The New Penguin Dictionary of

Geology

Philip Kearey

SECOND EDITION

PENGUIN BOOKS

To Jane, Eleanor, Georgina and Louisa

PENGUIN BOOKS

Published by the Penguin Group
Penguin Books Ltd, 80 Strand, London WC2R 0RL, England
Penguin Putnam Inc., 375 Hudson Street, New York, New York 10014, USA
Penguin Books Australia Ltd, 250 Camberwell Road, Camberwell, Victoria 3124, Australia
Penguin Books Canada Ltd, 10 Alcorn Avenue, Toronto, Ontario, Canada M4V 3B2
Penguin Books India (P) Ltd, 11 Community Centre, Panchsheel Park, New Delhi – 110 017, India
Penguin Books (NZ) Ltd, Cnr Rosedale and Airborne Roads, Albany, Auckland, New Zealand
Penguin Books (South Africa) (Pty) Ltd, 24 Sturdee Avenue, Rosebank 2196, South Africa

Penguin Books Ltd, Registered Offices: 80 Strand, London WC2R 0RL, England

www.penguin.com

First published 1996
Second edition 2001
2

Set in ITC Stone
Typeset by Rowland Phototypesetting Ltd, Bury St Edmunds, Suffolk
Printed in England by Clays Ltd, St Ives plc

Contents

Introduction

This dictionary contains 7703 entries which cover definitions of terms relevant to Geology. Because of the overlap which exists between the many branches of the Earth Sciences, for example between Geomorphology and Sedimentology or Palaeontology and Biology, some of the entries might be considered to be at the borders of Geology, but are included because of their importance to the subject.

For reasons of economy of space, I have concentrated on producing concise definitions of the entries rather than longer treatments. If further information should be required about any particular entry, it should be sought from the 432 references, arranged by topic, provided in the Bibliography, which lists modern works covering all aspects of Geology.

Also for reasons of space, I have omitted descriptions of individual geologists. Similarly, it would have been impossible to include definitions of, for example, every mineral ever described, every small local glacial event or every local stratigraphy. I hope that the reader will agree that I have selected the most important.

Entries and synonyms are distinguished by bold type, whilst those words which appear in the entries and are entries themselves appear in small capitals. Cross references to related terms are labelled 'See also' and references to contrasting terms are labelled 'Cf'. Abbreviations used in the text are as follows: Ma = million years, Ga = billion (10^9) years, L. = Lower, M. = Middle, U. = Upper, ~ = approximately.

Spelling in most cases follows the conventions used in Great Britain. Some, but not all, alternative spellings are given in American usage. Alternatives have not always been given for words containing once-ligatured diphthongs, e.g. 'ae'. I hope that this will not seriously inconvenience foreign readers.

All stratigraphic ages are given according to Harland, W. B., Armstrong, R. L., Cox, A. V., Craig, L. E., Smith, A. G. and Smith, D. G. (1990) *A Geologic Timescale 1990*, Cambridge University Press, Cambridge. A stratigraphic column, together with explanations of SI, cgs and imperial units, abbreviations used for very small and very large numbers, unit conversions, and facts about the Earth, will be found in the Appendix.

I would welcome suggestions for the inclusion of terms which do not appear in the present dictionary.

In this second edition I have taken the opportunity to add terms that did not appear in the first edition and to correct a few typographical errors. There are now almost a hundred new entries in this new edition, and the bibliography has been expanded.

Philip Kearey
Bristol, October 2000

Å Symbol for ÅNGSTRÖM.

A-subduction See AMPFERER SUBDUCTION.

aa A solidified LAVA FLOW with a very rough or clinkery surface.

Aalenian The lowest STAGE of the M. JURASSIC (DOGGER), 178.0–173.5 Ma.

AAS See ATOMIC ABSORPTION SPECTROSCOPY.

Ab Method of indicating PLAGIOCLASE FELDSPAR composition as a percentage of ALBITE, e.g. Ab_{15} indicates a composition of 15% ALBITE, 85% ANORTHITE.

abandonment The voluntary surrender of legal rights or title to a mining claim.

Abbé refractometer An instrument for determining the REFRACTIVE INDEX of MINERALS and liquids.

ablation The removal of detritus by wind action.

abnormal pressure (overpressure, geopressure) The pressure in a formation exceeding the HYDROSTATIC PRESSURE of a water column of DENSITY 1.114 Mg m^{-3}. Important in controlling the MIGRATION of fluids.

abrasion Mechanical EROSION by debris-charged wind, water or ice, which also removes the eroded material. See also ABLATION.

abrasion platform Any horizontal surface cut into a slope.

abrasive Any material suitable for grinding, polishing, scouring or cutting, e.g. DIAMOND, CORUNDUM, SAND, PUMICE.

absarokite A PORPHYRITIC BASALT with a small amount of ORTHOCLASE in the GROUNDMASS.

absolute age The age of a rock or FORMATION with respect to the present. Determined by RADIOMETRIC DATING methods. Cf. RELATIVE AGE.

absolute permeability The ability of a rock at 100% saturation to transmit a particular fluid. Cf. EFFECTIVE PERMEABILITY.

absolute plate motion The motion of a PLATE with respect to the interior of the Earth. Absolute motions can be determined by making use of the HOTSPOT REFERENCE FRAME. Cf. RELATIVE PLATE MOTION.

absolute temperature Temperature measured on the Kelvin scale, with respect to absolute zero (–273.15°C).

absorption The energy lost during the transmission of a waveform, excluding geometrical effects. For example, the loss of energy by a SEISMIC WAVE by conversion to heat.

Abukuma-type metamorphism METAMORPHISM at a high GEOTHERMAL GRADIENT and low pressure, characterized by the presence of ANDALUSITE and SILLIMANITE in PELITIC rocks.

abutment load The weight transferred to adjacent solid rock in a deep excavation.

abyss A very deep oceanic depression.

abyssal plain (basin plain, basin floor) A flat, generally smooth, sediment-covered, deep-ocean floor; usually floored by TURBIDITES.

abyssal zone (hadal zone) A deep-ocean ecological zone approximately from 2000 to 6000 m deep.

ac joint See CROSS JOINT.

Acadian orogeny An OROGENY affecting the northern parts of the APPALACHIAN FOLD BELT in the DEVONIAN, corresponding to the CALEDONIAN OROGENY of Europe. See also TACONIC OROGENY.

acanthite (Ag_2S) An ORE MINERAL of SILVER.

Acanthodiformes An order of subclass ACANTHODII, class OSTEICHTHYES, superclass PISCES; small, shark-like bony FISH. Range U. SILURIAN–L. PERMIAN.

Acanthodii A subclass of class OSTEICH-THYES, superclass PISCES; primitive FISH characterized by bony spines along the front edges of their fins and no bony internal skeleton. Range SILURIAN–PERMIAN.

accelerated erosion The increased rate of EROSION resulting from human activity.

accelerator mass spectrometry dating (AMS dating, accelerator radiocarbon dating) A RADIOCARBON DATING method in which a MASS SPECTROMETER is used to detect ^{14}C atoms. Makes use of smaller quantities of material than required for conventional RADIOCARBON DATING and extends its range to beyond 50 000 years.

accelerator radiocarbon dating See ACCELERATOR MASS SPECTROMETRY DATING.

accessory lithic A CLAST in a PYROCLASTIC rock formed of material torn from a VOLCANIC VENT's walls during a VOLCANIC ERUPTION.

accessory mineral A MINERAL comprising less than ~10% of a rock which is insignificant to nomenclature or classification. Cf. ESSENTIAL MINERAL.

accessory plate A plate of a specially cut MINERAL for use in a polarizing microscope, used in determining the character of a MINERAL in THIN SECTION.

accidental lithic A CLAST in a PYROCLASTIC rock plucked from the ground during the transport of TEPHRA.

accommodation structure A small STRUCTURE which allows a bed to fill all available space created during DEFORMATION.

accordion fold See CHEVRON FOLD.

accretion The process by which inanimate objects increase in size by the addition of material to their surfaces, e.g. accretion of continents, in which a CRATON increases in size by the welding to it of LITHOSPHERE brought into juxtaposition by SUBDUCTION.

accretion vein A MINERAL VEIN in which more than one phase of FRACTURE and infilling has occurred.

accretionary prism (accretionary wedge, subduction complex) A pile of sediments, characterized by repeated THRUST FAULTS, which accumulates on the landward side of the OCEAN TRENCH associated with a SUBDUCTION ZONE and grows by the OFFSCRAPING of sediment from the top of the downgoing PLATE by the leading edge of the overriding PLATE.

accretive plate margin See CONSTRUCTIVE PLATE MARGIN.

ACF diagram Triangular graph whose axes are Al_2O_3, CaO and $FeO+MgO$, used to illustrate chemical variation in a suite of rocks.

achnelith (Pelé's hair) Hair-like volcanic GLASS formed by LAVA exuding through a small orifice and blown by the wind.

achondrite A class of STONY METEORITE with no CHONDRULES.

achroite A white, potassium-rich variety of TOURMALINE.

acicular Needle-like.

acid clay A clay which releases hydrogen ions when in water suspension, e.g. FULLER'S EARTH.

acid rain Rainwater with a pH of less than 4 arising from the DISSOLUTION of gases produced naturally or from industrial processes.

acid rock See ACIDIC ROCK.

acidic rock (acid rock) An IGNEOUS ROCK with >10% free SILICA.

aclinic line See MAGNETIC EQUATOR.

acme zone (epibole, peak zone) A BIOZONE in which a particular species or genus is at its maximum abundance.

acmite See AEGIRINE.

acoustic basement The boundary between overlying sediments and underlying IGNEOUS or METAMORPHIC ROCKS, characterized by strong SEISMIC REFLECTIONS on a SEISMOGRAM.

acoustic impedance The product of SEISMIC VELOCITY and bulk DENSITY. The acoustic impedance contrast across a boundary determines the proportions of seismic energy transmitted and reflected at the boundary.

acoustic log A GEOPHYSICAL BOREHOLE LOG in which measurements are taken which utilize the properties of acoustic wave propagation.

acritarchs A diverse group of microorganisms with hollow, organic-walled vesicles, 5–500 μm in diameter. Probably produced by several groups of PROTISTS and useful in BIOSTRATIGRAPHY. Range PRECAMBRIAN–RECENT.

actinides The elements with atomic numbers 89–104, with properties similar to actinium.

actinolite $(Ca_2(Mg,Fe)_5Si_8O_{22}(OH)_2)$ A green AMPHIBOLE.

Actinopterygii A subclass of class OSTEICHTHYES, superclass PISCES; the ray-fin FISHES. Range DEVONIAN–RECENT.

activation analysis A technique of identifying stable isotopes by irradiating a sample with neutrons, charged particles or gamma rays. The induced radioactivity allows identification of isotopes from their characteristic radiation.

active continental margin A CONTINENTAL MARGIN that is also a SUBDUCTION ZONE. Cf. PASSIVE CONTINENTAL MARGIN.

Actonian A STAGE of the ORDOVICIAN, 444.5–444.0 Ma.

actualism See UNIFORMITARIANISM.

adakite A magnesium-rich DACITE/ANDESITE, possibly formed by the direct melting of subducting oceanic LITHOSPHERE at a SUBDUCTION ZONE where an OCEAN RIDGE is being subducted.

adamantine lustre A brilliant, sparkling LUSTRE arising from a MINERAL's being transparent and having a high REFRACTIVE INDEX, e.g. DIAMOND, CERUSSITE.

adamellite A coarse-grained, ACID IGNEOUS ROCK with PLAGIOCLASE comprising more than two thirds of the FELDSPAR present. ·

Adams-Williamson equation An equation describing the relationship between SEISMIC VELOCITY (v), GRAVITY (g) and the ADIABATIC change in DENSITY (ρ) within the Earth (of radius r).

$$\frac{d\rho}{dr} = \frac{\rho(r)g(r)}{\phi(r)}$$

where
$\phi = \alpha^2 - \frac{4}{3}\beta^2$ and α is the P WAVE velocity and β the S WAVE velocity.

adaptive divergence See ADAPTIVE RADIATION.

adaptive radiation (adaptive divergence) An evolutionary trend by which organisms increase in DIVERSITY as they adapt to occupy a large number of ecological environments.

addition rule See WEISS ZONE LAW.

Adelaidean orogeny An OROGENY affecting S. Australia from late PROTEROZOIC–ORDOVICIAN.

adhesion A process in which dry SAND blown onto a damp surface is held by the surface tension of water rising between the grains by capillary action.

adhesion lamination (quasi-planar adhesion stratification) A flat to low angle LAMINATION with a crinkly appearance and very good SORTING. Forms in well-sorted

SAND by ADHESION resulting from a strong wind acting on a slightly damp surface.

adhesion plane bed The bedform corresponding to ADHESION LAMINATION.

adhesion pseudo cross-lamination The cross-laminae forming in TABULAR sets when climbing ADHESION RIPPLES deposit well-sorted SAND.

adhesion ripple (aeolian micro-ridge, anti-ripplet) A millimetric scale, straight or sinuous ridge forming orthogonal to a unidirectional wind by ADHESION. The RIPPLES have steep upwind sides and shallow dipping LEE SIDES, and migrate downwind.

adhesion structure A sedimentary STRUCTURE formed by ADHESION.

adhesion wart A submillimetric to millimetric, irregular protuberance believed to form by ADHESION under strong, frequently shifting winds.

adiabatic Describing the relationship between pressure and volume as a substance expands or compresses without emitting or absorbing heat.

adinole An ARGILLACEOUS rock which has undergone ALBITIZATION during CONTACT METAMORPHISM.

adit A subhorizontal tunnel driven into a hillside.

admissible section A geological section in which the observed STRUCTURES maintain the same style below the surface. A BALANCED SECTION can only be constructed from such a section.

admission The substitution of a TRACE ELEMENT for a MAJOR ELEMENT with similar ionic radius during MAGMA CRYSTALLIZATION.

adobe A clay similar to LOESS.

adoral On the same side of the body as the mouth.

adsorption Adhesion to the surface of a material.

adularescence The milky-bluish sheen shown by MOONSTONE.

adularia ($KAlSi_3O_8$) A colourless, TRANSLUCENT variety of POTASH FELDSPAR.

advection Transport of heat, or any other physical property, as a result of movement of a fluid, such as occurs in a CONVECTION CURRENT.

adventive cone See PARASITIC CONE.

aegirine (acmite) ($NaFe^{3+}Si_2O_6$) A sodic PYROXENE commonly occurring as a late product in the CRYSTALLIZATION of alkaline MAGMA.

aegirine-augite A PYROXENE intermediate in composition between AEGIRINE and AUGITE.

aegirite An obsolete name for AEGIRINE.

aenigmatite (cossyrite) ($Na_2Fe^{2+}_5TiO_2$ $(Si_2O_6)_3$) A rare, titanium-bearing silicate.

aeolian (eolian) Pertaining to wind-driven processes.

aeolian bounding surface An EROSION surface truncating STRATA within wind-deposited sediments.

aeolian micro-ridge See ADHESION RIPPLE.

aeolian placer A PLACER DEPOSIT commonly formed by the reworking of a BEACH PLACER by the wind.

aeolian plane-bed A SLIP-FACED bedform formed on a subhorizontal surface in an AEOLIAN SAND SHEET.

aeolian sand sheet An extensive area of wind-deposited SAND, with only rare SLIP-FACED bedforms, commonly found around DUNE FIELDS.

aeolian stratification The BEDDING and LAMINATION formed as wind-blown sediment, usually of SAND grade, accumulates on a dry surface. See also ADHESION STRUCTURE.

aeolianite A cemented, wind-blown sediment. The CLASTS may be grains of QUARTZ, calcium carbonate, GYPSUM, etc. and the cement commonly calcium carbonate, although other water-soluble MINERALS such as GYPSUM have been described. The cement originates as grains within the SAND mobilized by infiltrating rainwater or from GROUNDWATER which subsequently evaporates and REPRECIPITATES the MINERAL. Commonly found along coasts within 40° of the equator.

aerial photography Successive photographs in the visible and very near infra-red bands taken by a downward-pointing camera mounted on an aircraft. Three-dimensional topography can be studied by observing sequential overlapping pairs of photographs through a STEREOSCOPE, which allows geological STRUCTURE, GEOMORPHOLOGY, vegetation, etc. to be assessed.

aerodynamic ripple A WIND RIPPLE similar in size to an IMPACT RIPPLE, but more sinuous and lying parallel to the wind. May be caused by wind vortices close to the surface.

aerolite See STONY METEORITE.

Aeronian A STAGE of the SILURIAN, 436.9–432.6 Ma.

aerosol A suspension or dispersion of solid particles in a gas.

AFM diagram A triangular graph with axes Al_2O_3, FeO and MgO, used to represent the composition of metamorphosed PELITES.

AFMAG **A**udio **F**requency **Mag**netic field method. An ELECTROMAGNETIC INDUCTION METHOD for the location of electrically conducting bodies in the subsurface which makes use of natural variations in the GEOMAGNETIC FIELD in the audiofrequency range known as SFERICS. Used both on land and in airborne versions.

aftershock An EARTHQUAKE in the same area as, and following, the 'main' EARTHQUAKE. Probably caused by the transfer of STRESS to FAULTS in the proximity of the FOCUS. The number of aftershocks depends on the MAGNITUDE of the 'main' EARTHQUAKE and decreases exponentially with time. Cf. FORESHOCK.

agalmatolite (pagodite) A compact variety of PYROPHYLLITE.

agate Concentric layers of CHALCEDONY with different colours and POROSITY.

age A fourth order geological time unit.

age of the Earth The oldest rocks yet found provide radiometric ages of 3.89 Ga, while METEORITES and Moon rocks give ages of 4.66 Ga. U-Pb studies suggest that the MANTLE became a closed system at the latter date, implying that the Earth formed at the same time as the other planets of the Solar system.

agglomerate A BRECCIA or CONGLOMERATE formed during volcanic activity.

agglutinate Welded splatter, commonly of BASALTIC composition, deposited ballistically in STROMBOLIAN or FIRE FOUNTAIN VOLCANIC ERUPTIONS. See also WELDED TUFF.

aggradation The process of building up a surface with sediment deposited by wind or water.

aggregate **1** A mass of rock fragments and/or MINERAL grains. **2** Any granular solid material used alone, e.g. as BALLAST, or mixed with a binding material, e.g. concrete.

aggressivity A measure of the capacity of water to dissolve calcium carbonate. An important parameter in the development of KARST.

agmatite MIGMATITE which appears to form a network of VEINS in the COUNTRY ROCK.

Agnatha A class of superclass PISCES; FISH with no biting jaws and a sucker-like mouth, including the hagfish and lampreys. Range ORDOVICIAN–RECENT.

Agnostida An order of subphylum TRILOBITA, phylum ARTHROPODA; small, eyeless TRILOBITES with a similar cephalon and pygidium, two thoracic segments and no facial sutures. Range L. CAMBRIAN–ORDOVICIAN.

agonic line The line joining locations of zero magnetic DECLINATION.

agrichnia A TRACE FOSSIL indicative of farming activity.

ahermatypic coral A CORAL with no symbiotic ALGAE, usually solitary, non-REEF forming and capable of living in deeper water than HERMATYPIC CORALS.

Aijiashan (Neichiashan) An ORDOVICIAN succession in China covering part of the ARENIG, the LLANVIRN, LLANDEILO, COSTONIAN and part of the HARNAGIAN.

Ainusian A CRETACEOUS succession in the far east of the former USSR equivalent to the CENOMANIAN.

air wave Seismic energy transmitted through the atmosphere from a SHOT.

airborne geophysical survey A MAGNETIC, ELECTROMAGNETIC, RADIOMETRIC or GRAVITY SURVEY undertaken alone or in concert from a fixed-wing aeroplane or helicopter. More cost-effective than a ground survey but with limited DEPTH OF PENETRATION.

airgun A mechanical marine SEISMIC SOURCE in which a burst of high pressure air is suddenly released into the water, forming a bubble which oscillates in a similar fashion to an explosion.

Airy hypothesis A mechanism of ISOSTATIC COMPENSATION in which surface topography is balanced by variation in the thickness of a CRUST of constant DENSITY. Mountain ranges would be underlain by crustal roots and ocean basins by anti-roots, i.e. elevated MOHO. Cf. PRATT HYPOTHESIS.

Aistopoda An order of subclass LEPOSPONDYLI, class AMPHIBIA; small, snake-like AMPHIBIANS. Range L. CARBONIFEROUS–early PERMIAN.

Akchagylian See APSHERONIAN.

åkermanite ($Ca_2MgSi_2O_7$) A FELDSPATHOID of the MELILITE group.

A'KF diagram A triangular graph with axes **A'** (($Al_2O_3+Fe_2O_3$)-(Na_2O+K_2O+CaO)), **K** (K_2O) and **F** ($FeO+MgO+MnO$) used to show the composition of a METAMORPHIC ROCK.

aklé A complex DUNE with interlocking, crescentic, generally parallel ridges forming orthogonal to the wind direction in areas of high SAND supply.

alabandite (MnS) A rare ORE MINERAL of manganese.

alabaster See GYPSUM.

alas A THERMOKARST feature consisting of a steep-sided, flat-floored depression where local melting of PERMAFROST has occurred.

alaskite A bimineralic, LEUCOCRATIC, GRANITIC rock composed of QUARTZ and ALKALI FELDSPAR.

Albertian A CAMBRIAN succession in the E. USA covering part of the LENIAN and the ST DAVID'S.

albertite A pitch-black, solid BITUMEN of the ASPHALTITE group.

Albian A STAGE of the CRETACEOUS, 112.0–97.0 Ma.

albite ($NaAlSi_3O_8$) The sodic end-member of the PLAGIOCLASE FELDSPARS, An_{0-10}.

albite twinning TWINNING which forms zebra-like stripes, seen in PLAGIOCLASE FELDSPARS in THIN SECTION.

albitite A variety of SYENITE composed almost wholly of ALBITE and produced by soda FENITIZATION.

albitization The METASOMATIC REPLACEMENT of an existing MINERAL, usually another FELDSPAR, by ALBITE as sodium ions are introduced into the rock.

alcove An arcuate, steep-sided valley on the side of a rock OUTCROP produced by water EROSION.

alcrete A DURICRUST of indurated BAUXITE formed when aluminium sesquioxides accumulate in the zone of WEATHERING.

Alcyonaria See OCTOCORALLIA.

Alexandrian An ORDOVICIAN/SILURIAN succession in North America covering part of the U. ORDOVICIAN, the RHUDDANIAN and part of the AERONIAN.

alexandrite A GEM variety of CHRYSOBERYL.

algae Non-taxonomic name for simple plants which did not differentiate into root, stem and leaves.

algal mat See STROMATOLITE.

algal ridge A morphological feature found on some REEF FLATS.

alginite (bituminite) A COAL MACERAL of the EXINITE group made up of coalified ALGAL remains, characteristic of BOGHEAD COAL.

Algoma-type iron formation A type of BANDED IRON FORMATION developed as oxide, carbonate and sulphide facies. Generally centimetres to hundreds of metres thick and less than a few kilometres in length. Characteristic of ARCHAEAN GREENSTONE BELTS and probably formed in an OCEAN TRENCH with a volcanic source of iron.

Algoman orogeny See KENORAN OROGENY.

aliasing A phenomenon in sampling a waveform when frequencies greater than half the sampling interval (the NYQUIST FREQUENCY) cause distortion of the low frequency part of the signal. Avoided by making use of an ANTI-ALIAS FILTER.

aliquot A small sample of a material ASSAYED to estimate the properties of the whole.

alkali basalt (olivine basalt) BASALT containing NORMATIVE NEPHELINE and OLIVINE.

alkali feldspar A general term for FELDSPAR of the K-Na solid solution, ORTHOCLASE/SANIDINE→ORTHOCLASE.

alkali flat See PLAYA LAKE.

alkali granite A GRANITE containing alkali AMPHIBOLE or PYROXENE.

alkali lake A lake in an arid region rich in dissolved sodium and potassium carbonate, sodium chloride and other alkaline salts.

alkali metal Lithium, sodium, potassium, rubidium, caesium and francium.

alkali-lime index The SiO_2 value at which $CaO=Na_2O+K_2O$ on a graph of these quantities against SiO_2 for a suite of related IGNEOUS ROCKS.

alkaline earth metals Calcium, strontium, barium and radium.

alkaline rock An IGNEOUS ROCK with a high proportion of alkalis (Na_2O+K_2O) relative to SILICA, i.e. SILICA-UNDERSATURATED, containing NORMATIVE NEPHELINE or LEUCITE and characterized by ALKALI FELDSPAR, FELDSPATHOIDS, alkali-rich PYROXENES and AMPHIBOLES, PHLOGOPITE and Zr, Nb, Rb, Ba, Ti and P enrichment.

allanite (orthite) $((Ca,Ce)_3(Fe^{2+},Fe^{3+})$ $Al_2O(SiO_4)(Si_2O_7)(OH))$ A SOROSILICATE found as an ACCESSORY MINERAL in many IGNEOUS ROCKS.

Alleghanian orogeny An OROGENY affecting the central and southern parts of the APPALACHIAN FOLD BELT in the CARBONIFEROUS–PERMIAN, corresponding to the VARISCAN OROGENY of Europe.

alleghanyite $(Mn_5(SiO_4)_2(OH)_2)$ An hydrated manganese silicate of the HUMITE group, the manganese analogue of CHONDRODITE.

allemontite (AsSb) A rare VEIN MINERAL comprising a SOLID SOLUTION of arsenic and antimony.

Aller A PERMIAN succession in NW Europe covering the lower part of the CAPITANIAN.

allochem Abbreviation of **allochem**ical constituent. An organized AGGREGATE of CALCITE, once used synonymously with grain or particle.

allochemical metamorphism METAMORPHISM in which there is removal or addition of material and the chemical composition of the rock is altered.

allochthon (heterochthon) A large structural unit, e.g. a SUSPECT TERRANE, originating at a long distance from its present position. Cf. AUTOCHTHON, PARAUTOCHTHON.

allocyclic Referring to controls on sedimentary accumulation external to a sedimentary system, such as climate, tectonic activity, sea level change and source area geology. Cf. AUTOCYCLIC.

allodapic Descriptive of sediment deposited by MASS FLOW, particularly LIMESTONE.

allogenic Descriptive of material originating from existing rock and transported to its present location. Cf. AUTHIGENIC.

allogenic stream A stream flowing through an area where it gains no DISCHARGE, occurring where DISCHARGE is derived from further up the CATCHMENT or where STREAM FLOW is supplemented by GROUNDWATER.

allopatric speciation The evolutionary divergence of geographically separated populations to form distinct species.

allophane ($Al_2Si_2O_5(OH)_2$) A white, amorphous CLAY MINERAL found along FRACTURES.

Allotheria A subclass of class MAMMALIA whose only order is the MULTITUBERCULATA.

allotriomorphic texture A TEXTURE of an IGNEOUS ROCK in which crystals exhibit a form related to surrounding, previously crystallized MINERALS.

allotropy The existence of an element in two or more forms.

alluvial Pertaining to river, or running fresh water, sedimentary deposits.

alluvial architecture The three-dimensional distribution of sedimentary FACIES in an ALLUVIAL DEPOSIT, governed by AUTOCYCLIC and ALLOCYCLIC controls.

alluvial deposit An accumulation of sediment deposited from running fresh water in a channel or on an alluvial, coastal or DELTAIC plain and comprising GRAVEL, SAND, MUD, COAL and chemical precipitates. The ALLUVIAL ARCHITECTURE is controlled by channel type, vegetation cover, channel density, source area geology, climate, TECTONIC setting and surface DEFORMATION.

alluvial fan An ALLUVIAL DEPOSIT with a semi-conical, downstream-broadening shape formed where the topographic gradient reduces and the transporting capacity is diminished as the width of flow increases, such as along mountain fronts, FAULT scarps, valley sides and GLACIER margins.

alluvial placer An ALLUVIAL DEPOSIT containing economic MINERALS.

alluviation The deposition of ALLUVIUM.

alluvium Sediment transported and deposited by running fresh water.

almandine ($Fe_2Al_2Si_3O_{12}$) A GARNET. Appreciable amounts of Mg and Mn generally replace Fe^{2+}.

almandine spinel See RUBY SPINEL.

Alminian An EOCENE succession in the former USSR equivalent to the PRIABONIAN.

alnöite An ALKALI BASALT, normally containing MELILITE.

alpha quartz (α-quartz, low quartz) The QUARTZ POLYMORPH stable below 573°C, common in all types of rock. Cf. BETA QUARTZ.

Alpides The mountain belt stretching from the Alps to the Himalaya.

Alpine-Himalayan orogeny An OROGENY affecting Europe and Asia, commencing in EOCENE times and caused by the closure of the TETHYS ocean.

Alportian The highest STAGE of the CARBONIFEROUS, 325.6–322.8 Ma.

alstonite ($CaBa(CO_3)_2$) An ORTHORHOMBIC CARBONATE MINERAL.

altaite (PbTe) A very rare ORE MINERAL of lead.

alteration Any chemical or mineralogical change in a rock caused by chemical or physical action.

alteration halo A rim of MINERALS formed in the WALL ROCK round a VEIN or ORE as the result of HYDROTHERMAL ALTERATION. MINERAL zoning in the halo indicates the changing nature of the HYDROTHERMAL SOLUTION during its passage.

alternation frequency map A SUBSURFACE FACIES MAP which allows distinction between a sequence of many alternating thin units and fewer thick units within the same lithology.

altiplanation terrace See CRYOPLAN-ATION.

Altonian A MIOCENE succession in New Zealand covering the upper part of the BUR-DIGALIAN.

alum A series of double sulphate ISO-MORPHS with potash alum approximating the formula $(KAl(SO_4)_2.12H_2O)$.

alum shale SHALE containing ALUM, usually formed by the WEATHERING of PYRITE-bearing SHALE.

alumstone See ALUNITE.

alunite (alumstone, löwigite) $(KAl_3(SO_4)_2(OH)_6)$ A MINERAL used in the production of ALUM.

alunitization The development of new ALUNITE by HYDROTHERMAL ALTERATION.

alveoles (honeycombs, stone lattice) Centimetric-scale small depressions formed by WEATHERING, making up a dense network of holes on gently sloping rock surfaces, possibly as a result of salt crystallization or other chemical or biological processes.

alvikite A medium- to fine-grained CARBONATITE.

amalgam (AgHg) A naturally occurring SILVER-MERCURY SOLID SOLUTION.

amazonite A green variety of MICROCLINE.

amazonstone A semi-precious GEM-quality variety of AMAZONITE.

amber (succinite) A MINERAL formed from fossilized resin.

amber ice Ice containing dispersed, fine-grained sediment with an AMBER-like appearance.

ambiguity problem A problem common to the interpretation of POTENTIAL FIELDS relating to the fact that the anomaly of any given body can be computed, but the derivation of the body responsible for any particular anomaly has no unique solution.

amblygonite $(LiAlFPO_4)$ A rare MINERAL found in GRANITE PEGMATITES.

Amblypoda An order of infraclass EUTHERIA, subclass THERIA, class MAMMALIA; large MAMMALS with broad-surfaced, low-crowned cheek teeth. Range U. PALAEOCENE–M. OLIGOCENE.

amesite A variety of CHLORITE rich in magnesium and aluminium.

amethyst A purple variety of QUARTZ.

Amgan A CAMBRIAN succession in Siberia covering parts of the SOLVAN and MENEVIAN.

Aminikan An ERA of the PRECAMBRIAN, 2200–1650 Ma.

Ammonoidea An order of class CEPHALO-PODA, phylum MOLLUSCA; MOLLUSCS with a coiled shell containing folded internal compartment walls. Range DEVONIAN–U. CRETACEOUS.

Amniota Vertebrate animals (REPTILES, MAMMALS and BIRDS) laying an egg covered by a tough shell and containing all water, food and facilities necessary for the complete development of the embryo. Cf. ANAMNIOTA.

Amonton's Law $\tau = \mu\sigma$, where τ = SHEAR STRESS, μ = COEFFICIENT OF FRICTION, σ = NORMAL STRESS. Relates to frictional sliding along a FAULT PLANE.

amorphous Descriptive of a MINERAL with no regular arrangement of atoms, i.e. non-crystalline.

amosite (brown asbestos) An ASBESTI-FORM variety of the AMPHIBOLE CUMMING-TONITE.

Ampferer subduction (A-subduction) SUBDUCTION during CONTINENTAL COLLISION in which the CRUST of the overriding PLATE becomes detached from the MANTLE part of the LITHOSPHERE so that the CRUST of the downgoing PLATE directly underthrusts that of the overriding PLATE.

Amphibia/amphibians A class of vertebrates capable of free existence on dry land but requiring water for egg laying and the maturing of larval forms prior to metamorphosis to the adult state. Range DEVONIAN–RECENT.

amphiboles SILICATE MINERALS with an

internal structure consisting of a double chain of linked silicate tetrahedra and cations occupying sites between oxygen ions at the edges of the chains. General chemistry $A_{0-1}B_2C_5D_8O_{22}(OH)_2$, where A = vacant or Na,K, B = Na,Ca,Mg or Fe^{2+}, C = Mg,Fe^{2+},Fe^{3+} or Al, D = Si or Al and some OH may be replaced by F,Cl or O^{2-}. They exhibit VITREOUS LUSTRE and perfect CLEAVAGE. They occur commonly in IGNEOUS and METAMORPHIC ROCKS.

amphibolite A non-FOLIATED, METAMORPHIC ROCK composed mainly of AMPHIBOLES, formed by the REGIONAL METAMORPHISM of BASIC IGNEOUS ROCKS.

amphibolite facies (epidote-amphibolite facies) A METAMORPHIC FACIES formed under moderate to high temperature and pressure.

amphibolitization See URALITIZATION.

amphidromic point A nodal point in the sea where no vertical tidal movement occurs, and tidal currents are most rapid.

Amphineura A class of phylum MOLLUSCA; MOLLUSCS whose surface is covered by seven or eight calcareous plates. Range CAMBRIAN–RECENT.

amphoteric With both acidic and alkaline properties.

AMS dating See ACCELERATOR MASS SPECTROMETRY DATING.

AMT survey See AUDIOMAGNETOTELLURIC SURVEY.

amygdale (amygdule) A VESICLE in LAVA or PYROCLASTIC ROCK filled with low-temperature MINERALS, e.g. CALCITE, QUARTZ.

amygdaloidal Containing AMYGDALES.

amygdule See AMYGDALE.

An A method of indicating PLAGIOCLASE FELDSPAR composition as a percentage of ANORTHITE, e.g. An_{15} indicates a composition of 15% ANORTHITE, 85% ALBITE.

anabranch A river channel pattern in which the width of islands is more than

three times the river width at average DISCHARGE.

anaerobic (anoxic, azoic) Descriptive of an environment in which the concentration of dissolved oxygen is too low to support METAZOAN life. Corresponds to an oxygen concentration less than 0.1 ml l^{-1}.

anagenesis An evolutionary change in a single species lineage.

analcime (analcite) ($NaAlSi_2O_6.H_2O$) A ZEOLITE or FELDSPATHOID.

analcite See ANALCIME.

analcitite An IGNEOUS ROCK in which abundant PLAGIOCLASE crystals have been replaced by ANALCITE during late stage DEUTERIC ALTERATION in which MAGMATIC fluids penetrated CLEAVAGES and FRACTURES in the PLAGIOCLASE.

Anamniota Vertebrate animals (FISH and AMPHIBIANS) which produce an egg without a tough shell. Cf. AMNIOTA.

Anapsida/anapsids A subclass of class REPTILIA; REPTILES lacking apertures in the skull behind the eyes. Range TRIASSIC–RECENT.

anastomosing Descriptive of a pattern characterized by branching and rejoining sinuous routes, e.g. an anastomosing river.

anatase (TiO_2) A relatively rare titanium OXIDE MINERAL.

anatectite See ANATEXITE.

anatexis (palingenesis) PARTIAL MELTING and fusion producing a melt of smaller proportion than the original rock, leaving an unmelted refractory residuum. The melt composition depends on phase relationships in the solid and the temperature and pressure conditions of melting. Cf. SYNTEXIS.

anatexite (anatectite) A rock formed by ANATEXIS.

anauxite ($Al_2(SiO_7)(OH)_4$) A CLAY MINERAL.

andalusite (Al_2SiO_5) A NESOSILICATE important in METAMORPHIC ROCKS.

Andean mountain belt An OROGENIC

BELT forming in response to the steady-state SUBDUCTION of oceanic LITHOSPHERE beneath CONTINENTAL LITHOSPHERE, typified by the South American Andes. The recognition of SUSPECT TERRANES in the Andes now casts doubt on this simple process. Cf. HIMALAYAN MOUNTAIN BELT.

Anderken An ORDOVICIAN succession in Kazakhstan covering part of the SOUDLEYAN, LONGVILLIAN and MARSHBROOKIAN.

Andernach lava See MAYAN LAVA.

andesine A PLAGIOCLASE FELDSPAR ($A\ n_{30-50}$).

andesite A VOLCANIC ROCK of INTERMEDIATE composition characteristic of the CALC-ALKALINE BASALT-ANDESITE-DACITE-RHYOLITE association. Commonly PORPHYRITIC with PHENOCRYSTS of zoned sodic PLAGIOCLASE, BIOTITE, HORNBLENDE and PYROXENE in a GROUNDMASS of the same MINERALS and more sodic PLAGIOCLASE and QUARTZ. The extrusive equivalent of DIORITE, grading into LATITE with increasing ALKALI FELDSPAR and into DACITE with increasing ALKALI FELDSPAR and QUARTZ. Typical of the SUBDUCTION ZONE environment.

andesite line A line separating ISLAND ARC and continental ANDESITIC rocks from the BASALTIC oceanic region.

andradite ($Ca_3Fe_2^{3+}Si_3O_{12}$) A GARNET. Fe is frequently partially replaced by Al. Typically formed by the METAMORPHISM of impure calcareous rocks and found in META-SOMATIC SKARNS.

anelasticity Time-dependent ELASTICITY.

Angaraland A large continent in the northern hemisphere which existed in CAR-BONIFEROUS and PERMIAN times, comprising the former USSR east of the Urals.

Angiospermopsida/angiosperms A class of division TRACHAEOPHYTA, kingdom plantae; flowering LAND PLANTS. Range CRE-TACEOUS–RECENT.

angle of contact The angle between a liquid surface and a solid with which it is in contact.

angle of draw The angle between the end of a subsurface working and the point on the surface to which its associated subsid-ence extends.

angle of internal friction (angle of shearing resistance) The angle relating the SHEAR STRESS required for sliding (e.g. along a FAULT) to the NORMAL STRESS on a plane. The inverse tangent of the angle defines the COEFFICIENT OF FRICTION.

angle of repose The maximum angle of slope that can be maintained by an accumu-lation of material.

angle of shearing resistance See ANGLE OF INTERNAL FRICTION.

anglesite ($PbSO_4$) A SECONDARY MINERAL formed by WEATHERING of lead ORES.

Ångström (Å) 10^{-10}m.

angular shear strain The SHEAR STRAIN produced in an original right angle during STRAIN, defined as the tangent of the change in angle.

angular velocity The rate of change of an angle between two lines. Used to quantify the rate of rotation of one PLATE with respect to another, as angular velocity is not depen-dent on the distance from the POLE OF ROTATION. Cf. TANGENTIAL VELOCITY.

angularity (roundness) The ratio between the average radius of circles drawn within the corners and edges of a CLAST and the radius of the largest circle which could be drawn within the CLAST; thus a measure of the sharpness of the corners of a CLAST.

anhedral Descriptive of a MINERAL with no crystal FORM developed.

anhydrite ($CaSO_4$) An EVAPORITE MINERAL.

Anika A SILURIAN succession in the Mirnyy Creek area of NE Siberia equivalent to the TELYCHIAN.

anilite (Cu_7S_4) A rare sulphide ORE MINERAL.

Anisian The lower STAGE of the M. TRIASSIC, 241.1–239.5 Ma.

anisotropic fabric A FABRIC in a rock which shows PREFERRED ORIENTATION, e.g. clay flakes in a MUDROCK which orient as they settle from SUSPENSION and on COMPACTION.

anisotropy The presence of a PREFERRED ORIENTATION.

ankaramite A PORPHYRITIC, PHENOCRYST-rich ALKALI OLIVINE BASALT with more PYROXENE than OLIVINE and a GROUNDMASS of AUGITE/TITANAUGITE MICROLITES, LABRADORITE and BIOTITE. Occurs in association with BASANITE and ALKALI BASALT.

ankerite $(Ca(Fe,Mg)(CO_3)_2)$ A CARBONATE MINERAL of the DOLOMITE group.

annabergite (nickel bloom) $(Ni_3(AsO_4)_2.8H_2O)$ A green SECONDARY MINERAL of nickel.

annealing The movement of grain boundaries and grain growth in a polycrystalline rock by which the rock undergoes RECOVERY from DEFORMATION. Driven by elevated temperature or occurring slowly over a long period.

Annelida Segmented worms with well-defined heads. Range PRECAMBRIAN–RECENT.

annite $(K_2Fe_6Si_6Al_2O_{20}(OH)_4)$ The ferrous iron end-member of the BIOTITE MICAS.

annular drainage A DRAINAGE pattern in which SUBSEQUENT STREAMS follow paths approximating portions of concentric circles. Often occurs around dissected DOMES or BASINS.

anomalous lead Lead with an isotopic ratio which gives an incorrect radiometric age. Cf. COMMON LEAD. See also B-TYPE LEAD, J-TYPE LEAD.

anorogenic Unrelated to TECTONISM.

anorthite $(CaAl_2Si_2O_8)$ The most calcium-rich PLAGIOCLASE FELDSPAR, An_{70-90}.

anorthoclase $((Na,K)AlSi_3O_8)$ An ALKALI FELDSPAR.

anorthosite A rock consisting of >90% PLAGIOCLASE FELDSPAR, virtually restricted to rocks of PRECAMBRIAN age.

anoxia The state of being ANOXIC.

anoxic See ANAEROBIC.

anoxic event A geographically widespread, stratigraphically limited interval of organic-rich deposition, whose CARBON assists in the removal of oxygen, e.g. in the TURONIAN.

antecedent drainage A drainage system maintaining its general direction by cutting through a localized uplifted area at a rate greater than the UPLIFT.

antecedent platform theory The theory that CORAL-ALGAL REEFS and ATOLLS grow upwards after colonizing submarine banks which have built up to a suitable depth, as CORALS can only survive in relatively shallow water. No longer widely accepted, as an explanation based on a combination of subsidence and GLACIAL CONTROL THEORY is now preferred.

anteconsequent stream A stream flowing as a consequence of an early UPLIFT but ANTECEDENT to later periods of UPLIFT. Cf. CONSEQUENT, INCONSEQUENT, INSEQUENT, OBSEQUENT, SUBSEQUENT and RESEQUENT STREAMS.

anthophyllite $((Mg,Fe)_7Si_8O_{22}(OH)_2)$ A white ORTHOAMPHIBOLE.

Anthozoa A class of phylum CNIDARIA which includes the CORALS. Range ORDOVICIAN–RECENT.

anthracite A high RANK COAL with a high CARBON content and a very low volatile content.

anthracitization The process of conversion of BROWN COAL to ANTHRACITE.

Anthracosauria An order of subclass LABYRINTHODONTIA, class AMPHIBIA; possible ancestors of the AMPHIBIA of CARBONIFEROUS age.

anthraxolite A hard black ASPHALT of the ASPHALTITE group found in VEINS and as masses in SEDIMENTARY ROCKS, such as OIL SHALE.

anthraxylon Vitreous COAL constituents derived from woody tissue.

anthropogeomorphology The study of man as a GEOMORPHOLOGICAL agent.

anti-alias filter A FILTER applied to a signal so as to avoid ALIASING.

anti-ripplet See ADHESION RIPPLE.

Antiarchii An order of class PLACODERMI, superclass PISCES; small, grotesque, normally freshwater, PLACODERM FISH. Range M. DEVONIAN–top DEVONIAN.

anticlinal trap An OIL TRAP consisting of a domed STRUCTURE dipping away from a central high point. May be generated by HALOKINESIS.

anticline A FOLD, CLOSING in any direction, in which the older rocks occupy the core.

anticlinorium A large composite ANTICLINE made up of smaller FOLDS. Cf. SYNCLINORIUM.

anticrack faulting See TRANSFORMATIONAL FAULTING.

antidune A bedform with a symmetrical shape whose crest is orthogonal to the flow direction, commonly found in fast, shallow flows. Rarely identified in ancient deposits.

antiferromagnetism Rock magnetism in which the magnetic lattices possess magnetizations which are equal and in opposite directions so that the rock has a zero external MAGNETIC FIELD.

antiform A FOLD CLOSING upwards for which no information is available on the YOUNGING direction.

antiformal stack An IMBRICATE STRUCTURE in which different amounts of THRUST DISPLACEMENTS relative to FAULT separations result in a DUPLEX dipping towards both HINTERLAND and FORELAND.

antigorite $(Mg_3Si_2O_5(OH)_4)$ A PHYLLOSILICATE of the SERPENTINE group.

antimonite See STIBNITE.

antimony (Sb) A rare VEIN MINERAL.

antimony glance See STIBNITE.

antiperthite An intergrowth of ALKALI FELDSPARS in which the sodium-rich phase predominates. Cf. PERTHITE.

antitaxial growth A VEIN fill growing from the centre towards the walls. Cf. SYNTAXIAL GROWTH.

antithetic In a direction opposite to the prevailing sense of VERGENCE or asymmetry, applied to STRUCTURES or FABRICS. Cf. SYNTHETIC.

antithetic fault A FAULT that DIPS in the opposite sense to a NORMAL FAULT with which it is associated.

Antler orogeny An OROGENY of DEVONIAN age in NW America.

antlerite $(Cu_3SO_4(OH)_4)$ A SECONDARY MINERAL formed by the ALTERATION of COPPER ORES.

Anura An order of subclass LISSAMPHIBIA, class AMPHIBIA; the frogs and toads. Range L. JURASSIC–RECENT.

apalhraun An Icelandic flow of AA.

apatite $(Ca_5(PO_4)_3(F,Cl,OH))$ A MINERAL important as source of phosphate for fertilizer.

apex An American mining law term denoting the outcrop of a VEIN reaching the surface or the shallowest point of a subsurface VEIN.

apex law An American mining law whereby the holder of mining rights to the outcrop or APEX of a VEIN may mine it down-DIP beyond the vertical projection of his CLAIM.

aphanitic texture (aphyric texture) A TEXTURE of an IGNEOUS ROCK comprising a MICROCRYSTALLINE GROUNDMASS with no PHENOCRYSTS. Forms by rapid crystallization of a melt close to its LIQUIDUS temperature with no large suspended crystals.

Aphebian The lower part of the PROTEROZOIC of Canada, ~2560–1800 Ma.

aphelion The location in the Earth's elliptical orbit where it is farthest from the Sun. Cf. PERIHELION.

aphotic zone The level in a water body where light is absent, which occurs at depths >200 m in the oceans. Cf. DIPHOTIC ZONE, PHOTIC ZONE.

aphyric texture See APHANITIC TEXTURE.

API gravity (API scale) The American Petroleum Institute standard for expressing the specific gravity of oil. API gravity = $141.5/(density)-131.5$.

API scale See API GRAVITY.

Aplacophora A subclass of class AMPHINE-URA, phylum MOLLUSCA; MOLLUSCS whose dorsal surface is covered by seven or eight non-calcareous plates, and thus with no fossil record.

aplite A fine- to medium-grained IGNEOUS ROCK occurring as thin (<20 cm) VEINS within coarser-grained PLUTONIC rocks. Has a distinctive SUBHEDRAL-ANHEDRAL FABRIC with a sugary TEXTURE. Forms by the squeezing out of residual GRANITIC MAGMA into FRACTURES or JOINTS formed at a late stage of cooling of the intrusion. Commonly associated with PEGMATITE. Without qualification aplite refers to GRANITIC composition, but it can range in composition from GABBROIC to GRANITIC.

Apoda An order of subclass LISSAMPHIBIA, class AMPHIBIA; legless, snake-like AMPHIBIANS. Range L. JURASSIC–RECENT.

apophyllite $(KCa_4(Si_4O_{10})_2F.8H_2O)$ A PHYLLOSILICATE.

apophysis A VEIN or protuberance connected to a larger intrusive body. Used to describe centimetric-scale branches of DYKES and SILLS.

Appalachian fold belt A FOLD BELT extending along the east coast of North America, affected by the ACADIAN OROGENY, ALLEGHENIAN OROGENY and TACONIAN OROGENY during the PALAEOZOIC.

apparent cohesion The cohesion of grains by the surface tension of PORE FLUID.

apparent conductivity The inverse of APPARENT RESISTIVITY.

apparent dip The angle of inclination of a plane to the horizontal measured in a plane not orthogonal to the STRIKE of the plane. Always less than the TRUE DIP.

apparent polar wander The pattern

traced by PALAEOMAGNETIC POLES of sequential age of a single LITHOSPHERIC block plotted assuming the block remained stationary. Cf. TRUE POLAR WANDER.

apparent resistivity A complex weighted mean of the ELECTRICAL RESISTIVITIES in the subsurface which varies with the location and separation of the measuring electrodes. Apparent resistivity is only equal to true RESISTIVITY for a uniform infinite half-space.

apparent velocity The velocity of SEISMIC WAVES measured from a segment of the graph of ARRIVAL TIME against RANGE. If the refractor is dipping or undulating, this does not represent a true velocity.

appinite An heterogeneous group of medium- to coarse-grained HORNBLENDE- and BIOTITE-rich IGNEOUS ROCKS forming small intrusions, varying from SUBALKALINE to ALKALINE in composition and containing prominent HORNBLENDE PHENOCRYSTS.

applied geomorphology The field of GEOMORPHOLOGY which applies GEOMORPHOLOGICAL knowledge and techniques to the solution of environmental problems and the support of ENVIRONMENTAL ENGINEERING projects.

applied geophysics (geophysical exploration) Methods of investigating the subsurface by taking measurements at or near the Earth's surface. Includes GRAVITY, MAGNETIC, ELECTRICAL, ELECTROMAGNETIC, SEISMIC and RADIOMETRIC SURVEYING.

apron A FAN of unconsolidated sediment at the base of a mountain or GLACIER.

Apsheronian (Akchagylian) A PLIOCENE succession on the Russian Platform covering the upper part of the PIACENZIAN.

Aptian A STAGE of the CRETACEOUS, 124.5–112.0 Ma.

aptychus A calcareous plate which represents a lower jaw plate of a MESOZOIC AMMONITE.

aquamarine A green/blue, GEM-quality variety of BERYL.

aquiclude A formation with a low PER-

MEABILITY, important in controlling flow in adjacent overlying and underlying permeable formations. Cf. AQUIFUGE, AQUITARD.

aquifer A geological unit containing sufficient saturated PERMEABLE rock to yield significant amounts of water.

aquifer loss The water lost by flow from the AQUIFER to the WELL when DRAWDOWN occurs.

aquifer test A test in which known quantities of water are added to or withdrawn from a WELL and subsequent changes in HEAD measured.

aquifuge A geological unit with no connected pores which neither absorbs nor transmits water. Cf. AQUICLUDE, AQUITARD.

Aquitanian The lowest STAGE of the MIOCENE 23.3–21.5 Ma.

aquitard A formation allowing the throughflow of water at a much slower rate than an AQUIFER. Cf. AQUICLUDE, AQUIFUGE.

Arachnida A class of subphylum CHELICERATA, phylum ARTHROPODA; the spiders and scorpions. Range SILURIAN–RECENT.

Araeoscelidia An order of subclass EURAPSIDA, class REPTILIA; primitive, extinct REPTILES. Range U. CARBONIFEROUS–M. PERMIAN.

aragonite ($CaCO_3$) A POLYMORPH of calcium carbonate. A major skeletal component of many modern invertebrates and so a major constituent of modern carbonate accumulations. Changes by NEOMORPHISM to CALCITE with increasing age.

aragonite compensation depth See CARBONATE COMPENSATION DEPTH.

Araneida An order of class ARACHNIDA, subphylum CHELICERATA, phylum ARTHROPODA; the spiders. Range ?CARBONIFEROUS–RECENT.

Aratauran A JURASSIC succession in New Zealand equivalent to the HETTANGIAN and SINEMURIAN.

arborescent See DENDRITIC.

archae- Ancient.

Archaean (Archaeozoic, Azoic, Cryptozoic) The older EON of the PRECAMBRIAN ranging from the formation of the Earth at ~4600 Ma to 2500 Ma.

Archaeocyatha Small, cup-like, calcareous, marine, BENTHONIC METAZOANS probably related to the PORIFERA. Range CAMBRIAN.

archaeomagnetism PALAEOMAGNETISM applied to archaeological phenomena.

Archaeornithes A subclass of AVES whose only representative is *Archaeopteryx*, the first BIRD, of M. KIMMERIDGIAN age.

Archaeozoic See ARCHAEAN.

archetype The hypothetical ancestral form of an organism in which all basic characteristics are established without specialization.

Archie's Formula (Archie's Law) $\rho = a \rho_w \phi^{-m}$ An empirical law relating the ELECTRICAL RESISTIVITY of a porous rock ρ to its POROSITY ϕ and the resistivity of the PORE FLUID ρ_w, where a and m are constants.

Archie's Law See ARCHIE'S FORMULA.

archipelago A group of islands.

Archosauria/archosaurs A subclass of class REPTILIA including the CROCODILES, DINOSAURS and PTEROSAURS. Range L. TRIASSIC–RECENT.

arenaceous Sandy or SAND-like.

Arenig An EPOCH of the ORDOVICIAN, 493.0–476.1 Ma.

arenite 1 SANDSTONE. 2 A SANDSTONE whose MATRIX makes up less than 15% of the rock by volume.

arête A narrow, steep-sided ridge formed during the convergence of CIRQUES.

arfvedsonite ($Na_3Fe^{2+}_4Fe^{3+}Si_8O_{22}(OH)_2$) A deep green AMPHIBOLE.

argentiferous Containing SILVER.

argentite (silver glance) (Ag_2S) A high temperature form of the ORE MINERAL ACANTHITE.

argentopyrite ($AgFe_2S_3$) A sulphide of the WURTZITE GROUP.

argillaceous Descriptive of a detrital SEDIMENTARY ROCK with particles <4 μm.

argillite A hard, slightly metamorphosed, ARGILLACEOUS rock.

argillization A type of WALL ROCK ALTERATION forming CLAY MINERALS in a COUNTRY ROCK containing mineralization. The important subtypes are ADVANCED ARGILLIC ALTERATION and INTERMEDIATE ARGILLIC ALTERATION.

argon-argon dating A technique of RADIOMETRIC DATING based on the irradiation of a potassium-bearing sample with thermal and fast neutrons to produce ^{39}Ar. Measurement of the ratio of ^{40}Ar to ^{39}Ar then allows the age to be calculated.

argyrodite (canfieldite) (Ag_8GeS_6) A grey germanium-SILVER double sulphide found in VEIN deposits associated with ARGENTITE and other SILVER MINERALS.

arkose A SANDSTONE with <15% MATRIX by volume containing a FELDSPAR:QUARTZ ratio of at least 25% and a FELDSPAR:rock fragment ratio of at least 50% by volume.

arkosic arenite A SANDSTONE comprising >25% FELDSPAR, with the FELDSPAR content exceeding that of rock fragments and <15% mud MATRIX (material <30 μm in diameter).

arkosic wacke (feldspathic greywacke, feldspathic wacke) A SANDSTONE comprising >5% of SAND grade particles, with the FELDSPAR content exceeding that of rock fragments and >15% by volume of MUD MATRIX (material <30 μm in diameter).

armalcolite ($(Fe,Mg)Ti_2O_5$) An OXIDE MINERAL with the same structure as PSEUDOBROOKITE, first found on the Moon, subsequently in METEORITES and on Earth. Indicative of a highly reducing environment.

Armorican orogeny See HERCYNIAN OROGENY.

armoured mud ball (clay ball, mud ball, mud pebble, pudding ball) An approximately spherical, centimetric-sized lump of cohesive sediment that has been gouged from a stream bed/bank. Often found in BADLANDS and along EPHEMERAL STREAMS, and also in tidal channels and on BEACHES.

armoured surface ARMOURING found in an equilibrium river channel, periodically disrupted during floods. Cf. PAVED BED.

armouring Heterogeneous river bed material forming a thin layer of coarse grains one to two grains thick that inhibits transportation of underlying finer material.

Arnold An OLIGOCENE succession in New Zealand comprising the BORTONIAN and KAIATAN.

Arnsbergian A STAGE of the CARBONIFEROUS, 331.1–328.3 Ma.

aromatics (C_nH_{2n-6}) Benzene-based compounds which occur in many CRUDE OILS up to a concentration of ~10%.

Arowhanan A CRETACEOUS succession in New Zealand covering part of the CENOMANIAN.

arrival time (onset time) The time at which a SEISMIC PHASE arrives at a detector.

arroyo (coulée, dry wash, wash) A steep-sided, EPHEMERAL STREAM channel found in deserts, in which several metres of poorly sorted sediment from occasional RUNOFF may accumulate.

arsenical pyrites See ARSENOPYRITE.

arsenopyrite (arsenical pyrites) (FeAsS) An arsenic MINERAL common in metallic sulphide ORES.

arterite A MIGMATITE with VEIN-like GRANITIC intrusions into the COUNTRY ROCK.

artesian aquifer (confined aquifer) An AQUIFER in which water is under sufficient pressure to drive it to the surface when penetrated by a WELL.

artesian basin A BASIN-shaped area of ARTESIAN WELLS.

artesian discharge The discharge from a WELL penetrating an ARTESIAN AQUIFER.

artesian head The hydrostatic head of an ARTESIAN AQUIFER.

artesian pressure The hydrostatic pressure of an ARTESIAN AQUIFER at the surface.

artesian spring A SPRING fed from an ARTESIAN AQUIFER.

artesian well A WELL fed from an ARTESIAN AQUIFER.

Arthrodira An order of class PLACODERMI, superclass PISCES; FISH with bony plated bodies and heavily armoured heads. Range DEVONIAN.

Arthropleurida A class of superclass MYRIAPODA, phylum UNIRAMIA; terrestrial, herbivorous uniramians with multisegmented limbs with a large, lobed plate at their base. Range DEVONIAN–U. CARBONIFEROUS.

Arthropoda/arthropods A diverse invertebrate phylum characterized by an exoskeleton and jointed appendages. Bilaterally symmetrical, with the body usually organized into specialized divisions, e.g. head, thorax, abdomen. Range L. CAMBRIAN–RECENT.

Articulata 1 A class of phylum BRACHIOPODA in which the VALVES are held together solely by muscles. Range L. CAMBRIAN–RECENT. **2** A subclass of class CRINOIDEA, subphylum PELMATOZOA; stemmed or stemless CRINOIDS with simple cups. Range L. TRIASSIC–RECENT.

artificial recharge The replenishment of an AQUIFER by pumping water into it from surface supplies or other AQUIFERS.

Artinskian A STAGE of the PERMIAN, 268.8–259.7 Ma.

Artiodactyla/artiodactyls An order of infraclass EUTHERIA, subclass THERIA, class MAMMALIA; hoofed MAMMALs including cattle, sheep and other even-toed herbivores. Range EOCENE–RECENT.

Arundian A STAGE of the CARBONIFEROUS, 345.0–342.8 Ma.

asbestiform With an extremely fibrous form.

asbestos A group of INDUSTRIAL MINERALS with ASBESTIFORM HABIT, including CHRYSOTILE, CROCIDOLITE and AMOSITE.

Asbian A STAGE of the CARBONIFEROUS, 339.4–336.0 Ma.

asbolane (asbolite, earthy cobalt) A type of WAD containing cobalt.

asbolite See ASBOLANE.

aseismic Free from EARTHQUAKES.

aseismic ridge An oceanic ridge lacking regular EARTHQUAKE activity, possibly formed by the passage of the LITHOSPHERE over a near-stationary HOTSPOT in the MANTLE, e.g. Iceland-Faeroes ridge.

aseismic slip Movement along a FAULT PLANE unaccompanied by EARTHQUAKE activity, generally at rates of less than 0.1 m s^{-1}. Cf. SEISMIC SLIP.

ash 1 A fine-grained volcanic material. **2** An incombustible inorganic residue remaining after the burning of COAL.

ash flow (pyroclastic flow) A concentrated dispersion of hot, juvenile, volcanic fragments (ASH, PUMICE, SCORIA) in a gas which moves in response to GRAVITY. Forms deposits ranging in volume from 0.01 km^3 to more than 1000 km^3.

ash-flow field (ash flow plain, ignimbrite plain, ignimbrite plateau) A plain of flat-lying, gently dipping IGNIMBRITES erupted from numerous, closely spaced CALDERAS, often swamping existing topography and forming subhorizontal PLATEAUX, sometimes covering areas of up to 20 000 km^2.

ash-flow plain See ASH-FLOW FIELD.

Ashgill The oldest EPOCH of the ORDOVICIAN, 443.1–439.0 Ma.

ashlar A block of BUILDING STONE with straight edges.

asparagus stone APATITE of a yellow-green colour resembling asparagus.

aspect ratio The ratio of width to height of an object, shape, etc.

asperity ploughing A mechanism of

SLICKENSIDE formation in which a FAULT surface is abraded by a small protuberance (asperity) to form a LINEATION known as a SCRATCH, GROOVE, TOOL TRACK or GUTTER.

asphalt (bitumen, pitch, tar) A solid member of the PETROLEUM group.

asphalt-based crude oil (black oil) CRUDE OIL dominantly composed of NAPHTHENIC compounds, comprising some 15% of world supplies.

asphaltene An AGGREGATE of high-molecular weight HYDROCARBONS occurring in ASPHALT.

asphaltic pyrobitumen A black, structureless ASPHALT insoluble in carbon disulphide and generally with <5% oxygen.

asphaltite A group name for the organic compounds ALBERTITE, ANTHRAXOLITE, GRAHAMITE, IMPSONITE, NIGRITE and UINTAITE.

assay The amount of metal or metals in an ORE.

assay boundary (assay limit, cut off limit) The boundary of economic MINERAL concentration of an OREBODY.

assay limit See ASSAY BOUNDARY.

assay plan A map showing the variation in GRADE and distribution of metals in and around an OREBODY.

Asselian The lowest STAGE of the PERMIAN, 290.0–281.5 Ma.

assemblage zone (coenozone) A BIOSTRATIGRAPHIC ZONE based on the temporal range of a group of species.

assimilation (contamination) The chemical and/or physical incorporation of rock into an unrelated MAGMA.

associated gas Gas produced during the formation of PETROLEUM and accompanying oil in an OILFIELD. Cf. NON-ASSOCIATED GAS.

asterism A property of some crystals which appear to contain a star-like shape, arising from the reflection of light from oriented inclusions.

asteroid A small, rocky or metallic inter-

planetary body (BOLIDE). A concentration of these bodies is found in the Asteroid Belt between Mars and Jupiter. Cf. COMET.

Asteroidea/asteroids A class of subphylum ELEUTHEROZOA, phylum ECHINODERMATA including starfish and brittle stars. Range L. ORDOVICIAN–RECENT.

asthenosphere A mechanically weak layer of the MANTLE immediately beneath the LITHOSPHERE, corresponding to the depth range within the Earth where the melting temperature is most closely approached. The top is near the surface beneath OCEANIC RIDGES, 120–180 km deep under old ocean basins and at least 250 km deep, if present at all, beneath the continents.

astraeoid A condition of massive CORAL in which the CORALLITE walls become thin or disappear but the septae remain.

Astrapotheria An order of infraclass EUTHERIA, subclass THERIA, class MAMMALIA; rhinoceros-sized South American ungulates with long bodies, short legs and protruding lower incisors. Range U. PALAEOCENE–MIOCENE.

astrobleme A terrestrial CRATER over 10 km in diameter formed by the impact of an extraterrestrial body.

astrophyllite $((K,Na)_3(Fe,Mn)_7(Ti,Zr)_2Si_8(O,OH)_{31}))$ A rare titanium-bearing silicate.

asymmetric valley A valley with sides of unequal slope. Possibly the result of the geological STRUCTURE as in UNICLINAL SHIFTING. They are common in PERIGLACIAL regions where they form as a result of differences in slope aspect and thus solar radiation received.

asymmetrical fold A FOLD which is not mirror symmetrical about the AXIAL PLANE, caused by differing LIMB length or thickness.

atacamite $(CuCl(OH)_3)$ A SECONDARY MINERAL of COPPER.

Atdabanian A STAGE of the CAMBRIAN, 560–554 Ma.

Atlantic-type coast (transverse coast, discordant coast) A coastline formed where the topography and geological STRUCTURE are orthogonal to it. Cf. PACIFIC-TYPE COAST.

Atokan A CARBONIFEROUS succession in the USA covering the lower part of the MOSCOVIAN.

atoll An irregular, subcircular, annular CORAL-ALGAL REEF surrounding a central LAGOON in oceanic waters.

atollon A small ATOLL lying on the flank of a larger one.

atomic absorption spectroscopy (AAS) A method of CHEMICAL ANALYSIS in which a solution of a sample is passed into a flame which atomizes it. The amount of light absorbed from a source of the wavelength of a particular element focused on the flame gives a measure of that element's concentration.

atomic orbital A descriptor of the spatial distribution of an electron in an atom.

Atrypida An order of class ARTICULATA, phylum BRACHIOPODA; BRACHIOPODS with biconvex, impunctate shells with a short, rounded hinge and pedicle foramen or notch. Range M. ORDOVICIAN–U. DEVONIAN.

attached groundwater The water retained on the pore walls in rocks above the WATERTABLE.

attapulgite $(2MgO_3SiO_2.4H_2O–Al_2O_3 5SiO_2.6H_2O)$ One of the MINERALS making up the PALYGORSKITE group of CLAY MINERALS. The constituent of FULLER'S EARTH.

attenuation The progressive reduction in amplitude of waves with increasing time or distance.

Atterberg limit The result of tests on cohesive soil that characterize changes from solid to plastic to liquid states.

attrital coal A US Geological Survey term used in the field description of COAL in terms of its FUSAIN, VITRAIN and ATTRITUS constituents, the latter being described according to its LUSTRE.

attrital-anthraxylous coal A lustrous COAL with the ratio of ANTHRAXYLON to ATTRITUS in the range 1:1 to 1:3.

attritus The microconstituents of COAL including spores, cuticles, resins and granular opaque matter.

aubrite An ENSTATITE-rich, calcium-poor ACHONDRITE.

audiomagnetotelluric survey (AMT survey) A MAGNETOTELLURIC SURVEY conducted using frequencies in the audio range.

augen gneiss A GNEISS with a planar or linear SHAPE FABRIC consisting of eye-like lensoid shapes, often resulting from the DEFORMATION of PORPHYRITIC coarse-grained IGNEOUS ROCKS or by the growth of PORPHYROBLASTS.

augen structure (rib structure) An eye-like marking on the surface of a JOINT.

augite $((Ca,Na)(Mg,Fe,Al)(Si,Al)_2O_6)$ The most common PYROXENE, dark green to black in colour.

aulacogen An inactive RIFT VALLEY, partially or completely filled with sediment, formed by the failure of one arm of a RIFT-RIFT-RIFT TRIPLE JUNCTION during CONTINENTAL SPLITTING.

aureole See METAMORPHIC AUREOLE.

auric See AURIFEROUS.

aurichalcite $((Zn,Cu)_3(CO_3)_2(OH)_3)$ A rare MINERAL formed by the ALTERATION of BASE METAL ORES.

auriferous (auric) Containing GOLD.

aurostibite $(AuSb_2)$ GOLD-antimony ORE.

Austin A CRETACEOUS succession on the Gulf Coast of the USA equivalent to the CONIACIAN and SANTONIAN.

austral Of southern hemisphere origin. Cf. BOREAL.

authigenic Originating in place. Usually applied to DIAGENETIC MINERALS forming in sediments after deposition. Cf. ALLOGENIC.

autobrecciated lava A viscous LAVA FLOW

whose congealed crust has been fragmented by continued movements of molten LAVA in the interior. The smooth-faced blocks of crust weld together if sufficiently hot, but otherwise remain entrained within the mobile LAVA.

autochthon A large structural unit, e.g. a NAPPE, which originated close to its present position. Cf. ALLOCHTHON, PARAUTOCHTHON.

autoclastic Descriptive of a rock which has fragmented *in situ*.

autocorrelation function A function describing the CORRELATION of a waveform with itself in order to measure similarity and periodicity along it.

autocyclic Referring to controls on sedimentation inherent to a sedimentary system, such as channel type, sediment discharge, DIAGENESIS. Cf. ALLOCYCLIC.

autointrusion The injection of MAGMA into fissures of its earlier crystallized rock.

autolith (cognate inclusion) A XENOLITH of early-formed material from the parental MAGMA of an IGNEOUS ROCK.

autometamorphism The process whereby METAMORPHIC changes take place by the action of residual fluids as an IGNEOUS BODY cools.

autometasomatism The METASOMATISM of newly crystallized IGNEOUS ROCK by its residual fluids.

automorphic See EUHEDRAL.

autosuspension The condition in a fluid, only approached in TURBIDITY CURRENTS, in which the hydraulic forces of motion provide sufficient energy to maintain a suspended load in the flow.

autotrophic Descriptive of organisms, such as CYANOBACTERIA, capable of creating their own food supply by photosynthesis.

autunite $(Ca(UO_2)_2(PO_4)_2.10-12H_2O)$ A uranium ORE MINERAL.

auxiliary plane The NODAL PLANE of an EARTHQUAKE orthogonal to the FAULT PLANE. See FOCAL MECHANISM SOLUTION.

avalanche A sudden, rapid movement of disaggregated ice, snow, earth or rock down a slope.

Avalonian orogeny An OROGENY affecting the region from Georgia, USA to Newfoundland, Canada from CAMBRIAN–ORDOVICIAN times.

aven See POTHOLE.

aventurescence The visual phenomenon exhibited by AVENTURINE which arises from the presence of small inclusions of coloured MINERALS.

aventurine A GEM variety of QUARTZ or, rarely, OLIGOCLASE, containing coloured inclusions, such as HAEMATITE (red) or chrome MICA (green).

average velocity For GROUNDWATER, the volume of water passing through a given area of rock in unit time divided by its POROSITY.

Aves (birds) A class of feathered vertebrates with an egg typical of the AMNIOTA. Possibly evolved from an ARCHOSAUR and first appearing in the fossil record as *Archaeopteryx* in the U. JURASSIC. Rare as FOSSILS because of the delicacy of their bones.

avulsion The process by which a channel changes course by a sudden diversion. The periodicity of avulsion is controlled by the rate of topographic change, fluctuations in discharge and occurrence triggering events, e.g. EARTHQUAKES.

awaruite (Ni_3Fe) A rare alloy of NATIVE METALS found in PERIDOTITES.

axial dipole A model in which the time-averaged GEOMAGNETIC FIELD is approximated by that of a single magnet aligned along the Earth's spin axis.

axial flattening (axial strain) STRAIN in which one PRINCIPAL STRAIN direction is much smaller than the similarly sized other two.

axial geocentric dipole A model of the GEOMAGNETIC FIELD in which it is approximated by that of an axial dipole at the Earth's centre.

axial modulus The ratio of longitudinal STRESS to longitudinal STRAIN when there is no lateral STRAIN.

axial plane The plane containing two crystallographic axes.

axial plane cleavage A set of CLEAVAGE planes subparallel to the AXIAL PLANE of a FOLD and related to the formation of the FOLD.

axial plane foliation A set of FOLIATION planes subparallel to the AXIAL PLANE of a FOLD and related to the formation of the FOLD.

axial rift (axial or median valley) A valley, 2–3 km deep and 30–50 km wide, found on slow-spreading OCEANIC RIDGES such as the mid-Atlantic Ridge.

axial strain See AXIAL FLATTENING.

axial valley See AXIAL RIFT.

axially symmetric extension EXTENSION in one PRINCIPAL STRAIN direction and equal SHORTENING in all orthogonal directions. See STRAIN ELLIPSOID.

axially symmetric shortening SHORTENING in one PRINCIPAL STRAIN direction and equal EXTENSION in all orthogonal directions. See STRAIN ELLIPSOID.

axinite $((Ca,Fe,Mn)_3Al_2(BO_3)_3(Si_4O_{12})(OH))$ A CYCLOSILICATE found in VUGS in GRANITES or their CONTACT METAMORPHIC zones.

axiolitic structure An intergrowth of elongate fibres of ALKALI FELDSPAR with CRISTOBALITE; the fibres have grown outwards from the sides of a linear FRACTURE in RHYOLITIC GLASS by solid state growth during DEVITRIFICATION, with the FRACTURE acting as the nucleus.

axis of rotation The axis passing through the centre of the Earth about which the motion of a PLATE can be described by a single angle of rotation. See EULER'S THEOREM.

axonometry The measurement of crystal axes.

azimuth The horizontal angle of a direction measured clockwise from north.

Azoic See ARCHAEAN.

azoic See ANAEROBIC.

azurite $(Cu(CO_3)_2(OH)_2)$ A blue, SECONDARY MINERAL of COPPER.

B

β**-quartz** See BETA QUARTZ.

B-type lead (Bleiberg-type lead) An ANOMALOUS LEAD whose isotopic ratios give an age older than its COUNTRY ROCKS, possibly because the lead was derived from lead remobilized from an older deposit.

back The roof of an underground working.

back-arc basin (back-arc sea, marginal basin/sea) An isolated marine BASIN behind a SUBDUCTION ZONE, formed either by BACK-ARC SPREADING or by a step-back in the location of UNDERTHRUSTING adding oceanic LITHOSPHERE to the leading edge of the overriding PLATE. Up to three generations may be developed behind a single SUBDUCTION ZONE.

back-arc sea See BACK-ARC BASIN.

back-arc spreading A process whereby new LITHOSPHERE is created by rifting and subsequent SEAFLOOR SPREADING in the overriding slab of a SUBDUCTION ZONE, to form a BACK-ARC BASIN.

backfill The refill introduced into mine or quarry workings to support the worked area, ameliorate subsidence or dispose of waste materials.

backfold (retrocharriage) A FOLD with a VERGENCE in the opposite sense to the majority of FOLDS in a region.

backlimb The limb of an ASYMMETRIC FOLD with the lower DIP. Cf. FORELIMB.

backreef facies A sedimentary facies developing on the landward side of a CORAL-ALGAL REEF after growth and CEMENTATION. cf. FOREREEF FACIES.

backset bed A CROSS-BED inclined against the flow direction.

backshore The area of a BEACH between normal high TIDE level and the highest point reached by marine action such as storm waves.

backswamp An area of waterlogged land adjacent to a river, containing fine-grained sediment and organic matter, common where AGGRADATION of FLUVIAL deposits along a river channel formed LEVÉES which are overtopped during peak DISCHARGE.

backthrust A low-angle REVERSE FAULT with a VERGENCE different from the majority of REVERSE FAULTS in the area.

backwash The GRAVITY-fed return flow of water after a wave breaks on a BEACH. An important factor in determining BEACH gradient. Cf. SWASH.

backwearing (parallel retreat) ERO-SIONAL hillslope retreat without change in slope morphology. Occurs where the rate of EROSION on different sections of the slope is proportional to the dip. Cf. DOWNWEARING.

bacteria (Schizomycophyta) Primitive, microscopic (usually < 1 μm) organisms; range ARCHAEAN to Present.

baddeleyite (ZrO_2) A zircon MINERAL, a by-product of working certain CAR-BONATITES.

badland A landscape produced by the extensive incision and EROSION of weakly COHESIVE rocks consisting of deep gullies and ravines separated by steep ridges, small MESAS and BUTTES. Usually devoid of vegetation which has been stripped by EROSION.

bafflestone An AUTOCHTHONOUS carbonate rock whose original components were bound organically during deposition, the organisms forming baffles to trap finer MATRIX material. Cf. BINDSTONE.

Bagenov A JURASSIC succession in W. Siberia equivalent to the TITHONIAN.

bajada A low-lying area of confluent PEDIMENT slopes and ALLUVIAL FANS at the base of mountains around a desert.

Bajocian A STAGE of the JURASSIC, 173.5–166.1 Ma.

Bakhchisaraian A PALAEOCENE/EOCENE succession in the former USSR covering parts of the THANETIAN and YPRESIAN.

Bala The youngest sub-PERIOD of the ORDOVICIAN, 463.9–439.0 Ma.

balanced section (restored section) A CROSS-SECTION in which folded and/or faulted STRATA are arranged into their form prior to DEFORMATION. Provided certain factors obtain, the balanced section can be used to check the veracity of the initial, deformed section, which should be an ADMISSIBLE SECTION.

balas ruby A red, GEM variety of SPINEL.

ball clay (pipe clay) A fine-grained CLAY comprising up to 70% KAOLINITE plus ILLITE, QUARTZ, MONTMORILLONITE, CHLORITE and 2–3% carbonaceous material, important to the ceramics industry.

ball-and-pillow structure A STRUCTURE caused by the WET SEDIMENT DEFORMATION of interbedded SAND and MUD, characterized by globular protrusions and isolated pillows of SANDSTONE which form by differential settling of unconsolidated SAND into MUD.

ballas A variety of INDUSTRIAL DIAMOND in the form of very hard, dense, globular aggregates of minute, concentric DIAMOND crystals.

ballast Crushed rock used for road beds or on railway tracks.

ballistic ripple (impact ripple) The most common form of WIND RIPPLE which runs straight and orthogonal to the wind direction with wavelengths from <10 mm to 20 m and heights from a few mm to 1 m.

ballstone A SEDIMENTARY ROCK comprising subspherical NODULAR masses, usually calcareous, in an ARGILLACEOUS MATRIX.

Baltica (Baltoscandia) A continent made up of modern NW Europe which existed from CAMBRIAN to early DEVONIAN times.

Baltoscandia See BALTICA.

banakite A potassic volcanic rock similar to SHOSHONITE but with more silica.

Banan A TRIASSIC succession in China equivalent to the CARNIAN.

band ratio The ratio of the strength of a signal in different wavelength bands, often used in REMOTE SENSING to help discriminate different terrains or features.

band silicate See INOSILICATE.

band theory An approach to the calculation of the electronic structure of solids in which atoms arranged in a lattice have many neighbours so that the ATOMIC ORBITALS interact to form a band of molecular orbitals whose energies lie in certain energy ranges. The proximity of these bands governs the facility with which electrons could be excited into an empty orbital, where their high mobility would allow a high THERMAL and ELECTRICAL CONDUCTIVITY.

banded coal (humic coal) A COAL of heterogeneous origin banded at a centimetric scale by organic layers of varying appearance and diverse origin.

banded iron formation (BIF, cherty iron formation, itabirite, jaspillite) An iron-rich sediment with layers of CHERT or SILICA and iron MINERALS (commonly HAEMATITE) 5–30 mm thick and laminated at millimetric or submillimetric scale. Virtually all are of PRECAMBRIAN age and of ALGOMA- or SUPERIOR-TYPE. Constitutes a very considerable economic RESOURCE.

bank erosion The EROSION of a river bank by the detachment of particles or rotational FAILURE following undercutting.

banket An early PROTEROZOIC GOLD-

uranium-bearing QUARTZ PEBBLE CONGLOM-
ERATE of PLACER origin of the Witwatersrand
Goldfield, South Africa.

bankfull discharge The maximum DIS-
CHARGE possible in a river channel without
overlapping the banks.

bannisterite ((K,Na,Ca)Mn$_{10}$Al$_2$Si$_{15}$O$_{44}$.
10H$_2$O) A dark brown, TRANSLUCENT,
hydrated manganese silicate.

Banquereau A CRETACEOUS/PALAEOCENE
succession on the Scotian shelf of Canada
covering parts of the MAASTRICHTIAN and
DANIAN.

bar 1 The non-SI unit of pressure, equivalent to 10 PASCALS. **2** A linear deposit of
SAND/GRAVEL, generally parallel to subparallel to a coastline or river channel.

bar finger sand An elongate deposit at a
high angle to the coast caused by the PRO-
GRADATION of distributary channel mouth
systems in which the channels are fixed by
cohesive MUDS of the DELTA front and interdistributary bays.

Barabin A JURASSIC succession in W. Siberia
covering part of the CALLOVIAN and the
OXFORDIAN.

barchan An isolated, crescentic DUNE, 0.3–
30 m in height, with a gentle windward
slope and a SLIP-FACE bounded by two arms
extending downwind. Capable of movement with little change of shape at rates of
5–10 m a^{-1}.

barite (baryte) (BaSO$_4$) An INDUSTRIAL
MINERAL important because of its high
DENSITY.

Barker Index of Crystals A once popular
method of identifying crystals by measurement of INTERFACIAL ANGLES.

barkevikite ((Na,K)Ca$_2$(Fe,Mg,Mn)$_5$
(Si,Al)$_8$O$_{22}$(OH,F)$_2$) An obsolete name for
alumino-ferro-HORNBLENDE.

barranca A steep-sided gully formed by
EROSION, similar to a DONGA.

barred basin A partially restricted BASIN
in which free water circulation is impeded

by a barrier. Often ANOXIC or the site of
EVAPORITE deposition.

barrel The traditional unit for expressing
a volume of oil, equivalent to 35 Imperial
gallons or 42 US gallons.

Barremian A STAGE of the CRETACEOUS,
131.8–124.5 Ma.

barrier beach (bay bar) A COASTAL BAR
permanently exposed above sea level.

barrier boundary An hydraulic boundary
of an AQUIFER, particularly one preventing
expansion of a CONE OF DEPRESSION.

barrier island An island, mainly sandy,
elongate parallel to the shore and separated
from it by a marsh or LAGOON.

barrier reef A coastal CORAL-ALGAL REEF,
separated from the shore by a LAGOON.

Barrovian metamorphism REGIONAL
METAMORPHISM, first recognized in Scot-
land, in which zones of increasing META-
MORPHIC GRADE are characterized by the
appearance of a suite of INDEX MINERALS:
CHLORITE, BIOTITE, GARNET, STAUROLITE, KYAN-
ITE, SILLIMANITE.

Barrovian zone A zone of BARROVIAN
METAMORPHISM.

Bartonian A STAGE of the EOCENE, 42.1–
38.6 Ma.

baryte See BARITE.

barytocalcite (BaCa(CO$_3$)$_2$) A calcium-
barium double carbonate found in lead
VEINS.

basal sapping The process of undercutting the base of a slope by EROSION along
a SPRING-line, SALT WEATHERING or GLACIAL
action.

basal tar mat A band of HEAVY OIL and
TAR at the oil-water interface of an OILFIELD
which can plug pores and inhibit production by the prevention of natural WATER
DRIVE from maintaining RESERVOIR pressure.

basalt An APHANITIC MAFIC IGNEOUS ROCK
comprising PLAGIOCLASE FELDSPAR more calcic than An_{50} and PYROXENE, perhaps with
NEPHELINE, OLIVINE or QUARTZ and with

ACCESSORY iron-titanium oxide. Originates commonly from the PARTIAL MELTING of MANTLE PERIDOTITE and makes up layer 2 of the OCEANIC CRUST.

basaltic hornblende   See OXY-HORN-BLENDE.

basaltic layer   An outmoded term for the lower CONTINENTAL CRUST. Cf. GRANITIC LAYER.

basanite   A SILICA-UNDERSATURATED ALKALI OLIVINE BASALT containing OLIVINE, CLINOPYROXENE and PLAGIOCLASE FELDSPAR with >10% FELDSPATHOIDS in the form of NEPHELINE or LEUCITE. Found in association with other ALKALINE IGNEOUS ROCKS.

base level   The lower limit of subaerial EROSIONAL activity, usually defined by the level of running water, but with sea level as the general base level for continents.

base metal   A mining industry term for COPPER, lead, TIN or zinc.

base station   A reference location used during a geophysical survey.

base surge   A cloud of turbulent solids, possibly including water vapour, which travels rapidly along the ground from a large explosion or VOLCANIC ERUPTION and eventually deposits laminated and CROSS-BEDDED TEPHRA. Common in PHREATOMAGMATIC ERUPTIONS.

baseflow   The subsurface RUNOFF made up of THROUGHFLOW and/or GROUNDWATER.

baselap   A term employed in SEISMIC STRATIGRAPHY to describe the termination of a sequence along its lower boundary.

basement   The surface beneath which no SEDIMENTARY ROCK is found.

Bashkirian   An EPOCH of the CARBONIFEROUS, 322.8–311.3 Ma.

basic rock   A QUARTZ-free IGNEOUS ROCK containing calcic FELDSPARS and 45–55% SILICA, often with PYROXENE and OLIVINE.

basin   **1** A SYNFORMAL STRUCTURE with a subcircular OUTCROP. **2** A topographic depression containing, or capable of receiving, sediment.

basin floor   See ABYSSAL PLAIN.

Basin Groups 1–9   An ERA of the PRECAMBRIAN, 4150–3950 Ma.

basin plain   See ABYSSAL PLAIN.

basin-and-range   Terrain in which NORMAL FAULTS create interspersed block mountains and BASINS, frequently containing lakes.

bassanite (hemihydrate) $(2CaSO_4.H_2O)$ An EVAPORITE MINERAL which may be an intermediate stage in the formation of ANHYDRITE by the dehydration of GYPSUM.

bastard coal (batt)   **1** Thin layers of COAL found in SHALES immediately above a COAL SEAM. **2** Any COAL with a high ASH content.

bastard ganister   A SANDSTONE similar to GANISTER but containing more interstitial material and often with incomplete secondary SILICIFICATION.

bastard rock   An impure SANDSTONE with thin lenticular partings of SHALE or COAL.

bastite (schillerspar)   A variety of SERPENTINE.

bastnaesite $((Ce,La)CO_3(F,OH))$ A fluorocarbonate of lanthanum and cerium, an ORE MINERAL of the RARE EARTH ELEMENTS.

batch melting model   A PARTIAL MELTING model in which the melt remains at the melting site, in equilibrium with the solid, before migrating as a single batch of MAGMA.

Bath Stone   A soft JURASSIC FREESTONE of honey coloured, OOLITIC LIMESTONE used as a BUILDING STONE.

batholith   A large COMPOSITE INTRUSION with a surface area >100 km^2 made up of multiple PLUTONS of GABBROIC to GRANITIC composition.

Bathonian   A STAGE of the JURASSIC, 166.1–161.3 Ma.

bathyal zone   A mid-water ecological zone approximately 200–2000 m deep.

batt   **1** See BASTARD COAL. **2** A British term

for a hardened CLAY other than FIRECLAY.
3 A British term for compact, black, fissile, carbonaceous SHALE often intercalated with thin layers of COAL or IRONSTONE.

battery ore A manganese ORE of GRADE suitable for the production of MnO_2 for use in dry cell batteries.

bauxite An earthy rock composed almost wholly of aluminium hydroxide, often formed by the intense chemical WEATHERING of existing rocks in the tropics under high rainfall. The principal ORE of aluminium.

bauxitization The process of intense tropical WEATHERING of rock in which all soluble components and IRON OXIDE are removed to leave BAUXITE.

bay bar See BARRIER BEACH.

bayou A swampy creek or backwater.

bc joint See LONGITUDINAL JOINT.

beach The site of accumulation of sediment deposited by WAVES and currents around a sea or lake margin.

beach cusp A regular LUNATE feature, 1–60 m in width, occurring in the upper SWASH ZONE of a SAND or GRAVEL BEACH and probably formed by WAVE action.

beach deposit The unconsolidated sediment of a BEACH, dominated by SWASH-produced sedimentary STRUCTURES.

beach placer deposit A PLACER DEPOSIT found on ancient and modern BEACHES, especially well developed by storm conditions and LONGSHORE DRIFT. Important MINERALS include CASSITERITE, DIAMOND, GOLD, ILMENITE, MAGNETITE, MONAZITE, RUTILE, XENOTIME and ZIRCON.

beach ridge A linear accumulation of wave-deposited sediment on a BEACH.

beachrock A BEACH DEPOSIT lithified by CEMENTATION by MINERALS (normally ARAGONITE) which form as sea water evaporates.

bearing capacity The maximum load per unit area that a surface can support in safety without SHEAR FAILURE.

Becke line The bright line at the margin of a grain visible when viewed in THIN SECTION which can be used to determine the REFRACTIVE INDEX relative to another grain or the mounting medium.

becquerel The SI unit of radioactivity, the activity in which there is a decay of 1 nucleus s^{-1}.

becquerelite $(CaU_6O_{19}.11H_2O)$ An hydrated uranium oxide, formed by the ALTERATION of URANINITE.

bed 1 The smallest unit used in LITHOSTRATIGRAPHY. See also FORMATION, GROUP, MEMBER, SUPERGROUP. **2** See BEDDING.

bed roughness The frictional resistance to flow in a sedimentary system.

bedding (bed, stratum) A centimetric-to metric-scale layer of SEDIMENTARY ROCK bounded above and below by BEDDING PLANES.

bedding mullion A MULLION STRUCTURE consisting of a polished or striated BEDDING PLANE.

bedding plane A distinct surface separating two BEDS which marks a break in the continuity of sedimentation caused, for example, by a period of EROSION or cessation of sediment supply.

bedding plane cleavage A CLEAVAGE parallel to the BEDDING.

bedding plane schistosity The PREFERRED ORIENTATION of planar MINERALS parallel to BEDDING in a METAMORPHIC ROCK.

bedding plane slip The relative DISPLACEMENT of two BEDS during FLEXURAL SLIP FOLDING. May produce SLICKENSIDE STRIATIONS on the BEDDING PLANE.

bedding plane thrust A THRUST FAULT parallel to the BEDDING.

bedding-cleavage intersection The intersection of BEDDING and CLEAVAGE which produces a LINEATION often parallel, or nearly parallel, to a related FOLD AXIS.

bedform The shape of the surface of a bed

of granular sediment produced by fluid flow over it.

bedform theory A group of hypotheses attempting explanation of the physical reasons for the initiation, development, stability and characteristics of natural BEDFORMS.

bedload The material transported along the floor of a flowing fluid by rolling, sliding or SALTATION.

bedrock Unweathered rock beneath unconsolidated material.

beekite A variety of CHALCEDONY.

beforsite A CARBONATITE in which the dominant CARBONATE MINERAL is DOLOMITE.

beidellite $((Ca,Na)_{0.3}Al_2(OH)_2(Al,Si)_4O_{10}.4H_2O)$ A CLAY MINERAL of the MONTMORILLONITE group.

Belbekian See BODRAKIAN.

Belemnitida/belemnites An order of subclass COLEOIDEA, class CEPHALOPODA, phylum MOLLUSCA; MOLLUSCS characterized by a bullet-shaped internal shell. Range CARBONIFEROUS–EOCENE.

Bell gravimeter An axisymmetric marine GRAVIMETER with improved accuracy over beam-based instruments.

bell A conical NODULE or CONCRETION in the roof of a COAL mine which may fall without warning.

bell hole The cavity left after the fall of a BELL.

bell pit An obsolete, shallow mining method in which material extraction led to a bell-shaped excavation.

bell-metal ore See STANNITE.

Bellerophontida An order of class MONOPLACOPHORA, phylum MOLLUSCA; non-septate, planispirally coiled MOLLUSCS. Range CAMBRIAN–TRIASSIC.

belt of no erosion The area of a hillslope, extending away from the INTERFLUVE, where surface RUNOFF is incapable of eroding the soil surface. Generally on bare soil and often in BADLAND areas.

bench mark A site of known elevation above SEA LEVEL DATUM.

Bendigonian An ORDOVICIAN succession in Australia covering the lower part of the early ARENIG.

benefication The separation of desired MINERALS from GANGUE during exploitation of a MINERAL deposit, by which the MINERAL is concentrated prior to refining.

Benioff zone See BENIOFF-WADATI ZONE.

Benioff-Wadati zone (Benioff zone) The active seismic zone in a SUBDUCTION ZONE, originally mapped as a single plane of EARTHQUAKE FOCI dipping beneath the over-riding PLATE. Recently recognized as two planes, the less pronounced upper zone delineating the top of the downgoing slab, the main lower zone 10–20 km below, in which EARTHQUAKES originate from internal DEFORMATION of the slab.

benitoite $(BaTiSi_3O_9)$ A rare CYCLOSILICATE found in SERPENTINE.

benmoreite An alkali LAVA intermediate in composition between MUGEARITE and TRACHYTE.

benthic See BENTHONIC.

benthonic (benthic) Living on the sea floor.

benthos Organisms living on the seafloor.

bentonite A clay mainly made up of MONTMORILLONITE formed by the ALTERATION of volcanic ASH.

bentonite clays See SMECTITE CLAYS.

beraunite $(Fe^{2+}Fe^{3+}_5(OH)_5(H_2O)_4(PO_4)_4.2H_2O)$ An hydrated iron phosphate found in deposits of iron ORE.

bergschrund A CREVASSE at the head of a GLACIER separating mobile and stationary ice, which develops when a GLACIER moves downslope.

Bering land bridge The land connection of BERINGIA during the CENOZOIC, which was the only route available into North America for MAMMALS.

Beringia Originally the land bridge that linked North America and Asia in the region of the Bering Sea during the last ICE AGE. Now also includes NE Siberia and W. Alaska.

berm 1 A triangular feature orthogonal to the shore with a subhorizontal top and a more steeply dipping seaward surface found on certain BEACHES. **2** A narrow, man-made shelf on a slope of an OPENCAST MINE or quarry.

Bernoulli equation An equation describing the conservation of energy in a steady flow of an ideal, frictionless, incompressible fluid: $p/\rho + gz + v^2/2$ is constant along any STREAMLINE, where p = fluid pressure, ρ = fluid density, g = GRAVITY, z = vertical height above a datum, v = fluid velocity.

Berriasian The lowest STAGE of the CRETACEOUS, 145.6–140.7 Ma.

berthierine A green variety of CHAMOSITE.

Bertrand lens An auxiliary lens of the petrological microscope between the eyepiece and objective used to view INTERFERENCE FIGURES.

bertrandite $(Be_4Si_2O_7(OH)_2)$ An hydrated beryllium silicate found in PEGMATITES and exploited as a source of beryllium.

beryl $(Be_3Al_2(Si_6O_{18}))$ A CYCLOSILICATE found in PEGMATITES. A source of beryllium; coloured varieties may be valued as GEMSTONES, see AQUAMARINE, EMERALD.

beryllonite $(NaBePO_4)$ A rare, colourless to yellow GEM.

bessemer ore A valuable iron ORE with a phosphorus content <0.045%.

Besshi-type deposit (Kieslager deposit) A PROTEROZOIC and PALAEOZOIC, volcanic-associated, MASSIVE SULPHIDE DEPOSIT. Commonly a COPPER-zinc ORE with or without GOLD and SILVER and probably originating within a RIFT in a deep BACK-ARC BASIN accompanied by BASALTIC volcanism.

beta (β) pole diagram The represen-

tation on a STEREOGRAM of the GREAT CIRCLE traces of successive positions of a folded surface, which intersect at the β pole or axis. Cf. PI (π) DIAGRAM.

beta quartz (β-quartz, high quartz) A POLYMORPH of QUARTZ stable in the range 573–867°C. Cf. ALPHA QUARTZ.

betafite $((Ca,Fe,U)_{2-x}(Nb,Ti,Ta)_2O_6 (OH,F)_{1-z})$ An hydrated uranium niobate, tantalate and titanate of the PYROCHLORE group found in PEGMATITES.

biaxial stress (plane stress) A STRESS system in which the maximum PRINCIPAL STRESS is greater than the intermediate and the minimum PRINCIPAL STRESS is zero.

bieberite (cobalt vitriol) $(CoSO_4.7H_2O)$ An hydrated cobalt sulphate found as STALACTITES and encrustations in workings containing other cobalt MINERALS.

BIF See BANDED IRON FORMATION.

bifurcation ratio A quantitative measure of the rate at which a stream network bifurcates.

billabong Australian term for a permanent/semi-permanent waterbody on a FLOODPLAIN isolated by a change in mainstream location, e.g. an OXBOW lake.

bimodal Descriptive of a frequency distribution which contains two peaks, i.e. two values significantly more common than the rest.

binary system A chemical system of only two components.

binding coal See CAKING COAL.

bindstone An AUTOCHTHONOUS carbonate rock whose original components were bound organically during deposition, with the organisms binding finer MATRIX material together. Cf. BAFFLESTONE.

Bingham plastic A fluid which requires an initial YIELD STRESS to be overcome before it moves, e.g. some LAVA FLOWS.

biocenosis (biocoenosis, life assemblage) An assemblage of FOSSILS which lived together before death and fossilization. Cf. THANATOCENOSIS.

biochron The length of time represented by a BIOZONE.

bioclast A grain in a LIMESTONE consisting of the skeletal material of organisms.

bioclastic limestone A LIMESTONE in which the predominant grains are BIOCLASTS.

biocoenosis See BIOCENOSIS.

biodegradation The decomposition of CRUDE OIL by AEROBIC bacteria carried into the RESERVOIR by METEORIC WATER.

bioerosion trace A TRACE FOSSIL consisting of drillings or raspings on a hard substrate.

biofacies A body of SEDIMENTARY ROCK characterized by specific and distinctive biological characteristics. Cf. LITHOFACIES.

biogeochemical cycle The movement of chemical elements from organism to physical environment to organism.

biogenic Descriptive of material produced by organisms or their activities.

biogenic sedimentary structure See TRACE FOSSIL.

bioherm A discrete lens-shaped mass of organic origin, e.g. a shell bank.

biohorizon 1 An interface between STRATA at which a BIOSTRATIGRAPHIC change has occurred. 2 A BIOSTRATIGRAPHIC MARKER BED.

biokarst (phytokarst) KARST resulting from biological action, generally small-scale and especially common in the coastal zone, e.g. rock surfaces bored and abraded by marine organisms.

biolimiting Descriptive of the elements N, P and Si which, because of biological activity, are almost totally depleted in surface waters relative to deep waters.

biolithite A LIMESTONE in which there is evidence for the binding of grains by organic processes.

biome A complex, regional grouping of animal populations which coexist in a particular environment.

biomicrite A BIOCLASTIC LIMESTONE with a MATRIX of MICRITE.

biomineralization The CALCIFICATION of organic tissue.

biophile Descriptive of an element typically found in organisms.

biopyribole See PYRIBOLE.

biosome A sediment mass formed under uniform biological conditions.

biostasy The state of stability of a landscape when conditions favour WEATHERING and soil formation rather than EROSION and DENUDATION. Cf. RHEXISTASY.

biostratigraphy The subdivision and correlation of sequences of STRATA using FOSSILS.

biostratinomy The study of the *post-mortem* history of an organism before burial.

biostrome A BIOHERM consisting of a bedded accumulation of skeletal material.

biotite $(K(Mg,Fe)_3(AlSi_3O_{10})(OH)_2)$ A brown/green PHYLLOSILICATE MICA.

biotitization A type of WALL ROCK ALTERATION caused by the percolation of HYDROTHERMAL FLUIDS. It gives rise to new, often fine-grained BIOTITE which may replace and PSEUDOMORPH primary BIOTITE as well as growing elsewhere in the COUNTRY ROCK. Often the main feature of POTASSIC ALTERATION.

biotope 1 An area of uniform ecology. 2 An environment in which a certain assemblage of organisms lives or lived. Cf. ECOTOPE.

bioturbation The disruption of sediments mainly by the burrowing activities of organisms, e.g. by feeding.

biozone See ZONE.

bird 1 See AVES. 2 An instrument towed behind an aeroplane containing a package of instruments.

birdseyes See FENESTRAE.

birefringence The property shown by a MINERAL possessing TWO REFRACTIVE INDICES, so that a double image is produced through it. Interference colours are caused when viewed in polarized light.

birnessite ($Na_4Mn_{14}O_{27}.9H_2O$) A manganese MINERAL found in MANGANESE NODULES.

BIRPS **B**ritish **I**nstitutions **R**eflection **P**rofiling **S**yndicate. The consortium responsible for deep SEISMIC REFLECTION studies in the UK and elsewhere.

Birrimian orogeny An OROGENY affecting GREENSTONE BELTS in W. Africa during the PROTEROZOIC.

bischofite ($MgCl_2.6H_2O$) An hydrated magnesium chloride found in EVAPORITE deposits.

bismite (BiO_3) MONOCLINIC bismuth trioxide. Cf. SILLÉNITE.

bismuth A rare NATIVE ELEMENT found in certain VEIN deposits.

bismuth glance See BISMUTHINITE.

bismuthinite (bismuth glance) (Bi_2S_3) A bismuth ORE MINERAL.

bismutite (($BiO)_2CO_3$) Bismuth carbonate, found as a SECONDARY MINERAL in the oxidized zones of VEINS and PEGMATITES containing PRIMARY bismuth MINERALS.

bisulcate Descriptive of a shell with two grooves or depressions.

bitter lake A lake rich in sulphates and carbonates.

bitumen See ASPHALT.

bituminite See ALGINITE.

bituminous brown coal See PITCH COAL.

bituminous coal The normal type of COAL, intermediate in RANK between BROWN COAL/LIGNITE and ANTHRACITE, consisting of a mixture of BANDED and SAPROPELIC COALS and generally rich in volatile HYDROCARBONS.

Bivalvia/bivalves (lamellibranchs, pelecypods) A class of phylum MOLLUSCA; MOLLUSCS inhabiting a shell made up of two VALVES joined dorsally by a ligament and closed by adductor muscles. Gills are used for both respiration and filter feeding. Range CAMBRIAN–RECENT.

Bizon A SILURIAN succession in the Mirnyy Creek area of NE Siberia equivalent to the LUDLOW.

black diamond A variety of massive, CRYPTOCRYSTALLINE DIAMOND valued as an ABRASIVE.

black gold 1 PLACER GOLD coated with dark material so that its yellow colour is not seen. See also COATED GOLD. **2** A colloquial term for oil.

black oil See ASPHALT-BASED CRUDE OIL.

black sand An ALLUVIAL or BEACH PLACER with a preponderance of dark MINERALS, usually MAGNETITE or ILMENITE.

black shale A black/dark grey MUDSTONE rich in organic CARBON (>5% by weight), generally formed in ANOXIC marine bottom waters.

black smoker A high temperature (<400°C) hydrothermal vent in which a plume of hydrothermal fluids issue from a vent on an OCEAN RIDGE or submarine VOLCANO. The sulphides produced include PYRRHOTITE, PYRITE, MARCASITE, CHALCOPYRITE and SPHALERITE. May be the origin of certain MASSIVE SULPHIDE DEPOSITS. Cf. WHITE SMOKER.

blackjack A dark, iron-rich variety of SPHALERITE.

Blackwater A PERMIAN succession in Queensland, Australia, covering part of the WORDIAN, the CAPITANIAN, the LONGTANIAN and part of the CHANGXINGIAN.

blairmorite A rare type of PHONOLITE containing PHENOCRYSTS of ANALCIME.

blanket bog A type of BOG, often of PEAT, which drapes upland areas and infills hollows in regions of high precipitation and low evapotranspiration.

blanket sand An horizontally extensive, thin layer of SANDSTONE.

-blast Indicative of *in situ* growth during METAMORPHISM.

blasto- Descriptive of the complete or partial REPLACEMENT of a TEXTURAL element by crystals growing during METAMORPHISM.

Blastoidea/blastoids A subclass of class CYSTOIDEA, subphylum PELMATOZOA, phylum ECHINODERMATA; cystoids consisting of a pentamerally symmetrical cup (CALYX) of thirteen plates borne on a stalk. Range M. ORDOVICIAN–late PERMIAN.

blastomylonite A MYLONITE containing crystals, of much larger size than the fine-grained GROUNDMASS, that grew during DEFORMATION.

bleaching A weak WALL ROCK ALTERATION process in which colour becomes much less intense.

Bleiberg-type lead See B-TYPE LEAD.

blende See SPHALERITE.

Blenheim A PERMIAN succession in Queensland, Australia, covering the middle part of the WORDIAN.

blind layer (hidden layer) A layer which cannot be detected using FIRST ARRIVALS in SEISMIC REFRACTION because its velocity is less than that of the overlying layer and does not generate a HEAD WAVE, or because it is too thin to generate a FIRST ARRIVAL.

blind lode A LODE which does not reach the surface.

blind thrust A THRUST FAULT which does not reach the surface.

blind valley A steep-sided, river-cut valley which terminates in a steep cliff, common in LIMESTONE areas with subterranean flow.

blind vein A VEIN which does not reach the surface.

block **1** An angular fragment of rock with a diameter >256 mm. Cf. BOULDER. **2** A piece of CRUST or LITHOSPHERE of some kilometres or more which moves as a coherent body during tectonic events.

block-and-ash flow Type of PYROCLASTIC FLOW comprising a mixture of large BLOCKS and finer ASH, resulting from collapse of a LAVA DOME under its own weight or overpressure of confined gases.

block caving A large-scale method of underground mining in which the ORE is undermined and thus collapses into the excavation where it is sufficiently broken up for handling. Surface subsidence is inevitable.

block faulting NORMAL FAULTING giving rise to FAULT BLOCKS.

blockfield (blockmeer, felsenmeer, stone field) An accumulation of coarse detritus on a level or gently sloping surface in a mountainous area, comprising local rocks broken up by frost action.

blocking temperature **1** The temperature, usually a few tens of °C less than the CURIE TEMPERATURE, at which a THERMAL MAGNETIC REMANENCE is acquired and becomes fixed in a grain for at least 20 minutes in the absence of an external MAGNETIC FIELD. **2** The temperature at which gases used for RADIOMETRIC DATING become locked in a rock.

blocking volume The volume of a grain at which a CHEMICAL REMANENT MAGNETIZATION becomes locked in a rock.

blockmeer See BLOCKFIELD.

bloodstone (heliotrope) A green variety of CHALCEDONY containing red flecks of JASPER.

blowhole A vertical fissure in a SEA CLIFF through which WAVES and TIDES cause water to fountain to the surface.

blowout A violent extrusion of gas and oil from a WELL.

Blue John A mauve-purple FLUORITE from Derbyshire, England once used for the manufacture of ornaments.

blue asbestos (crocidolite) ($Na_2Fe_5Si_8$ $O_{22}(OH)_2$) A blue ASBESTIFORM MINERAL with fibres of good flexibility. The most dangerous ASBESTOS to health.

blue ground (hardebank) Fresh, resistant KIMBERLITE that often outcrops. Cf. YELLOW GROUND.

blue vitriol See CHALCANTHITE.

blue-green algae See CYANOBACTERIA.

bluehole A circular, steep-sided hole in a CORAL-ALGAL REEF, probably formed by KARSTIC processes at times of low sea level.

blueschist A low temperature, high pressure, REGIONALLY METAMORPHOSED rock containing abundant GLAUCOPHANE.

blueschist facies (glaucophane schist facies) A low temperature, high pressure METAMORPHIC FACIES characterized by the presence of the blue MINERAL GLAUCOPHANE, typically found within a SUBDUCTION ZONE environment.

bluestone **1** A common name for the imported stones of the Stonehenge monument, England, comprising DOLERITE, RHYOLITIC LAVA, TUFF and SANDSTONE, which probably came from the Preseli Hills, SW Wales. **2** See CHALCANTHITE.

boart See BORT.

bodden An irregularly shaped coastal inlet found in the S. Baltic, formed by a rise in sea level over uneven topography.

Bodrakian (Belbekian) An EOCENE succession in the former USSR covering part of the LUTETIAN and the BARTONIAN.

body wave (seismic body wave) A SEISMIC WAVE which travels through the interior of the transmitting medium, e.g. P WAVE, S WAVE. Cf. SURFACE WAVE.

body wave magnitude A measure of EARTHQUAKE MAGNITUDE based on the amplitude of body waves. Cf. SURFACE WAVE MAGNITUDE.

boehmite (AlO.OH) An ORE MINERAL of aluminium found in BAUXITE.

bog (peatland, mire) A wetland comprising accumulations of semi-decomposed plant matter arising from precipitation rather than GROUNDWATER. Cf. FEN.

bog burst A sudden disruption of a BOG which releases water and PEAT that flow a considerable distance.

bog iron ore A soft, porous LIMONITE found in freshwater environments formed by PRECIPITATION from iron-rich waters. Currently of no economic relevance.

bogaz An elongate, deep ravine in a KARST area formed by the widening of large JOINTS by SOLUTION.

Bøggild intergrowth A MICROSCOPIC intergrowth formed by EXSOLUTION in PLAGIOCLASE FELDSPAR.

boghead coal (torbanite) A SAPROPELIC COAL composed largely of ALGAL matter.

Bohdalec An ORDOVICIAN succession in Bohemia covering the LONGVILLIAN, MARSHBROOKIAN, ACTONIAN, ONNIAN and part of the PUSGILLIAN.

Bohemian garnet A variety of PYROPE in which red crystals are developed.

Bohm lamellae MICROSCOPIC FLUID INCLUSIONS in planar arrays, possibly formed by the ANNEALING of DEFORMATION LAMELLAE.

bole A FOSSIL LATERITE within a LAVA FLOW.

bolide The general term for an extraterrestrial body, such as a COMET or ASTEROID.

bolide impact hypothesis A mechanism proposed as the cause of MASS EXTINCTIONS in the FOSSIL record because of the considerable change in environment resulting from the impact of a large body.

Bolindian An ORDOVICIAN succession in Australia covering part of the ONNIAN and the ASHGILL.

bolson A trough or BASIN in an arid or semi-arid region with a PLAYA at its lowest point and the focus of local RUNOFF. Often of TECTONIC origin.

bomb See VOLCANIC BOMB.

bonanza A mining term for a rich OREBODY containing PRECIOUS METALS.

bond clay A highly plastic clay used to

bond less plastic materials in ceramics manufacture.

bone turquoise (odontolite) A FOSSIL tooth or bone coloured blue by iron phosphate and used as a GEM.

boninite An OLIVINE- and PYROXENE-bearing, PLAGIOCLASE-poor PILLOW LAVA with quench-textured PYROXENE and ACCESSORY MAGNESIOCHROMITE in a glassy GROUNDMASS. Found in volcanic ISLAND ARCS.

book structure **1** A STRUCTURE in an ORE deposit in which ORE alternates with GANGUE. **2** Structure developed in MICA crystals in which the CLEAVAGE is so well developed that the individual foliae have the appearance of pages.

bookshelf faulting A form of shear deformation in which adjacent FAULT BLOCKS are tilted like books on a shelf, with STRIKE-SLIP movement between them being of the opposite sense to the regional SHEAR sense.

boomer A marine SEISMIC SOURCE producing SEISMIC WAVES in the band 100–10 000 Hz. A high voltage is discharged through a coil embedded in an epoxy-resin block, producing EDDY CURRENTS in a spring-loaded aluminium plate which rapidly repel it and create compressional waves in the water.

Boomerangian A CAMBRIAN succession in Australia covering the upper part of the MENEVIAN.

booming dune See SINGING SAND.

boracite ($Mg_3ClB_7O_{13}$) A MINERAL which is a source of BORAX.

borax ($Na_2B_4O_5(OH)_4.8H_2O$) An EVAPORITE MINERAL which is an important INDUSTRIAL MINERAL.

bord and pillar mining See ROOM AND PILLAR MINING.

bore See TIDAL BORE.

boreal Of northern hemisphere origin. Cf. AUSTRAL.

borehole A hole drilled for exploration or exploitation purposes.

borehole breakout An IN SITU STRESS MEASUREMENT by the determination of the ELONGATION of a circular borehole in order to determine the orientation and magnitude of near-surface, *in situ* STRESS.

borehole geophysics See WELL LOGGING.

borehole gravimeter A specialized GRAVIMETER used in boreholes to determine the DENSITY of the WALL ROCK over a given vertical interval.

borehole logging See GEOPHYSICAL BOREHOLE LOGGING.

bornhardt A large, domed INSELBERG.

bornite (erubescite, purple copper ore) (Cu_5FeS_4) An ORE MINERAL of COPPER.

borolanite A variety of NEPHELINE SYENITE found in Scotland.

bort (boart) A badly coloured or flawed DIAMOND with no value as a GEM.

Bortonian An EOCENE succession in New Zealand covering parts of the LUTETIAN and BARTONIAN.

boss A body of PLUTONIC IGNEOUS ROCK with a circular plan and steep sides.

botryoidal Occurring as an AGGREGATE with rounded surfaces.

bottoming The base of an OREBODY.

bottomset beds The basal units of CROSS-STRATIFIED beds formed on the LEE-SIDE of a structure.

boudin (boudinage) The segmentation of a COMPETENT layer of rock surrounded by less COMPETENT material into a series of parallel, elongate STRUCTURES with rectangular to elliptical sections, caused by EXTENSION parallel to the layer orthogonal to the boudin. May be preceded by the development of a PINCH-AND-SWELL STRUCTURE which is subsequently cut into boudins by FRACTURES.

boudinage See BOUDIN.

Bouguer anomaly A GRAVITY measurement corrected for all non-geological sources of variation of GRAVITY, i.e. latitude, elevation, topography and, where appropriate, EARTH TIDE and EÖTVÖS effects.

Bouguer correction The correction to a GRAVITY observation for the attraction of the rock mass between the point of measurement and the chosen datum level.

boulangerite ($Pb_5Sb_4S_{11}$) A sulphosalt found in low- to medium-temperature HYDROTHERMAL VEIN deposits.

boulder A rounded rock fragment with a diameter >256 mm. Cf. BLOCK.

boulder clay (till) GLACIAL debris deposited directly from ice, comprising a wide variety of grain sizes. See GLACIAL DEPOSITION.

boulder tracing A method of MINERAL exploration, usually in a glaciated terrain, involving the tracing of mineralized BOULDERS back to their source.

Bouma sequence A fivefold division of successive sediments deposited by a waning TURBIDITY CURRENT, from base to top: A) structureless, most coarse-grained; B) PLANE BED in coarse- to fine-grained SAND; C) CURRENT RIPPLE LAMINATED bed in fine SAND-SILT; D) PLANE BED in SILT; E) structureless to very fine-grained in MUD. All five units may not be developed.

boundary layer The marginal region of a flow where frictional resistance causes the velocity to decrease near the boundary and SHEAR STRESSES are developed in the fluid.

boundary mapping (contact mapping) A method of GEOLOGICAL MAPPING involving the following of a geological contact via a zigzag route, used when EXPOSURE is good or the contact follows a topographic or vegetational feature.

boundary stratotype A CHRONOSTRATIGRAPHIC division comprising a sequence of rocks with standard reference points which are particularly complete at the sequence boundary. See GEOLOGICAL TIME-SCALE.

boundstone A LIMESTONE in which the grains were bound by an organism or organisms. See also BAFFLESTONE, BINDSTONE.

bourne An intermittent SPRING in CHALK forming when the WATER TABLE rises sufficiently high for water to flow in a normally DRY VALLEY.

bournonite ($PbCuSbS_3$) A SULPHIDE MINERAL found in HYDROTHERMAL VEINS associated with COPPER and lead mineralization.

bow-tie effect A feature of an UNMIGRATED SEISMIC REFLECTION section in which reflection events cross each other. Arises over a SYNFORMAL feature whose radius of curvature is less than that of the wavefront because reflections from different parts of the curved surface are focused onto the same portion of the section. Can be removed by MIGRATION.

Bowen's reaction series A series of MINERALS crystallizing from a MAGMA of specific chemical composition in which any MINERAL formed early in the series will later react with the melt to form a new MINERAL further down the series.

bowenite A yellow/green variety of SERPENTINE, sometimes used as a substitute for JADE.

bowlingite See SAPONITE.

box fold A composite FOLD with two ANTIFORMAL HINGES lying between two SYNFORMAL HINGES or vice versa.

boxstone A hollow CONCRETION.

boxwork A honeycomb-like STRUCTURE commonly found in GOSSANS which forms when residual LIMONITE remains in the cavity resulting when a sulphide grain is OXIDIZED.

BP Before Present, by convention taken as 1950.

Brachiopoda/brachiopods A phylum of solitary, bilaterally symmetrical, unsegmented marine invertebrates with a bivalved shell and a complex feeding apparatus (the lophophore). Range CAMBRIAN–RECENT.

brachyanticline An elongate PERICLINAL DOME with varying axial PLUNGE.

brachydont Descriptive of a tooth with

low, short crowns and well developed roots with narrow canals.

brachysyncline An elongate PERICLINAL BASIN with varying axial PLUNGE.

Bradydonti An order of subclass ELASMO-BRANCHII, class CHONDRICHTHYES, superclass PISCES; sharks with powerful crushing teeth for eating hard-shelled prey. Range end DEVONIAN–PERMIAN.

Bragg Law A law controlling X-RAY DIF-FRACTION. $n\lambda = 2d_{hkl}\sin\theta$, where n is an integer, λ the X-ray wavelength, d_{hkl} the spacing of the (hkl) planes of the crystal and 2θ the angle between the incident and diffracted X-ray beams.

braid A multithread channel formed, for example, by the meltwater flow from a GLA-CIER in a SANDUR.

braid bar An accumulation of sediment causing flow to divide, eventually forming an island at most flow states.

braided river/stream A river/stream that divides and rejoins around BARS of a width similar to the channel width and with a SINUOSITY of 1–1.3.

braidplain A gently sloping, extensive region covered by BRAID BARS and channels.

brammalite A variety of ILLITE in which sodium is the inter-layer cation.

branch line The location in an IMBRICATE FAULT system where a FAULT forks and DIS-PLACEMENT is transferred to another FAULT.

Branchiopoda/branchiopods A class of subphylum CRUSTACEA, phylum ARTHRO-PODA; small, bivalved animals enveloped by a carapace. Range L. DEVONIAN–RECENT.

braunite ($(Mn_2O_3)_3MnSiO_3$) A massive ORE MINERAL of manganese.

bravoite ($(Ni,Fe)S_2$) A rare nickel ORE MINERAL.

brazilian emerald A green GEM variety of TOURMALINE.

brazilian peridot TOURMALINE or CHRYSO-BERYL with the green colour of PERIDOT.

brazilian ruby A red TOURMALINE or pink TOPAZ.

brazilian topaz A clear blue variety of TOPAZ valued as a GEM.

brazilianite ($NaAl_3(PO_4)_2(OH)_4$) A rare yellow/green GEM found in PEGMATITES.

breached anticline An ANTICLINE whose core has been eroded so that the FOLD LIMBS form SCARPS.

breaching thrust A THRUST FAULT that off-sets an existing, structurally higher FAULT or FOLD STRUCTURE.

bread-crust bomb A VOLCANIC BOMB with a cracked outer crust and VESICULAR interior.

break-back thrust A REVERSE FAULT in a PIGGYBACK THRUST SYSTEM that forms in an existing THRUST STRUCTURE rather than nearer to the FORELAND.

break-point bar A permanently sub-merged COASTAL BAR formed near the shore when steep, high energy WAVES break, depositing sediment onshore on the sea-ward side of the break point and offshore landward of it.

breaker A WAVE that enters shallow water and increases in height until it breaks.

breaker zone The BEACH zone in which WAVE energy is dispersed by breaking.

breast The FACE of a mine working.

breccia A RUDITE with angular CLASTS.

breccio-conglomerate A RUDITE inter-mediate between BRECCIA and CONGLOMER-ATE, i.e. with approximately equal numbers of angular and rounded CLASTS.

breithauptite (NiSb) A nickel ORE MINERAL.

breunnerite A variety of MAGNESITE con-taining about 9% FeO.

breviconic Descriptive of a short CEPHALO-POD shell which expands rapidly.

brewsterite ($(Sr,Ba,Ca)(AlSi_3O_8)_2.5H_2O)$) A rare strontium-barium ZEOLITE.

brick clay A clay suitable for making

bricks, ranging from unlithified clay to MUDROCKS.

brick earth LOESS reworked by a river.

Brigantian A STAGE of the CARBONIFEROUS, 336.0–332.9 Ma.

bright coal The brightest type of COAL on the scale bright coal, banded bright coal, BANDED COAL, banded dull coal, DULL COAL.

bright spot A local increase in the amplitude (i.e. strength) of REFLECTED SEISMIC WAVES, produced by the presence of a strong contrast in REFLECTION COEFFICIENT, often indicating the presence of HYDROCARBONS in a RESERVOIR. Cf. DULL SPOT.

brine A concentrated aqueous solution of sodium chloride, in nature containing the cations Na^+, Ca^{2+}, K^+ and Mg^{2+} but with Cl^- the dominant anion, which is capable of LEACHING metals from the rocks through which it passes.

Bristol diamond A QUARTZ-filled AMYGDALE occurring in the MESOZOIC of the Bristol district, England.

brittle The property of a material that deforms by FRACTURE without appreciable VISCOUS or PLASTIC DEFORMATION.

brittle mica A MICA group lacking the alkali content of BIOTITE and frequently with calcium replacing magnesium. Includes CHLORITOID, CLINTONITE, MARGARITE, OTTRELITE and STILPNOMELANE.

brittle mineral A TENACITY descriptor of a MINERAL that breaks and powders easily.

brittle silver ore A popular name for STEPHANITE.

brittle stars See OPHIUROIDEA.

brittle strength The STRENGTH shown by a material which fails by FRACTURE. Cf. ULTIMATE STRENGTH.

brittle-ductile shear zone A SHEAR ZONE across which DISPLACEMENT is transferred by a combination of BRITTLE and DUCTILE processes.

brittle-ductile transition The transition from DEFORMATION by FRACTURE (i.e. CATA-CLASIS) to DEFORMATION by CRYSTAL PLASTICITY. Often refers to the depth in the CRUST at which this transition occurs.

brochantite $(Cu_4SO_4(OH)_6)$ A SECONDARY MINERAL of COPPER.

brockram A PERMO-TRIASSIC BRECCIA of northern England.

bromargyrite A rare SUPERGENE MINERAL.

bronzite $((Mg,Fe)SiO_3)$ A brown-green ORTHOPYROXENE.

bronzitite An IGNEOUS ROCK containing BRONZITE and lesser AUGITE and calcic PLAGIOCLASE, found in layered intrusions.

brookite (TiO_2) A rare titanium OXIDE MINERAL.

brousse tigrée A banding of vegetation in which closely spaced trees alternate with sparser bands.

brown asbestos See AMOSITE.

brown clay See RED CLAY.

brown coal A soft, low RANK COAL, including LIGNITE and SUB-BITUMINOUS COAL, with a CALORIFIC VALUE <19.3 MJ kg^{-1} and a fixed carbon content of 46–60%.

brown iron ore See LIMONITE.

brucite $(Mg(OH)_2)$ An INDUSTRIAL MINERAL found in LIMESTONES and where magnesium silicates have been ALTERED.

Brunhes A MAGNETOSTRATIGRAPHIC EPOCH of NORMAL POLARITY in the PLEISTOCENE, 0.78 Ma–present.

Bryophyta A division of kingdom plantae; spore-producing plants including the mosses, liverworts and hornworts. Range U. CARBONIFEROUS–RECENT.

Bryozoa A phylum of largely marine, colonial, moss-like invertebrates in which the animals are commonly housed in calcareous skeletons which make up a colony. Range ORDOVICIAN–RECENT.

bubble pulse The unwanted seismic signal generated by the oscillation of the bubble formed by an underwater explosion such as the discharge of an AIR GUN.

Bubnoff unit A unit used to quantify the rate of ground loss or slope retreat at right angles to the surface, equal to 1 mm ka^{-1} and equivalent to 1 m^3 km^{-2} a^{-1}.

Buchan zones A series of METAMORPHIC ZONES characterized by the successive appearance of ANDALUSITE, CORDIERITE, STAUROLITE and SILLIMANITE, first described from NE Scotland.

buchite A fine-grained to glassy HORNFELS produced by THERMAL or CONTACT METAMORPHISM of a CLAY-rich rock.

buckle fold A FOLD resulting from BUCKLING.

buckling (layer parallel shortening) FOLDING caused by the compression of a layered rock in the direction of the layering. The folded layers maintain their original thickness so the FOLDS produced are PARALLEL or CONCENTRIC.

buddingtonite (NH$_4$AlSi$_3$O$_8$) A rare FELDSPAR.

Budnanium A SILURIAN succession in Bohemia comprising the KOPANINA-SCHICHTEN and PRIDOLI-SCHICHTEN.

Bugando-Toro-Kibalian orogeny (Kibalian orogeny) An OROGENY of PRECAMBRIAN age affecting central and E. Africa at ~2075–1700 Ma.

buhrstone See BURRSTONE.

building stone Any rock suitable for use in construction.

Bulitanian A PALAEOCENE succession on the west coast of the USA covering part of the THANETIAN.

bulk finite strain The total FINITE STRAIN of a volume of rock calculated assuming that the STRAIN is homogeneous.

bulk material A general name for material used in construction, including CLAYS for bricks, LIMESTONE and SHALE for cement, SAND and GRAVEL for concrete, BUILDING STONE, ROAD STONE, AGGREGATE and BALLAST.

bulk modulus (incompressibility) An ELASTIC MODULUS defined as the ratio of the HYDROSTATIC PRESSURE to the DILATION.

Bunter A PERMIAN/TRIASSIC succession in Germany equivalent to the CHANGXING and SCYTHIAN. **2** A traditional British name for the SCYTHIAN EPOCH of the Lower TRIASSIC.

Buntsandstein A PERMIAN–Lower TRIASSIC succession in NW Europe covering part of the CAPITANIAN, the LONGTANIAN and the CHANGXINGIAN.

buoyancy A term describing the gravitationally powered ascent of relatively low DENSITY materials, such as HYDROCARBONS and EVAPORITES.

Burdigalian A STAGE of the MIOCENE, 21.5–16.3 Ma.

Burgess Shale An EXCEPTIONAL FOSSIL DEPOSIT of M. CAMBRIAN age in British Columbia where some 120 genera are preserved in aluminosilicates.

burial diagenesis (mesogenesis) The physical, chemical and biological processes acting on a sediment during burial as it is removed from the surface until the onset of METAMORPHISM or structural INVERSION.

burial metamorphism METAMORPHISM characteristic of thick sedimentary/volcanic sequences as they are rapidly buried in a sedimentary BASIN or OCEANIC TRENCH.

burmite An AMBER-like MINERAL, a variety of RETINITE.

burnt alum A porous, friable material produced by dehydrating ALUM at dull red heat, used industrially for dyeing, water purification, etc.

burrstone (buhrstone) A siliceous rock quarried in N. France since at least the 16th century and used for MILLSTONES.

burst An upward movement of low velocity fluid away from the lower parts of the turbulent BOUNDARY LAYER.

Burzyan The oldest PERIOD of the RIPHEAN, 1650–1350 Ma.

Bushveldt Complex A very large DIFFERENTIATED igneous complex of PRECAMBRIAN age in South Africa containing vast reserves of chromium, PLATINUM group metals and iron.

Busk Method See TANGENT-ARC METHOD.

bustamite $((Mn,Ca,Fe)SiO_3)$ A pink to brown PYROXENOID.

butane (C_4H_{10}) A colourless PARAFFIN SERIES gas present in NATURAL GAS.

butte A small isolated hill capped by resistant rock, possibly representing the former land surface.

butte témoin A BUTTE situated on the scarp side of a CUESTA of which it is a remnant.

bysmalith A large igneous intrusion or PLUTON of subcylindrical shape which has forced up and fractured the overlying COUNTRY ROCK.

byssate Descriptive of a BIVALVE attached to the substrate by tough, horny threads.

bytownite A PLAGIOCLASE FELDSPAR, An_{70-90}.

C

c joint See CROSS JOINT.

Caen stone A light creamy yellow LIME-STONE of JURASSIC age used as a BUILDING STONE in N. France and England.

Caerfai The oldest EPOCH of the CAMBRIAN, 570.0–536.0 Ma.

cafemic Descriptive of an IGNEOUS ROCK containing calcium, iron and magnesium.

CAI See COLOUR ALTERATION INDEX.

Cairngorm stone A semi-precious, GEM variety of SMOKY QUARTZ.

caking coal (binding coal) Any COAL that softens, melts and agglomerates on heating and quenching to produce hard COKE.

calamine See SMITHSONITE.

calanque (calas) A coastal inlet with a gorge-like form, possibly a KARSTIC and FAULT-controlled valley partially drowned by marine transgression.

calaverite (AuTe₂) An ORE MINERAL of GOLD.

calc-alkaline rock A rock with a higher concentration of calcium (CaO) than alkalis (Na₂O+K₂O) compared to ALKALINE IGNEOUS ROCKS. PLAGIOCLASE is thus the dominant FELDSPAR. Such rocks define the BASALT-BASALTIC ANDESITE-ANDESITE-DACITE-RHYOLITE OROGENIC ANDESITE association which characterizes SUBDUCTION ZONE environments.

calc-flinta (calc-silicate hornfels) A hard, fine-grained, calc-silicate rock formed by the CONTACT METAMORPHISM of an impure LIMESTONE.

calc-schist A metamorphosed, ARGIL-LACEOUS LIMESTONE with SCHISTOSITY.

calc-silicate hornfels See CALC-FLINTA.

calc-tufa See TUFA.

Calcarea A class of phylum PORIFERA; SPONGES with calcareous spicules making up the skeleton. Range CAMBRIAN–RECENT.

calcarenite A LIMESTONE with SAND-grade sized grains.

calcareous aggregate An irregular mass of carbonate grains which have been stuck together, often by microbial processes.

calcareous algae (blue-green algae, cyanobacteria) ALGAE in which calcium carbonate derived from life processes provides a skeleton for the whole or part of the plant. Common in fresh and shallow marine water, where they contribute to REEF and BIOCLASTIC LIMESTONES and provide both environmental and BIOSTRATIGRAPHIC markers. Range CAMBRIAN–RECENT.

calcareous ooze (carbonate ooze) A PELAGIC, biogenic sediment above the CAR-BONATE COMPENSATION DEPTH covering very large areas of the ocean floor, comprising the skeletal remains of calcareous micro-organisms, e.g. COCCOLITHOPOROIDS, FORA-MINIFERS and PTEROPODS.

Calcichordata A class of phylum ECHINO-DERMATA which may be ancestral to the FISH. Range CAMBRIAN–DEVONIAN.

calciclastic Descriptive of a CLASTIC carbonate rock.

calcification **1** A PEDOGENETIC process involving the accumulation of CALCITE or

DOLOMITE in soils low in moisture. **2** The process whereby organic tissue is converted to CALCITE.

calcilutite A LIMESTONE with MUD-grade sized grains.

calcirudite A LIMESTONE with very coarse grains.

calcisiltite A LIMESTONE with SILT-grade sized grains.

calcisphere A small (<500 µm) sphere of CALCITE, commonly found in PALAEOZOIC LIMESTONES, comprising a MICRITE coating around a hollow or SPARITIC interior, believed to be of ALGAL origin.

calcite (CaCO₃) The most common CAR-BONATE MINERAL, the principal component of LIMESTONE. Cf. ARAGONITE.

calcite compensation depth See CAR-BONATE COMPENSATION DEPTH.

calcrete (caliche, duricrust) A powdery, nodular to highly indurated, near-surface terrestrial material mainly composed of calcium carbonate, resulting from CEMEN-TATION and the introduction of CALCITE into soil, sediment and rock by GROUNDWATER in arid to semi-arid regions.

caldera A depression in the Earth's surface with a diameter >1.5 km formed by collapse into a MAGMA CHAMBER that has been vacated by the eruption or migration of MAGMA.

Caledonian orogeny A mid-PALAEOZOIC OROGENY affecting NW Europe and NE North America.

Caledonides The OROGENIC BELT formed by the CALEDONIAN OROGENY.

caliche 1 A nitrate deposit of the Atacama Desert of South America. **2** See CALCRETE.

Californian jade See CALIFORNITE.

californite (Californian jade) A compact green variety of VESUVIANITE used as a GEM substitute for JADE.

caliper log A borehole logging tool which measures the diameter of the hole; this var-ies with the COMPETENCE of the horizons penetrated.

calläis An archaeological term for MIN-ERALS, mostly VARISCITE, TURQUOISE and MALACHITE, used to make small blue or green beads in Neolithic and Copper Age times in W. Europe.

Callovian A STAGE of the JURASSIC, 161.3–157.1 Ma.

calomel (Hg₂Cl₂) A white-grey mercury chloride, found in association with CINNABAR.

calorific value The amount of energy released by the burning of one kilogram of COAL.

calyx 1 The cup-shaped body of a PELMATO-ZOAN made up of a number of plates. **2** A bowl-shaped depression on the top of a CORAL skeleton.

cambering The warping and sagging of relatively COMPETENT STRATA which overlie less COMPETENT STRATA, such as of CLAY, which tend to flow towards adjacent valleys so that a convex top is produced.

Cambrian The lowest SYSTEM of the PALAEOZOIC, 570–510 Ma, during which there was a vast radiation of shelled inver-tebrates.

camera A chamber within a chambered MOLLUSC, e.g. a NAUTILOID.

Camerata A subclass of class CRINOIDEA, subphylum PELMATOZOA; CRINOIDS with a CALYX of variable form in which all plates are united by rigid sutures. Range ORDOVICIAN–PERMIAN.

Campanian A STAGE of the CRETACEOUS, 83.0–74.0 Ma.

camptonite An alkali LAMPROPHYRE con-taining PHENOCRYSTS of BARKEVIKITE and/or KAERSUTITE AMPHIBOLE, AUGITE, OLIVINE and/or BIOTITE/PHLOGOPITE in a GROUNDMASS of calcium-rich PLAGIOCLASE, AMPHIBOLE and PYROXENE with minor ALKALI FELDSPAR, FELD-SPATHOIDS, APATITE, iron-titanium oxides and CARBONATE MINERALS.

Canada balsam A mounting medium for

THIN SECTIONS, with a REFRACTIVE INDEX of 1.54.

Canadian 1 The oldest sub-PERIOD of the ORDOVICIAN, 510.0–493.0 Ma. **2** See IBEXIAN.

Canadian asbestos See CHRYSOTILE.

cancrinite ($Na_6Ca(CO_3)(AlSiO_4)_6.2H_2O$) A rare FELDSPATHOID.

Candelaria A TRIASSIC succession in Nevada, USA equivalent to the SCYTHIAN.

canfieldite See ARGYRODITE.

Canglanpu A CAMBRIAN succession in China covering parts of the TOMMOTIAN and ATDABANIAN.

cannel coal A type of SAPROPELIC COAL composed of spores or fine organic fragments.

canyon A steep-sided, deep valley cut by a river, often as a result of REJUVENATION.

cap rock 1 (roof rock, seal) An impermeable rock lying above and sealing a gas or oil RESERVOIR. **2** A sheath around and over a SALT DOME composed of GYPSUM- or ANHYDRITE-bearing LIMESTONE.

Cape asbestos See CROCIDOLITE.

Cape ruby See PYROPE.

capillary fringe (capillary zone) The zone immediately beneath the WATER TABLE into which water can be drawn by capillary action, typically 0.1–3 m thick.

capillary habit A HABIT of MINERALS composed of flexible, thread-like crystals.

capillary pyrite See MILLERITE.

capillary zone See CAPILLARY FRINGE.

Capitan A PERMIAN succession in the Delaware Basin, USA equivalent to the CAPITANIAN.

Capitanian A STAGE of the PERMIAN, 252.5–250.0 Ma.

caprock model A theory that WATERFALLS occur where soft rock is eroded from under a harder rock, but which does not apply in all cases.

capuliform Cap-shaped.

Caradoc An EPOCH of the ORDOVICIAN, 463.9–443.1 Ma.

carat (karat) A unit of weight of DIAMONDS equal to 0.2 g, different from the measure used to describe the number of parts of GOLD per 24 parts of an alloy.

carbon (C) An element occurring free as DIAMOND and GRAPHITE, and combined in CARBONATE MINERALS, HYDROCARBONS and gases.

carbon cycle The progression of CARBON through the surface, interior and atmosphere of the Earth.

carbon-14 dating See RADIOCARBON DATING.

carbona A term used in Cornwall, England for large masses of rich TIN ORE.

carbonaceous chondrite A type of STONY METEORITE containing CHONDRULES and CARBON.

carbonado A variety of INDUSTRIAL DIAMOND taking the form of a black, CRYPTOCRYSTALLINE mixture of DIAMOND and GRAPHITE or amorphous CARBON.

carbonate buildup A REEF complex in the geological record or any large STRATIGRAPHIC accumulation of LIMESTONES and DOLOMITES.

carbonate compensation depth (calcite compensation depth, CCD) The level in oceanic sediment at which the rate of DISSOLUTION of calcium carbonate is the same as the rate of supply. Calcareous sediment does not accumulate below the CCD. Its level is controlled by the fact that carbonate DISSOLUTION increases with increasing pressure and decreasing temperature. These parameters increase with depth, but there are many other controlling factors.

carbonate minerals Common MINERALS containing the carbonate anion (($CO_3)^{2-}$). About 60 carbonate minerals are known, but the most common are ARAGONITE, CALCITE, DOLOMITE , MAGNESITE, RHODOCHROSITE and SIDERITE.

carbonate ooze See CALCAREOUS OOZE.

carbonate platform A water-covered, extensive (10^2–10^3 km wide), flat area of CARBONATE BUILDUP developed when a CRATON drowns, especially during phases of high sea level.

carbonate ramp A CARBONATE BUILDUP with the form of a gently sloping (<1°) surface passing into progressively deeper water.

carbonate shelf A CARBONATE BUILDUP which is similar to, but less extensive (few to hundreds of km wide) than, a CARBONATE PLATFORM.

carbonate-apatite ($Ca_5F(PO_4CO_3,OH)_3$) A variety of APATITE.

carbonate-hosted base metal deposit A STRATA-BOUND MINERAL DEPOSIT of lead and zinc, with or without COPPER, FLUORITE and BARITE, which occurs in thick sequences of LIMESTONE or DOLOMITE. Accounts for the majority of lead and zinc production in the USA and Europe.

carbonation 1 The process of introducing carbon dioxide into water. **2** The process of chemical WEATHERING in which MINERALS are replaced by carbonates, commonly by reaction with carbonated water.

carbonatite An IGNEOUS ROCK containing >50% CARBONATE MINERALS. Occurs as LAVA FLOWS, DYKES and SILLS and commonly associated with ALKALINE IGNEOUS ROCKS within RIFT systems. Formed by the derivation of carbonate-rich fluids from ascending MAGMAS originating from the PARTIAL MELTING of MANTLE PERIDOTITE.

carbonatization A type of WALL ROCK ALTERATION in which DOLOMITE, ANKERITE and other CARBONATE MINERALS form in the WALL ROCKS of EPIGENETIC MINERAL DEPOSITS.

Carboniferous A SYSTEM of the PALAEOZOIC, 362.5–290.0 Ma.

carbonization The decomposition of organic matter so that only a thin film of CARBON remains, which may retain features of the original organism.

carborundum An artificial ABRASIVE which has largely replaced CORUNDUM.

Carlin-type deposit A type of DISSEMINATED GOLD DEPOSIT occurring in carbonate rocks, probably formed by the deposition of MINERALS leached from underlying rocks by HYDROTHERMAL FLUIDS.

Carlsbad twin A TWIN, commonly found in the FELDSPARS, in which the twin axis is the c crystallographic axis.

carnallite ($KMgCl_3.6H_2O$) An EVAPORITE MINERAL.

carnelian (cornelian, carnellian) A red variety of CHALCEDONY used as a semiprecious GEM.

carnellian See CARNELIAN.

Carnian A STAGE of the TRIASSIC, 235.0–223.4 Ma.

Carnivora An order of infraclass EUTHERIA, subclass THERIA, class MAMMALIA; meateating placental MAMMALS. Range PALAEOCENE–RECENT.

carnosaur A carnivorous, bipedal, SAURISCHIAN DINOSAUR.

carnotite ($(K_2UO_2)_2(VO_4)_2.3H_2O$) An ORE MINERAL of uranium.

Carpoidea A bizarre group of CALCITEplated invertebrates similar to ECHINODERMS but differing in their lack of radial symmetry. Range L. CAMBRIAN–CARBONIFEROUS.

carrollite ($CuCo_2S_4$) A SULPHIDE MINERAL of the THIOSPINEL GROUP.

carse An ALLUVIAL FLOODPLAIN beside a river or ESTUARY.

carstone (iron pan) A brown SANDSTONE with a LIMONITE cement.

cascade fold A FOLD arising from GRAVITY collapse. See GRAVITY COLLAPSE STRUCTURE.

case-hardening The INDURATION of the surface of porous rocks, caused by the infilling of pore spaces by MINERAL cements precipitated from evaporating METEORIC WATER, soil moisture or GROUNDWATER under tropical to sub-tropical conditions.

Cassian A TRIASSIC succession in the Alps covering the upper part of the LADINIAN.

cassiterite (SnO_2) The major ORE MINERAL of TIN.

Castile A PERMIAN succession in the Delaware Basin, USA equivalent to the lower part of the LONGTANIAN.

Castlemainian An ORDOVICIAN succession in Australia covering the upper part of the EARLY ARENIG.

cat's eye Any GEMSTONE exhibiting CHATOYANCY, e.g. SPINEL, RUBY.

cataclasis (cataclastic flow) DEFORMATION by FRACTURE, sliding and rolling of rigid particles without internal STRAIN in which the STRENGTH of the material increases with CONFINING PRESSURE and DILATANCY.

cataclasite A cohesive FAULT ROCK with a random FABRIC containing 10–50% fragments in a finer grained MATRIX, probably formed as a result of CATACLASIS. Cf. PROTOCATACLASITE, ULTRACATACLASITE.

cataclastic flow See CATACLASIS.

cataphorite See KATOPHORITE.

catastrophism The postulate that important changes in the physical environment arise from major events of high magnitude, low frequency and short duration. Essentially the converse of UNIFORMITARIANISM.

catchment The total area of a drainage BASIN.

catena (toposequence) A sequence of soils found successively along a hillslope which are related by similar parental material.

cathodoluminescence (CL) The luminescence generated in MINERALS when bombarded by electrons. The phenomenon is used in PETROGRAPHIC studies to elucidate DIAGENETIC trends in SEDIMENTARY ROCKS and the compositional zoning of crystals in IGNEOUS or METAMORPHIC ROCKS.

cation exchange capacity (CEC) The amount of exchangeable cations that a MINERAL or soil can absorb at a given pH, expressed in mg equivalents per 100 g of material. These cations are mainly held on the surfaces of COLLOIDS.

cattierite (CoS_2) A SULPHIDE MINERAL of the DISULPHIDE GROUP.

cauldron The STRUCTURE underlying a CALDERA, comprising the subsided block and any RING DYKES.

cauldron subsidence The subsidence of cylindrical blocks of COUNTRY ROCK with MAGMA filling the space created to form RING DYKES.

caunter lode See COUNTER LODE.

caunter vein See COUNTER VEIN.

causse A LIMESTONE PLATEAU characterized by closed depressions, CAVES and AVENS.

Cautleyian A STAGE of the ORDOVICIAN, 440.6–440.1 Ma.

cave A hole or fissure in rock, usually large enough for the entry of a person.

cave coral A SPELEOTHEM of CALCITE in the form of a CORAL.

cave drapery A SPELEOTHEM in the form of a sheet, formed by fluid trickling along a dipping ceiling.

cave marble See CAVE ONYX.

cave onyx (cave marble) A SPELEOTHEM of CALCITE or ARAGONITE in a compact, banded form resembling ONYX.

cave pearl A SPELEOTHEM of CALCITE or ARAGONITE formed by the accretion of layers around a nucleus, such as a SAND grain.

cavitation erosion EROSION caused by the TURBULENT FLOW of meltwater at high velocity over rough BEDROCK under a GLACIER.

cay An island on a CORAL-ALGAL REEF made up of unconsolidated carbonates of SAND and GRAVEL grade.

Cayugan A SILURIAN succession in North America covering part of the GORSTIAN, the LUDFORDIAN and the PRIDOLI.

CBED See CONVERGENT BEAM ELECTRON DIFFRACTION.

CCD See CARBONATE COMPENSATION DEPTH.

CDP See COMMON DEPTH POINT.

CEC See CATION EXCHANGE CAPACITY.

cedar-tree laccolith A series of LACCOLITHS of the same IGNEOUS ROCK stacked vertically.

celadonite $(K(Mg,Fe^{2+})(Fe^{3+},Al)Si_4O_{10}(OH)_2)$ A green MICA which is an early AUTHIGENIC mineral in marine SANDSTONES.

celerity The velocity with which a wave advances, the product of its frequency and wavelength.

celestine See CELESTITE.

celestite (celestine) $(SrSO_4)$ An INDUSTRIAL MINERAL of strontium.

celsian $(BaAl_2Si_2O_8)$ A barium FELDSPAR.

cement bond log A GEOPHYSICAL BOREHOLE LOG which tests the quality of the cement bond between a borehole casing and its surrounding rock by measuring the ATTENUATION of a seismic signal.

cementation A DIAGENETIC process whereby AUTHIGENIC MINERALS are precipitated into the pores of sediments, causing them to become consolidated or lithified.

cementstone An ARGILLACEOUS LIMESTONE suitable for cement manufacture.

Cenomanian A STAGE of the CRETACEOUS, 97.0–90.4 Ma.

Cenozoic (Kainozoic) The youngest ERA comprising the PALAEOCENE to QUATERNARY, 65.0–0 Ma.

central eruption A VOLCANIC ERUPTION from a single VOLCANIC VENT. Cf. FISSURE ERUPTION.

cephalaspids See MONORHINA.

Cephalopoda/cephalopods A class of phylum MOLLUSCA, characterized by a differentiated head with well-developed eyes and a planispirally coiled, external shell (except for subclass COLEOIDEA), internally partitioned by calcareous septae. Range CAMBRIAN–RECENT.

cerargyrite (chlorargyrite, horn silver) $(AgCl)$ A SECONDARY MINERAL found in the oxidized zone of SILVER VEINS.

cerianite $((Ce,Th)O_2)$ A very rare OXIDE ORE MINERAL of thorium.

cerioid Descriptive of a CORAL in which the CORALLITES are polygonal and packed together.

cerussite (white lead ore) $(PbCO_3)$ An important ORE MINERAL of lead.

cesium vapour magnetometer A portable MAGNETOMETER capable of very precise measurement of the GEOMAGNETIC FIELD, often used on satellites and aircraft both singly and in pairs as a MAGNETIC GRADIOMETER.

Cetacea An order of infraclass EUTHERIA, subclass THERIA, class MAMMALIA; the whales, dolphins and porpoises. Range EOCENE–RECENT.

ceylonite See PLEONASTE.

cgs The centimetre-gram-second system of units.

chabazite $(Ca_2Al_2Si_4O_{12}.6H_2O)$ A ZEOLITE found mainly in cavities in BASALTIC rocks in association with other ZEOLITES, CALCITE and QUARTZ.

Chadian A STAGE of the CARBONIFEROUS, 349.5–345.0 Ma.

Chaetetida An order of class DEMOSPONGEA, phylum PORIFERA; SPONGES with a meandroid or CERIOID basal skeleton with fibrous, tufted, CALCITIC or ARAGONITIC walls. Range CAMBRIAN–RECENT.

Chaetognatha A phylum comprising the arrow worms, first appearing in the CARBONIFEROUS.

chain silicate See INOSILICATE.

chain structure group A group of SULPHIDE MINERALS characterized by a structure comprising rings, the only member being REALGAR.

chain width error A PLANAR CRYSTAL DEFECT in which the DISPLACEMENT at the fault has a component parallel to the fault plane. Cf. STACKING FAULT.

chalcanthite (bluestone, blue vitriol) ($CuSO_4.5H_2O$) A SECONDARY MINERAL of COPPER.

chalcedony (SiO_2) A MICROCRYSTALLINE form of SILICA.

chalcocite (redruthite) (Cu_2S) A COPPER ORE MINERAL.

chalcophile Descriptive of an element with a strong affinity for sulphur.

chalcopyrite ($CuFeS_2$) A major ORE MINERAL of COPPER.

chalcosiderite ($CuFe_6(PO_4)_4(OH)_8.4H_2O$) A rare SECONDARY MINERAL.

Chalk An informal name for the U. CRETACEOUS of northern Europe.

chalk A very fine-grained, white, porous LIMESTONE containing COCCOLITHS, common in the CRETACEOUS of western Europe.

Chalmak A SILURIAN succession in the Mirnyy Creek area of NE Siberia covering part of the RHUDDANIAN and the AERONIAN.

chalybeate Impregnated with iron salts.

chalybite An old name for SIDERITE.

chamosite ((Fe^{2+},Mg,Fe^{3+})$_5$Al(Si_3Al)O_{10} (OH,O)$_8$) An ORE MINERAL of iron found in MINETTE or CLINTON IRONSTONES.

Chamovnicheskian A STAGE of the CARBONIFEROUS, 299.9–298.3 Ma.

Champlainian An ORDOVICIAN succession in the E. USA covering part of the late ARENIG, the LLANVIRN, LLANDEILO, COSTONIAN, HARNAGIAN, SOUDLEYAN and LONGVILLIAN.

Chandler wobble The oscillation of the location of the Earth's rotational axis, with a period of 435 days, an amplitude of ~100 marc s and a decay time of ~40 years. Possibly caused by changes in the Earth's rotation rate resulting from changes in the magnetic coupling between the CORE and MANTLE.

Changshan A CAMBRIAN succession in China covering parts of the MAENTWROGIAN and DOLGELLIAN.

Changxing A PERMIAN succession in China equivalent to the CHANGXINGIAN.

Changxingian The highest STAGE of the PERMIAN, 247.5–245.0 Ma.

channel capacity The maximum DISCHARGE possible within a river channel.

channel resistance The resistance to flow in a channel arising from particle resistance, the resistance of bedforms, e.g. RIPPLES, DUNES, flow around channel bends and the spill resistance caused by changes in flow pattern at high FROUDE NUMBERS.

channel roughness See MANNING EQUATION.

channel sample A sample from channels cut across the FACE of an exposed ORE.

channel storage The capacity of a channel network to contain a flood DISCHARGE.

channel-belt deposit A sediment deposited between successive AVULSIONS in a river channel.

char COKE in the form of a granular mass of porous powder which can be briquetted.

chargeability A measure of INDUCED POLARIZATION in time-domain INDUCED POLARIZATION methods, equal to the area under the time-decaying voltage curve over a given time period divided by the steady-state voltage before measurement.

charnockite An ORTHOPYROXENE-bearing QUARTZ-FELDSPAR rock formed at high temperature and pressure, commonly found in GRANULITE FACIES METAMORPHIC terrains.

Charophyta Large, bushy, green CALCAREOUS ALGAE, mainly found in freshwater ponds and useful BIOSTRATIGRAPHICALLY from the CRETACEOUS to OLIGOCENE, when they underwent great diversification.

chatoyancy A MINERAL property caused by

the surface reflection of light to give a silky, banded appearance.

chatter marks Centimetric-scale, curved cracks found on glaciated surfaces, formed by the pressure of irregularly moving BOULDERS.

Chattian The higher STAGE of the OLIGO-CENE, 29.3–23.3 Ma.

Chautauquan A DEVONIAN/CARBONIFER-OUS succession in E. North America equivalent to the FAMENNIAN and HASTARIAN.

Chebotarev sequence An idealized temporal sequence of changes in GROUNDWATER, which becomes more charged the longer it is in contact with the AQUIFER. With increasing depth the anion sequence is bicarbonate, sulphate, chloride and the cations change from calcium to sodium.

cheirographic coast A coastline cut by successive deep bays and promontories, caused by TECTONIC activity.

chelate A very strong, multifunctional, claw-like bond formed between an organic molecule and a metal in CHELATION.

chelation The removal by complexing of metal atoms or cations in organic ring compounds by the formation of CHELATES during WEATHERING.

Chelicerata/chelicerates A subphylum of phylum ARTHROPODA whose body is divided into a combined head/thorax with usually six appendages, the first used for feeding and the remainder for locomotion and feeding, and an abdomen. No antennae are present. Range CAMBRIAN–RECENT.

Chelonia An order of subclass ANAPSIDA, class REPTILIA; the turtles and tortoises. Range TRIASSIC–RECENT.

cheluviation The combination of water containing organic extracts with soil cations by CHELATION. The solution then moves downwards by ELUVIATION and transfers metals to a lower level.

chemical and instrumental analysis of minerals Techniques for the identification of MINERALS, quantification of their chemis-

try and structure and elucidation of their development and TEXTURES. The methods include WET CHEMICAL ANALYSIS, X-RAY DIF-FRACTION, X-RAY FLUORESCENCE, ELECTRON MICROPROBE and MASS SPECTROMETRY.

chemical remanent magnetization (CRM, crystalline remanent magnetiz-ation) A REMANENT MAGNETIZATION acquired as FERROMAGNETIC grains grow through their BLOCKING VOLUME in an external MAGNETIC FIELD.

chemostratigraphy The use of chemical signatures (e.g. CARBON, oxygen and strontium isotopes, amino acid residues) in stratal sequences for the purpose of correlation.

chenier plain A coastal plain consisting of PROGRADATIONAL, alternate, coast-parallel bands of coarse CLASTIC and muddy sediment, occurring in environments undergoing periodic EROSION.

cheralite $((Th,Ca,Ce,La,U,Pb)(PO_4,SiO_4))$ A green radioactive MINERAL of the MONAZITE group, rich in thorium.

Cheremshanskian A STAGE of the CARBON-IFEROUS, 318.3–313.4 Ma.

chert (SiO_2) A granular MICROCRYSTALLINE to CRYPTOCRYSTALLINE variety of QUARTZ.

chertification The SILICIFICATION of a sediment into CHERT by the contemporaneous DISSOLUTION and PRECIPITATION of siliceous tests.

cherty iron formation See BANDED IRON FORMATION.

Chesterian A CARBONIFEROUS succession in the USA covering parts of the VISÉAN and SERPUKHALIAN.

chesterite $((Mg,Fe)_{17}Si_{20}O_{54}(OH)_6)$ A BIO-PYRIBOLE formed by the ALTERATION of ANTHOPHYLLITE.

Chev'ynskiy A PERMIAN succession in the Timan area of the former USSR covering parts of the WORDIAN and CAPITANIAN.

chevron construction An approximate method of constructing the geometry of a non-planar FAULT from the shape of its folded HANGINGWALL, which assumes that HEAVE is constant and conserved.

chevron fold (accordion fold, zig-zag fold) A FOLD with straight LIMBS and a sharp HINGE.

chevron mark A 'V'-shaped SOLE MARK.

Chewtonian An ORDOVICIAN succession in Australia covering the middle part of the EARLY ARENIG.

Chézy equation $v = C/(Rs)$, where v = mean flow velocity in a channel, C = Chézy roughness coefficient, R = HYDRAULIC RADIUS, s = channel slope.

chiastolite A variety of ANDALUSITE containing dark carbonaceous inclusions in the form of a cross.

chicken-wire texture A TEXTURE resembling a mesh developed in the sulphates which represents the PENECONTEMPORANEOUS DOLOMITIZATION of SABKHAS. Once taken as evidence of an ancient SABKHA environment, it is now realized that the TEXTURE is a common feature of the DIAGENESIS of sulphates.

Chientangkiang See JIANGTANGJIANG.

Chile saltpetre $(NaNO_3)$ Sodium nitrate, found in Chilean CALICHE.

chilled margin The fine-grained, outer layer of an IGNEOUS BODY formed by rapid cooling.

Chilopoda A class of superclass MYRIAPODA, phylum UNIRAMIA; the centipedes. Range CRETACEOUS–RECENT.

chimney An OREBODY of this shape lying within an ORE-locating STRUCTURE.

china clay A soft, white, plastic CLAY composed of KAOLINITE with low iron, formed by the ALTERATION of rocks rich in FELDSPAR.

china stone 1 KAOLINIZED GRANITE with unaltered FELDSPAR. **2** Certain very fine-grained LIMESTONES with a smooth TEXTURE.

chine A narrow ravine or canyon running to the sea.

chip sample A sample of rock chips taken from a FACE.

Chiroptera An order of infraclass EUTHERIA, subclass THERIA, class MAMMALIA; the bats. Range M. EOCENE–RECENT.

chitonozoans Small bag-shaped animals with a smooth or ornamented surface enclosing a chamber which may be spherical, cylindrical or conic. Range early ORDOVICIAN–CARBONIFEROUS.

Chixia A PERMIAN succession in China equivalent to the SAKMARIAN and ARTINSKIAN.

chloanthite $((Ni,Co)As_{3-x})$ An ORE MINERAL of nickel and cobalt.

chlorapatite $(Ca_5(PO_4)_3Cl)$ A variety of APATITE.

chlorargyrite See CERARGYRITE.

chlorastrolite An obsolete name for the fibrous, green variety of PUMPELLYITE.

chlorite $((Mg,Fe)_3(Si,Al)_4O_{10}(OH)_2.$ $(Mg,Fe)_3(OH)_6)$ An important group of PHYLLOSILICATES found in METAMORPHIC, SEDIMENTARY and ALTERED IGNEOUS ROCKS.

chloritization A type of WALL ROCK ALTERATION in which CHLORITE $\pm$ QUARTZ or TOURMALINE forms as the result of the passage of HYDROTHERMAL SOLUTIONS.

chloritoid $((Fe,Mg)_2Al_4O_2(SiO_4)_2(OH)_4)$ A NESOSILICATE of the HUMITE group.

Choanichthyes A group of FISH comprising the CROSSOPTERYGII and DIPNOI, which share lungs, nares and paired fins with fleshy lobes. Range L. DEVONIAN–RECENT.

chocolate block/tablet boudinage BOUDINAGE in which a COMPETENT layer is segmented into equidimensional STRUCTURES by layer-parallel EXTENSION in two directions.

Chokierian A STAGE of the CARBONIFEROUS, 328.3–325.6 Ma.

Chokrakian A MIOCENE succession on the Russian Platform covering the upper part of the SERRAVALLIAN.

Chondrichthyes A class of superclass PISCES; the sharks, skates and rays. Range SILURIAN–RECENT.

chondrite A STONY METEORITE containing CHONDRULES.

chondrodite $(Mg_5(SiO_4)_2(F,OH)_2)$ A NESOSILICATE of the HUMITE group.

Chondrostei An order of subclass ACTINOPTERYGII, class OSTEICHTHYES, superclass PISCES; FISH with a partly cartilaginous skeleton. Range L. DEVONIAN–RECENT.

chondrule A small, globular mass of PYROXENE, OLIVINE and occasionally GLASS found in certain STONY METEORITES.

chonolith An igneous intrusion which cannot be classified because of its irregular form.

Choquette and Pray classification A scheme for the classification of POROSITY types in carbonate rocks.

Chordata/chordates A phylum comprising animals with a rod of flexible tissue, protected in later forms by a backbone and first appearing in the CAMBRIAN.

Chotec A DEVONIAN succession in the former Czechoslovakia covering the upper part of the EIFELIAN.

chott A large, seasonally flooded desert BASIN.

chrome iron ore See CHROMITE.

chrome spinel See PICOTITE.

chromite (chrome iron ore) $(FeCr_2O_4)$ A SPINEL OXIDE MINERAL, the major ORE MINERAL of chromium.

chromitite A rock in which CHROMITE is the dominant MINERAL.

chron A small unit of geological time.

chronohorizon A thin, characteristic STRATIGRAPHIC interval that can be used for accurate time correlation or as a time reference zone.

chronosequence The development of a soil with time.

chronostratigraphic unit A STAGE or ZONE which, independent of FACIES, represents a layer of specific age.

chronostratigraphy A GEOLOGICAL TIME SCALE represented by a sequence of rocks with standard reference points.

chrysoberyl $(BeAl_2O_4)$ A rare MINERAL found in GRANITIC rocks, MICA SCHISTS and SANDS used as a GEM, particularly in the varieties ALEXANDRITE and CAT'S EYE.

chrysocolla $(Cu_4H_4Si_4O_{10}(OH)_8)$ A SECONDARY MINERAL of COPPER found in the oxidized zone of COPPER deposits.

chrysolite An OLIVINE with the composition Fo_{70-90}.

chrysoprase An apple green variety of CHALCEDONY.

chrysotile (Canadian asbestos, serpentine asbestos) $(Mg_3Si_2O_5(OH)_4)$ A CLAY MINERAL; a fibrous variety of SERPENTINE, possibly with an ASBESTIFORM habit.

Churchillian orogeny An OROGENY affecting the central part of the Canadian SHIELD in the PROTEROZOIC from ~1900–1850 Ma.

chute bar A sedimentary deposit formed at the downstream end of a narrow channel with a swift current.

Cimmeridian See KUYALNITSKIAN.

Cincinnatian An ORDOVICIAN succession in the E. USA equivalent to the MARSHBROOKIAN, ACTONIAN, ONNIAN and ASHGILL.

cinder cone An accumulation of SCORIA close to a VOLCANIC VENT resulting from a STROMBOLIAN ERUPTION.

cinders VESICULAR LAPILLI composed of dark GLASS.

cinnabar (HgS) The major ORE MINERAL of MERCURY.

cinnamon stone See HESSONITE.

CIPW classification An early system for the classification of IGNEOUS ROCKS based on NORMATIVE COMPOSITIONS.

CIPW norm A procedure for recomputing the chemical composition of a rock into a group of hypothetical standard MINERALS for simplified comparison between rocks.

circumgranular fracture (grain boundary fracture) A FRACTURE in a granular material which goes round the grains. Cf. INTERGRANULAR FRACTURE, INTRAGRANULAR FRACTURE.

cirque (corrie, cwm) A horseshoe-shaped, steep-walled valley head caused by GLACIAL EROSION.

Cirripedia A subclass of class MAXILLOPODA, subphylum CRUSTACEA, phylum ARTHROPODA; the barnacles. Range U. SILURIAN–RECENT.

citrine (Indian topaz, Madagascar topaz, quartz topaz, yellow quartz) A clear yellow variety of QUARTZ, valued as a semi-precious GEM.

CL See CATHODOLUMINESCENCE.

cladistics A taxonomic system applied to the study of evolutionary relationships which proposes that a common origin can be demonstrated by shared characteristics. It assumes that each new TAXON develops by the splitting of an ancestral lineage into two daughter TAXA.

cladogenesis The branching into two derivative TAXA as inferred in CLADISTICS.

cladogram A diagram illustrating the branching sequences in a CLADISTIC evolutionary sequence.

Cladoselachii An order of subclass ELASMOBRANCHII, class CHONDRICHTHYES, superclass PISCES; FOSSIL sharks with an elongate body and two dorsal fins with a spine. Range DEVONIAN–CARBONIFEROUS.

Claiborne An EOCENE succession on the Gulf Coast of the USA covering part of the LUTETIAN and the BARTONIAN.

claim An area of land staked out by a person who has certain rights to explore for and exploit MINERALS therein.

Clairault's formula An equation showing how GRAVITY varies over the reference SPHEROID, the basis of the GRAVITY FORMULA.

Clapeyron equation An equation linking the temperature of a MINERAL phase change, the volume change and the pressure.

clarain A type of BANDED COAL with bright to semi-bright bands of finely laminated COAL with a silky LUSTRE, often containing thin VITRAIN bands alternating with duller ATTRITAL material.

Clarence A CRETACEOUS succession in New Zealand comprising the URUTAWAN, MOTUAN and NGATERIAN.

clarite A MICROLITHOTYPE OF COAL, consisting of VITRINITE and EXINITE group MACERALS, which makes up to 20% of BITUMINOUS COALS.

clarodurain A LITHOTYPE OF BANDED COAL with characteristics intermediate between CLARAIN and DURAIN.

clarofusain A LITHOTYPE OF BANDED COAL with characteristics intermediate between CLARAIN and FUSAIN.

clast A particle of rock or single crystal which has been derived by WEATHERING and EROSION. The basic building block of a CLASTIC SEDIMENT.

clast-supported conglomerate (framework-supported conglomerate, grain-supported conglomerate, particle-supported conglomerate) A RUDITE which contains >85% CLASTS, which are mostly in contact. Cf MATRIX-SUPPORTED CONGLOMERATE.

clastation The disintegration of a rock into CLASTS by chemical or physical means.

clastic rock A SEDIMENTARY ROCK made up of CLASTS, classified according to CLAST size.

clastic sediment A sediment made up of CLASTS.

clathrates HYDROCARBON-bearing ice molecules which make up GAS HYDRATES.

clay A sediment with particles <4 μm in size.

clay ball See ARMOURED MUD BALL.

clay dune An AEOLIAN DUNE composed predominantly of CLAY AGGREGATES rather than QUARTZ SAND grains.

clay ironstone A sediment composed mainly of SIDERITE occurring in thin NODULES in some ARGILLITES.

clay minerals PHYLLOSILICATE MINERALS based on composite layers constructed from components with tetrahedrally and octahedrally co-ordinated cations. Mainly occurring as MICROSCOPIC, platy particles in fine-grained AGGREGATES which have varying degrees of PLASTICITY when mixed with water. They are hydrous silicates, mainly of aluminium, magnesium, iron and potassium, which lose adsorbed and constitutional water at high temperature to yield refractory materials. The most important groups are ILLITES, KANDITES, SMECTITES and VERMICULITES.

clay pan (soil crust) A compact ILLUVIAL layer of clayey material found in the soil zone formed by LEACHING and ELUVIATION of the upper soil zone.

claystone A CLASTIC SEDIMENTARY ROCK with the composition of SHALE but without its characteristic LAMINATION and FISSILITY.

clear water erosion The EROSION of a channel by a river whose sediment load has been removed, e.g. by the construction of a dam.

cleat Closely spaced JOINTING found in COAL SEAMS, generally in two orthogonal sets normal to the BEDDING.

cleavage 1 A MINERAL property whereby it breaks along regular, crystallographic planes. **2** A FOLIATION formed by DEFORMATION at low METAMORPHIC GRADE, along which a rock splits preferentially.

cleavage fan A radiating pattern of CLEAVAGE surfaces.

cleavage mullion A MULLION STRUCTURE comprising polished or striated cylinders bounded by CLEAVAGE PLANES.

cleavage plane A plane along which FRACTURE occurs in a rock.

cleavage refraction A change of orientation of CLEAVAGE, usually across a layer boundary.

cleavage tetrahedron A figure showing the four types of CLEAVAGE: CRENULATION, FRACTURE, PRESSURE SOLUTION and SLATY.

cleavage transection The oblique intersection of CLEAVAGE and a FOLD AXIS.

cleavage vergence The horizontal direction of rotation through the acute angle from the CLEAVAGE to an earlier FABRIC within the plane normal to the CLEAVAGE/FABRIC intersection.

cleavelandite A white, platy variety of ALBITE.

cleveite A variety of PITCHBLENDE with uranium oxide and RARE EARTH ELEMENTS.

Clifdenian A MIOCENE succession in New Zealand equivalent to the LANGHIAN.

climatic geomorphology The branch of GEOMORPHOLOGY covering the formation of landforms by climate.

Climatiformes An order of subclass ACANTHODII, class OSTEICHTHYES, superclass PISCES; FISH with bony jaws and a skeleton midway between sharks and bony fish. Range U. SILURIAN–L. PERMIAN.

climb The movement of a DISLOCATION by the diffusion of atoms normal to the structural displacement at the DISLOCATION in the formation of a CRYSTAL DEFECT.

climbing adhesion ripple structure An ADHESION STRUCTURE forming when net deposition occurs by ADHESION to a surface kept damp by the capillary rise of water.

climbing ripple A RIPPLE which climbs a gentle slope.

climbing dune A DUNE which climbs a gentle slope.

clino- Inclined.

clinochesterite $((Mg,Fe)_{17}Si_{20}O_{54}(OH)_6)$ A rare PYRIBOLE, probably formed by the ALTERATION of ANTHOPHYLLITE.

clinochlore A variety of CHLORITE.

clinoenstatite ($MgSiO_3$) A CLINOPYR-OXENE.

clinoferrosilite ($FeSiO_3$) A CLINOPYROXENE.

clinoform A major sloping depositional surface.

clinohumite ($Mg_9(SiO_4)(F,OH)_2$) A NESO-SILICATE of the HUMITE group.

clinohypersthene (($Mg,Fe)SiO_3$) A CLINO-PYROXENE.

clinojimthompsonite (($Mg,Fe)_{10}Si_{12}O_{32}$ $(OH)_4$) A MONOCLINIC PYRIBOLE with triple silicate chains.

clinometer A field instrument used to measure the magnitude of DIP.

clinoptilolite (($Na,K,Ca_{0.5})_6(Al_6Si_{30}O_{72})$. $24H_2O$) A ZEOLITE.

clinopyroxene A MONOCLINIC PYROXENE exhibiting non-parallel EXTINCTION. Cf. ORTHOPYROXENE.

clinothem A rock unit formed of STRATA which PROGRADE gently seawards to deep water.

clinozoisite ($Ca_2Al_3O(SiO_4)Si_2O_7(OH)$) A SOROSILICATE of the EPIDOTE group.

clint (flachkarren) A TABULAR block of LIMESTONE in a LIMESTONE PAVEMENT.

Clinton ironstone (Clinton ore) An OOL-ITIC HAEMATITE-CHAMOSITE-SIDERITE rock with ~40–50% iron, forming lenticular beds 2–3m thick. Formed in shallow water conditions.

Clinton ore See CLINTON IRONSTONE.

clintonite (xanthophyllite) ($Ca(Mg, Al)_{3-2}Al_2Si_2O_{10}(OH)_2$) A PHYLLOSILICATE; a BRITTLE MICA.

clitter A type of BLOCKFIELD made up of a scatter of GRANITE BOULDERS.

close fold A FOLD with an INTERLIMB ANGLE of 30–70°.

cluse A steep-sided valley cutting through a mountain ridge.

colorimetry A method of wet CHEMICAL ANALYSIS in which reagents are added to a solution of the unknown to form coloured compounds. The intensity of the colour is proportional to the concentration of the unknown.

Cnidaria (Coelenterata/coelenter-ates) A large phylum of living and extinct organisms which are the lowest animals with definite tissues, including the STROMA-TOPOROIDEA, SCYPHOZOA and ANTHOZOA.

co-ignimbrite breccia See LAG BRECCIA.

Coahuila A CRETACEOUS succession on the Gulf Coast of the USA comprising the DUR-ANGO and NUEVO LEON.

coal A combustible, organoclastic, SEDI-MENTARY ROCK containing >50% by weight of carbonaceous material and moisture and composed mainly of lithified plant remains.

coal ball A spheroidal mass of, commonly, CALCITE, DOLOMITE, SIDERITE and PYRITE in a COAL SEAM.

coal basin A sedimentary BASIN containing important COAL SEAMS.

coal gas (town gas) A gas produced by the distillation of BITUMINOUS COAL, with >50% hydrogen and 10–30% METHANE and a CALORIFIC VALUE of ~18MJ m^{-3}, used in heating and lighting.

Coal Measures A STRATIGRAPHIC term for the U. CARBONIFEROUS of western Europe, equivalent to the WESTPHALIAN.

coal measures A series of STRATA contain-ing economically workable COAL SEAMS.

coal quality map A contour map showing changes in the nature of a COAL SEAM over a mine area or COALFIELD.

coal seam A BED of COAL.

coal tar TAR produced by the distillation of BITUMINOUS COAL.

coalfield A region rich in COAL deposits.

coalification The process by which moist, partially decomposed vegetation such as PEAT is transformed, in response to burial or TECTONIC activity, progressively into BROWN COAL, BITUMINOUS COAL and ANTHRACITE with an increase in CARBON content and

CALORIFIC VALUE and a decrease in volatiles and moisture.

coarse-tail grading A feature shown by a GRADED BED in which only the larger grain sizes are graded. Cf. DISTRIBUTION GRADING.

Coast Range orogeny An OROGENY affecting the Coast Mountains of British Columbia in the JURASSIC and early CRETACEOUS, approximately equivalent to the NEVADAN OROGENY of the USA.

coastal aquifer An AQUIFER extending beneath the sea and accessible to seawater.

coastal bar A linear accumulation of mainly SAND grade sediment lying submerged in the nearshore zone subparallel to the coastline.

coastal notches Horizontal, slot-like recesses formed by EROSION at the base of a SEA CLIFF, normally at the high-water mark, and the main factor in the undermining of the cliff.

coated gold NATIVE GOLD covered by a surface film of IRON OXIDE which gives it a rusty or tarnished appearance. See also BLACK GOLD.

coated grain A sedimentary particle made up of successive, concentric layers, e.g. an OOLITH.

coaxial 1 Descriptive of any progressive DEFORMATION during which the PRINCIPAL STRAIN axes do not rotate with respect to reference lines within the rock. **2** Descriptive of FOLDS with a common AXIAL direction.

cobalt bloom An old name for ERYTHRITE, a SECONDARY MINERAL of cobalt.

cobalt glance See COBALTITE.

cobalt pentlandite (Co_9S_8) A SULPHIDE MINERAL of the METAL EXCESS GROUP.

cobalt vitriol See BIEBERITE.

cobaltite (cobalt glance) ((Co,Fe)AsS) A rare ORE MINERAL of cobalt.

cobble A rock fragment 64–256 mm in diameter.

Coble creep CREEP accomplished by the diffusion of individual atoms along grain boundaries.

Coblencien A DEVONIAN succession in France and Belgium covering the PRAGIAN and part of the EMSIAN.

Coccolithophoridae/coccolithophoroids Small (<20 µm), unicellular, spheroidal, PLANKTONIC ALGAE covered in calcareous plates, responsible for extensive marine carbonate production. Range U. TRIASSIC–RECENT.

coccoliths Minute calcareous plates which cover COCCOLITHOPHORIDAE, which contribute greatly to the formation of CHALK and deep sea OOZES and which are useful in the fine-scale STRATIGRAPHY of PELAGIC deposits of JURASSIC to PLEISTOCENE age.

cockpit karst (cone karst, kegelkarst) A KARST landscape of humid tropical areas comprising star-shaped, closed depressions separated by deep residual hills.

cockscomb pyrites A TWINNED form of MARCASITE.

COCORP **Co**nsortium for **C**ontinental **R**eflection **P**rofiling. The organization that undertakes deep SEISMIC REFLECTION profiling in the USA.

coefficient of friction See COULOMB CRITERION.

coefficient of roughness See CHÉZY EQUATION.

coefficient of work hardening/softening The slope of the STRESS-STRAIN curve as a rock of significant PLASTICITY is stressed.

Coelenterata See CNIDARIA.

Coelolepida An order of subclass DIPLORHINA, class AGNATHA, superclass PISCES; poorly known, tiny FISH with a forked tail and small mouth. Range late SILURIAN–L. DEVONIAN.

Coelosauria Small (crow-sized), bipedal, THEROPOD DINOSAURS similar in form to *Archaeopteryx*. Range U. TRIASSIC–CRETACEOUS.

coenosteum The laminated skeleton of a STROMATOPOROID.

coenozone See ASSEMBLAGE ZONE.

coesite (SiO_2) A high pressure form of SILICA.

coffinite ($U(SiO_4)_{1-x}(OH)_{4x}$) A black, hydrous uranium silicate.

cognate inclusion See AUTOLITH.

cognate lithic A non-VESICULAR CLAST found in a PYROCLASTIC ROCK consisting of juvenile material derived from the same, or closely related, MAGMA as the VESICULAR CLASTS in the rock.

cohesion (cohesive strength) The strength of bonding between particles or surfaces; in rock mechanics specifically the inherent SHEAR STRENGTH of a plane across which there is no NORMAL STRESS.

cohesive strength See COHESION.

cokability index A measure of the suitability of a COAL (only certain CAKING COALS of specific RANK) for making COKE.

coke A dense, porous product of the CARBONIZATION of CAKING COAL in an oven.

coking coal A COAL suitable for making COKE.

col A gap in a WATERSHED commonly caused by GLACIAL EROSION.

colatitude ($90° − \phi$), where ϕ = latitude.

cold working See DISLOCATION GLIDE.

colemanite ($CaB_3O_4(OH)_3.H_2O$) A MINERAL which is a source of BORAX, found in arid areas and saline lakes.

Coleoidea A subclass of class CEPHALOPODA, phylum MOLLUSCA; MOLLUSCS with an internal shell or lacking a shell, including many squids and the octopodids. Range U. MISSISSIPPIAN–RECENT.

collapse breccia (solution breccia) Chaotic, angular fragments of rocks formed when an underlying EVAPORITE layer is removed by DISSOLUTION.

collinite A component of the MACERAL VITRINITE in which no microscopic cell structure is visible.

collision zone A linear belt where two continents or MICROCONTINENTS have collided as the result of their intervening ocean having been consumed by SUBDUCTION.

colloform banding A TEXTURE found in certain MINERAL deposits in which crystals have grown in a radiating and concentric fashion, possibly resulting from geochemical controls.

colloid A dispersion of extremely fine particles in suspension, possible because the ultra-small size of the particles ($1–10 \mu m$) allows the supporting forces to exceed the gravitational forces promoting settling.

collophane (collophanite) Massive, CRYPTOCRYSTALLINE varieties of APATITE making up the bulk of PHOSPHATE ROCK and FOSSIL bone.

collophanite See COLLOPHANE.

colluviation The process by which COLLUVIUM forms.

colluvium Sediment transported by weakly selective, non-FLUVIAL processes such as MASS-WASTING and SLOPE-WASH. Cf. ALLUVIUM.

colonial coral A CORAL in which the exoskeleton is built by several animals.

colonnade A volcanic layer with COLUMNAR JOINTING which is regular and vertical. Cf. ENTABLATURE.

Colorado Plateau-type deposit (roll-front-type deposit, sandstone uranium-type deposit, western states-type deposit) A uranium-rich, SANDSTONE-uranium-VANADIUM BASE METAL DEPOSIT.

Colorado ruby A fiery-red variety of PYROPE.

Colorado topaz A brown-yellow variety of TOPAZ.

colour alteration index (CAI) A calibration of the colour change of a CONODONT element with temperature, used for the assessment of THERMAL MATURATION of sedimentary BASINS and the heating during thermal and TECTONIC events.

colour index (colour ratio) The total percentage of MAFIC (Mg- and Fe-rich) MINERALS in the MODAL ANALYSIS of an IGNEOUS ROCK, used as an aid to classification.

colour ratio See COLOUR INDEX.

coloured stone mining Mining for GEMS other than DIAMONDS.

columbite (niobite) ((Fe,Mn)Nb$_2$O$_6$) An ORE MINERAL of niobium found in GRANITIC rocks and PEGMATITES.

columnar jointing The vertical JOINT pattern developed during the cooling of a large body of VOLCANIC ROCK which divides it into regular polygonal columns. Thick bodies may consist of alternating COLONNADES and ENTABLATURES.

comagmatic Descriptive of IGNEOUS ROCKS derived from the same MAGMA.

Comanche A CRETACEOUS succession on the Gulf Coast of the USA comprising the TRINITY, FREDERIKSBERG and WASHITA.

comb See COMBE.

comb structure A TEXTURE shown by an IJOLITE intruded by VEINS, mineralogically similar to the COUNTRY ROCK, in which PYROXENE, WOLLASTONITE and FELDSPAR exhibit a PRISMATIC to ACICULAR HABIT with long MINERAL axes orthogonal to the walls of the VEIN.

combe (coombe, comb, coomb) A small, steep-sided valley which may contain an EPHEMERAL STREAM.

combination trap A combined STRUCTURAL and STRATIGRAPHIC OIL or GAS TRAP.

combined gold (AuTe) GOLD in combination with tellurium.

comendite A fresh, unaltered, PERALKALINE, silicic, GLASSY ROCK with <~12.5% NORMATIVE FEMIC MINERALS.

comet A relatively small interplanetary body (BOLIDE) comprising meteoric dust and frozen ices of H$_2$O, CO$_2$, CO and HCHO occupying eccentric orbits round the Sun and which may have rarely impacted on the Earth. Cf. ASTEROID, BOLIDE.

Comley Lowest SERIES of the CAMBRIAN.

comminution The breaking down of material into a fine powder.

commissure A line or plane of junction, such as between the VALVES of a shell.

common depth point (CDP) The location on a reflector which produces SEISMIC REFLECTIONS for a number of different source-receiver combinations. CDP STACKING generally improves the SIGNAL TO NOISE RATIO of a SEISMIC RECORD.

common lead Lead formed by the addition of radiogenic lead to primeval lead, probably with a simple history. Cf. ANOMALOUS LEAD.

compactibility The property of a sediment that allows COMPACTION.

compaction The process of volume reduction and PORE FLUID expulsion within a sediment in response to increasing overburden load, commonly expressed by the change in POROSITY.

companion sand SAND which may be present in a PARNA.

comparator electromagnetic method An ELECTROMAGNETIC INDUCTION EXPLORATION method using an artificial, time-varying electromagnetic field.

compensated load force The force on a continental block arising from the ISOSTATIC compensation of a load, such as at a CONTINENTAL MARGIN or PLATEAU UPLIFT.

compensator electromagnetic method An ELECTROMAGNETIC INDUCTION METHOD using a constant transmitter-receiver separation in which the primary electromagnetic field is compensated so that the instrument responds only to the small secondary fields generated by subsurface conducting bodies.

competent Of high relative STRENGTH.

complex dune See DRAA.

complex twin See COMPOUND TWIN.

compliance tensor A complete description, in terms of 36 components of ELASTIC MODULI, of the relationship between STRESS and STRAIN in an ANISOTROPIC material.

composite intrusion A type of MULTIPLE INTRUSION composed of MAGMAS of contrasting composition.

composite seam A COAL SEAM made up of two or more different COAL beds juxtaposed when intervening STRATA were wedged out.

composite volcano See STRATOVOLCANO.

composition plane The plane along which two TWINNED crystals are joined, often the same as the TWIN PLANE.

compositional layering A set of layers of distinct, different composition, usually applied to IGNEOUS or METAMORPHIC ROCKS in which the origin of the layering is equivocal.

compositional maturity index The ratio of QUARTZ + CHERT grains to FELDSPARS + rock fragments in a SANDSTONE, providing a measure of maturity as the former pair is relatively resistant to breakdown compared to the latter pair.

compound twin (complex twin) A crystal in which at least two types of TWINNING has occurred.

compressibility The inverse of the BULK MODULUS.

concealed coalfield A COALFIELD hidden beneath other STRATA.

concentration deposit A particularly rich EXCEPTIONAL FOSSIL DEPOSIT, formed by winnowing, slow deposition rate, PLACERS and concentration traps.

concentration factor The degree of enrichment of an element or elements above their normal crustal abundance in the formation of an OREBODY.

concentric fold A FOLD in which the layer thickness perpendicular to its surface is constant and the layer boundaries form concentric arcs of circles.

conchoidal fracture A MINERAL FRACTURE producing smooth, curving surfaces similar to the interior surface of a shell.

conchoidal fringe joint A shorter FRACTURE cross-linking EN ECHELON FRACTURES.

Conchostracha An order of class BRANCHIOPODA, subphylum CRUSTACEA, phylum ARTHROPODA; small, bivalved invertebrates enveloped by a carapace. Range L. DEVONIAN–RECENT.

concordant With margins parallel to the BEDDING or FOLIATION of the COUNTRY ROCK.

concordant coast A coast parallel to the general trend of relief with a consequent straight, regular outline.

concordant intrusion An igneous intrusion which does not cut across the BEDDING or FOLIATION of the COUNTRY ROCK.

concretion A NODULE without a concentric structure.

concurrent range zone A ZONE based on the co-occurrence of two or more species.

condensate A very light CRUDE OIL with API GRAVITY >50°.

condensed sequence A series of STRATA much thinner than developed elsewhere, probably caused by a diminished sediment supply.

Condoblin A succession in E. Australia of middle DEVONIAN age.

Condylarthra An order of infraclass EUTHERIA, subclass THERIA, class MAMMALIA; extinct ungulate-like MAMMALS, transitional between the UNGULATA and INSECTIVORA. Range late CRETACEOUS–U. MIOCENE.

cone karst See COCKPIT KARST.

cone of depression The inverted conical form adopted by the WATER TABLE after the extraction of water by pumping.

cone sheet A type of small RING INTRUSION with margins which DIP inwards, probably towards the centre of the upper part of a MAGMA CHAMBER. Possibly emplaced during the UPLIFT of a central conical block associated with a pressured MAGMA CHAMBER.

cone-in-cone structure A MINERAL STRUC-TURE in the form of a series of nested, concentric cones.

confined aquifer See ARTESIAN AQUIFER.

confining pressure The radial STRESS applied to a cylindrical unit deformed by axially symmetric loading.

conformability See CONFORMITY.

conformable Descriptive of a continuous series of BEDS, i.e. with no UNCONFORMITIES.

conformity (conformability) The STRATIGRAPHIC continuity of STRATA.

congelifraction (frost shattering, frost splitting, frost weathering, frost wedging) The fragmentation of rock by the expansion and contraction of water freezing and melting in pores, cracks or fissures.

congeliturbation See CRYOTURBATION.

conglomerate A RUDITE with rounded CLASTS larger than 2 mm.

conglomerate test A PALAEOMAGNETIC test for determining the age of a REMANENT MAGNETIZATION. If the NATURAL REMANENT MAGNETIZATION of the CLASTS of a CONGLOM-ERATE are randomly oriented, the magnetization predates its formation and vice versa.

congruent melting Melting in which the melt has the same composition as the solid. Cf. INCONGRUENT MELTING.

Coniacian A STAGE of the CRETACEOUS, 88.5–86.6 Ma.

conical fold A FOLD which approximates the shape of a conical surface.

conjugate Referring to a pair, or two sets, of FAULTS, SHEAR ZONES, ASYMMETRIC FOLDS, KINK BANDS, JOINTS, etc with opposing DIPS (FAULTS) or VERGENCES (FOLDS).

conjunctive use The combined use of GROUNDWATER and surface water, e.g. in ARTIFICIAL RECHARGE.

connate water The water deposited with, and contained within, a sediment, which may be changed in composition during burial, e.g. to a BRINE.

Conodontophora/conodonts Marine animals with vertebrate affinities represented in the FOSSIL record by scattered elements of their mineralized feeding apparatus. Generally 0.2–2.0 mm in size and composed of APATITE. Range U. CAMBRIAN–uppermost TRIASSIC.

Conrad discontinuity A SEISMIC DIS-CONTINUITY seen in certain regions of the CONTINENTAL CRUST at a depth of 10–12 km marking the boundary between the UPPER and LOWER CONTINENTAL CRUSTS.

consequent stream A stream flowing in the direction of the original slope of the surface. Cf. ANTECONSEQUENT, INCON-SEQUENT, INSEQUENT, OBSEQUENT, SUB-SEQUENT and RESEQUENT STREAMS.

conservation deposit An EXCEPTIONAL FOSSIL DEPOSIT in which soft tissues or the impressions of soft tissues are preserved.

conservative plate margin A PLATE MARGIN at which LITHOSPHERE is neither created nor destroyed, i.e. a TRANSFORM FAULT. Cf. DESTRUCTIVE PLATE MARGIN, CON-STRUCTIVE PLATE MARGIN.

consolidation The processes of COM-PACTION of a loose material.

constant separation traversing (CST, electric profiling, electric trenching) A RESISTIVITY METHOD in which the electrodes are maintained at a fixed separation and moved along a profile to map lateral variations in RESISTIVITY and locate buried three-dimensional bodies of anomalous RESISTIVITY. Cf. VERTICAL ELECTRICAL SOUNDING.

constructive plate margin (accretive plate margin) A PLATE MARGIN at which LITHOSPHERE is created, i.e. an OCEAN RIDGE. Cf. CONSERVATIVE PLATE MARGIN, DESTRUCTIVE PLATE MARGIN.

contact aureole (metamorphic aureole) A centimetric- to kilometric-scale zone of CONTACT METAMORPHISM around an IGNEOUS BODY.

contact mapping (boundary mapping) A technique of geological mapping in which a geological contact is followed by zigzagging along it.

contact metamorphism The THERMAL METAMORPHISM developed in rocks adjacent to the contact with an IGNEOUS BODY, reflecting the high temperature of the intrusion and possibly the expulsion of HYDROTHERMAL SOLUTIONS from the intrusion and the mobilization of GROUNDWATER in the COUNTRY ROCKS.

contact resistance The electrical resistance between an electrode and the ground.

contact twin A TWIN whose members are separated by a COMPOSITION PLANE. Cf. INTERPENETRANT TWIN.

contaminated magma A MAGMA of MANTLE derivation into which CRUSTAL material has been added by ASSIMILATION.

contamination See ASSIMILATION.

content grading (distribution grading) A gradual change in the entire grain size distribution in a GRADED BED.

continent-continent collision (continental collision) The collision of two PLATES of continental LITHOSPHERE by the consumption of an intervening ocean by SUBDUCTION beneath one of them to form a HIMALAYAN-TYPE MOUNTAIN RANGE. The buoyancy of continental LITHOSPHERE allows only very limited SUBDUCTION, but the persisting driving forces may power further INDENTATION TECTONICS.

continental collision See CONTINENT-CONTINENT COLLISION.

continental crust The upper layer of continent-bearing LITHOSPHERE, bounded at the base by the MOHOROVICIC DISCONTINUITY at a depth of ~20–80 km, at which there is a downward increase in density from ~3.0 to ~3.3 Mg m^{-3}. Its average composition is between GRANODIORITE and QUARTZ DIORITE. Sometimes divided into upper and lower continental crust by the CONRAD DISCONTINUITY. Cf. OCEANIC CRUST.

continental drift The hypothesis, largely attributed to A. Wegener in the early 20th century, that the continental blocks move relative to each other on geological timescales. Lack of knowledge of the ocean basins prevented the proposal of a cogent mechanism. Superseded by the theories of SEA FLOOR SPREADING and PLATE TECTONICS.

continental platform See CONTINENTAL SHELF.

continental reconstruction A PALAEO-GEOGRAPHIC configuration of continental blocks at some time in the past, determined by jigsaw fitting of their geometric shapes, rotations about EULER POLES, fitting together OCEANIC MAGNETIC ANOMALIES from either side of an OCEANIC RIDGE and PALAEOMAGNETIC and PALAEOCLIMATIC data.

continental rise The gently dipping part of the continental margin between the CONTINENTAL SLOPE and ABYSSAL PLAIN.

continental shelf (continental platform) The region of relatively shallow seafloor between the coastline and the CONTINENTAL RISE. Underlain by normal CONTINENTAL CRUST and usually less than about 200–400 m deep.

continental slope The steep slope between the CONTINENTAL SHELF and the more gentle slope to the ABYSSAL PLAIN. Usually underlain by thinned CONTINENTAL CRUST.

continental splitting The initiation of a new ocean by RIFTING of a continent and subsequent SEAFLOOR SPREADING.

continuous reaction series (reaction series) The continuous change in composition of a SOLID SOLUTION MINERAL as it retains equilibrium with a cooling MAGMA. Cf. DISCONTINUOUS REACTION SERIES.

continuous velocity log (sonic log, velocity log) A GEOPHYSICAL BOREHOLE LOG in which the SEISMIC VELOCITIES of the wallrock are determined by the measurement of the travel time of an ultrasonic pulse from one end of a SONDE to a GEOPHONE at the other.

contour current A permanent or semipermanent THERMOHALINE ocean current flowing subparallel to a continental slope, responsible for the deposition of CON-TOURITES.

contourite A muddy or sandy sediment reworked and deposited by a CONTOUR CURRENT.

contracting Earth An early model of Earth behaviour based on its thermal contraction over geological time as the result of cooling. Discounted since the discovery of RADIOACTIVITY and techniques which show the Earth has not contracted significantly over its history. Cf. EXPANDING EARTH.

contraction The decrease in length of a line during DEFORMATION. Cf. ELONGATION, STRETCH.

contractional kink band (reversed kink band) A STRUCTURE in which there is DISPLACEMENT of the upper side of a KINK BAND with respect to the lower in a manner similar to a REVERSE FAULT. Cf. EXTENSIONAL KINK BAND.

contrast enhancement See CONTRAST STRETCHING.

contrast stretching (contrast enhancement) An image processing technique used in REMOTE SENSING to reveal more information from a signal having a limited range of contrast.

contributing area The area of a CATCH-MENT contributing to storm RUNOFF.

convection The roughly circulatory or turbulent flow within a fluid whereby heat is transferred from one area to another at low THERMAL GRADIENT and approximately constant VISCOSITY. Driven by thermally induced density differences arising from heating at the base of a liquid and cooling from above.

convergent beam electron diffraction (CBED) A type of TRANSMISSION ELECTRON MICROSCOPY used to determine the symmetry of small crystals.

convergent evolution The tendency of unrelated species to evolve similar structures, physiology or appearance due to the same external factors.

convergent fan (normal fan) A CLEAV-AGE FAN radiating from a focus on the concave side of a folded layer. Cf. DIVERGENT FAN.

convergent plate margin See DESTRUC-TIVE PLATE MARGIN.

convolute Descriptive of a coiled shell in which the outer whorls embrace the inner ones so that they are nearly invisible. Cf. EVOLUTE.

convolute lamination An internal sedimentary STRUCTURE in SAND-SILT grade material in which the LAMINATION is disturbed or distorted into regular-irregular wavelength and amplitude FOLDS. Produced by LIQUEFACTION with subordinate FLUID-IZATION.

convolution A mathematical operation in which two functions are combined, e.g. convolving a function representing REFLEC-TION COEFFICIENTS with one representing a SEISMIC WAVE produces a SEISMOGRAM.

Cooley-Tukey method (Fast Fourier transform, FFT) An algorithm developed in the late 1960s which performs rapid FOURIER TRANSFORMATION. The increase in speed over previous techniques made possible the transformation of long signals into the frequency domain, in which many signal processing methods are much simpler.

cooling joint A JOINT in an IGNEOUS BODY forming perpendicular to the cooling surface by thermal contraction.

coomb See COMBE.

coombe See COMBE.

coordination number The number of anions surrounding a particular cation and forming a COORDINATION POLYHEDRON. Used in the description of CRYSTAL STRUCTURE.

coordination polyhedron The shape adopted by the anions around a particular cation. Used with COORDINATION NUMBER in the description of CRYSTAL STRUCTURE.

copal Tree resin which has not yet fully fossilized to AMBER, and may be up to 4 Ma old.

copaline See HIGHGATE RESIN.

copalite See HIGHGATE RESIN.

Cope's rule The phyletic trend towards increasing body size as an organism evolves.

Cope-Osborne theory A system of homologizing tooth cusps across the REPTILE-MAMMAL boundary and within the MAMMALS which assumed, perhaps incorrectly, that the single-pointed reptilian tooth grew extra cusps in order to become a mammalian molar.

copper (Cu) A NATIVE ELEMENT.

copper glance A mining term for CHALCOCITE.

copper pyrites An old name for CHALCOPYRITE.

copperas See MELANTERITE.

coppice dune See NEBKHA.

coppice mound See NEBKHA.

coprolite A FOSSIL made up of faecal material.

coquimbite $(Fe_{2-x}Al_x(SO_4)_3.9H_2O)$ An hydrated COPPER sulphate found in some ORE deposits and FUMAROLES.

coquina A carbonate rock made up of mechanically sorted debris, particularly shells.

coral sand A SAND-grade sediment of carbonate grains derived from the EROSION of a CORAL-ALGAL REEF.

coral terrace A terrace of CORAL elevated above the water by UPLIFT, such as on a GUYOT, or by a fall in sea level.

coral-algal reef A marine, largely biogenic, accumulation of calcium carbonate formed from a complex, productive biosystem by CORALS and ALGAE plus BRYOZOANS, GASTROPODS and serpulid worms. Such reefs only develop in shallow (<165 m), warm (21°C), saline (30–40 ppt) water free of suspended sediment.

Corallinaceae The most important group of the marine RHYDOPHYTA (CALCAREOUS red ALGAE). Range SILURIAN–RECENT.

corallite The skeleton of a single CORAL polyp.

coralloid Coral-like.

corallum The skeleton of a colonial CORAL.

corals Marine, solitary or colonial, polypoid invertebrates of class ANTHOZOA, phylum CNIDARIA, represented as FOSSILS by their calcareous skeletons which have radial to sub-radial symmetry. Range ?late PRE-CAMBRIAN–RECENT.

cordierite $((Mg,Fe)_2Al_4Si_5O_{18}.nH_2O)$ An hydrated CYCLOSILICATE found in METAMORPHIC ROCKS.

core 1 The central part of the Earth extending from the GUTENBERG DISCONTINUITY with the MANTLE at 2900 km to the centre of the Earth at 6370 km. The composition is predominantly iron and nickel, but it must contain some other element(s) to reduce the DENSITY to its known value. The outer core, from about 2900–4980 km, is fluid, and the site of generation of the GEOMAGNETIC FIELD, while the inner core, below 5120 km, is solid. **2** A cylindrical specimen of rock recovered by drilling.

core complex See METAMORPHIC CORE COMPLEX.

corestone A large COBBLE or BOULDER of relatively pristine rock found within a deep WEATHERING profile.

Coriolis force The effect of the Earth's rotation which deflects a body of fluid or gas moving relative to the Earth's surface to the right in the northern hemisphere and the left in the southern hemisphere. It is at a maximum at high latitudes and zero at the equator.

cornelian See CARNELIAN.

corner frequency The upper frequency limit of the SEISMIC WAVES generated by an EARTHQUAKE. This can be 0.05 Hz for very large events.

corniche An organic protrusion, often of CALCAREOUS ALGAE, growing from a steep rock surface near sea level, which forms a narrow walkway at the foot of a SEA CLIFF.

Cornish stone A crushed, partially WEATH-ERED GRANITE with appreciable FELDSPAR and KAOLINITE, used in the manufacture of bone china.

cornstone A concretionary LIMESTONE, usually formed under arid conditions.

corona structure Concentric zone(s) of at least one MINERAL, usually in radial arrangement, surrounding another MIN-ERAL. Formed by reaction with, or as an over-growth on, the primary MINERAL.

corrasion The mechanical EROSION by material transported across a surface by water, wind, ice or MASS MOVEMENT, the effect on the rock being termed ABRASION.

correlation The process of comparing two signals, such as SEISMIC RECORDS or STRATI-GRAPHIC logs, to determine the extent of their similarity at various different offsets.

corrie See CIRQUE.

corrosion A general term for chemical WEATHERING. Cf. CORRASION.

corrugation A large, linear feature on a FAULT PLANE of unknown origin.

corundum (Al_2O_3) An OXIDE MINERAL exploited as a REFRACTORY.

cosmopolitan distribution (pandemic distribution) The worldwide distribution of an organism.

cossyrite See AENIGMATITE.

Costonian The lowest STAGE of the ORDOVI-CIAN, 463.9–462.3 Ma.

cotectic Allowing the CRYSTALLIZATION of two or more solid phases from a single liquid over a finite temperature decrease.

coterminous With shared boundaries.

cottonballs A colloquial term for fine, fibrous crystals of ULEXITE.

Cotylosauria/cotylosaurs An order of subclass ANAPSIDA, class REPTILIA; stem-REPTILES with skulls pierced only by nose and eye openings. Range L. CARBONIFEROUS.

coulée See ARROYO.

coulée flow A very thick (up to 100 m), relatively short, flow of AA.

couloir 1 A deep gorge or ravine in the side of a mountain, possibly a result of AVA-LANCHE EROSION. **2** A parallel depression between YARDANGS in a desert area.

Coulomb failure criterion (Coulomb-Mohr failure criterion, Coulomb-Navier failure criterion) $\tau = S + \mu(\sigma - p)$ where τ = SHEAR RESISTANCE, μ = COEFFICIENT OF FRICTION, S = COHESION, σ = NORMAL STRESS and p = PORE FLUID PRESSURE. Thus the SHEAR RESISTANCE to faulting is reduced by the PORE FLUID PRESSURE within the material.

counter lode (caunter lode) A LODE trending in a different direction from that of the usual direction of the district.

counter vein (caunter vein) A VEIN trending in a different direction from that of the usual direction of the district.

counterpoint bar A river BAR deposited on the concave side of a channel bend because of a change in flow conditions or meander pattern migration.

country rock (host rock) The rock into which MAGMA or MINERALIZATION is intruded or emplaced.

Couvinien A DEVONIAN succession in France and Belgium covering part of the EMSIAN and the EIFELIAN.

covelline See COVELLITE.

covellite (covelline, indigo copper) (CuS) An ORE MINERAL of COPPER.

cow-dung bomb A VOLCANIC BOMB of characteristic shape.

crabhole An abrupt depression in the ground surface of centimetric- to metric-scale diameter and 50–600 mm depth, found in sediments susceptible to vertical cracking and horizontal packing.

crack-seal mechanism A mechanism of VEIN filling by repeated cycles of EXTEN-SIONAL FRACTURE followed by CEMENTATION.

crackle brecciation FRACTURES, usually healed with small VEINS of QUARTZ and other

MINERALS, that form the STOCKWORK mineralization in many DISSEMINATED MINERAL DEPOSITS, possibly formed by the release of volatiles from the host MAGMA on RETROGRADE BOILING.

crag-and-tail A streamlined ridge comprising a hill of resistant rock and a tail of less COMPETENT material, formed by GLACIER action.

crater 1 A large, bowl-shaped depression on the Earth's surface at the summit or flank of a VOLCANO. 2 A circular depression on a planetary surface caused by METEORITE impact.

craton The stable interior of a continental PLATE, unaffected by PLATE MARGIN activity since the PRECAMBRIAN.

cratonization (stabilization) The process of the transformation of a MOBILE BELT into a CRATON as the TECTONIC and thermal activity of the MOBILE BELT ceases.

creep A slow, largely continuous process of DEFORMATION occurring below the ELASTIC LIMIT in response to prolonged STRESS, i.e. all DEFORMATION that is not wholly ELASTIC.

creep strain Represented by the empirical expression: $\varepsilon = \varepsilon_e + \varepsilon_1(t) + v_t + \varepsilon_3(t)$, where ε = creep strain, ε_e = instantaneous ELASTIC STRAIN, $\varepsilon_1(t)$ = TRANSIENT CREEP, v_t = STEADY STATE CREEP and $\varepsilon_3(t)$ = accelerating CREEP, where t is time.

creep strength The long-term STRENGTH of a material, i.e. its STRENGTH at low STRAIN RATE. Cf. INSTANTANEOUS STRENGTH.

crenulation Small-scale folding or kinking.

crenulation cleavage (strain-slip cleavage) A MICROSCOPIC-scale CLEAVAGE formed by the folding of an existing FABRIC, generally resulting from the buckling of a previous SLATY CLEAVAGE.

Creodonta An order of infraclass EUTHERIA, subclass THERIA, order MAMMALIA; the more ancient order of slow-moving, small-brained, carnivorous, placental MAMMALS. Range late CRETACEOUS–PLIOCENE.

crescent-and-mushroom structure A FOLD INTERFERENCE STRUCTURE with the form of successive crescents and domes.

crescentic bar A COASTAL BAR formed by wave interaction, 100–200 m in length, lying just seawards of low water with horns pointing landwards.

crest 1 The top of a working face of an openpit mine or quarry. 2 The highest part of an ANTICLINE.

crest line A line joining the topographically highest points on a folded surface.

crest plane A plane joining the CREST LINES of a folded surface.

Cretaceous The youngest PERIOD of the MESOZOIC, 145.6–65.0 Ma.

Cretaceous-Tertiary boundary (K-T boundary) The end of the MESOZOIC ERA (65.0 Ma) at which time there was a MASS EXTINCTION event, related by some to the environmental effects of BOLIDE impact.

crevasse A deep fissure in a GLACIER.

crevasse splay A single or repeated flood event resulting from the breaching of a LEVÉE, in which coarse sediment is deposited on the landward side.

crianite A BASIC IGNEOUS ROCK comprising intergrown FELDSPAR, TITANAUGITE, OLIVINE and ANALCITE.

Crinoidea/crinoids A class of subphylum PELMATOZOA, phylum ECHINODERMATA; ECHINODERMS comprising the stalked sea lilies and unstalked feather stars. Range M. CAMBRIAN–RECENT.

cristobalite (SiO_2) A high temperature form of SILICA.

critical angle The angle of incidence of a SEISMIC WAVE on the interface to a layer of higher SEISMIC VELOCITY at which the refracted wave travels along the interface as a HEAD WAVE.

critical crack extension force (fracture toughness) A measure of rock STRENGTH which is independent of experimental procedure.

critical distance The distance from a SEISMIC SOURCE at which the HEAD WAVE first appears at the surface. Cf. CROSS-OVER DISTANCE.

critical flow Flow occurring when the flow velocity in a channel is the same as the wave velocity generated by a disturbance; the FROUDE NUMBER is one.

critical reflection A SEISMIC REFLECTION at the CRITICAL ANGLE, at which high amplitude waves are generated.

critical refraction A SEISMIC REFRACTION for which the angle of refraction is 90° and a HEAD WAVE is generated.

critical resolved shear stress (Schmid factor) The component of SHEAR STRESS parallel to a slip plane in a SLIP SYSTEM which controls the activity along that plane.

critical velocity The minimum velocity of a fluid required to entrain a particle.

critically refracted wave See HEAD WAVE.

CRM See CHEMICAL REMANENT MAGNETIZATION.

crocidolite (Cape asbestos) $(NaFe_3^{2+}$ $Fe_2^{3+}Si_8O_{22}(OH)_2)$ A sodic AMPHIBOLE occurring with an ASBESTIFORM habit as BLUE ASBESTOS.

Crocodilia An order of subclass ARCHOSAURIA, class REPTILIA; the crocodiles, alligators, caimans and gharials. Range TRIASSIC–RECENT.

crocoite $(PbCuO_4)$ A rare MINERAL found in the oxidized zone of lead deposits.

Croixian A CAMBRIAN succession in the E. USA equivalent to the MERIONETH.

crop management factor A term in the UNIVERSAL SOIL LOSS EQUATION.

cross fold A FOLD whose AXIS trends at an angle to the general FOLD trend in a region.

cross fracture A surface marking of a JOINT surface.

cross joint A JOINT perpendicular to the causative FOLD AXIS.

cross lode A LODE intersecting a larger one.

cross vein A VEIN intersecting a larger one.

cross-bedding See CROSS-STRATIFICATION.

cross-correlation A measure of the CORRELATION between two signals.

cross-lamination A millimetric scale CROSS-STRATIFICATION.

cross-over distance The distance along a SEISMIC REFRACTION profile at which the DIRECT SEISMIC WAVE from the SHOT is overtaken by the CRITICALLY REFRACTED wave. Cf. CRITICAL DISTANCE.

cross-section A diagrammatic representation in the vertical plane of the geology along a profile, usually constructed from a GEOLOGICAL MAP.

cross-set The fundamental unit of CROSS-STRATIFICATION.

cross-slip The movement of a DISLOCATION out of its plane. Cf. DISLOCATION GLIDE.

cross-stratification (cross-bedding, current-bedding, false bedding) The characteristic BEDDING STRUCTURE produced by the migration of bedforms with inclined depositional surfaces.

crosscut 1 A mining tunnel cut through COUNTRY ROCK to intersect an ORE-bearing STRUCTURE. **2** A horizontal underground tunnel which cuts a VEIN or OREBODY at a high angle. Cf. DRIFT.

crossite $(Na_2(Mg,Fe^{2+})_3(Fe^{3+},Al)_2Si_8O_{22}$ $(OH)_2)$ A sodic AMPHIBOLE intermediate in composition between GLAUCOPHANE and RIEBECKITE.

Crossopterygii An order of subclass SARCOPTERYGII, class OSTEICHTHYES, superclass PISCES; lobe-finned, bony FISH, possibly the ancestors of the lungfish. Range M. DEVONIAN–RECENT.

crotovina See KROTOVINA.

crucible swelling number A parameter indicating the suitability of COAL for COKE manufacture, involving the measurement of the profile of the COKE residue after heating a small sample of COAL to 800°C under standard conditions.

crude oil The liquid components of PETROLEUM, a mixture of HYDROCARBONS in which the majority belong to the PARAFFIN and NAPHTHENE SERIES.

Crudine A DEVONIAN succession in E. Australia equivalent to the LOCHKOVIAN.

crush belt (crush zone) A narrow belt of CRUST in which the rock has been broken and crushed, often by FAULT movement.

crush breccia A cohesive FAULT ROCK with <10% MATRIX, containing angular fragments >5 mm in size, formed by the TECTONIC reduction in grain size of the faulted rock. Cf. FINE CRUSH BRECCIA, CRUSH MICROBRECCIA.

crush microbreccia A cohesive FAULT ROCK similar to a CRUSH BRECCIA, but with fragments <1 mm in size. Cf. CRUSH BRECCIA, FINE CRUSH BRECCIA.

crush zone See CRUSH BELT.

crushing strength A parameter of a BUILDING STONE which describes its mechanical STRENGTH in sustaining loads and STRESSES in service.

crust The outermost solid layer of the Earth, distinguished chemically from the underlying MANTLE beneath the MOHOROVICIC DISCONTINUITY. OCEANIC and CONTINENTAL CRUST are markedly dissimilar.

Crustacea/crustaceans A subphylum of phylum ARTHROPODA; mainly marine invertebrates whose body is normally divided into head, thorax and abdomen, the head bearing five pairs of limbs, two pairs of antennae and maxillae and one pair of mandibles. Range CAMBRIAN–RECENT.

crustiform banding Layers of different MINERAL composition formed during the infilling of open spaces, commonly developed in dilatant spaces along FAULTS and solution channels in KARST by the permeation of mineralizing solutions.

cryergic Descriptive of the work of ground ice.

cryoconite A tubular depression in GLACIER ice formed by melting where a dark particle absorbs solar energy.

cryogenic magnetometer (squid magnetometer) A MAGNETOMETER in which the detecting system is maintained at superconducting temperatures by liquid helium, thus providing greater speed and sensitivity than other types.

cryolite (Greenland spar) (Na_3AlF_6) A fluoride MINERAL found in PEGMATITE VEINS and exploited as a flux.

cryoplanation A NIVATION process forming a flat ALTIPLANATION TERRACE.

cryosphere The part of the Earth comprising ice and frozen ground.

cryoturbation (congeliturbation, geliturbation) The process by which PATTERNED GROUND forms in response to the mixing activities of ice.

Cryptic An ERA of the PRECAMBRIAN, 4560–4150 Ma.

cryptic layering The gradual change in composition of cumulate MINERALS of a SOLID SOLUTION through layers in a CUMULATE IGNEOUS ROCK.

cryptocrystalline Formed of crystals only visible under very high magnification.

cryptodome A dome-like area of UPLIFT formed by an intrusion of, usually, ANDESITIC or DACITIC MAGMA.

cryptomelane (KMn_8O_{16}) A MINERAL found in manganese ORES.

cryptoperthite A sub-optical PERTHITE formed by the very rapid cooling of ALKALI FELDSPAR.

cryptovolcano A circular structure of highly deformed STRATA lacking direct evidence of formation by volcanic activity.

Cryptozoic See ARCHAEAN.

crystal class A classification of a crystal according to its symmetry. 32 classes exist, each of which is further subdivided into 7 CRYSTAL SYSTEMS.

crystal defect A departure from the usual regular arrangement of atoms in a crystal, comprising POINT DEFECTS, LINE DEFECTS and PLANAR DEFECTS.

crystal field See CRYSTAL FIELD THEORY.

crystal field theory A theory describing how the energy levels of atoms are perturbed by a regular array of nearby neighbouring negative charges, which produce a CRYSTAL FIELD.

crystal form See FORM.

crystal group See CRYSTAL SYSTEM.

crystal growth The process of PRECIPITATION following NUCLEATION by which a crystal grows spontaneously and is removed from solution.

crystal habit The relative development of faces shown by a crystal; faces large on some crystals may be small or missing on others.

crystal lattice The regular atomic arrangement of a crystalline solid portrayed as a three-dimensional array of regularly spaced points, each of which has the same arrangement of atoms in the same arrangement around it.

crystal morphology The shape and appearance of crystals, especially the regularity and symmetry of their faces and varied facial development. Also includes the regular internal arrangement of atoms and the precise geometrical arrangement of crystal features.

crystal plastic deformation A SLIP SYSTEM specified by crystallographic planes and directions.

crystal settling A mechanism for the formation of a CUMULATE IGNEOUS ROCK by concentration as MINERALS settle out of a MAGMA.

crystal structure The regular arrangement of atoms making up a crystalline solid, normally described in terms of UNIT CELLS, COORDINATION NUMBERS and COORDINATION POLYHEDRA. Usually determined by X-RAY DIFFRACTION.

crystal symmetry One of seven groups of crystals classified according to common symmetry characteristics.

crystal system (crystal group) The classification of crystals according to their particular types of CRYSTAL SYMMETRY. The seven crystal systems are CUBIC, HEXAGONAL, MONOCLINIC, ORTHORHOMBIC, TETRAGONAL, TRICLINIC and TRIGONAL.

crystal tuff A TUFF with crystal fragments more abundant than LITHIC or GLASS fragments.

crystal-liquid fractionation (fractional crystallization) The most important mechanism of MAGMATIC DIFFERENTIATION in which crystals separate from a melt by settling or floating, by flow in a conduit or by the CRYSTALLIZATION of early MINERALS on the walls of the MAGMA CHAMBER.

crystalline massif A portion of CONTINENTAL CRUST made up of METAMORPHIC or IGNEOUS ROCKS that is stable relative to its surroundings.

crystalline remanent magnetization See CHEMICAL REMANENT MAGNETIZATION.

crystalline rock An imprecise term for an IGNEOUS or METAMORPHIC ROCK, used in contrast to a SEDIMENTARY ROCK.

crystallinity The degree to which a rock exhibits crystal development.

crystallite 1 A very small, often imperfect crystal. **2** A minute inclusion in a GLASSY ROCK, indicative of incipient CRYSTALLIZATION.

crystallization The gradual formation of crystals from a liquid.

crystalloblastic texture (hypidioblastic texture) A TEXTURE formed when several MINERALS crystallize simultaneously during METAMORPHISM. Any of the MINERALS may be an inclusion of any other.

crystallographic preferred orientation An alignment or PREFERRED ORIENTATION of the axes of symmetry of CRYSTAL LATTICES in a deformed rock.

crystallographic symmetry elements The symmetry systems which allow the repetition of a basic structural unit in crystals. These are mirror planes, 2-, 3-, 4- and 6-fold rotation axes, centres of symmetry and 3-, 4- and 6-fold inversion axes.

cubanite ($CuFe_2S_3$) A SULPHIDE MINERAL of the WURTZITE GROUP.

cubic packing One end-member of the different ways perfectly sorted spheres can be arranged, in which orthogonal lines join the sphere centres. A sediment with such packing would have 48% intergranular POROSITY. Cf. RHOMBOHEDRAL PACKING.

cubic system (isometric system) A CRYSTAL SYSTEM whose members have three mutually perpendicular axes of equal length.

cubichnia TRACE FOSSILS made up of nesting structures.

cuesta An asymmetrical ridge caused by the differential EROSION of gently inclined STRATA, consisting of a steep scarp face, a well-defined crest and a gentle back slope accordant with the local stratal DIP.

cuirass A well-cemented DURICRUST covering the land surface, protecting underlying unconsolidated material from EROSION.

Cullen stone See KOLN STONE.

culm A soft, sooty COAL found in the CARBONIFEROUS rocks of SW England.

cummingtonite ($(Mg,Fe)_7Si_8O_{22}(OH)_2$) An AMPHIBOLE.

cumulate igneous rock An IGNEOUS ROCK characterized by a framework of touching or interlocking crystals and grains interpreted to have been concentrated by gravitational settling during MAGMATIC DIVERSIFICATION processes.

Cunningham A DEVONIAN succession in E. Australia approximately equivalent to the EMSIAN and EIFELIAN.

cupola A small, dome-like protuberance on a larger igneous intrusion.

cupriferous COPPER-bearing.

cuprite (Cu_2O) A red, TRANSPARENT, SECONDARY COPPER OXIDE MINERAL.

cuprouranite See TORBERNITE.

Curie Law 'The SUSCEPTIBILITY of a PARAMAGNETIC substance is inversely proportional to the ABSOLUTE TEMPERATURE.'

Curie temperature (Curie isotherm, Curie point) The temperature above which MINERALS cannot exhibit FERROMAGNETIC behaviour and only PARAMAGNETISM is possible.

Curie-Weiss law 'At temperatures above the CURIE POINT, the SUSCEPTIBILITY of a FERROMAGNETIC substance is inversely proportional to the difference between the temperature and the CURIE TEMPERATURE.'

current lineation A LINEATION caused by the movement of sediment or fluid.

current ripple A sinuous, crested BEDFORM formed by the transport and deposition of, usually, SAND-sized particles by unidirectional fluid movement.

current-bedding See CROSS-STRATIFICATION.

curtain See DRAPERY.

curtain-of-fire A line of coalescing FIRE FOUNTAINS simultaneously erupting along a fissure (i.e. a FISSURE ERUPTION).

cuspate fold profile A FOLD shape with the FOLD CLOSURE in one direction having a broad, arcuate shape and in the other direction a cuspoid shape, arising across an interface separating layers of different COMPETENCE.

cut-off An abandoned reach of a river channel, often produced where a meander loop is detached from the active river channel when the neck of the loop is breached. May be occupied by an OXBOW LAKE.

cut-off grade The lowest GRADE of ORE that can be exploited economically from an OREBODY.

cut-off limit See ASSAY BOUNDARY.

cut-off line (cut-off point) The intersection of a feature with a FAULT surface.

cut-off point See CUT-OFF LINE.

cutinite A COAL MACERAL of the EXINITE group made up of the waxy cuticular coatings of leaves and other plant tissues.

cuvette A non-TECTONIC, depositional BASIN.

cwm See CORRIE.

cyanobacteria (blue-green algae, Cyanophyta) Organisms similar to CALCAREOUS ALGAE but PROKARYOTIC. Range CAMBRIAN–RECENT.

Cyanophyta See CYANOBACTERIA.

cycle of erosion (Davisian cycle of erosion, geographical cycle) The sequence of DENUDATIONAL processes existing between the initial UPLIFT of an area and its reduction to a plane or PENEPLAIN close to BASE LEVEL.

cyclic evolution The concept that evolution was initially rapid and then followed by a longer phase of more gradual change.

cyclographic trace The trace of a GREAT CIRCLE on a STEREOGRAPHIC PROJECTION.

cyclosilicate (metasilicate, ring silicate) A CRYSTAL STRUCTURE classification in which the COORDINATION POLYHEDRA are Si tetrahedra and these form rings when each tetrahedron shares two corners with adjacent tetrahedra.

Cyclostomata An order of subclass MONORHINA, class AGNATHA, superclass PISCES; the lampreys, hagfish and slime eels. Range ?ORDOVICIAN–RECENT.

cyclothem A sequence of beds deposited in a single cycle of sedimentation, e.g. in COAL-bearing STRATA the sequence is SANDSTONE, SHALE, FIRECLAY, COAL, SHALE. Cyclothems were believed to reflect a variety of TECTONIC, climatic or sedimentological controls.

cylindrical fold (cylindroidal fold) A FOLD in which the FOLD PROFILE PLANE has the same orientation everywhere along the HINGE and the FOLD PROFILE is everywhere the same.

cylindroidal fold See CYLINDRICAL FOLD.

cymatogeny The large-scale warping of the CRUST over tens to hundreds of kilometres with little DEFORMATION, at a smaller scale than EPEIROGENY and less intense than OROGENY.

cymophane (Oriental cat's eye) A variety of CHRYSOBERYL exhibiting CHATOYANCY.

Cyprus-type deposit A type of VOLCANIC-ASSOCIATED MASSIVE SULPHIDE DEPOSIT, often containing COPPER, GOLD and zinc, associated with BASALTIC PILLOW LAVAS.

cyrtoconic Descriptive of a coiled GASTROPOD shell in the form of a curved, tapering cone.

Cystoidea/cystoids A class or superclass of subphylum PELMATOZOA, phylum ECHINODERMATA; sessile animals with an ovoid theca comprising irregularly arranged plates and five food-gathering arms. Range L. CAMBRIAN–U. PERMIAN.

D

D Letter used to indicate a phase of DEFOR-MATION, subscripted to denote each separate phase.

D″ layer The lowest 200–300 km of the MANTLE; a heterogeneous layer where there may be interaction with metallic CORE material.

dachiardite ((Na,K)$_{1.5}$Mg$_2$(Al$_{5.5}$Si$_{30.5}$O$_{72}$).18H$_2$O) A ZEOLITE found in GRANITE PEG-MATITE.

Dachsteinkalk A TRIASSIC succession in the Alps equivalent to the NORIAN and RHAETIAN.

dacite A SILICA-OVERSATURATED, INTERMEDI-ATE-ACID, CALC-ALKALINE LAVA, intermediate between ANDESITE and RHYOLITE in COMPO-SITION with PHENOCRYSTS of PLAGIOCLASE, minor OLIVINE, PYROXENE, AMPHIBOLE, BIO-TITE and Fe-Ti oxide in a fine-grained sil-iceous GROUNDMASS. Occurs both as LAVAS and PYROCLASTIC ROCKS.

Dacryoconarida/dacryoconarids A group of small, shell-bearing, PELAGIC MOL-LUSCS abundant in the DEVONIAN, possibly related to the CEPHALOPODA.

dagala See KIPUKA.

Dala A CARBONIFEROUS succession in China covering part of the BASHKIRIAN and the MOSCOVIAN.

Daleje A DEVONIAN succession in the former Czechoslovakia covering parts of the EMSIAN and EIFELIAN.

dalmation coast A flooded CONCORDANT COAST characterized by chains of islands and long inlets formed by the inundation of coast-parallel ridges and valleys arising from TECTONIC subsidence and/or sea level rise.

Dalradian The upper SUPERGROUP of the PRECAMBRIAN in Scotland and Ireland.

Dalslandian orogeny (Gothian orogeny, Gothic orogeny, Sveconor-wegian orogeny) The continuation of the PROTEROZOIC GRENVILLIAN OROGENY at 1100–1050 Ma into S. Sweden and S. Norway.

dambo A linear depression without a well-defined stream channel found on old, gently sloping land surfaces, especially in the tropics.

danburite (Ca(B$_2$Si$_2$O$_8$)) A yellow TECTO-SILICATE, occurring as an ACCESSORY MINERAL in PEGMATITES.

Danian 1 The lowest STAGE of the PALAEO-CENE, 65.0–60.5 Ma. **2** A PALAEOCENE suc-cession on the west coast of the USA covering the lower part of the DANIAN.

dannemorite ((Fe,Mn,Mg)$_7$Si$_8$O$_{22}$(OH)$_2$) A rare, manganese-bearing AMPHIBOLE.

daphnite A variety of CHLORITE rich in iron and aluminium.

Darcy equation $Q = KA\,dP/dx$, where Q = volume of flow per unit time through a porous medium, K = HYDRAULIC CONDUC-TIVITY, A = cross-sectional area of medium, dP/dx = HYDRAULIC GRADIENT.

darcy The unit of PERMEABILITY; defined as the PERMEABILITY which allows a flow of 1 mm s^{-1} of a fluid of VISCOSITY 10^{-3} Pa s through an area of 100 mm² under a pressure gradient of 0.1 atm mm⁻¹.

dark ruby silver See PYRARGYRITE.

Darriwilian An ORDOVICIAN succession in Australia equivalent to the LLANVIRN.

Dasycladales A group of green CALCAREOUS ALGAE; erect marine plants with a central stem from which branches radiate. Range L. CAMBRIAN–RECENT.

Datang A CARBONIFEROUS succession in China equivalent to the VISÉAN.

datolite ($CaB(SiO_4)(OH)$) A NESOSILICATE found as a SECONDARY MINERAL in cavities in BASALTIC LAVAS.

Datsonian A CAMBRIAN/ORDOVICIAN succession in Australia covering parts of the DOLGELLIAN and TREMADOC.

datum A fixed reference point.

daughter element An element formed by the radioactive decay of an existing element.

Davisian Cycle of Erosion See CYCLE OF EROSION.

Dawson Canyon A CRETACEOUS succession on the Scotian shelf of Canada covering part of the CENOMANIAN, the TURONIAN, CONIACIAN and part of the SANTONIAN.

daya A small, SILT-filled, solutional depression on a LIMESTONE surface in arid areas of the Middle East and N. Africa.

dB See DECIBEL SCALE.

de Broglie relationship $\lambda = h/p$, where λ = electron beam wavelength, h = Planck's constant and p = electron momentum; the basis of ELECTRON MICROSCOPY.

de-asphalting The DEGRADATION of CRUDE OIL by the PRECIPITATION of ASPHALT, probably caused by the introduction of natural gas. Although the quality of the oil is improved, the RESERVOIR POROSITY may be reduced by plugging with BITUMEN.

dead line 1 The depth in an OREFIELD below which no economic mineralization is present. **2** The level below which no significant oil is likely to be present.

death assemblage See THANATOCENOSIS.

debris avalanche A rapid form of MASS MOVEMENT in a narrow channel down a steep slope.

debris fall The near free fall of WEATHERED material from a vertical or overhanging face.

debris flow A SEDIMENT GRAVITY FLOW process in which particles up to BOULDER size are supported principally by their buoyancy in, and the COHESIVE STRENGTH of, a sediment-water slurry.

debris streaking A mechanism of SLICKENSIDE formation in which products of the ABRASION of an asperity accumulate on either side of it in the sense of DISPLACEMENT.

debrite A sediment deposited from a DEBRIS FLOW.

Debye-Scherrer method An X-RAY POWDER DIFFRACTION method in which the spatial distribution of cones of diffracted rays is measured on a cylindrical film around the sample in a specialized camera.

Decapoda/decapods An order of subclass EUMALACOSTRACA, subphylum CRUSTACEA, phylum ARTHROPODA; ARTHROPODS with ten legs, e.g. lobsters and advanced crabs. Range U. DEVONIAN–RECENT.

decarboxylation The thermocatalytic decomposition of oxygen-containing KEROGEN to release carbon dioxide during BURIAL DIAGENESIS, causing an increase in carbon content and a loss of water.

decay constant The constant λ in the equation $-(dN/dt) = \lambda N$, which describes the rate of radioactive decay in terms of the number of radioactive atoms present (N) and the rate of change of that number with time (dN/dt). λ is related to the HALF-LIFE T by $T = 0.693/\lambda$.

decay series The sequence of DAUGHTER ELEMENTS produced by the radioactive decay of a parent element.

decaying ripple A type of WAVE RIPPLE.

Deccan traps MAASTRICHTIAN-PALAEOCENE age, voluminous THOLEIITIC BASALTS over 1200 m thick which cover an area of 500 000 km² in west-central India, probably the result of a very large HOTSPOT.

decibel scale (dB) A logarithmic scale used to measure the power or amplitude of a signal.

decke See NAPPE.

declination The horizontal angle between TRUE NORTH and MAGNETIC NORTH.

declivity A GEOMORPHOLOGICAL term for the gradient of a slope.

décollement (detachment) A structural discontinuity of STRAIN, folding or FOLD STYLE within the Earth, typically the undeformed surface between STRAINED or FAULTED areas or the boundary between ALLOCH-THONOUS and AUTOCHTHONOUS rocks.

deconvolution The process of undoing a previous FILTERING (CONVOLUTION) operation. An important processing technique applied to SEISMIC RECORDS which removes from the waveform the deleterious effects of its passage through the Earth during which it increases in length, providing a sharper indication of reflecting interfaces.

decussate texture A TEXTURE comprising a random arrangement of TABULAR or PRISMATIC CRYSTALS.

dedolomitization The partial or complete CALCIFICATION of DOLOMITE, caused by the reaction of EVAPORATED-RELATED COLLAPSE BRECCIAS with calcium-rich sulphatic waters derived by the DISSOLUTION of GYPSUM.

Deep Sea Drilling Project (DSDP) An international programme, planned by the JOINT OCEANOGRAPHIC INSTITUTES FOR DEEP EARTH SAMPLING, to drill the Earth in deep water, using the drilling ship *D/V Glomar Challenger*. Commenced in 1968 and terminated in 1983 when it was superseded by the OCEAN DRILLING PROGRAM.

deep lead A buried ALLUVIAL PLACER DEPOSIT.

deep weathering WEATHERING to a depth of tens of metres beneath the surface, proceeding as METEORIC WATER percolates through pores, JOINTS and fissures in the bedrock, dissolving and chemically altering the MINERALS present with only QUARTZ unaffected. Moisture eventually penetrates the solid rock along crystal boundaries and CORESTONES shrink, leading to the development of SAPROLITE. Common on flat landscapes in the humid tropics.

deep-focus earthquake An EARTHQUAKE with a depth of FOCUS greater than 300 km. Cf. SHALLOW-FOCUS EARTHQUAKE, INTERMEDI-ATE-FOCUS EARTHQUAKE.

deep-sea cone See SUBMARINE FAN.

deep-sea fan See SUBMARINE FAN.

deep-sea trench See OCEAN TRENCH.

deep-water fan See SUBMARINE FAN.

deerite $((Fe,Mn)_{13}(Fe,Al)_7Si_{13}O_{44}(OH)_{11})$ A black, hydrous iron-manganese silicate found in metamorphosed SHALE, siliceous IRONSTONE and impure LIMESTONE.

deflation The process by which particles are removed from the ground surface by wind action.

deflation lag Large particles remaining after the removal of finer material by DEFLATION.

deformation A geological process in which the application of FORCE causes a change in geometry, such as the production of a FOLD, FAULT or FABRIC, often associated with METAMORPHIC reactions.

deformation band A TABULAR zone of a CRYSTAL LATTICE differing in crystallographic orientation from the rest of the MINERAL grain.

deformation history (polyphase deformation) The chronological sequence of DEFORMATION events in a rock or region.

deformation lamellae Narrow planar zones in MINERAL grains such as QUARTZ, FELDSPAR and OLIVINE which are related to DEFORMATION of the CRYSTAL LATTICE.

deformation mechanism The means by which DEFORMATION is accomplished at a microscopic scale. The three basic mechanisms are CATACLASIS, INTRACRYSTALLINE PLASTICITY and DIFFUSIVE MASS TRANSFER.

deformation mechanism map A plot of the progression of DEFORMATION in coordinates of STRESS, temperature and grain size.

deformation path A line showing successive states in a DEFORMATION HISTORY, plotted in three-dimensional coordinates of, commonly, STRAIN and rotation. Cf. STRAIN PATH.

deformation twinning A systematic reorientation of part of a CRYSTAL STRUCTURE during DEFORMATION so that the deformed zone is geometrically related to the rest of the crystal. Common in CARBONATE MINERALS.

degradation The modification of CRUDE OIL by differential SOLUTION, DE-ASPHALTING or BIODEGRADATION, in which non-HYDROCARBONS and AROMATICS are generally removed preferentially.

delamination The process of CRUST DETACHMENT from the MANTLE during CONTINENTAL COLLISION, or when an upper crustal layer detaches from a lower layer by FLAKE TECTONICS.

delay time **1** In SEISMIC REFRACTION, the difference in ARRIVAL TIME of a refracted SEISMIC HEAD WAVE and the travel time if the wave had travelled the distance between SHOT and detector at the velocity of the refractor. Delay times allow the construction of the refractor geometry. **2** In EARTHQUAKE SEISMOLOGY, the difference in ARRIVAL TIME of an EARTHQUAKE-generated SEISMIC WAVE and the time predicted by a model of the velocity structure of the Earth, which may allow the model to be refined or a SEISMIC TOMOGRAPHIC analysis to be made.

delayed recovery The removal of TEMPORARY STRAIN (RECOVERY) a measurable time after removal of the deforming STRESS, characteristic of rocks showing VISCOELASTIC behaviour.

delayed runoff **1** RUNOFF that penetrates the subsurface before DISCHARGE. **2** The water stored temporarily as snow or ice.

delessite An iron-rich, oxidized variety of CHLORITE.

dell A small headwater valley, characteristically choked with sediment and the site of a SWAMP, often located at the head of a gorge on a PLATEAU.

Delmontian A NEOGENE succession on the west coast of the USA covering part of the MESSIAN, the ZANCLIAN and part of the PIACENZIAN.

delta A constructional, triangular-shaped, sediment body, up to thousands of square kilometres in area, where river systems interact with fresh to marine waters and deposit sediments as the flow volume expands and channel flow becomes unconfined.

demagnetization The removal of NATURAL REMANENT MAGNETIZATION by incremental increases in temperature or the strength of an applied MAGNETIC FIELD.

demantoid A green variety of ANDRADITE.

demoiselle See EARTH PILLAR.

Demospongea A class of phylum PORIFERA (SPONGES) in which the siliceous spicules have rays at 60° and 120°. Range CAMBRIAN–RECENT.

demultiplexing The recovery of information from a MULTIPLEXED signal.

dendritic (arborescent, dendroid) Descriptive of a branching, ramifying or dichotomizing form.

dendrochronology The dating of trees and ancient wood by measuring and counting annual growth rings for, usually, archaeological purposes.

dendroclimatology PALAEOCLIMATOLOGY using information obtained from tree growth rings.

dendrogeomorphology A branch of DENDROCHRONOLOGY used in the understanding of GEOMORPHOLOGICAL features, as tree rings are affected by events such as inclination, corrosion of bark, SHEAR, burial, exposure, inundation, climate change and nudation which may be the result of such processes as FAULTING, shoreline warping, volcanism, flooding, MASS WASTING, AVA-LANCHING, GLACIAL fluctuations, etc.

dendroid See DENDRITIC.

Dendroidea The more primitive of the GRAPTOLITHINA. Normally sessile and reaching a height of up to 100 mm. Range M. CAMBRIAN–early CARBONIFEROUS.

density Mass per unit volume (kg m^{-3} or Mg m^{-3}). Density estimation is important in both the REDUCTION and interpretation of GRAVITY SURVEY data and can be accomplished by direct measurement or *in situ*. The latter methods include GAMMA-GAMMA and GRAVIMETER GEOPHYSICAL BOREHOLE LOGGING, NETTLETON'S METHOD and using the relationship between SEISMIC VELOCITY (from SEISMIC REFRACTION data) and density.

density contrast The difference in DENSITY between a body of rock and its surroundings, which controls the magnitude and sign of the GRAVITY ANOMALY over it.

density current A current driven by the density of the fluid, e.g. TURBIDITY CURRENT.

density log See GAMMA-GAMMA LOG.

density of states The number of MOLECULAR ORBITALS whose energy lies within a given band.

dentate With tooth-like projections.

denudation Removal of material by WEATHERING plus EROSION.

denudation chronology The reconstruction of the EROSIONAL history of the Earth's surface using EROSIONAL remnants where the STRATIGRAPHY is incomplete.

depletion The loss of water from surface or GROUNDWATER RESERVOIRS at a greater rate than their RECHARGE.

depletion allowance The proportion of income from the exploitation of a deposit not subject to tax, in recognition that it will eventually become exhausted.

depletion drive (dissolved-gas drive, solution gas drive) The mechanism whereby oil in a RESERVOIR is driven towards a WELL by the action of dissolved gas in the oil.

depocentre The site of maximum thickness of sediment accumulation in a sedimentary BASIN over a particular period of time.

depositional environment-related diagenesis (eogenesis) DIAGENETIC processes occurring where the composition of the PORE-FLUIDS is mostly controlled by the overall physical, biological and chemical characteristics of the depositional system.

depositional remanent magnetization (DRM, detrital remanent magnetization) A NATURAL REMANENT MAGNETIZATION carried by unconsolidated sediment acquired during the deposition of magnetized grains.

depression spring A SPRING forming where the land surface cuts the WATER TABLE.

depth of compensation The depth at which ISOSTATIC effects cause the pressure due to the overlying rocks to be everywhere equal.

depth of penetration 1 In electrical surveying, the depth at which current flow becomes insignificant, representing the limit to which information is derived. **2** In other geophysical methods, the maximum depth to which reliable information is obtained.

depth zone A zone proposed in the outmoded concept of the existence of a primary correlation between METAMORPHIC processes and depth, in which there were considered to be a set of continuous zones parallel to the Earth's surface grading downwards. See also EPIZONE, MESOZONE, KATAZONE.

Derbyshire spar A popular name for FLUORITE or FLUORSPAR.

derived fossil See REMANIÉ FOSSIL.

deroofing The uncovering of a PLUTON by DENUDATION.

descloisite (PbZn(VO₄)OH) An important ORE MINERAL of vanadium, found in the oxidation zone of lead-zinc deposits.

desert armour See DESERT PAVEMENT.

desert mosaic See DESERT PAVEMENT.

desert pavement (desert armour, desert mosaic, stone pavement) A superficial layer of PEBBLES covering unconsolidated, finer-grained materials in arid regions which protects them from EROSION.

desert rose (rock rose) A cluster of platy crystals, commonly of BARYTES or GYPSUM, in the crude shape of a flower, created in arid climates by evaporation.

desert varnish A dark, shiny coating, <1 mm thick, found on PEBBLES in arid regions, generally composed of iron and manganese oxides and CLAYS. Much thinner than a WEATHERING RIND.

desiccation crack A MUD CRACK formed by the subaerial drying of mud.

desilication The removal of SILICA from a rock by chemical action.

desmine See STILBITE.

Desmoinsian A CARBONIFEROUS succession in the USA covering the upper part of the MOSCOVIAN.

Desmostyla An order of infraclass EUTHERIA, subclass THERIA, class MAMMALIA; aquatic, hippopotamus-sized MAMMALS. Range MIOCENE.

desquammation See SPHEROIDAL WEATHERING.

destructive plate margin A PLATE MARGIN at which oceanic LITHOSPHERE is destroyed by SUBDUCTION at an ISLAND ARC or ANDEAN-TYPE SUBDUCTION ZONE. Cf. CONSTRUCTIVE PLATE MARGIN, CONSERVATIVE PLATE MARGIN.

detachment See DÉCOLLEMENT.

detachment fault A FAULT which marks the DISPLACEMENT along a DETACHMENT horizon or DÉCOLLEMENT plane.

detailed grid mapping GEOLOGICAL MAPPING at scales from 1:100 to 1:10 in which the exposure is marked with a square grid and the detail in each square transferred to graph paper.

detrital Descriptive of a particle, generally of a resistant MINERAL, derived from an existing rock by WEATHERING and/or EROSION.

detrital remanent magnetization See DEPOSITIONAL REMANENT MAGNETIZATION.

deuteric (epimagmatic) Descriptive of ALTERATION arising from reaction between primary MAGMATIC MINERALS and HYDRO-THERMAL SOLUTIONS that separated from the MAGMA at a late stage in its solidification.

deuteric reaction The permeation of non-water bearing MINERALS through crystal FRACTURES in a newly crystallized MAGMA and their reaction with PRIMARY MINERALS.

DEVAL See DEVIATION FROM AXIAL LINEARITY.

development well A WELL used in the development of an oilfield and the production of oil rather than in exploration.

deviation from axial linearity (DEVAL) A small bathymetric offset of an OCEAN RIDGE crest.

deviatoric stress The STRESS remaining in a TRIAXIAL STRESS system when the mean STRESS is subtracted.

devitrification The development of crystals in a GLASSY ROCK.

devolatilization The loss of volatiles during COALIFICATION.

Devonian A PERIOD of the PALAEOZOIC, 408.5–362.5 Ma.

dewatering The removal of GROUNDWATER in order to lower the WATER TABLE so that work can take place in the dewatered area.

Dewey Lake A PERMIAN succession in the Delaware Basin, USA probably equivalent to the CHANGXINGIAN.

Dewu A CARBONIFEROUS succession in China equivalent to the SERPUKHALIAN.

dextral The sense of movement across a boundary, such as a FAULT, in which the side opposite the observer moves to the right. Cf. SINISTRAL.

diabanite An iron-rich, aluminium-poor variety of CHLORITE.

diabase An American term for DOLERITE.

diachronism A time-transgressive unit, e.g. a LITHOSTRATIGRAPHIC unit that varies in age.

diadochy The REPLACEMENT of an ion in a CRYSTAL LATTICE by another.

diagenesis All physical, chemical and biological processes that occur in a sediment after deposition and before METAMORPHISM, during which sedimentary assemblages and their interstitial PORE FLUIDS react and attempt to reach equilibrium with their evolving geochemical environment.

diallage An obsolete name for a CLINO-PYROXENE.

dialogite See RHODOCROSITE.

diamagnetism The fundamental, weak magnetism of all substances arising from the orbit of electrons around a nucleus. Superimposed by PARAMAGNETISM and FERROMAGNETISM in more highly magnetic substances.

diamict The general term for DIAMICTON and DIAMICTITE.

diamictite A terrigenous SEDIMENTARY ROCK with particle sizes ranging from CLAY to BOULDER size.

diamicton Unconsolidated DIAMICTITE.

diamond (C) A naturally occurring, high-pressure form of CARBON, valued as an INDUSTRIAL MINERAL because of its HARDNESS and as a GEM. Diamonds ultimately derive from KIMBERLITE and LAMPROITE and form PLACER DEPOSITS.

diaphthoresis See RETROGRADE METAMORPHISM.

diapir A volume of rock rising upwards buoyantly because of its low DENSITY relative to its surroundings and causing DEFORMATION of overlying STRATA. Rocks forming diapirs include EVAPORITES (commonly HALITE) and GRANITE.

diapirism The buoyant ascent of a DIAPIR.

Diapsida REPTILES with two openings in the cheek region of the skull, including lizards, snakes, CROCODILIANS, DINOSAURS and PTEROSAURS.

diaspore (AlO.OH) An aluminium ORE MINERAL found in BAUXITE.

diastem An UNCONFORMITY in which a pause in sedimentation is marked only by an abrupt change in lithology.

diastrophism A little-used term for large-scale TECTONIC DEFORMATION of the CRUST, e.g. EPEIROGENY, CYMATOGENY, OROGENY, FOLDING, FAULTING, UPLIFT, subsidence and PLATE movement.

diatom A MICROSCOPIC, unicellular, planktonic ALGA growing in water. Possesses a siliceous cell wall which may contribute to sediment.

diatomaceous earth See DIATOMITE.

diatomaceous ooze A PELAGITE made up of the siliceous TESTS of DIATOMS.

diatomite (diatomaceous earth, kieselguhr) Fine-grained, hydrated SILICA formed by the accumulation of the TESTS of DIATOMS.

diatreme A vertical PIPE or funnel-shaped igneous intrusion, 200–2000 m thick and up to 2 km deep, made up of a chaotic BRECCIA of blocks of COUNTRY ROCK, magmatic material and possibly MANTLE-derived XENOLITHS and XENOCRYSTS passing down into a DYKE. A FORCEFUL INTRUSION of a mixture of MAFIC MAGMA, VOLCANIC GASES and ACCIDENTAL LITHIC blocks and CLASTS.

dichotomous Regularly bifurcating.

dichroiscope An instrument for the determination of PLEOCHROISM.

dichroism See PLEOCHROISM.

dichroite A GEM-quality variety of CORDIERITE.

dickite ($Al_2Si_2O_3(OH)_4$) A CLAY MINERAL found in HYDROTHERMAL deposits.

differential stress The two-dimensional STRESS difference between the maximum and minimum STRESSES.

differential thermal analysis (DTA) A method used in the study of CLAY MINERALS in which a sample and an inert material are heated. When a temperature difference between the two is observed, the CLAY MINERAL is undergoing a reaction, and this behaviour can be used in its identification.

differentiation The separation of a MAGMA into two or more fractions.

differentiation index The sum of NORMATIVE QUARTZ, ALBITE, ORTHOCLASE, NEPHELINE, KALIOPHILITE and LEUCITE, which quantifies the degree of DIFFERENTIATION. The index increases with the quantity of FELSIC MINERALS.

diffluence A mechanism by which a GLACIER overflows an obstacle at its lowest point.

diffraction The radial scattering of energy at an abrupt discontinuity whose radius of curvature is smaller than the wavelength of the incident wave. Applicable to SEISMIC WAVES and the basis of the X-RAY DIFFRACTION method.

diffraction pattern The interference pattern formed in X-RAY DIFFRACTION by atomic planes in a crystal which is characteristic of the MINERAL.

diffractometer An instrument allowing the distribution and intensity of DIFFRACTED radiation to be measured, e.g. in X-RAY DIFFRACTION methods.

diffuse reflectance spectroscopy A technique for the quantitative measurement of STREAK.

diffusion The transport of matter by the mixing of molecules and ions by thermal agitation.

diffusion coefficient See FICK'S LAW OF DIFFUSION.

diffusion in sediments The tendency of random molecular and ionic movements in PORE-FLUIDS to reduce chemical gradients.

diffusive mass transfer A DEFORMATION MECHANISM involving the movement of material by the solid state DIFFUSION of vacancies, atoms, molecules or ions, controlled by temperature and STRAIN RATE.

digenite (Cu_9S_5) A COPPER ORE MINERAL.

dikaka A SAND DUNE covered by scrub or grass.

dike See DYKE.

dilatancy The volume increase of a granular material when it is subjected to STRESS and approaching FAILURE. Caused by changes in crack and pore distribution in the rock and important in EARTHQUAKE PREDICTION as it changes the ratio of P WAVE to S WAVE velocity, the ELECTRICAL CONDUCTIVITY, the ground level, the GROUNDWATER level and affects the acoustic emissions from MICROEARTHQUAKES.

dilatancy-diffusion theory An EARTHQUAKE PREDICTION model for the build-up of STRAIN prior to an EARTHQUAKE in which MICROCRACKS increase in number before locking up; an influx of water then leads to FAILURE. Cf. DILATANCY-INSTABILITY THEORY.

dilatancy-hardening The strengthening of a rock caused by a reduction in PORE-FLUID PRESSURE due to DILATANCY.

dilatancy-instability theory An EARTHQUAKE PREDICTION model for the build-up of STRAIN prior to an EARTHQUAKE in which MICROCRACKS increase in number and then avalanche to relax slightly the build-up of STRESS. Cf. DILATANCY-DIFFUSION THEORY.

dilatant zone The open space formed in the more steeply dipping part of a PINCH-AND-SWELL STRUCTURE arising from NORMAL FAULTING.

dilatation See DILATION.

dilation (dilatation) The ratio of change in volume to initial volume. Negative dilation indicates a volume decrease and vice versa.

dilation vein A VEIN formed by the infilling of an existing void caused by a FAULT or FISSURE.

dilution gauging A technique for the estimation of STREAM FLOW by introducing a tracer and timing its passage over a known distance.

diluvialism An early form of CATASTROPHISM which related the shaping of the landscape to Noah's Flood.

diluvium An obsolete term for unconsolidated sediment which could not be explained by FLUVIAL or marine action.

dim spot A reduction in amplitude of SEISMIC WAVES on a SEISMIC REFLECTION profile indicative of a reduction in REFLECTION COEFFICIENT. Cf. BRIGHT SPOT.

dimension stone A BUILDING STONE dressed into regularly shaped blocks.

dimorphism The state of an element or compound existing in two forms, e.g. DIAMOND and GRAPHITE.

Dinantian 1 The older sub-period of the CARBONIFEROUS, also called the MISSISSIPPIAN, 362.5–322.8 Ma. **2** A CARBONIFEROUS succession in NW Europe equivalent to the TOURNAISIAN and VISÉAN.

dinocyst An organic-walled vesicle of a DINOFLAGELLATE.

dinoflagellates MICROSCOPIC, unicellular PROTISTS with DINOCYSTS 30–60 μm across, of importance in the BIOSTRATIGRAPHY of the JURASSIC to PLEISTOCENE.

dinosaur A general term for the orders SAURISCHIA and ORNITHISCHIA, subclass ARCHOSAURIA, class REPTILIA which dominated the terrestrial ecology of the MESOZOIC, many of which achieved great size.

diopside ($CaMgSi_2O_6$) A white to light green CLINOPYROXENE.

dioptase (emerald copper) ($Cu_6(Si_6O_{18}).6H_2O$) A rare, green CYCLOSILICATE.

diorite A medium- to coarse-grained intrusive IGNEOUS ROCK containing PLAGIOCLASE more Ab-rich than Ab_{50} and <20% QUARTZ, AMPHIBOLE and/or PYROXENE. It is equivalent to ANDESITE in composition and grades into TONALITE with increased QUARTZ and into MONZONITE with >10% ALKALI FELDSPAR. Found in ISLAND ARC settings.

Diorite model A type of PORPHYRY COPPER DEPOSIT characterized by the presence of zones of POTASSIC ALTERATION, SERICITIZATION, INTERMEDIATE ARGILLIC ALTERATION and PROPYLITIC ALTERATION with increasing distance from the mineralization.

dip The inclination of a planar surface, measured in the vertical plane perpendicular to its STRIKE. Cf. APPARENT DIP.

dip fault A steep FAULT whose STRIKE is parallel to the DIP of the BEDDING.

dip isogon A line joining points of equal DIP in a folded sequence, which can be used to define the shape of a FOLD PROFILE and to distinguish different FOLD STYLES.

dip moveout The MOVEOUT generated by a dipping reflector. Cf. NORMAL MOVEOUT.

dip separation An offset of a planar feature in the vertical plane normal to the FAULT. This can be resolved into HEAVE and THROW. Cf. STRIKE SEPARATION.

dip slope 1 The topographic slope parallel to the DIP of the BEDDING and generally at an angle lower than it. **2** In a tilted FAULT BLOCK, the top surface which dips gently away from the fault scarp.

dip-angle system An ELECTROMAGNETIC INDUCTION METHOD based on the measurement of the orientation of the resultant of the primary (due to the source) and secondary (due to an electrically anomalous body) electromagnetic fields. Cf. PHASE COMPONENT SYSTEM.

dip-slip The movement parallel to the DIP of a planar surface such as a FAULT.

diphotic zone The level in a water body where sunlight is faint and little photosynthesis can take place. Cf. APHOTIC ZONE, PHOTIC ZONE.

Diplopoda A class of superclass MYRIA-

PODA, phylum UNIRAMIA; the millipedes. Range SILURIAN–RECENT.

Diplorhina (Pteraspidomorpha) A subclass of class AGNATHA, superclass PISCES; jawless FISH, perhaps similar to the modern hagfish. Range U. SILURIAN–U. DEVONIAN.

dipmeter log A GEOPHYSICAL BOREHOLE LOG in which formation DIP and STRIKE are measured by taking four MICRORESISTIVITY readings around the borehole.

Dipnoi An order of subclass SARCOPTERYGII, class OSTEICHTHYES, superclass PISCES; the lungfish. Range L. DEVONIAN–RECENT.

dipole field The MAGNETIC FIELD due to two MAGNETIC POLES of identical strength and opposite polarity. The GEOMAGNETIC FIELD approximates such a field. Also, the shape of the MAGNETIC FIELD produced by a current flowing in a loop at a large distance compared with the diameter of the loop.

dipyre (mNa$_4$(Al$_3$Si$_9$O$_{24}$)Cl+nCa$_4$(Al$_6$Si$_6$O$_{24}$) CO$_3$) A member of the SCAPOLITE series containing 20–50% MEIONITE, found in REGIONALLY METAMORPHOSED rocks.

direct runoff The RUNOFF of precipitation falling directly on saturated soil and unable to infiltrate.

direct seismic wave A SEISMIC BODY WAVE taking the most direct route from SEISMIC SOURCE to detector.

direct-shipping ore (lump ore) ORE requiring no BENEFICATION before transportation.

directional fabric The FABRIC in an IGNEOUS, SEDIMENTARY or METAMORPHIC ROCK comprising the alignment of MINERALS caused by motion during the formation of the rock.

dirt band A thin bed of inorganic rock in a COAL SEAM.

discharge See STREAM FLOW.

discharge hydrograph See HYDROGRAPH.

discoaster A COCCOLITH with the form of a STELLATE shield, usually built of six radiating rays.

disconformity (lacuna) An UNCONFORMITY marked by evidence of EROSION, across which there is no change in DIP. Cf. HIATUS.

discontinuous reaction series A sequence of MINERAL reactions which occur at specific temperatures as a MAGMA cools. The higher temperature species dissolve at that temperature and the new MINERAL remains until the next reaction temperature is reached. Cf. CONTINUOUS REACTION SERIES.

discordant 1 Descriptive of an IGNEOUS ROCK which cross-cuts BEDDING or FOLIATION. **2** Descriptive of UNCONFORMABLE STRATA.

discordant coast A coast where the structural grain runs transverse to the coastline. Cf. CONCORDANT COAST.

discovery well A WELL in which oil or gas was revealed in a new location.

dish structure A slightly concave-up STRUCTURE in SANDSTONE marked by a 0.2–2 mm thick, CLAY-rich coating produced by the upward escape of PORE FLUID.

disharmonic folds FOLDS whose FOLD STYLE changes from layer to layer, probably reflecting the different RHEOLOGIES of the folded layers.

dislocation A surface across which there is a loss of continuity, e.g. a FAULT, a termination of a half-plane in a CRYSTAL LATTICE.

dislocation climb A DEFORMATION process by the formation and movement of DISLOCATIONS in a CRYSTAL LATTICE such that they move out of their original lattice planes.

dislocation creep A type of CREEP in which parts of crystals glide past each other along crystalline DISLOCATIONS.

dislocation glide (cold working) A DEFORMATION process by the formation and movement of DISLOCATIONS within a CRYSTAL LATTICE such that each DISLOCATION remains in its own lattice plane.

dislocation line A line normal to the DISPLACEMENT of a structure at a DISLOCATION in a crystal.

dismicrite A LIMESTONE mainly composed

of MICRITE with patches or lenses of SPARITE, caused by disturbance of the original lime MUD by ALGAE or escaping gas.

dispersion The dependence of propagation velocity on wave frequency, characteristic of SURFACE SEISMIC WAVES. The general increase in velocity with depth means that lower frequencies travel faster than higher frequencies.

dispersive pressure The fluid condition in which grains are supported above a bed in a dispersed state due to grain collisions and interactions, which give rise to a viscous force with a strong vertical component.

displaced terrane (exotic terrane) A provably ALLOCHTHONOUS TERRANE.

displacement The relative distance moved across a line or plane.

displacement plane The plane at right angles to the walls of a SHEAR ZONE, containing the SHEAR DIRECTION.

displacement pressure The smallest capillary pressure required to force HYDROCARBONS into the largest interconnecting pores of a water-wet rock.

displacement vector See SLIP VECTOR.

displacive transformation The rapid transformation of one POLYMORPH into another by the expansion, distortion or rotation of COORDINATION POLYHEDRA without bonds being broken. Cf. RECONSTRUCTIVE TRANSFORMATION.

disseminated deposit A generally low-GRADE deposit in which ORE MINERALS are dispersed throughout a COUNTRY ROCK, e.g. DIAMONDS in KIMBERLITE, PORPHYRY COPPER DEPOSIT.

dissipative beach A low gradient BEACH protected by a BAR which absorbs and dissipates much wave energy. Cf. REFLECTIVE BEACH.

dissolution A DIAGENETIC process by which a solid is dissolved in an aqueous PORE FLUID leaving behind a pore space in the COUNTRY ROCK.

dissolved gas drive See DEPLETION DRIVE.

distal Descriptive of a feature far from its source. Cf. PROXIMAL.

disthene See KYANITE.

distribution coefficient See PARTITION COEFFICIENT.

distribution grading (content grading) A feature shown by a GRADED BED in which there is a gradual change in the entire grain size distribution. Cf. COARSE-TAIL GRADING.

disulphide group A group of SULPHIDE MINERALS characterized by the presence of anion pairs such as S_2^{2-}, AsS^{2-}.

diurnal variation The daily variation in the GEOMAGNETIC FIELD affecting all the GEOMAGNETIC ELEMENTS. Cf. SECULAR VARIATION.

divergent erosion The difference between EROSION in low latitudes, where chemical WEATHERING affects PLANATION SURFACES to a greater extent than steeper slopes, and mid-latitudes, where the opposite effects obtain.

divergent evolution The evolutionary radiation of organisms.

divergent fan (reverse fan) A CLEAVAGE FAN which radiates from a focus on the convex side of the folded layer. Cf. CONVERGENT FAN.

diversity A measure of taxonomic variety in terms of species, genera, etc.

divide See WATERSHED.

diving seismic wave A SEISMIC WAVE that follows a curved path between the SEISMIC SOURCE and detector, caused by a progressive increase in SEISMIC VELOCITY with depth.

Dix formula A formula by which the INTERVAL VELOCITY of a SEISMIC WAVE can be calculated for a given depth interval between reflectors.

djurleite ($Cu_{1.96}S$) A SULPHIDE MINERAL derived from a CHALCOCITE structure.

Dobrotiva An ORDOVICIAN succession in Bohemia equivalent to the early LLANDEILO.

dog-tooth spar A form of CALCITE in the shape of a canine tooth.

Dogger The middle EPOCH of the JURASSIC, 178.0–157.1 Ma.

dogger A metric-scale flattened, ovoid, calcareous or ferruginous CONCRETION in SAND or CLAY.

dolerite (diabase) A fine- to medium-grained MAFIC IGNEOUS ROCK, mineralogically and chemically equivalent to BASALT, commonly forming minor intrusions.

Dolgellian A STAGE of the CAMBRIAN, 514.1–510.0 Ma.

doline (shakehole) A circular to oval, simple closed depression found in KARST TERRAIN, formed by solution, CAVE collapse, PIPING or subsidence.

dolocrete A DURICRUST cemented by DOLOMITE.

dololithite A DOLOSTONE comprising detrital fragments of DOLOMITE derived by WEATHERING from an existing rock.

dolomite $(CaMg(CO_3)_2)$ A CARBONATE MINERAL found in MAGNESIAN LIMESTONE, formed by DOLOMITIZATION.

dolomitization The formation of DOLOMITE or a DOLOSTONE by REPLACEMENT of the calcium of a calcium carbonate precursor by magnesium.

dolostone A rock made up of DOLOMITE.

domain A subdivision of a larger area or volume that is more homogeneous than the whole. See also MAGNETIC DOMAIN.

dome 1 A volcanic feature formed by the accumulation of MAGMA above a VOLCANIC VENT. **2** An ANTIFORM with a circular to subcircular OUTCROP pattern.

dome dune A circular to subcircular DUNE, 0.1–1 km in diameter, with a poorly developed or absent SLIP-FACE.

dome-and-basin structure A FOLD INTERFERENCE STRUCTURE formed by early upright FOLDS whose AXES and AXIAL PLANES make a large angle with later FOLDS.

domichnia TRACE FOSSILS formed from dwelling structures.

dominance diversity A term used in PALAEOECOLOGY to describe the relative abundance of taxa within a sample, calculated in different ways, such as by use of the SHANNON-WEAVER DOMINANCE DIVERSITY EQUATION. Cf. EQUITABILITY.

donga A GULLY or ARROYO, particularly in southern Africa, formed by the EROSION of surficial deposits by RUNOFF and PIPING.

doorstopper technique See OVERCORING.

dormant volcano A volcano which is currently inactive but which has erupted within historical time. Cf. EXTINCT VOLCANO.

Dorogomilovskian A STAGE of the CARBONIFEROUS, 298.3–295.1 Ma.

Dott classification A classification scheme for SANDSTONES.

double zigzag structure A FOLD INTERFERENCE STRUCTURE formed by overfolded early FOLDS whose AXIAL PLANES make a large angle with later FOLDS, but whose AXES make a small angle with these, and where the later flow direction is oblique to the earlier AXIAL PLANES.

down-plunge projection (down-plunge view) The reconstructed profile of a FOLD STRUCTURE constructed at right angles to the PLUNGE of the FOLD AXIS.

down-plunge view See DOWN-PLUNGE PROJECTION.

downhole geophysical survey See GEOPHYSICAL BOREHOLE LOGGING.

downthrow The DISPLACEMENT of one side of a DIP-SLIP FAULT relative to the other.

downthrown Descriptive of the side of a FAULT with relative downward movement. Cf. UPTHROWN.

Downtonian A SERIES of the early DEVONIAN in British STRATIGRAPHY, approximately contemporaneous with the GEDINNIAN.

downward continuation The computation, from the POTENTIAL FIELD measured at a certain level, of what the field would be at a lower level. Based on the solution of LAPLACE'S EQUATION. Cf. UPWARD CONTINUATION.

downwash A process involved in the formation of GRÈZE LITÉE in which a half-fluid mixture of fine sediment is spread over a stony layer.

downwearing The EROSION of the slopes at the top of a hill or ESCARPMENT more rapidly than the lower slopes. Cf. BACKWEARING.

draa (complex dune, 'uruq) A large DUNE formed by a regional wind pattern.

drag fold 1 A MINOR or PARASITIC FOLD, probably formed by SHEAR in an INCOMPETENT layer between two COMPETENT layers folded by FLEXURAL SLIP. **2** A FOLD produced by FAULT DRAG.

drag force The FORCE exerted by a fluid on a surface in a direction parallel to the flow which controls settling behaviour and, with LIFT FORCE, the initiation of sediment movement.

drainage coefficient The amount of RUNOFF per unit area in 24 hours.

drainage density The average length of stream channel per unit area of a drainage BASIN, giving a measure of the degree of FLUVIAL dissection.

drainage network An hierarchical system of channel links within a drainage BASIN.

drainage ratio The ratio of RUNOFF to precipitation over a given period of time.

drainage well A WELL used to drain excess superficial water into an AQUIFER.

drapery (curtain) A TABULAR or folded SPELEOTHEM that hangs from CAVE ceilings or wall projections with a curtain-like appearance.

dravite $(NaMg_3Al_6B_3Si_6O_{27}(OH,F)_4)$ A brown, manganese-bearing TOURMALINE, sometimes used as a GEM.

drawdown The loss of head of pressure around a WELL which is being pumped.

dreikanter A VENTIFACT of PEBBLE size moulded into a three-faceted form by wind ABRASION. Cf. EINKANTER, ZWEIKANTER.

Dresbachian A CAMBRIAN succession in the E. USA covering the lower part of the MAENTWROGIAN.

Dreuss A JURASSIC succession in Utah/Idaho, USA, covering part of the CALLOVIAN.

driblet See SPATTER.

drift 1 An horizontal underground tunnel following a VEIN or parallel to the STRIKE of an OREBODY. Cf. CROSSCUT. **2** A gradual change in the reading of a stationary geophysical instrument, such as a GRAVIMETER, with time. **3** Unconsolidated superficial sediment. **4** An accumulation of sediment on the ocean floor transported by ocean currents.

drill hole A metalliferous mining term for borehole.

drill string All equipment within a DRILL HOLE during drilling.

drilling fluid See DRILLING MUD.

drilling mud (drilling fluid) An oil- or water-based fluid containing CLAY, lime or BARITE forced down a DRILL HOLE during drilling to cool and lubricate the bit, seal the sides of the borehole and prevent BLOWOUTS.

dripstone See STALAGMITE.

drive A tunnel driven along or near an ORE deposit.

DRM See DEPOSITIONAL REMANENT MAGNETIZATION.

dromorthid One of a family of huge, extinct, flightless Australian birds. Range 24 Ma to 50 000 years or less.

dropstone A CLAST dropped through the water column into soft sediment, typically released from ice.

dross Inferior or worthless COAL.

drowned placer See BEACH PLACER.

drumlin A rounded hummock of GLACIAL TILL.

druse A cavity into which EUHEDRAL crystals in the COUNTRY ROCK project.

drusy Containing cavities, often lined with crystals.

dry gas Natural gas composed almost entirely of METHANE. Cf. WET GAS.

dry lake See PLAYA LAKE.

dry placer A PLACER that cannot be exploited for lack of water.

dry valley A valley seldom, if ever, occupied by a stream.

dry wash See ARROYO.

DSDP See DEEP SEA DRILLING PROJECT.

DTA See DIFFERENTIAL THERMAL ANALYSIS.

ductile A TENACITY descriptor of a MINERAL that can be drawn into a wire.

ductile deformation DEFORMATION resulting in MACROSCOPICALLY continuous STRAIN; DEFORMATION in which there is a large, non-elastic, PERMANENT STRAIN before FAILURE.

ductile flow FLOW at a MACROSCOPIC scale.

ductile stringer A FAULT GOUGE feature in a FAULT ROCK made up of hard inclusions drawn out in the direction of SHEAR.

ductility The phenomenon of deforming by DUCTILE DEFORMATION.

duff Fine COAL of too low a CALORIFIC VALUE for direct sale.

Dulankara An ORDOVICIAN succession in Kazakhstan covering parts of the MARSH-BROOKIAN, ACTONIAN and ONNIAN.

dull coal The dullest type of COAL. See BRIGHT COAL.

dumortierite $(Al_7O_3(BO_3)(SiO_4)_3)$ A NESO-SILICATE found in aluminium-rich META-MORPHIC ROCKS.

dune 1 An accumulation of unconsolidated material (SAND, CLAY, GYPSUM or carbonate) shaped by the wind into a distinguishable landform. **2** A bedform resulting from transport and deposition in a current under a particular range of flow conditions.

dune grass A grass used in DUNE STABILIZATION in a temperate environment with adequate rainfall.

dune ridge A DUNE inundated by the sea and separated from the land to enclose a LAGOON.

dune stabilization The artificial prevention of EROSION or immobilization of a DUNE for engineering purposes.

dune-bedding The large-scale CROSS-STRATIFICATION developed in DUNES.

dungannonite A CORUNDUM-bearing DIORITE containing NEPHELINE.

Dunham classification A classification scheme for LIMESTONES.

dunite (peridotite) A medium- to coarse-grained ULTRAMAFIC ROCK comprising >90% OLIVINE, often of MANTLE origin.

Duntroonian An OLIGOCENE succession in New Zealand covering the middle part of the CHATTIAN.

duplex An IMBRICATE STRUCTURE in which FAULTS branch from an underlying FLOOR THRUST and join a common, higher-level ROOF THRUST.

durability Resistance to WEATHERING.

durain A LITHOTYPE of BANDED COAL made up of hard, grey-black bands with a dull to GREASY LUSTRE.

Durango A CRETACEOUS succession on the Gulf Coast of the USA equivalent to the BERRIASIAN, VALANGINIAN and HAUTERIVIAN.

duricrust A hard, MINERAL-cemented crust occurring in WEATHERED material or the soil zone, commonly composed of ALCRETE, CAL-CRETE, DOLOCRETE, FERRICRETE, GYPCRETE, SALCRETE or SILCRETE. Formed by the mobilization and deposition of chemicals during DEEP WEATHERING.

duripan A SILICA-cemented soil layer.

durite A MICROLITHOTYPE OF COAL made up of INERTINITE and EXINITE MACERALS.

Durlston Beds A CRETACEOUS succession in England covering the lower part of the BERRIASIAN.

duroclarain A LITHOTYPE OF BANDED COAL intermediate between DURAIN and CLARAIN.

dust Solid particles <0.08 mm in diameter suspended in the air, originating from many possible sources and sometimes deposited as LOESS.

dust storm A phenomenon in which DUST reduces visibility to <1000 m.

Dvur An ORDOVICIAN succession in Bohemia covering parts of the PUSGILLIAN and CAUTLEYAN.

Dyfed The middle sub-PERIOD of the ORDO-VICIAN, 476.1–472.7 Ma.

dyke (dike) A sheet-like, near-vertical, minor igneous intrusion that cuts across horizontal to gently dipping planar structures in the COUNTRY ROCK.

dyke swarm A set of DYKES, generally sub-parallel, with a common origin.

dynamic correction The correction for MOVEOUT time on a SEISMIC REFLECTION section, applied before STACKING a COMMON DEPTH POINT gather.

dynamic equilibrium (dynamic homeostasis) A self-regulating system in which any change in the energy status results in a change in the system variables to regain equilibrium.

dynamic homeostasis See DYNAMIC EQUILIBRIUM.

dynamic metamorphism METAMORPHISM caused by intense localized STRESS.

dynamic recrystallization The formation of new crystal species as a result of DEFORMATION or TECTONISM.

dynamothermal metamorphism REGIONAL METAMORPHISM at high temperature and pressure.

dyscrasite (Ag_3Sb) A SILVER ORE MINERAL.

dyscrystalline Descriptive of a poorly crystalline IGNEOUS ROCK.

dystrophic lake A lake poor in nutrients and oxygen and rich in undecomposed plant matter. Cf. EUTROPHIC LAKE, OLIGOTROPHIC LAKE.

E

e plagioclase An intermediate PLAGIO-CLASE with a fine-scale, slab-like MICROTEX-TURE detectable by X-RAY or ELECTRON DIFFRACTION.

Eagle Ford A CRETACEOUS succession on the Gulf Coast of the USA covering part of the CENOMANIAN and the TURONIAN.

Early Imbrian A PERIOD of the HADEAN, 3850–3800 Ma.

Early Llandeilo A STAGE of the ORDOVICIAN, 468.6–467.0 Ma.

Early Llanvirn A STAGE of the ORDOVICIAN, 476.1–472.7 Ma.

Earth movement A general term for DEFORMATION of the CRUST.

Earth Observation System A multi-satellite project between NASA, the European Space Agency and other countries planned to give improved information on the global LITHOSPHERE, hydrosphere and atmosphere.

Earth Resources Technology Satellite (ERTS satellite) The original name of a LANDSAT SATELLITE used in REMOTE SENSING.

Earth tide DEFORMATION, on a centimetric scale, of the solid Earth by the gravitational attractions of the Moon and Sun. Provides information on the internal RIGIDITY of the Earth and requires correction in GRAVITY REDUCTION.

earth flow A rapid type of MASS MOVEMENT of unconsolidated material down a slope. Usually occurs due to an increase in PORE FLUID pressure, which reduces the friction between particles.

earth hummock A type of PATTERNED GROUND in which rounded hummocks form an irregular net-like pattern as the result of FROST HEAVING.

earth pillar (demoiselle) A column of earthy material capped by a BOULDER which protects it from EROSION, typical of BADLAND and MORAINIC areas.

Earth's magnetic field See GEOMAGNETIC FIELD.

earthquake A sudden release of accumulated STRESS along a subsurface DIS-CONTINUITY.

earthquake engineering (engineering seismology) The study of hazards arising from EARTHQUAKES to man-made structures.

earthquake focus (focus, hypocentre) The location of origin of an EARTHQUAKE, usually assumed to be a point but in reality usually part of a FAULT PLANE of finite lateral extent.

earthquake intensity (intensity) A subjective measure of the strength of the effects of an EARTHQUAKE at and around the EPICENTRE, expressed on a scale of I to X or XII. Cf. EARTHQUAKE MAGNITUDE. See also MERCALLI SCALE.

earthquake magnitude (magnitude) A measure of the amount of energy released by an EARTHQUAKE estimated from the amplitude and frequency of the SEISMIC WAVES it produces corrected for distance and expressed on a logarithmic scale, e.g. LOCAL MAGNITUDE, BODY WAVE MAGNITUDE. Cf. EARTHQUAKE INTENSITY. Cf. RICHTER SCALE.

earthquake mechanism A model of the

causative motion at an EARTHQUAKE FOCUS. See ELASTIC REBOUND THEORY, FOCAL MECHANISM SOLUTION.

earthquake prediction An attempt to make a short-term estimate of the time, place and MAGNITUDE of an EARTHQUAKE; also long-term forecasting of the probability of EARTHQUAKES of a given MAGNITUDE in a given time for a given region. Many prediction methods have been devised but, as yet, only some six successful predictions have been made.

earthquake swarm A prolonged series of EARTHQUAKES of small to moderate MAGNITUDE without a single 'main' event.

earthy cobalt See ASBOLANE.

earthy lustre The non-metallic LUSTRE of porous MINERAL AGGREGATES such as CLAYS.

Eastonian An ORDOVICIAN succession in Australia covering the HARNAGIAN, SOUDLEYAN, LONGVILLIAN, MARSHBROOKIAN, ACTONIAN and part of the ONNIAN.

ecdysis The moulting process by which some animals shed the exoskeleton or outer skin.

Echinodermata/echinoderms A phylum of marine invertebrates with a spiny CALCITE endoskeleton, a water vascular system and pentameral symmetry. Range L. CAMBRIAN–RECENT.

Echinoidea/echinoids A class of subphylum ELEUTHEROZOA, phylum ECHINODERMATA in which the body is enclosed in a globular to discoid TEST of interlocking calcareous plates which carry movable appendages. Range ORDOVICIAN–RECENT.

echo dune A depositional DUNE forming on a steeper slope than a CLIMBING DUNE.

echo sounder (fathometer) An instrument for measuring water depth by determining, near the sea surface, the travel time of an acoustic pulse reflected from the sea bed.

echogram (fathogram) A graph of seafloor topography made by an ECHO SOUNDER.

eckermannite $(Na_3(Mg,Li)_4(Al,Fe)Si_8O_{22}$ $(OH,F)_2)$ A rare blue-green, alkali AMPHIBOLE found in some PLUTONIC ALKALINE IGNEOUS ROCKS.

eclogite A coarse-grained METAMORPHIC ROCK comprising pink, PYROPE-rich GARNET, green OMPHACITE ± KYANITE, of deep-seated origin.

eclogite facies A METAMORPHIC FACIES of high pressure and medium temperature, characterized in BASALTIC rocks by the presence of OMPHACITE and PYROPE.

economic basement The level below which there is minimal probability of finding an economic MINERAL RESOURCE.

economic geology Geological studies for the exploration and exploitation of materials which can be profitably utilized by man.

economic yield The maximum rate at which water can be extracted from an AQUIFER without damaging water quality or creating a deficiency.

ecostratigraphy The study of the occurrence and evolution of FOSSIL communities through time, especially with application to STRATIGRAPHIC correlation.

ecosystem The interdependence of species with themselves and their environment.

ecotone The narrow transition zone between different communities.

ecotope The habitat of an organism. Cf. BIOTOPE.

ectinite A METAMORPHIC ROCK developing without introduction or loss of its component MINERALS.

eddy current Loops of electric current induced to flow in a conducting body by a time-varying MAGNETIC FIELD. The measurement of the MAGNETIC FIELDS generated by eddy currents in subsurface conductors forms the basis of the ELECTROMAGNETIC INDUCTION METHODS of GEOPHYSICAL EXPLORATION.

eddy viscosity The component of resistance to DEFORMATION in a fluid arising from the generation of eddies in a TURBULENT FLOW.

edelopal A variety of OPAL with a very brilliant play of colours.

edenite $(NaCa_2Mg_5AlSi_7O_{22}(OH)_2)$ A variety of HORNBLENDE.

Edentata An order of infraclass EUTHERIA, subclass THERIA, class MAMMALIA; toothless MAMMALS including aardvarks, armadillos and sloths. Range U. PALAEOCENE–RECENT.

edge coal A COAL SEAM of high inclination.

edge dislocation A LINE DEFECT in a crystal formed when an additional half-plane of atoms is inserted and the surrounding atoms adjust position to accommodate it.

edge water The water in a saturated RESERVOIR rock surrounding an oil pool.

Ediacara The younger EPOCH of the VENDIAN, 590–570 Ma.

Ediacara fauna Complex animals of PROTEROZOIC age (~670 Ma) with unusual soft-body preservation found in New South Wales, Australia.

Edrioasteroidea/edrioasteroids A class of subphylum ELEUTHEROZOA, phylum ECHINODERMATA; ECHINODERMS with a discoidal to cylindrical exoskeleton of irregular, flexible, polygonal plates. Range L. CAMBRIAN–L. CARBONIFEROUS.

EDX See ENERGY DISPERSIVE X-RAY ANALYSIS.

effective elastic thickness An expression of FLEXURAL RIGIDITY (usually of the LITHOSPHERE) in terms of the thickness of an ideal elastic PLATE behaving in the required manner.

effective permeability The ability of a rock to allow the passage of a fluid in the presence of other fluids, e.g. oil and water.

effective porosity The percentage of a given mass of rock or soil consisting of inter-connecting interstices.

effective stress The difference between applied NORMAL STRESS and PORE FLUID PRESSURE.

efficiency of a water well The ratio of AQUIFER LOSS to total DRAWDOWN of a WELL.

effusive eruption A VOLCANIC ERUPTION characterized by a lack of explosive activity, caused by a MAGMA low in volatiles.

effusive igneous body See EXTRUSIVE IGNEOUS BODY.

Egyptian blue A fused mixture of QUARTZ, LIME and COPPER ORE ground to fine powder and used in antiquity as a pigment.

Eh Oxidation potential, a measure of the electron concentration of a system in internal equilibrium.

Eifelian A STAGE of the DEVONIAN, 386.0–380.8 Ma.

einkanter A VENTIFACT with one face. Cf. DREIKANTER, ZWEIKANTER.

EIS See ENVIRONMENTAL IMPACT STATEMENT.

ejecta The solid material thrown from a VOLCANO or impact CRATER.

Ekman layer The thickness over which an EKMAN SPIRAL is representative.

Ekman spiral A graphical representation of current velocity distribution with depth caused by applied wind STRESS on the water surface. The spiral shape arises from the CORIOLIS EFFECT and frictional resistance of the underlying water.

Ekman transport The net DISPLACEMENT from an EKMAN SPIRAL.

elaeolite (eleolite) A massive variety of NEPHELINE.

Elasmobranchii A subclass of class CHONDRICHTHYES, superclass PISCES; sharks and related FISH. Range M. DEVONIAN–RECENT.

elasmosaurs Extinct, long-necked, aquatic REPTILES.

elastic bitumen See ELATERITE.

elastic constants (elastic moduli) Constants that define the elastic properties of an isotropic medium and which control the SEISMIC WAVE VELOCITY of the medium.

elastic deformation (elasticity) DEFOR-

MATION that is instantaneously and totally recoverable.

elastic limit (yield point, yield stress) The STRESS above which ELASTIC BEHAVIOUR is no longer followed.

elastic mineral A TENACITY descriptor indicating a MINERAL that bends and returns to its initial shape on release of pressure.

elastic moduli See ELASTIC CONSTANTS.

elastic rebound theory A model for the mechanism of an EARTHQUAKE whereby forces progressively distort a body of rock until its STRENGTH is exceeded and energy is suddenly released in a catastrophic event.

elastic strain The change in shape resulting from ELASTIC DEFORMATION.

elastic wave A wave which vibrates its host medium without causing permanent DEFORMATION, e.g. a SEISMIC WAVE.

elastic-plastic Descriptive of a material which undergoes ELASTIC DEFORMATION below the YIELD STRESS, at which it behaves with PLASTICITY. Cf. RIGID-PLASTIC.

elastica A type of FOLD PROFILE in which the FOLD ANGLE is negative, e.g. an ANTICLINE with limbs converging downwards.

elasticity 1 See ELASTIC DEFORMATION. **2** Of an ARTESIAN AQUIFER, referring to the presence of compressed water, so more is present than would be the case under atmospheric pressure.

elastoviscous deformation (Maxwell model) DEFORMATION comprising ELASTIC and VISCOUS behaviour in series, i.e. the application of STRESS causes instantaneous ELASTIC STRAIN followed by flow at constant STRAIN RATE.

elaterite (elastic bitumen) Solid BITUMEN resembling dark brown rubber, elastic when fresh.

elbaite A lithium-rich variety of TOURMALINE, which may be green, pink or blue.

electric calamine See HEMIMORPHITE.

electric drilling See VERTICAL ELECTRICAL SOUNDING.

electric log A GEOPHYSICAL BOREHOLE LOG in which ELECTRICAL RESISTIVITY is measured down a borehole using a variety of different electrode configurations to study the nature of the zone invaded by DRILLING MUD FILTRATE and the pristine formations present.

electric profiling See CONSTANT SEPARATION TRAVERSING.

electric trenching See CONSTANT SEPARATION TRAVERSING.

electrical conductivity The reciprocal of ELECTRICAL RESISTIVITY, unit S m^{-1}.

electrical resistivity The resistance in ohms between the opposite faces of a unit cube of material, unit ohm m.

electrode configuration The arrangement of electrodes in electrical surveying. See SCHLUMBERGER CONFIGURATION, WENNER CONFIGURATION.

electrode polarization See OVERVOLTAGE.

electrolytic polarization 1 The build-up of charge on metal electrodes connected to a DC source, which can be avoided by the use of NON-POLARIZING ELECTRODES. **2** See MEMBRANE POLARIZATION.

electromagnetic induction methods (electromagnetic methods, EM methods) GEOPHYSICAL EXPLORATION methods based on energizing the subsurface with time-varying electromagnetic fields. These induce EDDY CURRENTS to flow in subsurface conductors which generate their own fields and which can be detected at the surface, e.g. VLF, INPUT®, AFMAG.

electromagnetic methods See ELECTROMAGNETIC INDUCTION METHODS.

electron microprobe (microprobe) An instrumental analytical technique in which the chemistry of small phases and intergrowths are examined by use of a focused electron beam in a vacuum acting on a polished section.

electron microscopy The use of a beam of high-energy electrons to form images of the surface or internal structure of a material, based on the DE BROGLIE RELATIONSHIP.

electron spin resonance (ESR) A dating method based on the detection of unpaired electrons resulting from ionizing radiation and/or heating.

electrum. A natural GOLD–SILVER alloy.

eleolite See ELAEOLITE.

Eleutherozoa A subphylum of phylum ECHINODERMATA, distinguished by the absence of a stalk. Range ORDOVICIAN–RECENT.

elevation correction The correction applied to GRAVITY and MAGNETIC SURVEY data to compensate for the varying elevations of observations.

elevation energy The potential energy of a mass of water with respect to its elevation above a datum.

Ellesmerian orogeny An OROGENY in DEVONIAN times affecting the Canadian Arctic.

elongation The relative change in length of a line with respect to its original length during DEFORMATION.

elongation lineation A LINEATION formed by a set of parallel, elongate objects in a deformed rock.

Elsonian orogeny An OROGENY during the PROTEROZOIC from 1500–1400 Ma affecting the eastern Canadian SHIELD.

Elster glaciation The first of the four GLACIAL periods of northern Europe in the QUATERNARY.

elutriation A natural process by which CLASTIC particles are separated by grain size, either in water or PYROCLASTIC FLOWS.

eluvial placer A PLACER formed by the CREEP of material down a slope.

eluviation The movement of soil material through the soil zone, resulting from THROUGHFLOW or LEACHING.

eluvium *In situ* WEATHERED BEDROCK.

elvan A Cornish mining term for a DYKE cutting GRANITE.

EM methods See ELECTROMAGNETIC INDUCTION METHODS.

Embrithopoda An order of infraclass EUTHERIA, subclass THERIA, class MAMMALIA; an order whose only representative is *Arsinotherium*, a huge, horned form. Range L. OLIGOCENE.

Embry and Clovan classification An expanded form of the DUNHAM CLASSIFICATION of LIMESTONES.

emerald A deep green GEM variety of BERYL.

emerald copper See DIOPTASE.

emery A natural ABRASIVE comprising CORUNDUM and MAGNETITE, formed by the THERMAL METAMORPHISM of ferruginous BAUXITE.

emplacement The intrusion of an IGNEOUS ROCK body into an envelope of COUNTRY ROCK.

Emsian A STAGE of the DEVONIAN, 390.4–386.0 Ma.

emu Electromagnetic unit, the CGS unit of magnetism.

en echelon An arrangement of parallel lines or planes in which each is of finite length and displaced obliquely from its neighbours in a consistent sense.

enantiomorphism The existence of two chemically identical crystals which are mirror images.

enantiotrophy The conversion of one POLYMORPH to another at a critical temperature and pressure. Cf. MONOTROPY.

enargite (Cu_3AsS_4) An ORE MINERAL of COPPER.

endellite An American term for HALLOYSITE.

enderbite A CHARNOCKITIC rock comprising QUARTZ, ANTIPERTHITE, HYPERSTHENE and MAGNETITE.

endichnia TRACE FOSSILS occurring within the preserving bed.

endlichite A variety of VANADINITE containing arsenic.

endogenetic (endogenic, endogenous) Originating within the Earth. Cf. EXOGENETIC.

endogenic See ENDOGENETIC.

endogenous See ENDOGENETIC.

endometamorphism See ENDOMORPHISM.

endomorphism (endometamorphism) The ALTERATION of the composition of a MAGMA by the ASSIMILATION of COUNTRY ROCK.

endoskarn A SKARN DEPOSIT formed by the REPLACEMENT of an intrusion. Cf. EXOSKARN.

endrumpf A PENEPLAIN reduced to a flat or gently undulating landscape by EROSION.

energy dispersive X-ray analysis (EDX) An X-RAY DIFFRACTION technique in which the whole X-ray spectrum is examined in a single measurement, which has a higher detection limit than WAVELENGTH DISPERSIVE ANALYSIS.

engineering geological map A map showing units defined by their engineering properties.

engineering geology The application of geological information, techniques and principles to the design, construction and maintenance of engineering works.

engineering seismology See EARTHQUAKE ENGINEERING.

englacial (intraglacial) Within ice.

enhanced oil recovery (tertiary oil recovery) Techniques used for the recovery of oil after normal pumping and reinjection are no longer effective, by such methods as steam injection to mobilize viscous oil.

ensialic Within or on CONTINENTAL CRUST.

enstatite ($MgSiO_3$) The magnesian endmember of the ORTHOPYROXENE SOLID SOLUTION series.

enstatitite A PYROXENITE composed almost entirely of ENSTATITE .

entablature A volcanic layer with fan-like JOINTS, commonly alternating with COLONNADES.

enterolithic Descriptive of a SEDIMENTARY STRUCTURE in the form of ropy folds, formed by the crumpling of an EVAPORITE resulting from its swelling on HYDRATION.

entrainment equivalence Descriptive of grains in a sediment bed when they begin to move at identical fluid SHEAR STRESS.

enveloping surface The surface which would join the CREST or TROUGH LINES of a set of FOLDS.

environmental engineering geology The branch of APPLIED GEOMORPHOLOGY which covers the study of features and processes in relation to environmental management and engineering.

environmental impact statement (EIS) A summary of information about the impact an action, such as an engineering project, will have or has had on the environment.

Eocambrian The late PRECAMBRIAN, approximately equivalent to the RIPHEAN.

Eocene An EPOCH of the PALAEOGENE, 56.5–35.4 Ma.

eogenesis See DEPOSITIONAL-ENVIRONMENT-RELATED DIAGENESIS.

eolian See AEOLIAN.

eon A large division of geological time comprising a number of ERAS.

eonothem The largest CHRONOSTRATIGRAPHIC unit.

Eosuchia An order of subclass LEPIDOSAURIA, class REPTILIA; a brigade of poorly known fossil REPTILES which may or may not be related. Range U. PERMIAN–U. TRIASSIC.

Eötvös balance An early form of GRAVIMETER.

Eötvös correction The correction applied to GRAVITY SURVEY data collected from a moving platform (ship or aeroplane) to take account of the component of E–W motion which reinforces or decreases the centripetal force arising from the Earth's rotation, which, in turn, decreases the gravitational attraction of the Earth.

epeiric sea (epicontinental sea) A shallow inland sea.

epeirogenesis See EPEIROGENY.

epeirogeny (epeirogenesis) Very large-scale TECTONIC movements which cause UPLIFT/subsidence of the CONTINENTAL and OCEANIC CRUST without significant DEFORMATION, REGIONAL METAMORPHISM or intrusion.

ephemeral stream A stream or river which does not flow at all times of the year. Cf. PERENNIAL STREAM.

epi- Upon, above.

epibole See ACME ZONE.

epicentral angle The angle subtended at the centre of the Earth by the EPICENTRE of an EARTHQUAKE and the location of its detection.

epicentral distance The great circle distance between epicentre and recorder. Approximately the EPICENTRAL ANGLE multiplied by 111.1, in kilometres.

epicentre The location on the surface vertically above the FOCUS of an EARTHQUAKE.

epichnia A TRACE FOSSIL occurring on the top of the preserving bed.

epiclastic Descriptive of sedimentary material redeposited from an existing sediment.

epicontinental On a continent.

epicontinental sea See EPEIRIC SEA.

epidiorite A METAMORPHIC ROCK with a GRANULAR TEXTURE derived from a BASIC IGNEOUS ROCK but containing the same MINERALS as DIORITE.

epidosite A METAMORPHIC ROCK comprising EPIDOTE and QUARTZ.

epidote (pistacite) $(Ca_2(Al,Fe)Al_2(SiO_4)$ $(SiO_7)(O,OH)_2)$ A group of apple-green, hydrated calcium aluminosilicate SOROSILICATES.

epidote-amphibolite facies See AMPHIBOLITE FACIES.

epidotization An ALTERATION process whereby the FELDSPAR of, commonly, a BASIC IGNEOUS ROCK is ALBITIZED with the separation of EPIDOTE and ZOISITE.

epifaunal Descriptive of an organism which lives on the floor of an ocean, lake or river, either attached to a larger organism or free-moving. Cf. INFAUNAL.

epigene At or near the Earth's surface.

epigenesis Changes affecting SEDIMENTARY ROCKS after compaction, excluding WEATHERING and METAMORPHISM.

epigenetic Descriptive of a deposit forming after its COUNTRY ROCK. Cf. HYPOGENE.

epigenetic drainage See SUPERIMPOSED DRAINAGE.

epilimnion The upper, oxygenated, circulating layer of a stratified lake, from <10m to >50 m thick. Cf. HYPOLIMNION.

epimagmatic See DEUTERIC.

epimorph The natural cast of a MINERAL.

epistilbite $(CaAl_2Si_6O_{16}.5H_2O)$ A white to colourless ZEOLITE.

epitaxy A cement overgrowth different in MINERALOGY from the main grain.

epithermal deposit An EPIGENETIC DEPOSIT formed at low temperatures (50–200°C) near the Earth's surface (<1500 m).

epizone A DEPTH ZONE of moderate temperature and low HYDROSTATIC PRESSURE. See also MESOZONE, KATAZONE.

EPMA Electron probe microanalysis.

epoch A third order geological time unit.

epsilon cross-stratification A lateral, ACCRETIONARY, sedimentary STRUCTURE with gently inclined BEDDING surfaces dipping approximately perpendicular to the PALAEOCURRENT direction.

Epsom salts See EPSOMITE.

epsomite (Epsom salts) $(MgSO_4.7H_2O)$ An EVAPORITE MINERAL, also found as an efflorescence on the walls of CAVES or workings.

equal-angle net See WULFF NET.

equal-area net See SCHMIDT NET.

equal-area projection See SCHMIDT NET.

equant See EQUIDIMENSIONAL.

equidimensional (equant) With the same, or nearly the same, dimensions in all directions.

equigranular Descriptive of a TEXTURE in which grains are all of about the same size.

equipotential surface A surface on which the POTENTIAL is constant.

era A first order geological time unit.

erathem A first order CHRONOSTRATI-GRAPHIC unit.

erg 1 (koum, sand sea) A region covered by DUNES or SAND SHEETS. **2** The CGS unit of energy or work.

Erian A DEVONIAN succession in E. North America covering the lower part of the GIVETIAN.

erionite $(K_2NaCa_{1.5}Mg(Al_9Si_{27}O_{72}).28H_2O)$ A ZEOLITE found in FRACTURES in RHYOLITIC and BASALTIC rocks.

erodibility The resistance of a soil to the entrainment and transport of its particles by an agent of EROSION, controlled by its mechanical and chemical properties.

erosion The process whereby particles are detached from rock or soil and transported away, the principal agents being ice, wind and water.

erosion control practice factor A term in the UNIVERSAL SOIL LOSS EQUATION.

erosion surface A flat plain resulting from EROSION and representing the final phase of a CYCLE OF EROSION.

erosional sheltering A mechanism of SLICKENSIDE formation whereby debris is deposited in the direction of slip behind an asperity on a FAULT surface.

erosivity A measure of the potential ability of an eroding agent, such as rainfall or wind, to cause EROSION, based on its kinetic energy.

Erqiao A TRIASSIC succession in China equivalent to the RHAETIAN.

erratic A stone transported by a GLACIER and deposited far from its point of origin.

ERTS satellites See EARTH RESOURCES TECH-NOLOGY SATELLITE.

erubescite See BORNITE.

erythrite (cobalt bloom) $(Co_3(AsO_4)_2.8H_2O)$ A pink SECONDARY MINERAL of cobalt.

escape tectonics See INDENTATION TEC-TONICS.

escarpment (scarpslope) The steeper slope of a CUESTA. Cf. DIPSLOPE.

esker (osar) An elongated ridge of strati-fied GRAVEL, probably formed by streams flowing beneath or on a GLACIER.

essential mineral A MINERAL whose pres-ence or absence determines the name of a rock. Cf. ACCESSORY MINERAL.

essexite An ALKALINE GABBRO composed of PLAGIOCLASE, HORNBLENDE, BIOTITE and TITANAUGITE with minor ALKALI FELDSPAR and NEPHELINE.

estuary A partly enclosed body of water open to the sea where fresh and sea water intermix.

Etalian A TRIASSIC succession in New Zea-land equivalent to the ANISIAN.

etchplain A broad, erosional land surface in the tropics formed by the DEEP WEATHER-ING of CRYSTALLINE ROCKS where EROSION has not caused deep incision.

eu- Well, good, rich in.

euclase $(BeAl(SiO_4)(OH))$ A rare beryllium CYCLOSILICATE.

eucrite 1 An obsolete term for a coarse-grained, commonly OPHITIC, BASIC GABBRO of deep-seated origin containing PLAGIO-CLASE (near BYTOWNITE), ORTHO- and CLINO-PYROXENE and OLIVINE. **2** A stony METEORITE of BASALTIC composition.

eucryptite (LiAlSiO$_4$) An INOSILICATE formed as an ALTERATION product of SPODUMENE.

eucrystalline Descriptive of an IGNEOUS ROCK which is well-crystallized.

eudialite See EUDIALYTE.

eudialyte (eudialite) (Na$_4$(Ca,Fe, Ce,Mn)$_2$ZrSi$_6$O$_{17}$(OH,Cl)$_2$) A pink-red or yellow-brown, complex, hydrated sodium-calcium-iron zirconosilicate found in some ALKALINE IGNEOUS ROCKS.

eugeosyncline A GEOSYNCLINE with abundant MAGMATIC activity.

euhedral (automorphic, idiomorphic) Descriptive of a grain with a fully developed CRYSTAL FORM.

eukaryote An organism with cells possessing a distinct nucleus. Cf. PROKARYOTE.

Euler pole (pole of rotation) See EULER'S THEOREM.

Euler's theorem 'Any point on the surface of a sphere can be moved to any other point by a single rotation about a specific axis, the points at which the axis intersect the surface being known as EULER POLES.' Used in describing PLATE movements across the globe.

eulysite A METAMORPHIC ROCK containing iron and manganese silicates, e.g. HEDEN-BERGITE, FAYALITE, ALMANDINE, SPESSARTINE.

Eumalacostraca A subclass of class MALA-COSTRACA, subphylum CRUSTACEA, phylum ARTHROPODA, characterized by a non-bivalve shell, six abdominal somites and a tail formed by the last somite and telson, including shrimps, crabs and lobsters. Range DEVONIAN–RECENT.

euphotic zone See PHOTIC ZONE.

Euryapsida A subclass of class REPTILIA, characterized by only one temporal open-ing; possibly not a natural grouping. Range PERMIAN–U. CRETACEOUS.

Eurypterida/eurypterids A class of sub-phylum CHELICERATA, phylum ARTHROPODA; large (up to 2 m), predatory invertebrates with a small prosoma whose last appendage may be modified as a swimming paddle, a long episthosoma of 12 somites and a telson in the form of a spine or paddle. Range ORDOVICIAN–PERMIAN.

eustasy Global change in sea level.

eutaxitic texture A TEXTURE of flattened glassy discs (FIAMME) in an ASHY MATRIX seen in WELDED TUFFS.

eutectic Descriptive of a mixture of at least two substances which have crystallized simultaneously.

Eutheria (placentals) An infraclass of subclass THERIA, class MAMMALIA; the placental MAMMALS. Range CRETACEOUS–RECENT.

eutrophic lake A lake containing much dissolved plant nutrient and a seasonal lack of oxygen in the lowest layer due to significant amounts of decaying organic matter. Cf. DYSTROPHIC LAKE, OLIGOTROPHIC LAKE.

eutrophism The process whereby a lake is rejuvenated by an increase in plant nutrients so that ALGAE bloom on the surface, preventing light penetration and oxygen absorption.

euxenite ((Y,Er,Ce,La,U)(Nb,Ti,Ta)$_2$ (O,OH)$_6$) A massive, brown-black, niobate, tantalate and titanate of yttrium, erbium, cerium, thorium and uranium found in GRANITE PEGMATITES.

euxinic Descriptive of an environment of restricted water circulation where ANAER-OBIC conditions obtain.

evaporative drawdown The loss in BRINE volume and its lowering in level as evaporation takes place in an EVAPORITE BASIN.

evaporative pumping The upward movement of GROUNDWATER towards a DEFLATION surface caused by severe evaporation at the sediment-air interface.

evaporite A rock made up of MINERAL(s) formed by PRECIPITATION from concentrated BRINES.

evaporite basin A low-lying area where evaporation exceeds fluid input so that BRINES are sufficiently concentrated for the PRECIPITATION of EVAPORITE MINERALS to take place.

evaporite deposit A deposit from which various salts can be recovered, principally ANHYDRITE, BORAX, CARNALLITE, CELESTINE, COLEMANITE, GYPSUM, HALITE, KERNITE, nitrates, SYLVITE and TRONA.

evaporite-related collapse breccia Chaotic, angular rock fragments forming when an underlying EVAPORITE layer is removed by DISSOLUTION.

event stratigraphy The recognition, study and correlation of the effects of significant physical events, e.g. marine TRANSGRESSIONS, VOLCANIC ERUPTIONS, in the expectation that truly synchronous HORIZONS could be defined.

event stratinomy The EVENT STRATIGRAPHY of individual BEDS.

evolute Descriptive of a coiled shell in which all whorls are exposed. Cf. CONVOLUTE.

evolution path The trend in ORGANIC MATTER DIAGENESIS as KEROGEN decreases in oxygen and hydrogen content.

ex- Out of, not having.

Excelsior A TRIASSIC succession in Nevada, USA equivalent to the ANISIAN and LADINIAN.

exceptional fossil deposit (fossil lagerstätte) A deposit in which FOSSILS are exceptionally well preserved, often including soft tissues, or exceptionally rich.

excess mass The difference in mass between an anomalous body of rock (e.g. an OREBODY) and the COUNTRY ROCK which would otherwise occupy its volume. Can be estimated from GRAVITY ANOMALIES using GAUSS' THEOREM.

exchangeable sodium percentage A property of certain CLAYS referring to the percentage of sodium that is readily lost by DISSOLUTION. If this is high the CLAY is very susceptible to EROSION.

exfoliation (desquamation, onionskin weathering, spheroidal weathering) The DEGRADATION of BOULDERS by the spalling of surface layers, millimetres to a few metres in thickness, probably arising from the release of LITHOSTATIC PRESSURE on exhumation, by WEATHERING or the growth of salt crystals just below the surface of the rock.

exhalite A chemical deposit, principally on the seafloor, formed mainly from HYDROTHERMAL exhalations such as BLACK SMOKERS.

exichnia A TRACE FOSSIL occurring as infillings of the preserving BED with another substrate.

exinite A COAL MACERAL group made up of small organic particles such as algae, spores, cuticles, etc. and relatively high in hydrogen and volatiles.

Exmoor A PERMIAN succession in Queensland, Australia, covering part of the ARTINSKIAN, the KUNGURIAN and part of the UFIMIAN.

exogenetic (exogenic, exogenous) Originating at or near the Earth's surface. Cf. ENDOGENETIC.

exogenic See EXOGENETIC.

exogenous See EXOGENETIC.

exoskarn A SKARN DEPOSIT developed in metasediments. Cf. ENDOSKARN.

exotic terrane See DISPLACED TERRANE.

expanding Earth A model suggesting the Earth has expanded significantly over geological time, explaining the relative movements of the continents by the dismemberment of a once complete shell of CONTINENTAL CRUST. Modern measurements have shown that the Earth has not expanded at the required rate. Cf. CONTRACTING EARTH.

expansive soil A soil which shrinks and swells with changing moisture content, such as one containing MONTMORILLONITE.

exploitation well A WELL sunk in a proven deposit.

exploratory well A WELL sunk in the hope of finding new oil or gas accumulations. Cf. DEVELOPMENT WELL.

explosion breccia An IGNEOUS BRECCIA formed by explosive volcanic activity.

explosive index The percentage of PYRO-CLASTIC material in the products of a VOL-CANIC ERUPTION.

exposed coalfield A COALFIELD in which COAL crops out at the surface. Cf. CONCEALED COALFIELD.

exposure A surface where *in situ* rock is seen free from soil or vegetation cover.

exposure mapping (outcrop mapping) A type of GEOLOGICAL MAPPING in which every EXPOSURE is visited and its limits plotted, along with relevant topographic features.

exsolution The development of two or more compositionally different phases from a SOLID SOLUTION, usually as cooling takes place.

extended Griffith failure criterion (Griffith-Murrell failure criterion) An extension of the two-dimensional GRIFFITH FAILURE CRITERION into three dimensions.

extension An increase in the length of a line during DEFORMATION.

extension joint (tension joint) A JOINT forming by TENSILE FAILURE perpendicular to the least PRINCIPAL STRESS.

extensional cleavage A set of planes oblique to an existing planar structure so that DISPLACEMENTS on the CLEAVAGE cause net EXTENSION parallel to the existing planes.

extensional crenulation cleavage A CLEAVAGE formed when one set of SHEAR BANDS is developed more strongly than the other.

extensional fault A FAULT across which EXTENSION has occurred.

extensional fault system A set of related FAULTS on which individual DISPLACEMENTS give a net EXTENSION in the system as a whole.

extinct volcano A VOLCANO that has not erupted within historical time.

extinction 1 The disappearance of a group of organisms. **2** The phenomenon of least illumination at a particular orientation of a crystal when viewed in THIN SECTION by a microscope with polarized light and crossed polars.

extraformational conglomerate A CONGLOMERATE whose CLASTS originate mostly outside the BASIN of deposition.

extraversion The mechanism whereby the SUPERCONTINENT proposed in the SWEAT HYPOTHESIS rifted and the resulting continents moved apart.

extrusion tectonics See INDENTATION TEC-TONICS.

extrusive igneous body (effusive igneous body) An IGNEOUS BODY emplaced at the surface.

exudatinite A COAL MACERAL of the EXINITE group of secondary origin, found in cavities in other MACERALS, which was soft and mobile at some stage during COALIFICATION.

eyot a small island on the bend of a river.

F Letter used to indicate a phase of FOLI-ATION formation, subscripted to denote each separate phase.

f joint See EN ECHELON FRACTURE.

Fa A method of indicating OLIVINE composition as a percentage of FAYALITE, e.g. Fa$_{10}$ indicates a composition of 10% FAYALITE, 90% FORSTERITE.

Fa Lang A TRIASSIC succession in China equivalent to the LADINIAN.

fabric The pervasive features of a rock.

fabric cross Orthogonal FABRIC axes of MONOCLINIC symmetry.

fabric diagram (petrofabric diagram) A STEREOGRAM showing the components of a PETROFABRIC.

fabric domain A region of a rock that is homogeneous with respect to the orientation of a FABRIC ELEMENT.

fabric element A feature of a rock that contributes to its FABRIC, e.g. CLEAVAGE, FRACTURE, LINEATION, grain shape, grain boundaries and CRYSTALLOGRAPHIC orientations.

fabric skeleton Lines linking the highly populated parts of a STEREOGRAM showing FABRIC ELEMENTS, which define PREFERRED ORIENTATIONS.

fabric symmetry The class of symmetry shown by a FABRIC ELEMENT.

face A mining and quarrying term for the exposed rock surface, excluding the BACK or floor.

face method A mapping method used in quarries and OPENPIT MINES in which geological data are plotted in plan view or at a reference datum measured accurately near the TOE, and data from higher levels are estimated.

facet An element of the surface of a crystal or cut GEM.

facies All lithological and palaeontological features of a particular SEDIMENTARY ROCK, from which depositional environment may be inferred.

facies map A map illustrating lateral changes in the lithology of a FORMATION, GROUP or SYSTEM within a sedimentary BASIN, which allows complex STRATIGRAPHIC data to be presented and which may be used to construct PALAEOGEOGRAPHIC or palaeoenvironmental conditions.

facing The direction in which BEDS become younger. See also YOUNGING.

fahl ore See TETRAHEDRITE.

fahlband A band of METAMORPHIC ROCK carrying disseminated sulphides that are more abundant than ACCESSORY MINERALS but too few to form an OREBODY.

fahlerz See TETRAHEDRITE.

failed rift See AULACOGEN.

failure Loss of STRENGTH.

failure criterion The relationship between PRINCIPAL STRESSES which gives the condition for FAILURE.

failure strength (failure stress) The STRESS at which FAILURE occurs.

failure stress See FAILURE STRENGTH.

fairfieldite $(Ca_2(Mn^{2+},Fe^{2+})(PO_4)_2.2H_2O)$ A white, hydrated phosphate of calcium and manganese found in GRANITE PEGMATITES.

fall velocity See SETTLING VELOCITY.

falling dune A steep-sided STATIONARY DUNE forming on the LEE SLOPE of an obstacle such as a hill where SAND collects.

false bedding See CROSS-STRATIFICATION.

false ruby See PYROPE.

false-colour composite A colour image of REMOTE SENSING data made by combining images of several SPECTRAL BANDS, some of which are beyond the visible spectrum.

famatinite (Cu_3SbS_4) A COPPER sulphosalt found in low- to medium-GRADE COPPER deposits.

Famennian The highest STAGE of the DEVONIAN, 367.0–362.5 Ma.

Famennien A DEVONIAN succession in France and Belgium covering parts of the FRASNIAN and FAMENNIAN.

fan A slope of detritus increasing in width down the slope.

fan shooting A method of SEISMIC REFRACTION surveying in which the detectors are sited at similar distances from the SEISMIC SOURCE from which they radiate out in an arc. Provides information on the three-dimensional form of a refractor.

fanglomerate A RUDITE deposited in an ALLUVIAL FAN.

faro A small, elongate REEF enclosing a LAGOON up to 30 m deep, forming on the rim of a BARRIER REEF or ATOLL.

fasciculate Descriptive of a compound CORAL whose CORALLITES, although associated, are sufficiently spaced to avoid mutual interference.

Fast Fourier Transform (FFT) See COOLEY-TUKEY METHOD.

fathogram See ECHOGRAM.

fathometer See ECHO SOUNDER.

faujasite $((Na_2,Ca,Mg)_{32}(Al_{64}Si_{128}O_{384}).$ $256H_2O)$ A rare colourless or white ZEOLITE.

fault A discontinuity surface across which there has been SHEAR DISPLACEMENT.

fault block A body of rock partly or completely defined by FAULTS and differing in elevation from its surroundings.

fault breccia A non-foliated, incohesive FAULT ROCK with >30% visible fragments surrounded by a MATRIX.

fault drag The bending of a marker across a FAULT.

fault gouge An incohesive FAULT ROCK with <30% visible fragments surrounded by a MATRIX.

fault inlier An INLIER created by a FAULT crossing a valley.

fault plane The plane along which a FAULT acts.

fault plane solution See FOCAL MECHANISM SOLUTION.

fault propagation The process by which a FAULT extends along its length.

fault reactivation The re-use of a FAULT in a later phase of DEFORMATION.

fault rock A rock produced by the action of a FAULT.

fault set See MULTIPLE FAULTS.

fault zone A TABULAR volume containing many FAULTS and FAULT ROCKS.

fault-bend fold A FOLD produced in the HANGINGWALL by the movement of a FAULT over a non-planar FAULT surface.

faunule A FOSSIL fauna from a small area or STRATIGRAPHIC range.

fayalite (Fe_2SiO_4) The iron-bearing end-member of the OLIVINE SOLID SOLUTION series.

FDSN See FEDERATION OF DIGITAL SEISMIC NETWORKS.

feather edge The intersection line on a map of a STRATIGRAPHIC boundary with a higher boundary such as an UNCONFORMITY, which marks the zero ISOPACHYTE of the STRATA between the boundaries.

feather fracture See EN ECHELON FRACTURE.

feather joint (pinnate fracture) A minor JOINT adjacent to a larger FRACTURE and intersecting it at an acute angle.

feather ore A PLUMOSE or ACICULAR form of JAMESONITE.

feather structure See PLUME STRUCTURE.

Federation of Digital Seismic Networks (FDSN) A global network of SEISMOMETERS based on digital recording which has superseded the WORLD-WIDE STANDARDIZED SEISMOGRAPH NETWORK.

Feixianguan A TRIASSIC succession in China covering the GRIESBACHIAN and part of the NAMMALIAN.

feldspars Framework aluminosilicates of sodium, potassium and calcium, the most abundant MINERAL group in the CRUST. Common feldspars are SOLID SOLUTIONS of the three end-member components ANORTHITE, ALBITE and ORTHOCLASE. Combinations predominantly of ALBITE/ANORTHITE are termed PLAGIOCLASE and combinations of ALBITE/ORTHOCLASE termed ALKALI FELDSPAR.

feldspathic greywacke See ARKOSIC WACKE.

feldspathic wacke See ARKOSIC WACKE.

feldspathization See POTASSIC FENITIZATION.

feldspathoids (foids) A group of aluminosilicate MINERALS with a variety of framework structures, similar to the FELDSPARS but containing less SILICA. Characteristic of UNDERSATURATED ALKALINE IGNEOUS ROCKS.

felsenmeer See BLOCKFIELD.

felsic Containing at least one of the light-coloured MINERALS FELDSPAR, LENAD or SILICA as the major component of the MODE. Cf. MAFIC.

felsitic Descriptive of GRANULAR, CRYPTO-CRYSTALLINE AGGREGATES formed by the DEVITRIFICATION of GLASS.

femic Ferromagnesian.

fen A mire resulting from GROUNDWATER rather than precipitation.

fenester (fenster) A TECTONIC 'window' through which rocks below a THRUST SHEET have been exposed by EROSION.

fenestrae (birdseyes) Millimetric-sized, rounded to planar to irregular voids in a SEDIMENTARY ROCK which may be partly or completely infilled by sediment or a cement. Formed by desiccation or air entrapment.

Fengshan A CAMBRIAN succession in China covering the upper part of the DOLGELLIAN.

fenite COUNTRY ROCK which has been subject to METASOMATISM by the EMPLACEMENT OF ALKALINE or CARBONATITE INTRUSIVE ROCKS.

fenitization The process of forming a FENITE. See POTASSIC FENITIZATION, SODIC FENITIZATION.

fenster See FENESTER.

feral relief A landscape in which valley sides are deeply dissected by INSEQUENT STREAMS, related to rapid RUNOFF.

ferberite $(FeWO_4)$ The iron end-member of the WOLFRAMITE MINERAL series.

fergusite An ALKALINE SYENITE comprising large PSEUDOLEUCITE crystals in a MATRIX of AEGIRINE-AUGITE, OLIVINE, APATITE, SANIDINE and IRON OXIDES.

fergusonite $((Y,Ce,Nb)NbO_4)$ An ORE MINERAL of the RARE EARTH ELEMENTS found in GRANITE PEGMATITES.

fermentation The ANAEROBIC process by which bacteria metabolize oxygen-containing organic matter, liberating hydrogen and carbon dioxide, which takes place during shallow organic matter DIAGENESIS.

ferrallitization A PEDOGENETIC process involving the accumulation of sesquioxides of iron and aluminium under humid tropical conditions.

ferricrete (laterite) An iron-rich DURI-CRUST often formed in DEEP WEATHERING profiles in humid tropical and subtropical conditions.

ferrierite $((Na,K)Mg_2(Al_{5.5}Si_{30.5}O_{72}.18H_2O)$ A ZEOLITE.

ferriferous (ferruginous) Containing iron.

ferrimagnetism A form of FERROMAG-NETISM, typical of most natural crustal magnetic MINERALS, in which MAGNETIC DOMAINS are oppositely magnetized but of different strength, giving rise to strong SPONTANEOUS MAGNETISM.

ferrimolybdite $(Fe_2(MoO_4)_3.8H_2O)$ A bright yellow, soft molybdate produced by the ALTERATION of MOLYBDENITE.

ferro- Iron-bearing.

ferroactinolite $(Ca_2Fe_5Si_8O_{22}(OH)_2)$ A dark green AMPHIBOLE.

ferrocarbonatite A CARBONATITE containing SIDERITE.

ferrogabbro A GABBRO containing iron-rich OLIVINE.

ferrohastingsite $(NaCa_2(Fe^{2+})_4AlAl_2Si_6$ $O_{22}(OH)_2)$ A variety of sodic, iron-rich HORN-BLENDE.

ferrohedenbergite $((Ca,Fe)Si_2O_6)$ A CLINO-PYROXENE found in QUARTZ SYENITES, GRANO-PHYRES and FERROGABBROS.

ferrohortonolite An OLIVINE with the composition Fo_{30-10}.

ferromagnesian Containing iron and magnesium.

ferromagnetism Strong magnetic be-haviour resulting from internal quantum-mechanical exchange or super-exchange forces which cause electron spins to become coupled but quench the coupling between the magnetization associated with the electron orbits. This results in a strong SPONTANEOUS MAGNETIZATION.

ferropseudobrookite $(FeTi_2O_5)$ A rare iron-titanium OXIDE MINERAL.

ferrosilite $(Fe_2Si_2O_6)$ A PYROXENE found cementing SANDSTONES and PELITES.

ferruginous See FERRIFEROUS.

fersmannite $(Na_4Ca_4Ti_4(SiO_4)_3(O,OH,F)_3)$ A rare, titanium-bearing NESOSILICATE.

fetch The extent of open water across which a WAVE-generating wind blows, determining the height and energy, and hence the EROSIONAL and depositional potential, of the WAVES.

FFT See FAST FOURIER TRANSFORM.

fiamme The irregular, flattened, GLASSY discs in a WELDED TUFF which determine EUTAXITIC TEXTURE.

fibre growth A SLICKENSIDE formation mechanism in which elongate crystals grow on a FAULT surface in the movement direction.

fibroblastic A type of METAMORPHIC FABRIC in which the grains are of equal size and fibrous HABIT due to solid-state CRYSTALLIZ-ATION during METAMORPHISM.

fibrolite See SILLIMANITE.

fibrous fracture The FRACTURE seen in MINERALS giving a fibrous appearance.

fibrous texture A TEXTURE with the appearance of a mass of fibres, as shown by ASBESTOS.

Fick's law of diffusion A law controlling DIFFUSION IN SEDIMENTS: $J_i = -D_i dC/dx$, where J_i = mass of component i transported per unit area per unit time, D_i = DIFFUSION COEFFICIENT, dC/dx = concentration gradient.

field capacity A descriptor of a soil in which gravitationally driven draining has ceased.

Filicopsida A class of division TRACHAEO-PHYTA, kingdom PLANTAE; the ferns and their relatives. Range U. SILURIAN–RECENT.

filiform Thread-like.

Fillipovskiy See IREN'SKIY.

filter pressing A possible mechanism of MAGMA DIFFERENTIATION in which melt separates from a crystal-rich MAGMA by draining or by being pressed out, perhaps by the overlying weight of the MAGMA body.

filtering The process of modifying a waveform, usually to suppress or enhance certain information, such as the improvement of a SEISMIC RECORD by the removal of NOISE frequencies.

fine crush breccia A rock similar to a CRUSH BRECCIA, but with fragments between 1 and 5 mm in size. Cf. CRUSH BRECCIA, CRUSH MICROBRECCIA.

fineness An expression of the quality of NATIVE GOLD in ppt, i.e. 1000 is pure GOLD.

finger lake A long, narrow lake in a deep trough, probably the result of GLACIAL EROSION.

finite rotation The actual rotation about a EULER POLE required to bring two linear features (e.g. OCEAN RIDGES) together. Cf. INSTANTANEOUS ROTATION.

finite strain (total strain) The total change in shape of a deformed body relative to its shape before DEFORMATION.

fiord See FJORD.

fire damp A combustible gas contained in COAL comprising a mixture of METHANE and other HYDROCARBONS.

fire fountaining A type of VOLCANIC ERUPTION in which low VISCOSITY MAGMA erupts continuously and rises a few hundred metres as an incandescent jet, typical of HAWAIAN ERUPTIONS.

fire opal A variety of OPAL with a brilliant orange colour.

fireclay (refractory clay) An UNDERCLAY rich in KAOLINITE with commercial application as a refractory because of its low MICA and iron content.

firn See NÉVÉ.

first arrival (first break) The earliest seismic signal to be recorded from a particular SEISMIC SOURCE.

first break See FIRST ARRIVAL.

first order fold A FOLD larger than a SECOND ORDER FOLD and which folds the ENVELOPING SURFACE of SECOND ORDER FOLDS.

fish **1** See PISCES. **2** An instrument package towed behind a ship.

fishscale dune pattern See AKLÉ.

fissility The property of a fine-grained rock which has surfaces of weakness along which it splits easily. A more general term than CLEAVAGE in that it implies no causative mechanism.

fission track dating A dating method based on measuring the concentration of fission tracks generated during the decay of ^{238}U. Also used to study thermal history and rates of UPLIFT.

fissure eruption The eruption of MAGMA at several points along an elongate volcanic conduit. Cf. CENTRAL ERUPTION.

fissure vein An old term for VEIN.

fixed carbon The CARBON left after all volatiles have been expelled from COAL.

fjord (fiord) A deep, narrow, GLACIAL trough inundated by the sea.

flachkarren See CLINT.

flagstone A SANDSTONE containing MICA, which enhances its FISSILITY.

flake graphite A commercial term for flat, platy grains of GRAPHITE disseminated through a METAMORPHIC ROCK.

flake mica (ground mica, scrap mica) Fine-grained MICA that is a by-product of certain processing operations and which is used in various industrial processes.

flake tectonics The process at a SUBDUCTION ZONE of the detachment or DELAMINATION of an UPPER CRUSTAL layer from a subducting PLATE and its emplacement on the overriding PLATE.

flame structure A STRUCTURE at the base of a BED comprising upward-pointing fingers or wedges of sandy or silty sediment penetrating a finer-grained substrate because of load pressure. Often associated with LOAD CASTS.

Flandrian transgression The rise in sea level during HOLOCENE and late PLEISTOCENE times caused by the melting of ICE SHEETS of the last GLACIAL.

flap A type of RECUMBENT SYNCLINE formed by GRAVITY COLLAPSE.

flaser A MUD lens preserved in the trough of a ripple. There is gradation from flasers to LINSEN, both of which form in aqueous environments in which slack water and WAVE activity alternate.

flaser bedding A type of HETEROLITHIC BEDDING comprising discontinuous, curved lenses of MUD or SILT that were deposited in troughs or draped over the RIPPLES in CROSS-LAMINATED SANDS.

flaser gabbro A CATACLASTIC GABBRO in which CHLORITE or MICA flakes curl around AUGEN of QUARTZ and FELDSPAR.

flash A water-filled depression caused by surface subsidence.

flat **1** An horizontal to subhorizontal REPLACEMENT OREBODY. **2** The part of a FAULT that does not cut across datum surfaces, such as BEDDING. Cf. RAMP.

flattening (flattening strain) A shape change in which there is CONTRACTION along only one PRINCIPAL STRAIN direction.

flattening strain See FLATTENING.

flatiron A small, steep-sided, triangular MESA.

flatjack A thin metallic membrane inserted in rock to measure *in situ* STRESS.

flexible mineral A TENACITY descriptor of a MINERAL that will bend and remains bent on release of pressure.

flexural flow A FOLD MECHANISM in which there is layer-parallel SHEAR in the FOLD LIMBS without distortion in the HINGE ZONE.

flexural rigidity A term describing the resistance of an elastic beam, such as the LITHOSPHERE, to FLEXURE.

flexural slip A FOLD MECHANISM in which there is FLEXURAL FLOW by discontinuous layer-parallel SHEAR distributed between the rock layers and INTERLAYER SLIP on their bounding surfaces.

flexure The bending of an elastic beam or sheet in order to support a sediment or other load. The LITHOSPHERE can act in this way to support a mountain range, ICE SHEET or sediment load.

Flinn diagram A graph used to illustrate and analyse the shapes of STRAIN ELLIPSOIDS. Cf. HSU DIAGRAM.

flint (SiO_2) A term used for MICROCRYSTAL-LINE SILICA found in the CHALK, equivalent to CHERT in other rocks.

flint clay A MICROCRYSTALLINE rock composed mainly of KAOLIN which forms a very hard, non-plastic FIRECLAY.

flinty crush-rock An old term for PSEUDO-TACHYLITE or ULTRAMYLONITE; a fine-grained to CRYPTOCRYSTALLINE CATACLASTIC rock appearing similar to FLINT and often showing intrusive relationship to the COUNTRY ROCK.

float ELUVIAL material.

floatstone A coarse-grained LIMESTONE with MATRIX-supported CLASTS, of which >10% exceed 2 mm in size.

flocculation The aggregation of CLAY particles into randomly oriented lumps.

flood basalt An extrusion of low VISCOSITY BASALTIC MAGMA of very large volume.

flood lava A LAVA FLOW contributing to a PLATEAU LAVA.

flood routing The estimation of the shape of a HYDROGRAPH anywhere along a river during flooding.

floodplain An area of land periodically inundated by floodwater.

floor thrust (sole thrust) The basal THRUST from which FAULTS branch in an IMBRICATE FAN.

Floran A CAMBRIAN succession in Australia covering parts of the SOLVAN and MENEVIAN.

flos ferri A variety of ARAGONITE resembling CORAL.

flotation (froth flotation) A method of concentrating MINERALS by selective flotation in which the MINERAL attaches to bubbles blown through a mixture of ground ORE, water and a frothing agent, and rises to form a surface froth.

flour gold The very finest-grained PLACER GOLD.

flow The permanent DEFORMATION that has a continuous STRAIN distribution.

flow banding (flow layering) Layering produced by FLOW in an IGNEOUS or META-MORPHIC ROCK.

flow cleavage A type of SLATY CLEAVAGE in which the rock's FABRIC is believed to reflect RECRYSTALLIZATION accompanied by solid-state FLOW, so that original sedimentary BEDDING may be lost.

flow competence The maximum particle size capable of being transported in the BEDLOAD.

flow folding (shear folding) The formation of a FOLD by SHEAR or bulk FLOW of rock in a direction oblique or normal to the layering.

flow foliation (flow lineation) A more general term than FLOW BANDING for describing a line or surface produced by FLOW.

flow layering See FLOW BANDING.

flow lineation See FLOW FOLIATION.

flow regime A classification of flows based on the frictional resistance experienced. See also LOWER FLOW REGIME, UPPER FLOW REGIME.

flow separation The detachment of the BOUNDARY LAYER from a surface through generation of adverse pressure gradients close to that surface, commonly arising from an abrupt change in bed geometry. Important in the generation of TURBULENT FLOW.

flow visualization Techniques used to make flow structure visible to direct observation or photography, e.g. smoke in an airflow, dye in a stream.

flower structure 1 A STRUCTURE produced by local changes in direction in a STRIKE-SLIP

FAULT SYSTEM where FAULTS have opposed DIPS, leading to alternate zones of elevated and depressed blocks arising from local zones of TRANSPRESSION and TRANSTENSION. **2** A coherent flow pattern within, for example, a channel; an area of large-scale, turbulent recirculation.

flowstone A type of SPELEOTHEM deposited by water flowing over the walls or floor of a CAVE.

floxoturbidite A poorly graded sediment deposited from a TURBULENT GRAVITY-induced FLOW.

fluid inclusion An inclusion of fluid inside a crystal, which can be used to determine the pressure of formation of the crystal and hence the depth of formation.

fluidity The inverse of VISCOSITY.

fluidization A process whereby the vertical escape of fluid from a GRANULAR AGGREGATE exerts sufficient drag to support the grains against GRAVITY.

flume 1 A deep, narrow gorge containing a turbulent stream. **2** An apparatus used to reconstruct and study fluid flow and sediment transport.

fluorapatite ($Ca_5(PO_4)_3F$) The commonest variety of APATITE.

fluorescence The property of emitting light when light of a certain wavelength is absorbed.

fluorine test dating A dating method for bone by determining its fluorine content, which increases with time as it replaces calcium.

fluorinite A COAL MACERAL of the EXINITE group believed to originate from plant oils and fats.

fluorite (CaF_2) A MINERAL used as a flux in smelting.

fluorspar Material with sufficient FLUORITE to be exploited commercially.

flushed zone The annulus around a borehole from which PORE FLUIDS have been flushed out and replaced by DRILLING MUD FILTRATE.

flute A type of SOLE MARK; a feature of turbulent EROSION with a bulbous depression upstream.

fluting A type of differential EROSION in which the surface of a coarse-grained rock is made ridged or corrugated.

fluvial (fluviatile) Pertaining to a river or stream.

fluviatile See FLUVIAL.

fluvioglacial See GLACIOFLUVIAL.

fluviokarst A LIMESTONE landscape produced by the combined action of FLUVIAL EROSION and LIMESTONE DISSOLUTION.

fluxgate magnetometer A continuous-reading MAGNETOMETER for measuring the strength of the GEOMAGNETIC FIELD.

fluxion structure A finely lensoid or banded STRUCTURE produced by the ELONGATION of grains of contrasting composition at high SHEAR STRAINS, characteristic of MYLONITE and ULTRAMYLONITE.

flysch A thick sedimentary deposit deposited by a TURBIDITY CURRENT and originating from the EROSION of rapidly rising fold mountains in the early stages of OROGENY. Cf. MOLASSE.

Fo A method of indicating OLIVINE composition as a percentage of FORSTERITE, e.g. Fo_{10} indicates a composition of 10% FORSTERITE, 90% FAYALITE.

focal depth The depth to the FOCUS of an EARTHQUAKE.

focal mechanism solution (fault plane solution) The identification of the nature of the faulting responsible for an EARTHQUAKE and the orientation of the NODAL PLANES from recordings at SEISMOGRAPHS distributed around the globe.

focal sphere An hypothetical sphere centred on an EARTHQUAKE FOCUS which facilitates representation of FOCAL MECHANISM SOLUTION data on a STEREOGRAM.

focus See EARTHQUAKE FOCUS.

fodinichnia TRACE FOSSILS formed from feeding structures.

foidite A general term for a VOLCANIC ROCK with >60% FELDSPATHOIDS by volume of the FELSIC constituents.

foidolite A general term for a PLUTONIC rock with >60% FELDSPATHOIDS by volume of the FELSIC constituents.

foids See FELDSPATHOIDS.

fold A curved or angular shape of an originally planar geological surface.

fold amplitude Half the distance between the ENVELOPING SURFACES of the CRESTS and TROUGHS of a FOLD TRAIN.

fold angle See FOLD INTERLIMB ANGLE.

fold axial plane A planar surface defined by the successive positions of FOLD HINGES through a layered sequence.

fold axial surface A curved plane defined by the successive positions of FOLD HINGES through a layered sequence, the more general form of FOLD AXIAL PLANE.

fold axial trace The line of intersection of a FOLD AXIAL SURFACE with the topographic or some other defined surface.

fold axis The orientation of a FOLD HINGE.

fold belt A large-scale group of related FOLDS, probably forming part of an OROGENIC BELT.

fold class A classification of FOLDS based on DIP ISOGONS: in Class 1 they converge downwards, in Class 2 they are parallel and in Class 3 they diverge downwards.

fold closure The direction in which a FOLD HINGE lies with respect to its LIMBS or AXIAL PLANE; e.g. a FOLD may close eastwards or downwards.

fold core The part of a folded layer closest to the FOLD HINGE zone.

fold crest A FOLD HINGE zone which is concave downwards. Cf. FOLD TROUGH.

fold culmination An elevated zone on a FOLD HINGE or CREST which is of variable height, or a high point on a folded surface with DOME-AND-BASIN STRUCTURE.

fold depression The location on a FOLD where the FOLD HINGE LINE PLUNGE of a NON-CYLINDRICAL FOLD causes the HINGE zone to be depressed.

fold hinge The location of greatest curvature of a folded surface.

fold hinge line The line in a folded surface linking the points of maximum curvature.

fold inlier An INLIER created by a FOLD crossing a valley.

fold interference structure A STRUCTURE formed when a folded surface is deformed by a later set of FOLDS. See DOME-AND-BASIN STRUCTURE, DOUBLE ZIGZAG STRUCTURE, CRESCENT-AND-MUSHROOM STRUCTURE.

fold interlimb angle The angle between FOLD LIMBS.

fold limb The part of a FOLD between the HINGES.

fold mechanism A method whereby a FOLD forms, i.e. the DEFORMATION distribution and history during folding.

fold mullion A MULLION STRUCTURE formed by FOLD HINGES.

fold nappe A large ASYMMETRIC FOLD STRUCTURE with a subhorizontal AXIAL SURFACE and HINGE LINE, i.e. a RECUMBENT FOLD.

fold nose The HINGE zone of a FOLD.

fold orientation The direction of a FOLD in three-dimensions, defined by the trend of the AXIAL SURFACE and the PLUNGE.

fold plunge The PLUNGE of a FOLD AXIS.

fold profile The trace of a FOLD in a plane perpendicular to the FOLD HINGE at any point along its length.

fold style Various geometric features of a FOLD PROFILE considered useful in classification and indicating FOLD MECHANISM.

fold symmetry The similarity in size and shape of FOLD LIMBS.

fold system A group of related FOLDS, possibly of different size and geometry, that formed together.

fold test A PALAEOMAGNETIC test for determining the age of a REMANENT MAGNETIZATION in a FOLD. If there is less scatter of directions after correcting for the folding than before, the magnetization is prefolding.

fold tightness The definition of a FOLD in terms of the INTERLIMB ANGLE. See GENTLE FOLD, OPEN FOLD, ISOCLINAL FOLD.

fold train An isolated surface curved to form alternating concave upward and downward regions.

fold trend The orientation of a FOLD AXIS or FOLD HINGE in the horizontal plane.

fold trough A FOLD HINGE zone which is concave upwards. Cf. FOLD CREST.

fold wavelength The separation of two adjacent FOLD CRESTS or TROUGHS measured along the length of the FOLD TRAIN.

folding frequency See NYQUIST FREQUENCY.

foliation A repeated or penetrative planar feature in a rock which may be defined by FABRIC, compositional layering or pervasive FRACTURE. Most commonly used for METAMORPHIC FABRICS, e.g. CLEAVAGE, SCHISTOSITY, GNEISSOSITY.

Folk classification A scheme for the classification of LIMESTONES.

fool's gold A colloquial term occasionally applied to PYRITE or CHALCOPYRITE or a mixture which could be mistaken for GOLD.

footwall The wall lying beneath a horizontal or inclined FAULT or OREBODY.

footwall ramp A RAMP in which truncation of BEDDING or other datum surface is seen in the FOOTWALL.

Foraminiferida/forams An order of single-cell PROTOZOANS characterized by an ectoplasm of fine, granular pseudopodia and an endoplasm enclosed in a TEST of varying composition, with a single chamber or many chambers connecting through an opening. Of major importance to BIOSTRA-TIGRAPHY and environmental analysis of DEVONIAN to RECENT sediments.

force That which produces motion in a body; mass times acceleration. SI unit: NEW-TON, the force required to give a mass of 1 kg an acceleration of 1 m s^{-2}.

forced oscillation The vibration in a solid or quasi-solid not arising from a natural resonance.

forceful emplacement (forceful intrusion) An EMPLACEMENT in which HOST ROCK is actively deformed.

forceful intrusion See FORCEFUL EMPLACEMENT.

forearc basin An elongate BASIN between the TRENCH and volcanic region of a SUBDUCTION ZONE.

foreberg An elongate hill in the FORELAND of a mountain belt formed when an ALLUVIAL FAN has been faulted and warped by foreland-propagating THRUSTS.

foredeep basin See FORELAND BASIN.

foredune A linear, coastal DUNE found behind and parallel to the BACKSHORE zone of a BEACH.

foreland The undeformed marginal region bordering an OROGENIC BELT.

foreland basin (foredeep basin) A TEC-TONIC BASIN on the FORELAND of an OROGENIC BELT and genetically related to it, ascribed to FLEXURE caused by the load of the belt.

foreland thrust belt A THRUST BELT of discrete THRUST FAULTS in the FORELAND.

forelimb The limb of an ASYMMETRIC FOLD with the greater DIP. Cf. BACKLIMB.

forensic seismology See NUCLEAR EXPLOSION SEISMOLOGY.

forereef facies A central sedimentary FAC-IES of a CORAL-ALGAL REEF after growth and CEMENTATION.

foreset bed The inclined surface within a CROSS-LAMINATED bed. Cf. TOESET BED, TOPSET BED.

foreshock A small to moderate MAGNITUDE EARTHQUAKE which precedes a large, shallow EARTHQUAKE. Cf. AFTERSHOCK.

form (crystal form) A set of crystal faces which are equivalent to each other by the POINT GROUP symmetry or which can be generated by operating the symmetry elements of the POINT GROUP on one starting face. See also GENERAL FORM, SPECIAL FORM.

form line A line on a map indicating the general direction of the STRIKE of a FOLD.

form surface A planar surface that intersects the ground surface as a FORM LINE, used in structural mapping.

formation A grouping of BEDS used in LITHOSTRATIGRAPHY; the smallest unit mappable on a reasonable scale. See also BED, MEMBER, GROUP, SUPERGROUP.

formation evaluation The integration of physical data on rocks during oil exploration, commonly achieved by GEOPHYSICAL BOREHOLE LOGGING.

fornacite (furnacite) ($(Pb,Cu)_3((Cr,As)O_4)_2(OH)$) A basic, copper-lead chromarsenate found associated with DIOPTASE.

forsterite (Mg_2SiO_4) The magnesium-bearing end member of the OLIVINE SOLID SOLUTION series.

forsterite marble (ophicalcite) A MARBLE-like rock produced by the CONTACT METAMORPHISM of SILICA-bearing, DOLOMITIC LIMESTONE.

forward problem A type of problem in geophysics in which the expected behaviour of a model is calculated for comparison with observations. Cf. INVERSE PROBLEM.

foskorite A MAGNETITE-OLIVINE-APATITE rock found in some carbonate-ALKALINE igneous complexes, sometimes mined for its iron and phosphorus content.

fossil 1 The trace of an organism buried naturally and subsequently preserved permanently. **2** Of great age.

fossil assemblage An association of FOS-

SILS without ecological relations. Cf. FOSSIL COMMUNITY.

fossil community An association of ecologically interrelated FOSSILS. Cf. FOSSIL ASSEMBLAGE.

fossil fuel A fuel originally of organic origin.

fossil placer A lithified PLACER.

foundry sand A refractory, cohesive, porous SAND suitable for forming moulds for metal castings.

fourchite An INTRUSIVE IGNEOUS ROCK comprising ESSENTIAL titanium-bearing AUGITE and KAERSUTITE in a MATRIX of ANALCITE or GLASS.

fourier analysis A means of expressing a waveform in terms of a combination of sine waves.

fourier transform A method of changing between a waveform expressed as a time-variable amplitude to one expressed as a frequency-variable amplitude. Many FILTERING operations are more conveniently and efficiently accomplished in the frequency domain.

fowlerite A zinc-rich variety of RHODONITE.

foyaite A variety of NEPHELINE SYENITE with equal amounts of NEPHELINE and POTASH FELDSPAR and a subordinate MAFIC MINERAL, e.g. AEGIRINE.

fractal An irregular geometrical shape with the same degree of irregularity on all sides which appears the same when examined from far away or nearby. Many phenomena in geology, such as FAULT systems, behave as fractals.

fractional crystallization See CRYSTAL-LIQUID FRACTIONATION.

fractionation factor (α) A term used in determining temperature and environment from stable isotopes of CARBON and oxygen in sediments, e.g. for CALCITE: $10^3 \ln\alpha = 2.78(10^6 T - 2) - 2.89$, where T is temperature in Kelvin and $10^3 \ln\alpha$ is approximately the difference between the oxygen isotope composition of the MINERAL and that of the fluid from which it formed.

fractography The study of FRACTURE surfaces, which provides information on FRACTURE propagation.

fracture 1 In STRUCTURAL GEOLOGY, a discontinuity across which there has been separation, e.g. JOINT, FAULT. 2 In MINERALOGY, the breaking of MINERALS which does not occur along particular crystallographic directions, e.g. CONCHOIDAL FRACTURE, FIBROUS FRACTURE, HACKLY FRACTURE, SPLINTERY FRACTURE, UNEVEN (IRREGULAR) FRACTURE.

fracture cleavage A CLEAVAGE defined by closely spaced FRACTURES.

fracture porosity A POROSITY developed as a result of the presence of FRACTURES in a rock.

fracture toughness A measure of the resistance of a material to BRITTLE FAILURE by the spreading of cracks.

fracture zone 1 A zone of past or present TRANSFORM FAULT movement within the OCEANIC CRUST, generally represented by a topographic depression up to several thousands of kilometres in length and sometimes accompanied by flanking ridges. 2 A zone of fractured rock, especially when acting as an AQUIFER.

fragipan A brittle, acidic, cemented horizon of CLAY, SILICA, iron, aluminium or organic matter between the soil and the underlying BEDROCK, often the result of PERIGLACIAL activity.

fragmental Descriptive of a TEXTURE of a SEDIMENTARY ROCK in which the CLASTS are broken particles.

fragmentation level The level in a VOLCANIC VENT at which VESICLES can explosively tear apart the MAGMA to form PYROCLASTS suspended in up-rising gas.

framboid Spherical aggregate of microcrystals with a diameter < 150 μm.

framestone An AUTHOCHTHONOUS, organically bound LIMESTONE in which the organisms formed a framework during deposition.

francolite A variety of APATITE.

Franconian A CAMBRIAN succession in the E. USA covering parts of the MAENTWROGIAN and DOLGELLIAN.

franklinite $((Zn,Fe,Mn)(Fe,Mn)_2O_4)$ A rare, zinc-rich SPINEL.

Frasch process A method of extracting subsurface NATIVE SULPHUR by melting it with hot water and blowing the melt to the surface.

Frasnian A STAGE of the DEVONIAN, 377.4–367.0 Ma.

Frasnien A DEVONIAN succession in France and Belgium covering parts of the GIVETIAN and FRASNIAN.

Frederiksberg A CRETACEOUS succession on the Gulf Coast of the USA covering part of the ALBIAN.

free-air anomaly A GRAVITY measurement to which a LATITUDE and FREE AIR CORRECTION ($\pm$ an EARTH TIDE and an EÖTVÖS CORRECTION) have been applied. Provides an indication of the degree of ISOSTATIC compensation of a broad feature. Cf. BOUGUER ANOMALY.

free-air correction The correction applied to a GRAVITY measurement to account for the decrease in GRAVITY with height in free air. Cf. BOUGUER CORRECTION.

free face The wall of an OUTCROP which is too steep for debris to rest on it.

free oscillation The resonation of a body, such as the Earth, at particular harmonics when excited by an event, such as an EARTHQUAKE of MAGNITUDE ≥ 7.5. The oscillations provide information on the physical properties of the interior.

free water An American term for HELD WATER.

free-milling gold GOLD in an OREBODY in NATIVE form that is easily amalgamated or cyanided for recovery.

freestone A fine-grained stone that can be cut and worked without fracturing.

freeze-thaw A type of WEATHERING in response to alternate freezing and thawing.

freibergite A SILVER-bearing ORE MINERAL; a variety of TETRAHEDRITE.

French chalk A variety of TALC with a compact form used to mark cloth or remove grease stains.

Frenkel crystal defect A CRYSTAL DEFECT formed when an atom is transferred from its normal position to an immediately adjacent interstitial site not normally occupied.

friable Descriptive of a material that can be disintegrated into grains by finger pressure.

fringe joint See EN ECHELON FRACTURE.

fringing reef A type of CORAL-ALGAL REEF bordering an island or continent with a flat surface exposed at low tide.

frondescent mark A type of SOLE MARK that was modified by flowage.

frontal ramp A RAMP trending normal to the tectonic transport direction in a THRUST SYSTEM.

frost heave A type of MASS MOVEMENT in which soil is moved upwards by the migration of water which expands on freezing.

frost shattering See CONGELIFRACTION.

frost splitting See CONGELIFRACTION.

frost weathering See CONGELIFRACTION.

frost wedging See CONGELIFRACTION.

froth flotation See FLOTATION.

Froude number (Fr) A dimensionless number which quantifies the influence of GRAVITY on a flow: $Fr = U/\sqrt{(gL)}$, where $U =$ mean flow velocity, $g =$ GRAVITY, $L =$ a length term, usually mean flow depth. When $Fr < 1$, the influence of GRAVITY is pronounced and the flow is SUBCRITICAL, when $Fr > 1$ inertial forces predominate and the flow is SUPERCRITICAL and $Fr = 1$ indicates a STANDING WAVE or HYDRAULIC JUMP.

fuchsite A chrome-rich variety of MUSCOVITE.

fucoid An old term for a TRACE FOSSIL formed from a burrow.

fugichnia TRACE FOSSILS formed from escape structures.

fulgurite A hollow tube of GLASS (predominantly LECHATELIERITE) formed by the action of lightning on QUARTZ SAND.

fulje A deep parabolic depression between closely interlocking DUNES.

Fuller's earth An absorbent CLAY composed of calcium MONTMORILLONITE (in the US ATTAPULGITE and SEPIOLITE) used in decolouring oils etc.

fumarole A VOLCANIC VENT, distinct from those erupting MAGMA, from which VOLCANIC GASES escape.

functional morphology An attempted interpretation of the function of organs or structures in FOSSILS.

fundamental strength The maximum STRESS a material can sustain indefinitely at a given temperature and CONFINING PRESSURE.

Fungi One of the three kingdoms of multicellular organisms, along with the META-PHYTA and METAZOA; organisms which feed by ingesting organic matter. Range PRECAMBRIAN–RECENT.

furnacite See FORNACITE.

fusain A LITHOTYPE OF BANDED COAL composed of soft, friable material similar to charcoal.

fusibility The property of being capable of conversion from solid to liquid by heating.

fusiform Shaped like a spindle.

fusinite A COAL MACERAL of the INERTINITE group with a high REFLECTANCE.

fusite A MICROLITHOTYPE OF COAL composed of FUSINITE.

fusoclarain A LITHOTYPE OF BANDED COAL intermediate between FUSAIN and CLARAIN.

G

G See GRAVITATIONAL CONSTANT.

g See GRAVITY.

gabbro A coarse-grained, BASIC IGNEOUS ROCK composed of PLAGIOCLASE with An_{50}, PYROXENE with CLINOPYROXENE > ORTHOPYROXENE and ACCESSORY OLIVINE, QUARTZ and NEPHELINE.

Gabbs A TRIASSIC succession in Nevada, USA equivalent to the NORIAN and RHAETIAN.

gadolinite $((YFeBe_2SiO_4)_2O_2)$ A RARE EARTH ELEMENT ORE MINERAL found in PEGMATITES.

gahnite (zinc spinel) $(ZnAl_2O_4)$ A zinc-aluminium SPINEL found in crystalline SCHISTS, CONTACT METAMORPHOSED LIMESTONES, high temperature REPLACEMENT ORE deposits, GRANITE PEGMATITES and PLACER DEPOSITS.

gaining stream A watercourse that receives water from an AQUIFER because its valley cuts the WATER TABLE. Cf. LOSING STREAM.

Gal The CGS unit of GRAVITY, 1 Gal = 1 cm s^{-2}.

galaxite $(MnAl_2O_4)$ A rare, black manganese-aluminium SPINEL.

galena (lead-glance) (PbS) The major ORE MINERAL of lead.

galena group A group of SULPHIDE MINERALS characterized by a HALITE structure, with CUBIC close PACKING of anions in planes parallel to (111) and both cations and anions in regular octahedral six-fold coordination.

gallery An horizontal tunnel or passage in a mine.

Gallic A division of the CRETACEOUS, 131.8–88.5 Ma.

gamma (γ) The CGS subunit of MAGNETIC FIELD strength, $1\gamma = 10^{-5}$ GAUSS $= 10^{-9}$ TESLA.

gamma-gamma log (density log) A GEOPHYSICAL BOREHOLE LOG in which the scatter of artificial gamma rays provides a measure of the DENSITY of the adjacent wallrock.

gamma ray log A GEOPHYSICAL BOREHOLE LOG which measures natural gamma radiation, which is most prevalent in the presence of CLAY MINERALS.

gamma ray spectrometer An instrument which measures gamma rays in a similar way to a SCINTILLATION COUNTER, but which can discriminate between rays produced by uranium, thorium and potassium-40.

gangue The unwanted material with which ORE MINERALS are associated.

ganister A highly leached SANDSTONE or SILTSTONE beneath a COAL SEAM.

Gard notation A method of describing the unit cell relationship in POLYTYPISM.

garnets Cubic MINERALS with the general formula $A_3^{2+}B_2^{3+}Si_3O_{12}$, where A = magnesium, iron, manganese or calcium and B = aluminium, iron or, rarely, chromium. Characteristic of METAMORPHIC ROCKS, but also found in some IGNEOUS ROCKS and as detrital grains in sediments. Common garnets are ALMANDINE, ANDRADITE, GROSSULAR, PYROPE, SPESSARTINE and UVAROVITE.

garnierite $((Ni,Mg)_3Si_2O_5(OH)_4)$ A green, nickel-rich, SERPENTINE GROUP MINERAL.

gas cap The gas above an oil accumulation in an oil RESERVOIR.

gas cap drive The pressure exerted by a GAS CAP as oil is removed from a RESERVOIR which drives it towards the WELL.

gas hydrate An accumulation of CLATHRATES formed under low temperature and high pressure.

gas pool An accumulation of NATURAL GAS in a single RESERVOIR and GAS TRAP.

gas trap A stratal arrangement that can trap gas in the same manner as an OIL TRAP.

gas-oil ratio (GOR) The volume of gas as it exists in a RESERVOIR relative to the volume of oil.

Gasteropoda See GASTROPODA.

gastrolith A stone found in the stomach of an animal to aid food processing or to counteract its buoyancy when in water.

Gastropoda/gastropods (Gasteropoda) A class of phylum MOLLUSCA in which the anterior part of the foot is developed into a head and a helically coiled shell protects the organism. Range CAMBRIAN–RECENT.

Gault A CRETACEOUS succession in England covering the upper part of the ALBIAN.

Gauss **1** A MAGNETOSTRATIGRAPHIC EPOCH of NORMAL POLARITY in the NEOGENE, 3.34–2.42 Ma. **2** The CGS unit of MAGNETIC FIELD strength. Cf. TESLA.

Gauss' theorem 'The outward flux of the force of attraction over any closed surface in a gravitational field is equal to 4π times the mass enclosed by the surface.' Used to determine EXCESS MASS from GRAVITY measurements.

gaylussite $(Na_2Ca(CO_3)_2.5H_2O)$ An EVAPORITE MINERAL precipitated from saline lakes.

geanticline A broad area of ANTICLINAL UPLIFT.

Gebbie A PERMIAN succession in Queensland, Australia, covering the upper part of the ARTINSKIAN.

Gedinnine A DEVONIAN succession in France and Belgium equivalent to the LOCHKOVIAN.

gedrite $(Na_{0.5}(Mg,Fe)_2(Mg,Fe)_{3.5}(Al,Fe^{3+})_{1.5}$ $Si_6Al_2O_{22}(OH)_2)$ An AMPHIBOLE found in METAMORPHIC and METASOMATICALLY altered ROCKS.

gehlenite $(Ca_2Al_2SiO_7)$ A FELDSPATHOID of the MELILITE group.

Geiger counter (Geiger-Müller counter) An instrument for detecting beta rays by their ionization of a gas.

Geiger-Müller counter See GEIGER COUNTER.

geikielite $(MgTiO_3)$ A rare, titaniumbearing OXIDE MINERAL.

gelifluction A type of SOLIFLUCTION taking place in PERIGLACIAL environments underlain by PERMAFROST.

geliturbation See CRYOTURBATION.

gem (gemstone) A MINERAL or organic material with an intrinsic value because of its beauty, durability or rarity.

gemstone See GEM.

general diagenetic equation An equation whose solution can quantify the rates of bacterial decomposition, COMPACTION, DIFFUSION, BIOTURBATION, ADSORPTION, IONEXCHANGE, DISSOLUTION and PRECIPITATION.

general form A crystal FORM which has no particular relationship with the symmetry operators present in the SPECIAL FORM.

generalized reciprocal method A method of interpreting REVERSED SEISMIC REFRACTION PROFILES in terms of an undulating refractor, with more general application than the PLUS-MINUS METHOD.

geniculate twin (knee twin) A TWIN in which the TWIN PLANE changes the crystal shape so that it has the appearance of a knee joint.

gentle fold A FOLD with an INTERLIMB ANGLE of >120°.

geo A narrow, linear cleft running inland from a SEA CLIFF.

geo- Pertaining to the Earth.

geobarometry The use of FLUID INCLUSIONS or the chemistry of MINERAL systems to determine the pressure of deposition or of a subsequent event.

geobotanical survey GEOCHEMICAL prospecting utilizing metallophilic plants, plant poisoning and the identification of anomalous metallic concentrations in plant tissue.

geochemical anomaly An abnormal concentration of elements with respect to the background level.

geochemistry The study of the chemistry of the Earth's constituents.

geochronology The study of geological time.

geochronometry The measurement of geological time.

geocline A succession of STRATA with a low, uniform DIP.

geocronite $(Pb_5(Sb,As)_2S_8)$ A white sulphosalt MINERAL of lead.

geode A cavity lined with crystals which project towards its centre.

geodesy The science of the precise measurement and mapping of the Earth's surface.

geodynamics The study of the dynamic processes which affect, or have affected, the solid Earth, generally those related to TECTONICS.

geoelectric section A one-dimensional section obtained by VERTICAL ELECTRICAL SOUNDING or a similar ELECTROMAGNETIC INDUCTION METHOD showing how ELECTRICAL RESISTIVITY varies with depth; it can often be interpreted geologically.

geographical cycle See CYCLE OF EROSION.

geoid The EQUIPOTENTIAL SURFACE of the gravitational field of the Earth represented by the sea level surface, usually taken as the DATUM plane in GRAVITY REDUCTION.

geologic time-scale An absolute time-scale made up of standard STRATIGRAPHIC divisions based on rock sequences.

geology The study of the solid Earth.

geomagnetic correction The correction to magnetic survey data for the variation with latitude and longitude of the GEOMAGNETIC FIELD, often applied using the INTERNATIONAL GEOMAGNETIC REFERENCE FIELD.

geomagnetic dipole field The 80% of the GEOMAGNETIC FIELD that can be ascribed to the field of a single, fictitious, magnetic dipole. Cf. GEOMAGNETIC NON-DIPOLE FIELD.

geomagnetic elements (magnetic elements) Descriptors of the strength and orientation of the GEOMAGNETIC FIELD in terms of its total field, vertical field, horizontal field, INCLINATION and DECLINATION.

geomagnetic event (geomagnetic excursion) A short period of constant geomagnetic polarity, usually <10 000 years.

geomagnetic excursion See GEOMAGNETIC EVENT.

geomagnetic field The MAGNETIC FIELD of the Earth.

geomagnetic non-dipole field The part of the GEOMAGNETIC FIELD (~20%) not accounted for by the GEOMAGNETIC DIPOLE FIELD.

geomagnetic polarity time-scale A time-scale constructed by making use of changes in polarity of the GEOMAGNETIC FIELD that occur at intervals of from several per Ma to ~50 Ma. The timescale is complete from 160 Ma.

geomagnetic reversal See MAGNETIC REVERSAL.

geomagnetic variation See DIURNAL VARIATION, SECULAR VARIATION.

geomagnetism The study of the GEOMAG-

NETIC FIELD and its use in GEOPHYSICAL EXPLORATION for magnetic MINERALS.

geomorphic Concerning the form of the Earth or its surface features.

geomorphological threshold A change in a landform initiated by changes in the morphology of the landform itself with time.

geomorphology The study of the form of the ground surface and the processes which shape it.

geopetal fabric A FABRIC formed by the partial infilling of the bottom part of a cavity with sediment under the influence of GRAVITY.

geophone An instrument used on land to detect the arrival of seismic energy, commonly by using a moving coil technique to convert ground movement into a varying voltage. Cf. HYDROPHONE.

geophysical anomaly A perturbation from the norm in a measured field, usually resulting from a change in the physical properties of the subsurface, e.g. GRAVITY ANOMALY, MAGNETIC ANOMALY. In order to be able to recognize such anomalies, all non-geological sources of variation in the field are removed by a REDUCTION process.

geophysical borehole logging (borehole logging, downhole geophysical survey, geophysical well logging, well logging, wire-line logging) The recording of the properties or characteristics of the rock formations traversed by measuring apparatus in a borehole, which largely obviates the necessity of the expense of coring. The principal techniques utilized are ELECTRIC LOGS, INDUCTION LOGS, SELF POTENTIAL LOGS, RADIOACTIVITY LOGS, SONIC LOGS, TEMPERATURE LOGS and DIPMETER LOGS.

geophysical exploration See APPLIED GEOPHYSICS.

geophysical well logging See GEOPHYSICAL BOREHOLE LOGGING.

geophysics The application of the methods and techniques of physics to the study of the Earth and the processes affecting it.

geopressure See ABNORMAL PRESSURE.

geopressuring See OVERPRESSURING.

Georgiev A JURASSIC succession in W. Siberia equivalent to the KIMMERIDGIAN.

geostrophic current A current in which the pressure gradient and CORIOLIS FORCES are in balance.

geosyncline An elongate trough with a great thickness of sediment which subsequently becomes an OROGENIC BELT. Now an obsolete term since such features can be put into their PLATE TECTONIC setting.

geotectonic Relating to major Earth STRUCTURES, such as OROGENIC BELTS, CRATONS, BASINS, and the processes forming them.

geotherm A curve showing the variation of temperature with depth.

geothermal Concerning the flow of heat from the interior of the Earth to the surface.

geothermal brine A saline solution within GEOTHERMAL SYSTEMS and HOT SPRINGS, the major constituents being Na, K, Ca, Mg and Cl, possibly with other metals which may become of economic importance.

geothermal energy Energy obtained from a GEOTHERMAL SYSTEM by pumping out hot water or pumping water through hot rocks and back to the surface (HOT DRY ROCK CONCEPT).

geothermal field An area of high HEAT FLOW that can produce GEOTHERMAL ENERGY.

geothermal gradient The rate of change of temperature with depth in the Earth.

geothermal system A circulating GROUNDWATER system activated by a high GEOTHERMAL GRADIENT.

geothermometry The determination of the temperature of formation of a MINERAL deposit. This can be accomplished by the use of FLUID INCLUSIONS in MINERALS such as FLUORITE, BARYTE or QUARTZ, inversion points in POLYMORPHS, the resolution of EXSOL-UTION TEXTURES, equilibrium MINERAL ASSEM-BLAGES and stable isotopes.

gersdorffite (nickel arsenic glance) (NiAsS) A rare ORE MINERAL of nickel.

geyser A vent from which hot water and steam are violently and periodically ejected at the surface in a volcanic area, caused by the heating of GROUNDWATER by subsurface MAGMA.

geyserite (sinter) A variety of OPAL found in GEYSERS and HOT SPRINGS.

ghost reflection A SEISMIC MULTIPLE generated by a SEISMIC REFLECTION at the base of the WEATHERED LAYER.

Ghyben-Herzberg relationship A relationship allowing the determination of the thickness of a lens of fresh GROUNDWATER overlying salt water from their densities and the height of the top of the lens above sea level.

giant gas field A field with NATURAL GAS reserves exceeding a volume of $\sim 140 \times 10^9$ m³.

gibber A desert plain covered with a layer of PEBBLES or BOULDERS.

gibbsite (hydrargillite) (Al(OH)₃) An ORE MINERAL of aluminium found in BAUXITE deposits.

Gilbert A MAGNETOSTRATIGRAPHIC EPOCH of REVERSED POLARITY in the PLIOCENE-UPPERMOST MIOCENE, 5.9–3.6 Ma.

Gilbert delta A steep-fronted DELTA formed where there are steep nearshore slopes, inflow velocities are large and outflows are dominated by inertia. Usually forms in an area of little reworking by waves or TIDE.

gilbertite A fluorine-rich variety of MUS-COVITE.

gilgai Micro-relief characterized by small hummocks separated by shallow troughs.

gilsonite See UINTAITE.

Gilyakian A CRETACEOUS succession in the

far east of the former USSR equivalent to the TURONIAN.

gipfelflur A summit plane shown by mountain peaks of similar elevation.

girdle distribution A pattern of points around a GREAT CIRCLE on a STEREOGRAM.

Gisbornian An ORDOVICIAN succession in Australia covering part of the late LLANVIRN, the LLANDEILO and part of the COSTONIAN.

gismondine (gismondite) (CaAl₂Si₂O₈.4H₂O) A rare ZEOLITE found in AMYG-DALES in BASALTIC LAVA.

gismondite See GISMONDINE.

Givetian A STAGE of the DEVONIAN, 380.8–377.4 Ma.

Givetien A DEVONIAN succession in France and Belgium covering the lower part of the GIVETIAN.

glaci- (glacio-) Pertaining to GLACIER ice.

glacial **1** Adjective referring to a GLACIER. **2** A period of glaciation.

glacial control theory A theory that suggests that marine platforms abraded during low sea levels associated with glaciation were subsequently sites of CORAL-ALGAL REEF growth.

glacial deposit A sedimentary feature deposited by a GLACIER, composed mainly of TILL. Includes BRAIDS, DRUMLINS, ESKERS, KAMES, MORAINE and SANDURS.

glacial erosion EROSION by GLACIER ice by the processes of FROST SHATTERING, meltwater flow and PLUCKING to form features such as ARÊTES, CIRQUES, FJORDS, HANG-ING VALLEYS, HORNS, KNOCK-AND-LOCHAN topography and ROCHES MOUTONNÉES.

glacial plucking See PLUCKING.

glacial proximal trough A trough eroded by GLACIER ice, running water or wind which increases in velocity around a rock body.

glacial rebound The ISOSTATIC REBOUND that occurs during and after the melting of a GLACIER.

glacier A mass of ice and snow which

deforms and FLOWS under its own weight if sufficiently thick.

glacimarine Referring to sediments produced by the interaction of a GLACIER with the sea.

glacio- See GLACI-.

glaciofluvial (fluvioglacial) Referring to GLACIAL meltwater activity.

glaciolacustrine Concerning GLACIAL lakes or their deposits.

glaciotectonism The production of landforms and STRUCTURES by the DEFORMATION of soft rock and DRIFT as a consequence of the movement of GLACIER ice.

glacis An EROSIONAL PEDIMENT, sometimes in several generations, in arid regions.

glance A mining term for an OPAQUE MINERAL of high reflectivity.

glass Amorphous matter formed by the rapid cooling of MAGMA.

glass sand A SAND with even-sized grains which, because of a high SILICA content (95–99.8%) and low alumina content (<4%), is suitable for glass making.

glassy rock A rock consisting partly or wholly of GLASS, usually formed when MAGMA is quenched too rapidly for CRYSTALLIZATION to occur, but also forming in a FAULT ROCK by DYNAMIC METAMORPHISM (PSEUDOTACHYLITE) or by THERMAL METAMORPHISM (BUCHITE).

Glauber salt See MIRABILITE.

glauberite ($Na_2Ca(SO_4)_2$) Sodium-calcium sulphate, an EVAPORITE MINERAL.

glauchroite ($CaMnSiO_4$) A rare manganese-bearing OLIVINE.

glaucodot ($(Co,Fe)AsS$) A rare ORE MINERAL of cobalt.

glauconite ($(K,Na,Ca)_{0.5-1}(Fe^{3+},Al, Fe^{2+},Mg)_2(Si,Al)_4O_{10}(OH)_2.nH2O$) A CLAY MINERAL found as an AUTHIGENIC MINERAL in SEDIMENTARY ROCKS.

glaucony A green, marine, sedimentary FACIES characterized by the presence of GLAUCONITE which develops on CONTINENTAL MARGINS and bathymetric highs.

glaucophane ($Na_2Mg_3Al_2Si_8O_{22}(OH)_2$) A blue-black AMPHIBOLE occurring in METAMORPHIC ROCKS.

glaucophane schist facies See BLUESCHIST FACIES.

Gleedonian A STAGE of the SILURIAN, 425.4–424.0 Ma.

glei See GLEY.

gleittbretter See MICROLITHON.

glendonite A STELLAR PSEUDOMORPH of CALCITE after IKAITE formed under GLACIAL temperatures.

gley (glei) A waterlogged soil horizon formed by GLEYING.

gleying A PEDOGENETIC process involving waterlogging and the development of ANAEROBIC conditions so that organic decomposition is slow and ferrous iron forms.

glide (translation gliding) The movement of a DISLOCATION through a CRYSTAL LATTICE along a SLIP PLANE.

glide twinning TWINNING that occurs when a STRESS forces part of a crystal to move into a TWIN-related orientation.

glimmerite An ULTRABASIC ROCK composed mainly of MICA.

Global Positioning System (GPS) A constellation of artificial satellites which allows accurate three-dimensional positioning by radio interferometry. Used for navigation, in measuring the rate of PLATE movements and in studies of continental DEFORMATION.

Globigerina ooze A PELAGITE composed of the TESTS of the FORAMINIFERA *Globigerina*.

globulite A tiny, spheroidal CRYSTALLITE found in GLASSY ROCKS.

glomerocryst A cluster of PHENOCRYSTS.

glomeroporphyritic texture The TEXTURE of a rock containing GLOMEROCRYSTS.

GLORIA **G**eological **LO**ng **R**ange **I**nclined **A**sdic, a long-range SIDE-SCAN SONAR system.

glory hole A large open pit from which ORE is, or has been, extracted.

gloss coal The highest RANK LIGNITE; black and compact with a CONCHOIDAL FRACTURE and glossy LUSTRE.

Glossopteris flora A cold-climate flora widespread in GONDWANALAND before its breakup at ~180 Ma.

gmelinite $((Na_2,Ca)(Al_2Si_4O_{12}).6H_2O)$ A ZEOLITE found chiefly in the AMYGDALES of BASALTIC LAVAS associated with other ZEOLITES.

gnamma A BASIN on a rock surface, especially IGNEOUS ROCK and SANDSTONE, produced by WEATHERING.

gneiss A METAMORPHIC ROCK characterized by GNEISSOSITY.

gneissic layering The compositional layering found in a GNEISS.

gneissoid With a GNEISS-like TEXTURE or STRUCTURE unrelated to METAMORPHIC processes.

gneissosity A FOLIATION of compositional layering or lensoid structure found in high-GRADE METAMORPHIC ROCKS and deformed IGNEOUS ROCKS, caused by the DEFORMATION of an existing TEXTURE or STRUCTURE, METAMORPHIC SEGREGATION or both.

Gnetopsida A class of division TRACHAEOPHYTA, kingdom plantae; an artificial grouping of LAND PLANTS. Range L. JURASSIC–RECENT.

goethite $(\alpha FeO.OH)$ A 'rust-like' hydrated oxide of iron produced by the WEATHERING of iron MINERALS.

gold (Au) A NATIVE METAL, the commonest ORE MINERAL of gold.

gold dust Fine specks of GOLD found in PLACER DEPOSITS.

golden beryl A clear yellow, GEM variety of BERYL.

Goldschmidt's rule 'The number of naturally occurring MINERALS in a rock is equal to the number of components where a given MINERAL ASSEMBLAGE is stable over a range of temperature and pressure.'

Gondwana See GONDWANALAND.

Gondwanaland (Gondwana) A southern SUPERCONTINENT, comprising Africa, Malagasy, India, Sri Lanka, Australasia, Antarctica and South America, which probably formed over 2000 Ma ago and began to split some 180 Ma ago.

Goniatitida/goniatites A subclass of class CEPHALOPODA, phylum MOLLUSCA; AMMONOIDS with a trilobed suture. Range M. DEVONIAN-end PERMIAN.

goniometer An instrument for the accurate measurement of the INTERFACIAL ANGLES of crystals.

gonnardite $(Na_2Ca((Al,Si)_5O_{10})_2.6H_2O)$ A rare ZEOLITE found in VESICLES in BASALTIC rocks.

GOR See GAS-OIL RATIO.

Gorstian A STAGE of the SILURIAN, 424.0–415.1 Ma.

goshenite A colourless GEM variety of BERYL.

goslarite (white copperas) $(ZnSO_4.7H_2O)$ A rare hydrated zinc sulphate formed by the decomposition of SPHALERITE and found as a wall coating in lead mines.

gossan (iron hat) A mass of LIMONITE and GANGUE resulting from the OXIDATION by percolating surface waters of OUTCROPS of sulphide deposits, generally characterized by BOXWORK.

Gothian orogeny See DASLANDIAN OROGENY.

Gothic orogeny See DASLANDIAN OROGENY.

Gotlandian An old term for the SILURIAN once used in continental Europe.

gouge **1** The CLAY filling of a VEIN. **2** See FAULT GOUGE.

GPR See GROUND-PENETRATING RADAR.

GPS See GLOBAL POSITIONING SYSTEM.

grab sample A random, possibly hurried, sample of mineralized ground with no statistical validity, taken simply to check the type of mineralization.

graben A valley formed by the DOWN-THROWING of a FAULT BLOCK between NORMAL FAULTS.

grade 1 Of COAL, a little used classification based on the amount of ASH (i.e. impurity) present. **2** In GEOMORPHOLOGY, descriptive of the condition of DYNAMIC EQUILIBRIUM, e.g. the exact balance between sediment load in a channel and the amount that can be moved by the river. **3 (tenor)** Of ORE, the concentration of a metal in an OREBODY. **4** See METAMORPHIC GRADE.

graded bed A layer in which there is a gradual, abrupt or step-wise vertical and/or lateral change in the grain size distribution. Develops in AEOLIAN or fluid systems in response to changes in flow velocity or sediment supply which allow the deposition of different grain size populations.

grading See SORTING.

gradiometer An instrument measuring the horizontal or vertical gradient of a POTENTIAL FIELD, most commonly applied to magnetic surveys.

grahamite A member of the ASPHALTITE group with a high DENSITY and high FIXED CARBON.

grain boundary fracture See CIRCUM-GRANULAR FRACTURE.

grain boundary gliding A form of CREEP by movement along grain margins.

grain boundary sliding A DEFORMATION mechanism involving DISPLACEMENT of grains relative to each other along grain margins.

grain fall lamination A LAMINATION produced in an AEOLIAN system by the fall of SAND from suspension in zones of FLOW SEP-ARATION, i.e. as FORESET and TOESET stratification.

grain flow The GRAVITY-driven movement of a sediment supported by grain-grain con-tact in which the grains move separately and non-cohesively.

grain size coarsening The RECRYSTALLIZ-ATION of aggregates of grains into a larger grain size, commonly occurring when temperature increases.

grain surface texture A TEXTURAL property reflecting the ABRASIVE and CORROSIVE processes which affected a CLAST during ERO-SION, transport, deposition and DIAGENESIS.

grain-supported conglomerate See CLAST-SUPPORTED CONGLOMERATE.

grainstone A little-used term for a CLASTIC ROCK.

grammatite See TREMOLITE.

Grampian event A major episode of DEFORMATION in Scotland in the CAMBRIAN.

granite A coarse-grained IGNEOUS ROCK composed of >20% QUARTZ and FELDSPAR of which PLAGIOCLASE and ALKALI FELDSPAR are present in approximately equal amounts.

granite porphyry (porphyritic micro-granite) A GRANITIC rock containing PHENOCRYSTS, usually of FELDSPAR and/or QUARTZ, in a medium- to fine-grained GROUNDMASS.

granite-greenstone terrain A type of ARCHAEAN CRUST composed of associated GRANITE BATHOLITHS and GREENSTONE BELTS, typically in the GREENSCHIST FACIES.

granitic With the MINERAL composition of GRANITE.

granitic layer An outmoded term for the upper CONTINENTAL CRUST. Cf. BASALTIC LAYER.

granitization A postulated METASOMATIC process whereby a rock is converted to GRANITE.

granitoid A term for any GRANITIC rock.

granoblastic texture The TEXTURE of a METAMORPHIC ROCK with equal-sized grains.

granodiorite A coarse-grained IGNEOUS ROCK composed of >20% QUARTZ and FELD-

SPAR of which PLAGIOCLASE makes up >67% of the total FELDSPAR.

granophyre A fine- to medium-grained, commonly PORPHYRITIC, acidic, FELSIC ROCK characterized by a GROUNDMASS containing intergrown QUARTZ and ALKALI FELDSPAR.

granophyric texture A small-scale GRAPHIC TEXTURE.

Granton Shrimp Bed An EXCEPTIONAL FOSSIL DEPOSIT of L. CARBONIFEROUS age near Edinburgh, Scotland, where the soft parts of CONODONTS, shrimps, worms and hydroids are preserved in phosphate.

Grantsville A TRIASSIC succession in Nevada, USA equivalent to the ANISIAN and LADINIAN.

granular texture A TEXTURE comprising MINERAL grains of approximately equal size.

granulation The grinding, during DEFORMATION, of crystals into smaller, equal-sized grains, which can rotate relative to each other.

granule A particle of 2–4 mm diameter.

granulestone A CLASTIC ROCK made up of GRANULES.

granulite A METAMORPHIC ROCK formed in the high-temperature, high-pressure GRANULITE FACIES, characterized by a MINERAL ASSEMBLAGE of PLAGIOCLASE and PYROXENE ± GARNET, QUARTZ, anhydrous aluminosilicates, ALKALI FELDSPAR, CALCITE and FORSTERITE-rich OLIVINE, commonly with a CRYSTALLOBLASTIC FABRIC. Common in ARCHEAN SHIELD areas and probably formed at a high METAMORPHIC GRADE in the lower CONTINENTAL CRUST.

granulite facies A METAMORPHIC FACIES found at great depths and temperatures >650°C in which the characteristic MINERAL ASSEMBLAGE of a BASIC IGNEOUS ROCK is CLINOPYROXENE ± HYPERSTHENE + PLAGIOCLASE.

granulite-gneiss belt A very high-GRADE METAMORPHIC ROCK association of the ARCHAEAN made up of QUARTZITE, LIMESTONE, PELITE, BANDED IRON FORMATION, AMPHIBOLITE and layered ANORTHOSITE complexes.

granulitic texture (intergranular texture) A TEXTURE comprising crystals of AUGITE and/or OLIVINE between laths of PLAGIOCLASE.

grapestone A rock formed of OOIDS which have been agglutinated, often by microbial processes, into irregular masses.

graphic texture An intergrown TEXTURE of QUARTZ and ALKALI FELDSPAR visible in hand specimen, probably formed by the simultaneous CRYSTALLIZATION of both MINERALS.

graphite (C) A soft, grey-black, low-pressure form of CARBON.

graphitization The conversion of amorphous CARBON to GRAPHITE by heat.

Graptolithina/graptolites A class of extinct colonial animals, probably of phylum HEMICHORDATA, composed of a chitinous exoskeleton with individuals inhabiting tubes (thecae) which overlap in single or double rows (stipes) along branches originating from a small, conical cup (sicula). Range M. CAMBRIAN–M. DEVONIAN.

gravel A sediment of variable composition with a grain size larger than SAND, i.e. 2 mm (4.75 mm in engineering terminology)–20 mm.

gravimeter (gravity meter) A small, portable instrument for measuring changes in GRAVITY.

gravitational constant (G, universal gravitational constant) The constant of proportionality in NEWTON'S LAW OF GRAVITATION, 6.67×10^{-11} $m^3kg^{-1}s^{-2}$.

gravitational load (gravitational pressure) The pressure exerted by GRAVITY on a subsurface area.

gravitational potential (U) The scalar defined: $U = GM/r$, where G = the GRAVITATIONAL CONSTANT and r the distance to a gravitating body of mass M. The first derivative of U in any direction gives the gravitational field in that direction.

gravitational pressure See GRAVITATIONAL LOAD.

gravity (acceleration due to gravity, g) The acceleration experienced by an object under the influence of the Earth's mass according to NEWTON'S LAW OF GRAVITATION. The measurement of small variations in gravity forms the basis of a GEOPHYSICAL EXPLORATION method.

gravity anomaly The variation in the Earth's gravitational field caused by a geological feature of anomalous DENSITY.

gravity collapse structure A STRUCTURE produced by movement down a slope under the influence of GRAVITY.

Gravity Formula CLAIRAULT'S FORMULA with internationally agreed constants, which describes how GRAVITY varies around the reference SPHEROID and is used to compute the LATITUDE CORRECTION in the REDUCTION of GRAVITY data.

gravity gliding (gravity sliding) The movement of a body of rock down an inclined plane in response to GRAVITY. The plane is usually a DETACHMENT HORIZON or DÉCOLLEMENT parallel to the BEDDING with a LISTRIC FAULT at the PROXIMAL end to allow separation of the body.

gravity log A GEOPHYSICAL BOREHOLE LOG in which a specialized GRAVIMETER is read remotely down a borehole in order to provide DENSITY information on the formations it penetrates.

gravity meter See GRAVIMETER.

gravity settling The GRAVITY-driven settling of heavy MINERALS to the floor of a MAGMA chamber.

gravity sliding See GRAVITY GLIDING.

gravity survey A survey in which variation in GRAVITY is measured in order to locate and define subsurface bodies of anomalous DENSITY.

gravity tectonics TECTONICS in which the primary driving mechanism is downslope sliding under the effect of GRAVITY.

gravity unit (gu) The SI subunit used for GRAVITY measurements; $1\ gu = 1\ \mu m\ s^{-2} = 0.1$ MILLIGAL.

gravity wave An aqueous or airborne flow in response to GRAVITY.

gray wethers See SARSEN.

graywacke See GREYWACKE.

greasy lustre A LUSTRE in which a MINERAL appears to have an oily coating, caused by the scatter of light on a MICROSCOPICALLY rough surface.

great circle A circle corresponding to a circumference of a sphere. Cf. SMALL CIRCLE.

Green-Ampt equation An empirical equation which describes the INFILTRATION process.

greenalite $((Fe,Mg)_3Si_2O_5(OH)_4)$ An hydrated iron SERPENTINE MINERAL exploited as an ORE.

Greenhouse effect The heat retention (and consequent warming) of the Earth by the absorption of short-wave solar energy and reradiation at longer (thermal) wavelengths to which the atmosphere is opaque. Gases such as carbon dioxide, water vapour, methane and chlorofluorocarbons are important opaque 'greenhouse gases'.

Greenland spar See CRYOLITE.

greenockite (CdS) A yellow ORE MINERAL of cadmium.

greensand A SANDSTONE with a green colour because of the presence of GLAUCONITE.

Greensand A STRATIGRAPHIC term. See UPPER GREENSAND, LOWER GREENSAND.

greenschist A green, SCHISTOSE, METAMORPHIC ROCK coloured green due to the presence of CHLORITE, EPIDOTE or ACTINOLITE.

greenschist facies A METAMORPHIC FACIES of low-GRADE REGIONAL METAMORPHISM of temperature range 300–500°C in which the MINERAL ASSEMBLAGE of BASIC ROCKS is ALBITE + EPIDOTE + CHLORITE + ACTINOLITE, which replace PRIMARY MINERALS.

greenstone 1 A general term for a dark-green, altered, low- to medium-GRADE, metamorphosed BASIC IGNEOUS ROCK such as SPILITE or DOLERITE, the green colour reflecting the GREENSCHIST FACIES MINERAL ASSEMBLAGE. **2** A loose term used by archaeologists for a group of BASIC IGNEOUS and METAMORPHIC ROCKS such as AMPHIBOLITE, SCHIST, GABBRO and EPIDIORITE, sometimes including SERPENTINITE and JADE.

greenstone belt A linear to irregular volcano-sedimentary belt, 10–15 km thick, occurring within GRANITE-GREENSTONE TERRAIN and containing MAFIC-ULTRAMAFIC volcanics with subordinate GREYWACKE, BANDED IRON FORMATION, CHERT and carbonate. Probably originated in a PRECAMBRIAN BACK-ARC BASIN.

greenstone-granite terrain See GRANITE-GREENSTONE TERRAIN.

greisen An AGGREGATE of QUARTZ and white MICA (usually MUSCOVITE or LEPIDOLITE) with ACCESSORY CASSITERITE, FLUORITE, RUTILE, TOPAZ and TOURMALINE, formed by the ALTERATION of GRANITE by HYDROTHERMAL SOLUTIONS.

greisen deposit A GREISEN with associated tin-tungsten mineralization.

greisenization **1** The process of GREISEN formation. **2** Fluorine METASOMATISM.

Grenville orogeny See GRENVILLIAN OROGENY.

Grenvillian orogeny (Grenville orogeny) An OROGENY during the Proterozoic at ~1000 Ma which affected the eastern part of the Canadian SHIELD, equivalent to the DASLANDIAN OROGENY in Europe.

grey copper ore See TETRAHEDRITE.

grey gneiss BIMODAL TRONDHJEMITE.

grey wethers See SARSEN.

greywacke (graywacke) Old term for an immature SANDSTONE with >15% CLAY MINERALS.

grèze litée A bedded SCREE of angular rock debris whose DIP is parallel to the BEDROCK slope, probably formed by NIVATION and DOWNWASH processes.

Griesbachian The lowest STAGE of the TRIASSIC, 245.0–243.4 Ma.

Griffith crack (Griffith flaw) A MICROSCOPIC to submicroscopic crack, flaw or inclusion around which TENSILE STRESSES concentrate, leading to crack propagation and FAILURE.

Griffith failure criterion A relationship between PRINCIPAL STRESSES which gives the condition for FAILURE, which assumes that FAILURE occurs when the potential energy of a GRIFFITH CRACK remains constant or decreases as it increases in length. $\sigma \geq \sqrt{(2E\lambda/\pi C)}$, where σ = TENSILE STRESS, E = YOUNG'S MODULUS, λ = surface energy of crack, C = crack half-length.

Griffith flaw See GRIFFITH CRACK.

Griffith-Murrell failure criterion See EXTENDED GRIFFITH FAILURE CRITERION.

grike (gryke, kluftkarren) A trough separating CLINTS in a LIMESTONE PAVEMENT, developing in response to SOLUTIONAL widening.

grit A hard, coarse-grained SANDSTONE.

groove See TOOL TRACK.

groove cast (groove mark) A millimetric- to decimetric-scale, elongated, sublinear scour or SOLE MARK, formed when CLASTS (e.g. PEBBLES, shells, etc.) or tools (e.g. icebergs) moving in a current erode a track into soft, COHESIVE sediment.

groove mark See GROOVE CAST.

grossular $(Ca_3Al_2Si_3O_{12})$ A GARNET, frequently found in metamorphosed impure calcareous rocks.

ground mica See FLAKE MICA.

ground roll The large amplitude oscillation of the ground in the vicinity of an explosion, caused by RAYLEIGH WAVES.

ground truth The confirmation, or otherwise, of an interpretation of REMOTE SENSING data by actual examination of the ground area under consideration.

ground-penetrating radar (GPR) A GEOPHYSICAL EXPLORATION method using radar waves reflected back from subsurface discontinuities to depths of a few tens of metres to produce an output similar to a SEISMIC REFLECTION SEISMOGRAM, but resulting from changes in the dielectric constant of subsurface layers rather than their ACOUSTIC IMPEDANCE.

groundmass (matrix) The fine-grained material of a rock in which larger bodies may be set.

groundwater The water in porous rocks beneath the WATER TABLE.

group A grouping of FORMATIONS used in LITHOSTRATIGRAPHY. See also BED, MEMBER, FORMATION, SUPERGROUP.

group velocity The velocity of the energy in a signal suffering DISPERSION. Cf. PHASE VELOCITY.

growan A partly WEATHERED or decomposed IGNEOUS ROCK, particularly GRANITE, probably forming by ALTERATION during METAMORPHISM or DEEP WEATHERING.

growing ripple See VORTEX RIPPLE.

growth fabric A FABRIC characteristic of the manner in which the rock formed, rather than the effects of DEFORMATION.

growth fault (synsedimentary fault) A FAULT along which movement is contemporaneous with sedimentation.

growth fibre A fibrous crystal growing in response to STRAIN by a CRACK-SEAL mechanism.

growth twinning TWINNING taking place during crystal growth by the continual addition of clusters of atoms.

grunerite ($Fe_7Si_8O_{22}(OH)_2$) A light brown AMPHIBOLE, commonly found in metamorphosed, iron-rich sediments.

grus An accumulation of poorly sorted, angular rock fragments formed during the WEATHERING of CRYSTALLINE ROCKS, particularly GRANITE.

gu See GRAVITY UNIT.

gryke See GRIKE.

Guadelupian A STAGE of the PERMIAN in

the Delaware Basin, USA equivalent to the WORDIAN and CAPITANEAN.

Guan Ling A TRIASSIC succession in China equivalent to the ANISIAN.

guano A deposit of calcium phosphate formed by the reaction of bird or bat excreta with LIMESTONE, once heavily worked for fertilizer.

gudmundite (FeSbS) A SULPHIDE MINERAL of the DISULPHIDE GROUP with an ARSENOPYRITE structure.

Gulf **1** The younger EPOCH of the CRETACEOUS, 97.0–65.0 Ma. **2** A CRETACEOUS succession on the Gulf Coast of the USA comprising the WOODBINE, EAGLE FORD, AUSTIN, TAYLOR and NAVARRO.

gull A fissure or crack, possibly sediment-filled, found on ESCARPMENTS and caused by CAMBERING.

gully erosion (gullying) The EROSION of steep-sided channels and small ravines in poorly consolidated superficial material or BEDROCK by EPHEMERAL STREAMS, often prompted by the reduction or removal of vegetation cover.

gullying See GULLY EROSION.

gumbo A certain type of soil which yields a sticky mud when wet.

Günz Glaciation The first of the GLACIAL periods of the QUATERNARY of the Alps.

Gushan A CAMBRIAN succession in China covering parts of the MENEVIAN and MAENTWROGIAN.

Gutenberg discontinuity A SEISMIC DISCONTINUITY between the CORE and MANTLE at 2900 km depth.

gutter A linear depression in a sediment surface caused by the EROSION associated with vortices in the flow.

gutter cast A type of SOLE MARK in the form of a linear to sinuous, 'U'-shaped depression <10 cm wide, formed by fluid scouring parallel to the flow direction.

guyot A flat-topped, submarine peak, originating as a VOLCANO above sea level and subjected to subaerial and marine planation, subsequently sinking below the surface. Origin related to the passage of an oceanic PLATE over a HOTSPOT.

Gymnocodiaceae A group of CALCAREOUS ALGAE with vegetative structure similar to green ALGAE, but with sporangia resembling red ALGAE. Range PERMIAN–CRETACEOUS.

Gymnospermopsida/gymnosperms A class of kingdom plantae, the 'naked seeded' plants. Range CARBONIFEROUS–RECENT.

gypcrete A DURICRUST composed of GYPSUM.

gypsite An earthy variety of GYPSUM found coating GYPSUM outcrops in arid regions.

Gypsum Springs A TRIASSIC/JURASSIC succession in Utah/Idaho, USA, covering the U. TRIASSIC, HETTANGIAN, SINEMURIAN, PLIENSBACHIAN, TOARCIAN, AALENIAN and part of the BAJOCIAN.

gypsum (alabaster) ($CaSO_4.2H_2O$) An important EVAPORITE MINERAL.

gyroconic Descriptive of a GASTROPOD shell which coils loosely.

gyrolite ($Ca_2Si_3O_7(OH)_2.H_2O$) An hydrated calcium silicate, often found in AMYGDALES with APOPHYLLITE.

gyromagnetic remanent magnetization A REMANENT MAGNETIZATION produced by rotating a sample in an alternating MAGNETIC FIELD.

gyttja (nekron mud) A rapidly accumulating, organic MUD formed in an EUTROPHIC LAKE.

Gzelian The youngest EPOCH of the CARBONIFEROUS, 295.1–290.0 Ma.

H

habit 1 The relative development of individual crystal forms and faces. **2** The typical appearance of an organism.

Hackberryfrio An OLIGOCENE succession on the Gulf Coast of the USA covering part of the RUPELIAN and the CHATTIAN.

hackle mark A type of PLUME STRUCTURE.

hackly fracture A FRACTURE with jagged, sharp-edged surfaces.

hackmanite A variety of SODALITE which changes from deep red to pale green on exposure to light.

hadal zone See ABYSSAL ZONE.

hade The angle made with the vertical by a FAULT PLANE.

Hadean An ERA of the PRECAMBRIAN, 4560–3800 Ma.

Hadrynian The upper part of the PROTEROZOIC of Canada, ~1000–570 Ma.

haematite (hematite) (Fe_2O_3) A major ORE MINERAL of iron, also found as an ACCESSORY MINERAL in many rocks.

haematization (hematization) The REPLACEMENT of the hard parts of an organism by HAEMATITE. Cf. PYRITIZATION.

hagatalite A variety of ZIRCON containing significant quantities of the RARE EARTH ELEMENTS.

Hagedoorn method See PLUS-MINUS METHOD.

hailstone-generated structure A sedimentary STRUCTURE formed by hail impacting on unconsolidated sediment. See also RAINDROP-GENERATED STRUCTURE.

hairpin An abrupt change in the direction of an APPARENT POLAR WANDERING PATH, usually indicative of a CONTINENT-CONTINENT COLLISION.

half-graben 1 A wedge-shaped zone comprising a dipping FAULT on one side of a block of tilted STRATA. **2** In common usage, a valley formed by movement on a single NORMAL FAULT.

half-life The time taken for the mass of a radioactive element to decrease by a half. See also DECAY CONSTANT.

half-spreading rate The rate of movement of a PLATE on one side of an OCEAN RIDGE, i.e. half the PLATE SPREADING rate for symmetric spreading.

half-width The horizontal distance over which a spatial function decays to half its maximum value. Particularly used for MAGNETIC and GRAVITY ANOMALIES, as it can be used to estimate the LIMITING DEPTH of the causative body.

halite (rocksalt) (NaCl) A common EVAPORITE MINERAL.

hälleflinta A metamorphosed VOLCANIC or PYROCLASTIC ROCK with a FLINT-like appearance.

halloysite ($Al_4Si_4O_{10}(OH)_8.8H_2O$) A CLAY MINERAL composed of an irregular series of KAOLINITE layers and interlayer water.

halmyrolysis An early stage of DIAGENESIS or decomposition of sediment on the seafloor.

haloclasty The disintegration of rock by the action of salts, which cause disruption by CRYSTALLIZATION, HYDRATION or thermal expansion.

halocline The interface between dense seawater and lighter freshwater in an ESTUARY or between waters of different SALINITY within a water column.

halokinesis The mobilization and flow of subsurface salt, which can give rise to SALT PILLOWS and SALT DIAPIRS.

halotrichite (iron alum) $(Fe^{2+}Al_2(SO_4)_4.22H_2O)$ A rare hydrated iron-aluminium sulphate forming fibrous crystals.

Hamagian A STAGE of the ORDOVICIAN, 462.3–457.5 Ma.

hambergite $(Be_2(OH,F)BO_3)$ A beryllium borate found in PEGMATITES.

hammada A flat to gently dipping, bare rock surface in a desert, sometimes with a thin cover of GRAVEL or a BOULDER LAG, forming from the WEATHERING of surface rock.

Hammer chart A graticule used in conjunction with a topographic map in the calculation of TOPOGRAPHIC CORRECTIONS in GRAVITY REDUCTION.

hamra A red, sandy soil with a high CLAY content found in desert regions.

hanging valley A tributary valley whose floor is at a higher level than the main valley, caused by the latter's deepening by GLACIAL EROSION.

hangingwall The wall and body of rock above an inclined or horizontal FAULT or OREBODY.

hangingwall ramp A RAMP in which BEDDING or DATUM surfaces are seen in the HANGINGWALL.

harbour bar A COASTAL BAR located across a river mouth or harbour.

hard coal A high RANK COAL with a CALORIFIC VALUE >23.86 MJ kg^{-1}, including most BITUMINOUS COAL and ANTHRACITE.

hard pan A layer of iron oxyhydroxides above the WATER TABLE formed by the REPRECIPITATION of MINERALS LEACHED from the overlying VADOSE ZONE.

hardebank Resistant KIMBERLITE that does not break up when exposed. Cf. YELLOW GROUND.

hardground An indurated surface resulting from the synsedimentary lithification of seafloor sediment and representing a decrease in sedimentation rate.

hardness 1 A MINERAL property determined by reference to MOHS' SCALE. **2** A property of water which prevents the formation of a lather with soap, caused by the presence of dissolved alkaline earth salts and usually expressed in ppm. Temporary hardness is caused by calcium bicarbonate and is removed by boiling; permanent hardness is due to other calcium and magnesium salts such as sulphates and is removed by ION-EXCHANGE processes or detergents.

Harker diagram A VARIATION DIAGRAM which shows the chemical relationships of a suite of rocks by plotting constituents against SILICA content.

harmotome $(Ba(Al_2Si_6O_{16}).6H_2O)$ A ZEOLITE found in mineralized VEINS.

hartite $(C_{20}H_{34})$ A naturally occurring HYDROCARBON.

harzburgite An ULTRAMAFIC/ULTRABASIC ROCK composed of >40% OLIVINE, 5–60% ORTHOPYROXENE and <5% CLINOPYROXENE. Forms the residue when BASALTIC MAGMA is derived from MANTLE LHERZOLITE, so is found in ULTRAMAFIC XENOLITHS in ALKALI BASALTS and in the MANTLE part of OPHIOLITES.

Hastarian The lowest STAGE of the CARBONIFEROUS, 362.5–353.8 Ma.

Hastings Beds A CRETACEOUS succession in England covering part of the BERRIASIAN and the VALANGINIAN.

hastingsite $(NaCa_2Fe_4(Al,Fe)Al_2Si_6O_{22}(OH)_2)$ A calcic AMPHIBOLE found in IGNEOUS and METAMORPHIC ROCKS.

Haumurian A CRETACEOUS succession in

New Zealand covering the upper part of the CAMPANIAN and the MAASTRICHTIAN.

Hauptdolomit (Stinkschiefer) A PERMIAN succession in NW Europe covering the lower part of the WORDIAN.

Hauterivian A STAGE of the CRETACEOUS 135.0–131.8 Ma.

haüyne $((NaCa)_{4-8}(AlSiO_4)_6(SO_4)_{1-2})$ A FELDSPATHOID found in PHONOLITE and related rocks.

Hawaiian eruption A VOLCANIC ERUPTION characterized by FIRE FOUNTAINING and the quiet extrusion of MAFIC LAVA, typically in an OCEAN ISLAND setting.

hawaiite An ALKALINE VOLCANIC ROCK containing ANDESINE PLAGIOCLASE (An_{30-50}), PYROXENE, iron-titanium oxide ± minor QUARTZ or NEPHELINE.

hawleyite (CdS) A SULPHIDE MINERAL of the SPHALERITE GROUP.

haycockite $(Cu_4Fe_5S_8)$ A SULPHIDE MINERAL of the SPHALERITE GROUP.

HDR See HOT DRY ROCK.

head A superficial deposit formed under PERIGLACIAL conditions.

head difference The difference in height of the water surface in two bodies of water.

head loss A lowering of the water level, e.g. in an AQUIFER.

head of water (hydraulic head) The height of the water surface above a particular point, giving a measure of the water pressure at that point.

head wave (critically refracted wave, seismic head wave) The refracted SEISMIC WAVE incident at the CRITICAL ANGLE so that it propagates along the boundary between two media at the higher SEISMIC VELOCITY. Its measurement is the basis of the SEISMIC REFRACTION method.

headwall erosion EROSION of the headwall of a ravine after GULLYING has been initiated, which can occur at very rapid rates.

heat flow (heat flux) The heat flowing from the Earth's interior through a unit area at the surface, generally in units of mW m^{-2}.

heat flow unit (HFU) The CGS unit of 1 cal cm^{-2} s^{-1} = 41.8 mW m^{-2}.

heat flux See HEAT FLOW.

heave 1 A group of MASS MOVEMENT processes including FROST HEAVE, FREEZE-THAW and CAMBERING. **2** The horizontal separation resulting from movement on a FAULT.

heavy metal pollution The artificial introduction of heavy metals (As, Cd, Cu, Pb, Ni, Ag) into the environment in quantities that will adversely affect it.

heavy mineral A MINERAL with a specific gravity greater than 2.85, i.e. which sinks in bromoform.

heavy oil Oil with an API GRAVITY <22°.

heavy oil sand A TAR SAND which can be exploited by techniques such as cyclic high pressure steam injection.

heavy spar A colloquial term for BARYTES.

heazlewoodite (Ni_3S_2) A SULPHIDE MINERAL of the METAL EXCESS GROUP.

hectorite $(Na_{0.3}(Al,Mg,Li)_3Si_4O_{10}(OH)_2.$ $4H_2O)$ A lithium-bearing CLAY MINERAL of MONTMORILLONITE type.

hedenbergite $(CaFeSi_2O_6)$ A CLINOPYROXENE found chiefly in CONTACT METAMORPHOSED LIMESTONES, iron-rich METAMORPHIC ROCKS and IGNEOUS ROCKS.

held water The water retained above the WATER TABLE by capillary action.

helictic structure (helictitic structure) Helical trails of MINERAL inclusions in a PORPHYROBLAST formed during METAMORPHIC RECRYSTALLIZATION, which may represent the remnants of an existing FOLD FABRIC enclosed by the PORPHYROBLAST or the rotation of the PORPHYROBLAST in the MATRIX during METAMORPHISM.

helictite A small calcium carbonate SPELEOTHEM growing in curved or spiral shapes in LIMESTONE CAVES.

helictitic structure See HELICTIC STRUCTURE.

Helikian The middle part of the PROTEROZOIC of Canada, ~1800–1000 Ma.

heliodor A clear yellow variety of BERYL coloured by iron, used as a GEM.

helioplacoids The most primitive of the ECHINODERMATA, classified in the stem group of both the PELMATOZOA and ELEUTHEROZOA. Range L. CAMBRIAN.

heliotrope A green and red variety of CHALCEDONY, sometimes of GEM quality.

helium isotopes Helium, a noble gas, has a low atomic weight and this causes only a short residence time on Earth. All helium in the atmosphere must arise from radioactive decay, and this provides a means of measuring the passage of geological time.

helluhraun An Icelandic LAVA FLOW of PAHOEHOE.

hematite See HAEMATITE.

hematization See HAEMATIZATION.

hemi- Half.

Hemichordata A phylum which includes the acorn worms, related to the CHORDATA, and originating in the M. CAMBRIAN.

hemicrystalline Descriptive of a rock composed of both crystals and GLASS in approximately equal proportions.

hemihydrate See BASSANITE.

hemimorphism A trait of certain crystals in which no symmetry element is present to cause repetition of the upper hemisphere faces in the lower hemisphere.

hemimorphite (electric calamine) ($Zn_4Si_2O_7(OH)_2.H_2O$) A minor ORE MINERAL of zinc, sometimes found in the oxidized zone of other zinc ORES.

hemipelagic deposit See HEMIPELAGITE.

hemipelagite (hemipelagic deposit) A generally fine-grained, structureless sediment of decimetric thickness formed by the settling of grains from suspension and from low density, low VISCOSITY TURBIDITY CURRENTS, NEPHELOID LAYERS and other ocean currents.

Herangi A JURASSIC succession in New Zealand comprising the ARATAURAN and URUROAN.

Hercynian orogeny See VARISCAN OROGENY.

hercynite ($FeAl_2O_4$) A SPINEL.

Heretaungan An EOCENE succession in New Zealand covering parts of the YPRESIAN and LUTETIAN.

Herkimer diamond A QUARTZ-filled cavity occurring in DOLOMITE from Herkimer, New York State, USA.

Hermann-Maugin symbol A shorthand notation for describing the symmetry of a crystal.

hermatypic Descriptive of a REEF-building organism.

hermatypic coral A CORAL with symbiotic ALGAE, living in water <50 m deep. Cf. AHERMATYPIC CORAL.

herringbone cross-bedding A type of CROSS-LAMINATION in which the FORESETS alternate in direction as a result of reversing currents.

herringbone structure See PLUME STRUCTURE.

Hervy A succession in E. Australia of late DEVONIAN age.

hessite (Ag_2Te) A rare ORE MINERAL of SILVER.

hessonite (cinnamon stone) A yellow-brown variety of GROSSULAR containing iron.

Heterian A JURASSIC succession in New Zealand covering the lower part of the KIMMERIDGIAN.

hetero- Different from.

heterochthon See ALLOCHTHON.

Heterocorallia A subclass of class ANTHOZOA, phylum CNIDARIA; a very small group of CORALS which may be related to the

RUGOSA or a separate group. Range late DEVONIAN–late CARBONIFEROUS.

heterogeneous nucleation An EXSOLUTION phenomenon in which exsolved phases appear at internal CRYSTAL DEFECTS, e.g. grain boundaries, DISLOCATIONS. Cf. HOMOGENEOUS NUCLEATION.

heterogeneous simple shear SIMPLE SHEAR in which the relationship between DISPLACEMENT and distance changes abruptly. Cf. HOMOGENEOUS SIMPLE SHEAR.

heterogeneous strain STRAIN in which the relationship between DISPLACEMENT and distance changes abruptly. Cf. HOMOGENEOUS STRAIN.

heterolithic bedding A fine interbedding of SAND and MUD formed in an area of variable current flow. The three main types are FLASER BEDDING, WAVY BEDDING and LENTICULAR BEDDING.

heterolithic unconformity See NONCONFORMITY.

heteromorphism The CRYSTALLIZATION of almost identical MAGMAS into different MINERAL ASSEMBLAGES by their possession of different cooling profiles.

Heterostrachi An order of subclass DIPLORHINA, class AGNATHA, superclass PISCES; the oldest known jawless FISH, with heavy armour. Range U. CAMBRIAN–DEVONIAN.

Hettangian The lowest STAGE of the JURASSIC, 208.0–203.5 Ma.

heulandite ($CaAl_2Si_7O_{18}.6H_2O$) A ZEOLITE found in cavities in BASIC IGNEOUS ROCKS.

Hexacorallia See SCLERACTINIA.

Hexactinellida A class of phylum PORIFERA (SPONGES) with OPALINE SILICA spicules arranged in rays at 90°. Range CAMBRIAN–RECENT.

hexagonal pyrrhotite (Fe_9S_{10}, $Fe_{11}S_{12}$) A SULPHIDE MINERAL of the NICKEL ARSENIDE GROUP.

hexagonal system A CRYSTAL SYSTEM whose members have three lateral axes of equal length intersecting at an angle of 60° to each other and reaching the edges of the six vertical faces, and a vertical axis of different length perpendicular to the other three.

Hexapoda/hexapods A superclass of phylum UNIRAMIA; the INSECTS. Range DEVONIAN–RECENT.

hexastannite ($Cu_2Fe_2SnS_6$) A SULPHIDE MINERAL of the WURTZITE GROUP.

HFU See HEAT FLOW UNIT.

hiatus A CHRONOSTRATIGRAPHIC gap caused by non-deposition. Cf. DISCONFORMITY.

hidden layer See BLIND LAYER.

hiddenite (lithia emerald) A green GEM variety of SPODUMENE.

high energy window A period during the mid-HOLOCENE when wave energy was higher than at present.

high quartz See BETA QUARTZ.

high sulphur crude oil CRUDE OIL with >1.7% by weight of sulphur (not in the form H_2S).

high-grade gneiss terrain One of the two main types of ARCHAEAN CRUST. Cf. GRANITE-GREENSTONE TERRAIN.

Highgate resin (copalite, copaline) A colloquial name for AMBER from Highgate, N. London.

hillebrandite ($Ca_2SiO_4.H_2O$) Hydrated dicalcium silicate found as fibrous masses in THERMALLY METAMORPHOSED impure LIMESTONES.

Hilt's law 'The RANK of COAL increases with depth.'

Himalayan mountain belt A broad mountain chain formed by CONTINENTAL COLLISION, either a single collision or a series involving several SUSPECT TERRANES, possibly followed by INDENTATION TECTONICS. Cf. ANDEAN MOUNTAIN BELT.

hinge fault A FAULT whose DISPLACEMENT decreases to zero at one end, allowing it to act as a hinge.

hinge line The locus of points of maximum curvature of a surface.

hinterland 1 The region from which the surface rocks have been translated in an OROGENIC BELT. Cf. FORELAND. **2** An area of land drained by a river system. **3** The area supplying sediment to a depositional system.

Hirnantian The highest STAGE of the ORDOVICIAN, 439.5–439.0 Ma.

Hjulström curve An empirical curve showing the critical bed condition for EROSION of river channel bed deposits in terms of mean flow velocity.

hogback A long ridge with a sharp crest and steep slopes on both flanks, produced by the differential EROSION of steeply inclined STRATA.

Holkerian A STAGE of the CARBONIFEROUS, 342.8–339.4 Ma.

hollandite ($Ba_2Mn_8O_{16}$) An ORE MINERAL of manganese found at or near the contacts between lithium-rich PEGMATITES and BASIC COUNTRY ROCKS.

holmquistite ($Li_2(Mg,Fe)_3(Al,Fe^{2+})_2Si_8O_{22}(OH)_2$) An AMPHIBOLE found in lithium-rich PEGMATITES.

holo- Complete.

holoblast A MINERAL formed entirely during METAMORPHISM.

Holocene (Recent) The youngest EPOCH of the QUATERNARY, 0.01 Ma–present.

Holocephali A subclass of class CHONDRICHTHYES, superclass PISCES; the 'rabbit fishes' or 'rat fishes'. Range L. CARBONIFEROUS–RECENT.

holocrystalline Descriptive of a rock which is completely crystalline.

holohedral Descriptive of a completely developed crystal.

holohyaline Descriptive of a rock which is composed entirely of GLASS.

holokarst Any LIMESTONE landscape with a fully developed range of KARST features.

hololeucocratic Descriptive of an IGNEOUS ROCK with a COLOUR INDEX <5.

holomictic The status of a lake which undergoes complete water circulation, destroying any seasonal stratification. Cf. MEROMICTIC.

Holostei An order of subclass CHONDROSTEI, class OSTEICHTHYES, superclass PISCES; FISH characterized by heavy rhombic scales, a swim bladder, reduced, unjointed fin rays, a short, mobile jaw and a symmetrical tail. Range U. PERMIAN–RECENT.

Holothuroidea A class of subphylum ELEUTHEROZOA, phylum ECHINODERMATA; worm-like invertebrates with a non-rigid CALCITIC skeleton, poorly preserved as FOSSILS. Range SILURIAN–RECENT.

holotype The single, type specimen of a species.

homeomorph 1 A crystal similar to another in CRYSTAL FORM and HABIT, but with a different composition. **2** An organism similar to another but deriving from different ancestors.

homeomorphism The relationship between crystal HOMEOMORPHS.

homeomorphy The general similarity between species which are different in detail.

hominids Man-like animals coming to dominance in the QUATERNARY.

homoclinal With uniform DIP.

homoclinal ridge An EROSIONAL feature with a form intermediate between a HOGBACK and CUESTA.

homogeneous nucleation An EXSOLUTION mechanism of NUCLEATION of the second phase in which the nuclei are uniformly distributed, which takes place within grains in the absence of CRYSTAL DEFECTS. Cf. HETEROGENEOUS NUCLEATION.

homogeneous simple shear SIMPLE SHEAR in which the relationship between DISPLACEMENT and distance is linear or continuously changing. Cf. HETEROGENEOUS SIMPLE SHEAR.

homogeneous strain STRAIN in which the relationship between DISPLACEMENT and distance is linear or continuously changing. Cf. HETEROGENEOUS STRAIN.

homology The similarity between parts of different organisms as a result of evolutionary differentiation of the parts from the same ancestor. Cf. HOMOPLASY.

homoplasy The similarity between parts of different organisms as a result of CONVERGENT EVOLUTION rather than a common ancestor. Cf. HOMOLOGY.

homopycnal flow The immediate, three-dimensional mixing of river and basinal water of equal DENSITY, accompanied by considerable sediment deposition. Cf. HYPERPYCNAL FLOW, HYPOPYCNAL FLOW.

homotaxis Indicative that STRATA in different areas with the same FOSSIL assemblage are not necessarily of the same age due to migration taking a finite time.

honestone See WHETSTONE.

honeycombs See ALVEOLES.

hoodoo A pillar of rock or weakly consolidated sediment formed by differential EROSION, found in arid to semi-arid regions.

Hooke's law 'The extension of an elastic spring is proportional to the FORCE applied to it.'

hopper crystal A crystal with recessed faces caused by more rapid growth at its edges.

horizon A time-plane with distinctive characteristics.

horizonation The formation of horizons in a SOIL PROFILE as the result of PEDOGENETIC processes.

horn A GLACIAL EROSION feature formed when a mountain is eroded from all sides to leave a pyramidal peak.

horn lead See PHOSGENITE.

horn silver See CERARGYRITE.

hornblende $((Ca,Na)_{2-3}(Mg,Fe,Al)_5$ $Si_6(Si,Al)_2O_{22}(OH)_2)$ An important, widespread, monoclinic calcic AMPHIBOLE.

hornblendite A PLUTONIC IGNEOUS ROCK composed almost completely of HORNBLENDE.

hornfels A fine- to medium-grained PELITIC rock, possibly containing PORPHYROBLASTS of BIOTITE, ANDALUSITE and CORDIERITE, commonly with relict sedimentary/TECTONIC structures, forming as a result of THERMAL METAMORPHISM and solid-state RECRYSTALLIZATION in a CONTACT AUREOLE.

hornfels facies An imprecisely defined METAMORPHIC FACIES produced by CONTACT METAMORPHISM.

hornito A small chimney or cone of LAVA SPATTER on the surface of PAHOEHOE.

hornstone A very fine-grained PYROCLASTIC ROCK.

horse A package of rocks surrounded and isolated by FAULTS in an IMBRICATE STRUCTURE.

horse-tailing A feature made up of a number of mineralized FRACTURES with the general appearance of a horse's tail, which is sometimes found at, or near, the end of a VEIN.

horsetail fault See LISTRIC FAN.

horst A FAULT BLOCK elevated by a series of parallel to subparallel, outward-dipping, CONJUGATE STEP FAULTS with the same sense of DISPLACEMENT. Cf. GRABEN.

Horton equation An empirical equation which describes the INFILTRATION process.

Hortonian overland flow INFILTRATION-excess OVERLAND FLOW which occurs when rainfall exceeds the INFILTRATION CAPACITY of the soil surface.

hortonolite An OLIVINE with the composition Fo_{50-30}.

host grain-controlled replacement REPLACEMENT that is selective or specific to sites of high surface free energy in the host MINERAL or grain, typical of solutions of low chemical reactivity. Cf. PRECIPITATE CON-TROLLED REPLACEMENT.

host rock See COUNTRY ROCK.

hot dry rock (HDR) A method of extracting GEOTHERMAL energy from poorly permeable rocks, such as high heat-production GRANITE, by drilling twin bore-holes, fracturing the rock between them by the HYDROFRAC TECHNIQUE and using the boreholes for the ingress of cold water and egress of heated water.

hot spring A surface seepage from a natural GEOTHERMAL SYSTEM.

hot-working DISLOCATION CLIMB taking place at an elevated temperature.

hotspot A heat source of limited areal extent arising from a MANTLE PLUME persisting for tens of millions of years and remaining nearly stationary with respect to the MANTLE.

howieite $(Na(Fe,Mn)_{10}(Fe,Al)_2Si_{22}O_{31}(OH)_{13})$ A black, hydrated sodium-manganese-iron silicate found in metamorphosed SHALE, siliceous IRONSTONE and impure LIMESTONE.

HREE Heavy RARE EARTH ELEMENT.

Hsu diagram A method of illustrating the shapes of STRAIN ELLIPSOIDS. Cf. FLINN DIAGRAM.

Huashiban A CARBONIFEROUS succession in China covering the lower part of the BASH-KIRIAN.

hübnerite See HUEBNERITE.

Hudsonian orogeny An OROGENY affecting the Canadian SHIELD in the PRECAMBRIAN at ~1750–1800 Ma.

huebnerite (hübnerite) $(MnWO_4)$ An end-member of the WOLFRAMITE MINERAL series.

hum A residual LIMESTONE hill, often rising from a POLJE.

humic coal See BANDED COAL.

humification A biological, PEDOGENETIC process in which organic matter is converted into HUMUS.

humite $(Mg_7(SiO_4)_3(F,OH)_2)$ A NESOSILICATE similar to OLIVINE found in thermally metamorphosed LIMESTONES.

hummocky cross-stratification A BEDDING STRUCTURE consisting of gently curved, low-angle CROSS-STRATIFICATION, possibly forming under storm conditions or by a FLOW REGIME with a unidirectional component.

humus The organic part of soil.

Hunsrück Slate An EXCEPTIONAL FOSSIL DEPOSIT of DEVONIAN age in the Rhineland, Germany where ARTHROPODS, MOLLUSCS, ECHINODERMS, FISH and plants are preserved in PYRITE.

huntite $(CaMg_3(CO_3)_4)$ An EVAPORITE MINERAL.

Huobachong A TRIASSIC succession in China equivalent to the NORIAN.

Huronian An ERA of the PRECAMBRIAN, 2450–2200 Ma.

hush A stream valley artificially deepened for MINERAL exploration.

Huttenlocher intergrowth A suboptical intergrowth between ALBITE and ANORTHITE.

Huygen's principle 'A point on an advancing WAVEFRONT can be regarded as the centre of a fresh disturbance.' Used to predict the path of a WAVEFRONT of, e.g., a SEISMIC WAVE.

hyacinth 1 (jacinth) A red-brown variety of ZIRCON valued as a GEM. **2** A brown variety of GROSSULAR.

hyaline Glassy.

hyalite A clear, colourless variety of OPAL with a globular or BOTRYOIDAL surface.

hyaloclastite A poorly sorted, non-bedded, generally BASALTIC, VOLCANIC ROCK composed of quench-fragmented GLASS with very blocky and angular non-VESICULAR fragments. Formed when shards of glassy LAVA crusts come into contact with water or ice.

hyalocrystalline (hypocrystalline, hypohyaline) Descriptive of material composed of both GLASS and crystals.

hyalophane $((K,Ba)(Al,Si)_2Si_2O_8)$ A rare barium FELDSPAR.

hyalopilitic Descriptive of a form of PILOTAXITIC TEXTURE in which the crystals are embedded in a glassy MATRIX.

hyalosiderite An OLIVINE with the composition Fo_{70-50}.

hybrid fracture See MIXED MODE FRACTURE.

hybrid joint A type of JOINT intermediate between a SHEAR JOINT and an EXTENSION JOINT.

hybrid rock A rock formed by the ASSIMILATION of WALL ROCK or XENOLITHS into a MAGMA.

hybridization The process whereby valence shell ATOMIC ORBITALS are rearranged to point in specific spatial directions to allow the construction of complex molecules involving polyvalent atoms.

hydrargillite See GIBBSITE.

hydration The ALTERATION of a material by the addition of water, e.g. in WEATHERING. Cf. HYDROLYSIS.

hydration layer dating A dating method for OBSIDIAN artefacts based on the phenomenon of OBSIDIAN's absorbing water at a rate dependent on its chemical composition and the ambient temperature.

hydraulic conductivity A measure of the ease with which a fluid flows and the ease with which a porous rock allows its passage. See DARCY EQUATION.

hydraulic equivalence A term which expresses the size of a grain of given DENSITY in the equivalent size of a QUARTZ grain with an identical SETTLING VELOCITY, enabling the behaviour of diverse grains to be compared.

hydraulic fracture (hydrofracture) FRACTURE caused by PORE FLUID PRESSURE, either natural or induced.

hydraulic geometry The study of channel form in relation to the external controls of DISCHARGE and sediment.

hydraulic gradient The loss of pressure with distance along the direction of flow when water flows along a constant diameter, horizontal pipe, i.e. the ratio of the HEAD LOSS between two points to their separation.

hydraulic head See HEAD OF WATER.

hydraulic jump An abrupt increase in flow depth arising where a fast flow with a FROUDE NUMBER >1 changes rapidly to a slow flow with a FROUDE NUMBER <1, as happens at abrupt changes in bed relief in alluvial channels.

hydraulic mining A mining technique in which material is blasted out of the ground by high-pressure water jets.

hydraulic radius The ratio of a stream's wetted perimeter length to the cross-sectional flow in a channel, a measure of the efficiency of a section in conveying flow.

hydraulically rough A term indicating that sediment particles or other irregularities on a bed have a diameter about five times the thickness of the VISCOUS SUBLAYER of a flow, encouraging TURBULENCE and vertical mixing. Cf. HYDRAULICALLY SMOOTH.

hydraulically smooth A term indicating that sediment particles or other irregularities on a bed have a diameter less than the thickness of the VISCOUS SUBLAYER of a flow, and TURBULENCE is probably not generated. Cf. HYDRAULICALLY ROUGH.

hydroboracite $(CaMgB_6O_8(OH)_6.3H_2O)$ A borate MINERAL forming prismatic or ACICULAR crystals.

hydrocarbon A solid, liquid or gas made up of compounds of carbon and hydrogen in varying proportions.

hydrocarbon saturation The condition of a RESERVOIR in which the PORE FLUID has been reduced to the minimum possible level by HYDROCARBONS. This can be determined using RESISTIVITY LOGGING.

hydrocarbon trap See GAS TRAP, OIL TRAP.

hydrocerrussite $(Pb_3(CO_3)_2(OH)_2)$ A colourless, hydrated lead carbonate occurring as an encrustation on lead MINERALS.

hydrofrac technique A method whereby artificial HYDRAULIC FRACTURE is induced in rocks penetrated by a borehole by sealing off part of it and pumping in fluid until fracturing occurs.

hydrofracture See HYDRAULIC FRACTURE.

hydrogeology The study of subsurface water (GROUNDWATER).

hydrogeomorphology (fluvial geomorphology) The study of landforms produced by FLUVIAL processes.

hydrograph The graph of the DISCHARGE of a river or stream plotted against time for a particular point.

hydrogrossular $(Ca_3Al_2O_8(SiO_4)_{1-x}(OH)_{4x})$ A hydrous GARNET.

hydrohaematite (hydrohematite, turgite) A natural mixture of HAEMATITE and GOETHITE.

hydrohematite See HYDROHAEMATITE.

hydroisostasy An ISOSTATIC reaction to the removal or loading of a mass of water.

hydrolith A rock formed by PRECIPITATION from water.

hydrological cycle All processes encompassing the evaporation of water from the sea, its fall as precipitation on land and thence its flow through AQUIFERS or watercourses back to the sea.

hydrology The study of the distribution, conservation and use of water in the Earth and its atmosphere.

hydrolysis The ALTERATION of a MINERAL by the addition of hydroxyl (OH^-) ions, e.g. during WEATHERING. Cf. HYDRATION.

hydrolyzate A sediment concentrating in fine-grained ALTERATION products which contain elements (Al, K, Si, Na) which are easily hydrolyzed. Found in CLAYS, SHALES and BAUXITES.

hydromagnesite $(Mg_5(CO_3)_4(OH)_2.4H_2O)$ Hydrated magnesium hydroxide and carbonate found in AMORPHOUS masses and rarely as crystals in SERPENTINITE.

hydromica See ILLITE.

hydromuscovite See ILLITE.

hydrophane A variety of OPAL, OPAQUE with a PEARLY LUSTRE when dry and TRANSPARENT when in water.

hydrophone An instrument used to detect seismic energy in water by converting pressure variations into a varying voltage with a piezoelectric material. Cf. GEOPHONE.

hydrostatic The condition when the pressure or stress acting on a buried rock is equivalent to that of a water column of the same depth. Cf. LITHOSTATIC.

hydrostatic gradient The systematic increase in PORE FLUID pressure with burial depth caused by the weight of the overlying water column up to sea level or the WATER TABLE.

hydrostatic pressure See PRESSURE.

hydrostatic stress See PRESSURE.

hydrothermal alteration The ALTERATION, ranging from minor colour changes to complete RECRYSTALLIZATION, produced by HYDROTHERMAL SOLUTIONS, often found alongside VEINS or OREBODIES from zones a few centimetres thick to several times the thickness of the OREBODY.

hydrothermal deposit A MINERAL deposit formed by PRECIPITATION from a HYDROTHERMAL SOLUTION, not necessarily of economic importance.

hydrothermal solution A hot, aqueous solution of high SALINITY responsible for many types of MINERAL and ORE deposit, e.g. VEINS, STOCKWORKS, MASSIVE SULPHIDE DEPOSITS. Originates from surface water, GROUNDWATER, sea water, METEORIC WATER, formation water, METAMORPHIC water and MAGMATIC water.

hydrothermal vent The location of the exhalation of HYDROTHERMAL SOLUTIONS, such as are found along OCEAN RIDGE systems, including BLACK SMOKERS and WHITE SMOKERS. Probably the source of many MASSIVE SULPHIDE DEPOSITS.

hydrotroilite COLLOIDAL hydrous ferrous sulphide formed by the reaction of SULPHUR, produced by the bacterial reduction of sulphates, with iron during the DIAGENESIS of organic matter, which slowly converts to PYRITE.

hydrovolcanic eruption See PHREATIC ERUPTION.

hydroxyapatite ($Ca_5(PO_4)_3(OH)$) A phosphate MINERAL akin to APATITE.

hydrozincite (zinc bloom) ($Zn_3(CO_3)_2(OH)_6$) A SECONDARY MINERAL produced by the ALTERATION of zinc ORE.

hygromatophile element See INCOMPATIBLE ELEMENT.

hyolithelminthid A small, conical or tubular FOSSIL composed of phosphate, of uncertain taxonomic status, found in the TOMMOTIAN.

hyp- See HYPO-.

hypabyssal Descriptive of an IGNEOUS ROCK crystallizing nearer the surface than a PLUTONIC rock and further from the surface than a VOLCANIC ROCK.

hyper- Exceeding.

hypermelanic Descriptive of a rock with 90–100% dark MINERALS.

hyperpycnal flow The water entering a BASIN on the floor of a lake or ocean which, because of its relatively high DENSITY, flows as a bottom-hugging current and causes sediment to travel a large distance from its point of entry. Cf. HOMOPYCNAL FLOW, HYPOPYCNAL FLOW.

hypersolvus granite A GRANITE which crystallized above the SOLVUS temperature and thus contains only one type of ALKALI FELDSPAR. Cf. SUBSOLVUS GRANITE.

hypersthene (($Mg,Fe)SiO_3$) A common ORTHOPYROXENE.

hypersthenite A coarse-grained ULTRA-BASIC rock mainly composed of HYPERSTHENE.

hypichnia TRACE FOSSILS found on the base of the preserving BED.

hypidioblastic texture See CRYSTALLOBLASTIC TEXTURE.

hypidiomorphic See SUBHEDRAL.

hypidiotopic fabric A FABRIC characterized by the presence of some MINERALS which exhibit their crystal form.

hypo- (hyp-) Below.

hypocentre See EARTHQUAKE FOCUS.

hypocrystalline See HYALOCRYSTALLINE.

hypogene Caused by an ascending HYDROTHERMAL SOLUTION. Cf. EPIGENETIC.

hypohyaline See HYALOCRYSTALLINE.

hypolimnion The lower, cold layer of a stratified lake undisturbed by diurnal or seasonal mixing. Tends to ANOXIA, allowing the preservation of organic material. Cf. EPILIMNION.

hypopycnal flow The water entering a BASIN on the floor of a lake or ocean which, because of its relatively low DENSITY, takes the form of a buoyant surface plume or jet. Cf. HOMOPYCNAL FLOW, HYPERPYCNAL FLOW.

hypothermal deposit An EPIGENETIC DEPOSIT formed at temperatures in the range 300–600°C at depths of 3–15 km. Usually found as FRACTURE-fill and REPLACEMENT bodies near deep-seated ACID PLUTONS in deeply eroded PRECAMBRIAN or PALAEOZOIC terrain. Important ORES of GOLD, TIN and tungsten are of this type.

hypsographic curve A graph of the distribution of elevation and depth with reference to SEA LEVEL. Cf. HYPSOMETRIC CURVE.

hypsometric curve A graph of the percentage elevation and depth distribution on the continents and oceans. Cf. HYPSOGRAPHIC CURVE.

hypsometry The measurement of the

land surface or seafloor with respect to a given DATUM, normally mean sea level.

hysteresis The difference between the paths followed with time during loading and unloading, e.g. of elastic loading, magnetization.

I

I In EARTHQUAKE SEISMOLOGY, a P WAVE that has travelled through the inner CORE.

Iapetus The ocean assumed to have lain between North America and Europe/Africa ~500 Ma ago, whose closure was responsible for the CALEDONIAN OROGENY.

Ibexian (Canadian) A CAMBRIAN/ORDOVICIAN succession in the E. USA covering part of the DOLGELLIAN, the TREMADOC and the early ARENIG.

ice age 1 A long period of GLACIATION, e.g. the PERMO-TRIASSIC ice age. **2** The GEOCHRONOLOGICAL equivalent of a CHRONOSTRATIGRAPHIC STAGE, e.g. the CENOZOIC ice ages.

ice cap A GLACIER composed of a small ICE SHEET of <50 000 km² which buries the landscape.

ice mound A PERMAFROST-related landform, such as a PINGO or PALSA.

ice rafting The transport of material by floating ice.

ice sheet A GLACIER of >50 000 km², composed of a flattened dome which buries the landscape.

Iceland spar A variety of clear, colourless CALCITE.

icelandite Intermediate member of THOLEIITIC series similar to ANDESITE, but with more iron and less alumina.

Ichang See YICHANG.

ichnite A TRACE FOSSIL comprising a footprint.

ichnofacies A model for a palaeoenvironment based on TRACE FOSSILS.

ichnofauna The organisms responsible for TRACE FOSSILS.

ichnofossil See TRACE FOSSIL.

ichnology The study of TRACE FOSSILS or their recent counterparts.

Ichthyopterygia A class of the REPTILIA whose sole order is the ICHTHYOSAURIA. Range M. TRIASSIC–U. CRETACEOUS.

Ichthyosauria/ichthyosaurs An order of subclass ICHTHYOPTERYGIA, class REPTILIA; long-snouted marine REPTILES. Range M. TRIASSIC–U. CRETACEOUS.

Ichthyostegalia An order of subclass LABYRINTHODONTIA, class AMPHIBIA; one of the earliest body FOSSILS of the AMPHIBIANS, very similar to their CROSSOPTERYGII FISH ancestors. Range L. CARBONIFEROUS–U. DEVONIAN.

ICPES See INDUCTIVELY COUPLED PLASMA EMISSION SPECTROMETRY.

ICPMS See INDUCTIVELY COUPLED PLASMA EMISSION MASS SPECTROMETER.

idaite (~Cu_3FeS_4) A SULPHIDE MINERAL with a COVELLITE structure of the LAYER SULPHIDES GROUP.

Idamean A CAMBRIAN succession in Australia covering parts of the MAENTWROGIAN and DOLGELLIAN.

iddingsite An ALTERATION product of OLIVINE comprising GOETHITE, QUARTZ, MONTMORILLONITE group CLAY MINERALS and CHLORITE.

idioblastic texture A TEXTURE of a META-

MORPHIC ROCK in which grains show fully developed crystal forms.

idiomorphic See EUHEDRAL.

idiomorphic fabric See IDIOTOPIC FABRIC.

idiotopic fabric (idiomorphic fabric) A FABRIC in a SEDIMENTARY ROCK in which most constituent crystals are EUHEDRAL.

idocrase See VESUVIANITE.

IGC International Geological Congress.

IGCP International Geological Correlation Program.

IGF See INTERNATIONAL GRAVITY FORMULA.

igneous body A volume of IGNEOUS ROCK with discrete boundaries with the surrounding COUNTRY ROCK into which it was emplaced.

igneous breccia A RUDITE formed of igneous material.

igneous foliation A FOLIATION in an IGNEOUS ROCK defined by compositional variation or changes in crystal size or shape, possibly produced by DIFFERENTIATION or flow in a MAGMA.

igneous rock A rock which has solidified from molten or partially molten material.

ignimbrite A poorly sorted, PYROCLASTIC ROCK, comprising mainly PUMICE and ASH, possibly with broken PHENOCRYSTS and dismembered vent wall material, of large volume (1 km^3–2000 km^3).

ignimbrite plain See ASH-FLOW FIELD.

ignimbrite plateau See ASH-FLOW FIELD.

IGRF See INTERNATIONAL GEOMAGNETIC REFERENCE FIELD.

IGSN See INTERNATIONAL GRAVITY STANDARDIZATION NET.

ijolite A PLUTONIC rock with >90% NEPHELINE and MAFIC MINERALS, usually PYROXENE, and also AMPHIBOLE, SPHENE, APATITE and MELANITE. Normally has a normal igneous TEXTURE, particularly SUBOPHITIC and COMB-STRUCTURE. Forms concentric intrusions and DYKES in continental areas.

ikaite ($CaCO_3.6H_2O$) A chalk-like CARBONATE MINERAL forming submarine pillars.

Ikskiy A PERMIAN succession on the eastern Russian Platform equivalent to the ARTINSKIAN.

Ilibeyskiy A PERMIAN succession in the Timan area of the former USSR equivalent to the L. SAKMARIAN.

Illinoian glaciation The third of the GLACIATIONS of the QUATERNARY in North America.

illite (hydromica, hydromuscovite) ($K_{1.5}Al_2(Al_{1.5}Si_{2.5}O_{10})(OH)_2$) A common CLAY MINERAL.

illitization The ALTERATION of a MINERAL to AUTHIGENIC ILLITE.

illuviation The accumulation of material in the lower soil zone by the LEACHING and ELUVIATION of fine-grained material and water-soluble MINERALS from the upper soil zone and their downward transport.

ilmenite (titaniferous iron ore) ($FeTiO_3$) An iron-titanium OXIDE ORE MINERAL, also occurring as an ACCESSORY MINERAL in many rocks.

ilvaite ($CaFe_3^{2+}Fe^{3+}O(Si_2O_7)(OH)$) A SOROSILICATE chiefly found as a product of contact METASOMATISM.

imaginary component (out-of-phase component, quadrature) The part of the secondary electromagnetic field that is 90° out of phase with the primary field in ELECTROMAGNETIC INDUCTION METHODS. It is large in the presence of a good conductor and at low EM frequencies.

imbibition The absorption of a fluid by a granular rock by capillary action.

imbricate Overlapping like tiles.

imbricate fan An IMBRICATE STRUCTURE in which FAULTS branch from a FLOOR THRUST and terminate in FOLD or STRAIN zones in the overlying STRATA.

imbricate fault One of a number of closely spaced FAULTS in an IMBRICATE STRUCTURE.

imbricate slice The rock between IMBRI-CATE FAULTS in an IMBRICATE STRUCTURE.

imbricate stack An IMBRICATE STRUCTURE in a THRUST BELT.

imbricate structure A set of subparallel, overlapping slices of rock bounded by closely spaced FAULTS which join and form one FAULT at depth, developed at all scales.

imbrication A FABRIC in a CLASTIC ROCK resulting from the alignment of CLASTS in which the plane containing the long and intermediate axes dips at a small angle (<20°) upstream.

immature oil Heavy, low API GRAVITY, NAPHTHENE- or ASPHALT-rich oil, which has not undergone sufficient cracking to convert it into a light oil.

immiscible Descriptive of two or more liquids that are incapable of mixing to form a single liquid and separate into two phases. Cf. MISCIBLE.

impact hypothesis See BOLIDE IMPACT HYPOTHESIS.

impact ripple See BALLISTIC RIPPLE.

impactite A rock formed by BOLIDE impact.

impactogen A continental RIFT VALLEY system in the DISTAL region of a CONTINENT-CONTINENT COLLISIONAL OROGEN resulting from tensional STRESSES associated with INDENTATION TECTONICS.

impsonite A member of the ASPHALTITE group.

in situ **combustion** A technique of recovering heavy oil from a RESERVOIR by combustion, which breaks it down to COKE and light oil, the latter being pushed towards producing WELLS. Used when primary methods of recovery have failed.

in situ **mining** The extraction of MINERALS without physical mining, such as in the FRASCH PROCESS.

in situ **stress measurement** The direct measurement of stress *in situ* using a BORE-HOLE BREAKOUT LOG, a FLATJACK, the HYDRO-FRAC TECHNIQUE, an INCLUSION STRESS METER or OVERCORING.

in-phase EM component See REAL COMPONENT.

in-sequence thrust A THRUST in which movement on the FAULT folds or tilts an existing FAULT surface of the same phase of DEFORMATION.

Inarticulata A class of phylum BRACHIO-PODA, characterized by the absence of hinging between the VALVES. Range TOMMOTIAN–RECENT.

inclination The angle between the total MAGNETIC FIELD and the horizontal.

inclined extinction See OBLIQUE EXTINCTION.

inclined fold (tilted fold) A FOLD whose AXIAL PLANE is inclined.

inclusion trail A linear array of MICRO-SCOPIC inclusions within a crystal, commonly fluids, IRON OXIDES and iron hydroxides.

incompatible element (hygromatophile element) An element that is difficult to substitute into the crystal structure of a rock-forming MINERAL because of size, charge or valency requirements, which consequently is less likely to crystallize out of a MAGMA and often concentrates in PEG-MATITIC or HYDROTHERMAL FLUIDS. The MANTLE is depleted in such elements and the CRUST enriched.

incompetent bed A layer which flows during DEFORMATION.

incompressibility See BULK MODULUS.

incongruent melting Melting in which there is dissociation or reaction with the melt so that one crystalline phase is converted to another plus a liquid of different composition. Cf. CONGRUENT MELTING.

incongruent solution DISSOLUTION yielding a solution with different proportions from those in the solid.

inconsequent stream A stream which does not follow major land surface or geological features. Cf. ANTECONSEQUENT, CONSEQUENT, INSEQUENT, OBSEQUENT, RESEQUENT and SUBSEQUENT STREAMS.

indentation tectonics (escape tectonics, extrusion tectonics) DEFORMATION affecting the LITHOSPHERE of the overriding PLATE in a CONTINENT-CONTINENT COLLISION, in which it is dissected by a series of long TRANSCURRENT FAULTS, whose geometry depends on the shape of the indenting continent and which allow further convergence. Indentation tectonics was probably the cause of the continued DEFORMATION of Asia and rotation of Indochina by the convergence of India with Asia during the formation of the Himalaya.

index fossil A FOSSIL species whose abundance characterizes a specific HORIZON.

index mineral A MINERAL whose first occurrence marks the limit of a METAMORPHIC ZONE.

indialite $((MgFe)_2Al_4Si_5O_{18}.nH_2O)$ A CYCLOSILICATE that is the high-temperature POLYMORPH of CORDIERITE.

Indian topaz 1 See CITRINE. 2 A yellow variety of CORUNDUM.

indicated reserve See ORE RESERVE.

indicator boulder An ERRATIC whose origin can be used to determine the source area of TILL.

indicatrix A three-dimensional geometrical figure that represents geometrically the different vibration directions in a MINERAL in terms of an ellipsoid whose axes are proportional to the REFRACTIVE INDEX of a ray travelling parallel to it.

indicolite (indigolite) A blue, sodium-rich variety of TOURMALINE.

indigo copper See COVELLITE.

indigolite See INDICOLITE.

Indigskiy A PERMIAN succession in the Timan area of the former USSR equivalent to the L. ASSELIAN.

Induan See UST'KEL'TERSKAYA.

induced magnetization The phenomenon whereby a material is made to be magnetic by the presence of an external MAGNETIC FIELD, which is lost on removal of the field.

induced polarization A GEOPHYSICAL EXPLORATION method based on the phenomenon that metallic MINERALS can store electric charge, which is gradually released when the energizing current is removed. The only method capable of detecting DISSEMINATED DEPOSITS.

induction log A GEOPHYSICAL BOREHOLE LOG in which the RESISTIVITY structure of the wallrock is determined by inducing current to flow in it by ELECTROMAGNETIC INDUCTION, used when the DRILLING MUD is insulating or the borehole is cased and ELECTRIC LOGS cannot be made.

Inductively Coupled Plasma Emission Mass Spectrometer (ICPMS) An analytical technique in which an aerosol of the sample is dissociated into its constituent ions in a high temperature plasma before being injected into a device for analysing their masses.

Inductively Coupled Plasma Emission Spectrometry (ICPES) An analytical technique for the determination of MAJOR, minor and TRACE ELEMENTS. An aerosol of the sample is injected into a plasma of high-temperature, ionized argon. The wavelengths of the radiation emitted by the atoms of the sample are diagnostic of the elements present.

induration Hardening.

industrial diamond A DIAMOND of less than GEM quality, used for drilling, cutting, lapping etc.

industrial mineral A MINERAL of economic importance in itself rather than because of the element(s) it contains.

inertinite A group of COAL MACERALS which have high CARBON and low hydrogen contents and are hard with a high relief.

inertite A MICROLITHOTYPE OF COAL group comprising INERTINITE MACERALS, of which

FUSITE, FUSINITE and SEMIFUSITE are the most common.

inertodetrinite A COAL MACERAL consisting of broken fragments of INERTITE and SCLEROTINE.

infaunal Descriptive of an organism that burrows into the substrate. Cf. EPIFAUNAL.

inferred reserve See ORE RESERVE.

infiltrability See INFILTRATION CAPACITY.

infiltration The entry of water into the soil, usually by downward flow through the surface.

infiltration capacity (infiltrability) The maximum rate at which water can enter the soil by INFILTRATION.

infiltration rate The volume flux of water flowing into the soil per unit area of surface.

infrared spectrometry A method used to study the hydroxyl groups of CLAY MINERALS.

infrastructure The part of an OROGENIC BELT which was deformed and metamorphosed at deep levels of the CRUST. Cf. SUPRASTRUCTURE.

infusorial earth See TRIPOLITE.

injection complex An association of IGNEOUS or METAMORPHIC ROCKS which are intimately intermixed.

injection gneiss A layered GNEISS, usually of GRANITIC or GRANODIORITIC composition, formed by MAGMA injection along parallel STRUCTURES in the COUNTRY ROCK. A type of MIGMATITE.

injection structure A STRUCTURE in a sediment formed when one layer forces into another.

injection well A WELL in a gas field or OILFIELD through which water, gas or steam is injected into the RESERVOIR to maintain its pressure and thus enhance recovery.

inland sabkha See PLAYA LAKE.

inlier An area of older rocks surrounded by younger rocks. Cf. OUTLIER.

inosilicate (band silicate, chain silicate) A CRYSTAL STRUCTURE classification in which the COORDINATION POLYHEDRA are Si tetrahedra and these form chains when each tetrahedron shares two corners with adjacent tetrahedra.

INPUT© **IN**duced **PU**lse **T**ransient. An ELECTROMAGNETIC INDUCTION METHOD using a transient primary field to produce a decaying secondary field when the primary is absent, so that it can be accurately monitored.

Insectivora An order of infraclass EUTHERIA, subclass THERIA, class MAMMALIA; MAMMALS which include INSECTS in their diet, probably the basic stock from which the other EUTHERIA were derived. Range U. CRETACEOUS–RECENT.

insects Members of superclass HEXAPODA, phylum UNIRAMIA. Range DEVONIAN–RECENT.

inselberg A large, steep-sided OUTCROP rising abruptly from a flat landscape, formed as a residual produced by the PARALLEL RETREAT of BEDROCK slopes or as a remnant on a land surface affected by DEEP WEATHERING.

insequent stream A stream developed as a result of indeterminate features. Cf. ANTECONSEQUENT, INCONSEQUENT, OBSEQUENT, RESEQUENT and SUBSEQUENT STREAMS.

insolation The heat received from the sun.

insolation weathering (thermoclastis) The shattering or disintegration of surface rock by the rapid expansion and contraction resulting from large temperature fluctuations, probably not a process of major importance.

inspissation The process of drying of oil that has reached the surface.

instantaneous rotation The relative rotation of a PLATE about its EULER POLE at a given instant in time. Cf. FINITE ROTATION.

instantaneous strength The short-term STRENGTH of a material. Cf. CREEP STRENGTH.

intensity See EARTHQUAKE INTENSITY.

intensity of magnetization The

strength of the magnetic behaviour of a material, which can be expressed as an INDUCED MAGNETIZATION and possibly a REMANENT MAGNETIZATION.

inter- Between.

inter-arc basin An elongate BASIN between the outer and inner arcs of an ISLAND ARC, usually filled with volcanic sediments.

interconnectedness An index of the proportion of individual sediment bodies of a specific type that are in touch with each other in a given succession.

interdistributary bay A bay between distributaries in a DELTA.

interdune The topographically low, generally flat ground between AEOLIAN DUNES, often with a GRAVEL surface.

interfacial angle The angle between faces of a crystal, measured with a GONIOMETER.

interference figure The pattern of coloured curves and black areas seen in the THIN SECTION of a MINERAL through polarized light and crossed Nichols.

interference ripple A BEDDING surface sedimentary STRUCTURE comprising RIPPLES in at least two orientations at a high angle forming a polygonal pattern, produced by simultaneous or sequential currents in different directions.

interflow 1 (subsurface stormflow) Rapid subsurface flow within the soil layer. **2** Flow at an intermediate level within a water body, often at a PYCNOCLINE.

interfluve An area of high ground between two adjacent river valleys.

interglacial A phase of relatively warm temperatures between GLACIALS, during which ICE SHEETS retreated, forest replaced TUNDRA and sea level rose.

intergranular displacement DISPLACEMENT along an adjacent grain boundary during DEFORMATION. Cf. INTRAGRANULAR DISPLACEMENT, TRANSGRANULAR DISPLACEMENT.

intergranular fracture FRACTURE taking place along adjacent grain boundaries. Cf. CIRCUMGRANULAR FRACTURE, INTRAGRANULAR FRACTURE.

intergranular texture See GRANULITIC TEXTURE.

intergrowth An interlocking arrangement of crystals arising from simultaneous CRYSTALLIZATION or EXSOLUTION.

interlayer slip DEFORMATION accomplished by DISPLACEMENT along BEDDING or FOLIATION surfaces.

intermediate argillic alteration A type of low grade WALL ROCK ALTERATION in which PLAGIOCLASE is altered to KAOLIN and MONTMORILLONITE group MINERALS.

intermediate rock A rock with <10% QUARTZ plus either PLAGIOCLASE in the range An_{10-50}, or an ALKALI FELDSPAR or both ALKALI and PLAGIOCLASE FELDSPARS.

intermediate sulphur crude oil CRUDE OIL with 0.6–1.7% sulphur.

intermediate-focus earthquake An EARTHQUAKE with a depth of FOCUS between 70 km and 300 km. Cf. SHALLOW-FOCUS EARTHQUAKE, DEEP-FOCUS EARTHQUAKE.

internal boudinage A type of BOUDINAGE in which there is the development of sinusoidal thickening and thinning within an ANISOTROPIC, homogeneous material, resulting from layer-parallel EXTENSION in which an instability occurs.

internal wave A WAVE occurring along a density interface within a body of stratified fluid, e.g. a THERMOCLINE.

International Geomagnetic Reference Field (IGRF) A complex formula representing the observed or predicted GEOMAGNETIC FIELD for a given EPOCH and used to compute the GEOMAGNETIC CORRECTION in the REDUCTION of MAGNETIC SURVEY measurements.

International Gravity Formula (IGF) A formula used for the LATITUDE CORRECTION in the REDUCTION of GRAVITY SURVEY data collected during the period 1930 to 1967. Now superseded by the GRAVITY FORMULA.

International Gravity Standardization Net (IGSN) A world-wide network of locations where the absolute value of GRAVITY is known, used to convert the relative readings of GRAVIMETERS into absolute values.

International Program for Ocean Drilling (IPOD) A phase of the DEEP SEA DRILLING PROGRAM. Cf. OCEAN DRILLING PROGRAM.

International Seabed Authority The authority responsible, under the United Nations Law of the Sea, for the administration of the seabed and its resources beyond 200 miles of shorelines.

interpenetrant twin A TWINNING phenomenon in which two individuals appear to have grown through each other.

interpluvial A relatively dry phase between PLUVIALS of the PLEISTOCENE and HOLOCENE.

intersection cleavage A CLEAVAGE which crosses another planar feature and thus creates a LINEATION with it.

intersection lineation A LINEATION produced by the intersection of two planes, e.g. CLEAVAGE and BEDDING.

intersection point The point on the longitudinal profile of an ALLUVIAL FAN above which there is incision into older FAN deposits and below which there is deposition.

intersertal texture The TEXTURE of a random network of FELDSPAR laths whose gaps are filled with GLASS or very small crystals.

interstadial A phase of relative warmth during a major GLACIAL, of insufficient magnitude and/or duration to be classed as an INTERGLACIAL.

interstice entrapment A PLACER formation mechanism involving the trapping of particles between grains of the stream bed.

interstitial Within the pores or between the grains or crystals of a rock.

interstitial crystal defect A CRYSTAL

DEFECT in which an alien atom occupies a small gap between atoms in the regular arrangement.

intertidal bar A SANDWAVE or SWASH BAR which migrates rapidly onshore due to SWASH action and is destroyed by storm activity, in which it is welded to the upper foreshore to form a BERM.

interval velocity The SEISMIC VELOCITY over a given vertical interval, which may comprise units with different velocities.

intra- Inside.

intraclast See INTRAFORMATIONAL CLAST.

intracrystalline plasticity DEFORMATION by the movement of DISLOCATIONS through a CRYSTAL LATTICE.

intrafolial fold A FOLD consisting solely of the HINGE of an ASYMMETRICAL FOLD which has been rotated and stretched by continued SIMPLE SHEAR.

intraformational clast (intraclast) A CLAST formed by the EROSION of sediment soon after deposition and incorporated into a slightly younger deposit. Cf. LITHOCLAST.

intraglacial See ENGLACIAL.

intragranular displacement DISPLACEMENT taking place within a grain during DEFORMATION. Cf. INTERGRANULAR DISPLACEMENT, TRANSGRANULAR DISPLACEMENT.

intragranular fracture A FRACTURE in a granular material which takes place within the grains. Cf. CIRCUMGRANULAR FRACTURE, INTERGRANULAR FRACTURE.

intragranular glide GLIDE in which the CRYSTAL LATTICE either side of the SLIP PLANE remains undistorted.

intramicrite A LIMESTONE with >10% INTRACLASTS in a MICRITE MATRIX.

intrasparite A LIMESTONE with >10% INTRACLASTS in a SPARITE MATRIX.

intratelluric crystal A crystal that has grown slowly at depth before more rapid secondary CRYSTALLIZATION.

intrenched meander A meander in a

river channel that is incised into the landscape as a result of EROSION.

intrusive igneous body An IGNEOUS BODY emplaced at depth.

Inverian event An OROGENIC phase affecting NW Scotland at ~2400 Ma during the later stages of the SCOURIAN.

inverse grading (negative grading, reverse grading) An upward increase in mean grain size in a sedimentary BED, implying the sedimenting grains were independently mobile and deposited under the influence of grain-grain interactions.

inverse problem A type of problem in geophysics in which the best estimate of a model is inferred from the observations. Cf. FORWARD PROBLEM.

inversion 1 See INVERSE PROBLEM. 2 In STRUCTURAL GEOLOGY, the reversal of the sense of vertical movement of a FAULT, block or region, such as the transformation of a sedimentary BASIN into a mountain by OROGENESIS.

inversion twinning A TWINNING phenomenon resulting from a phase transition during which CRYSTAL SYMMETRY is lowered and the TWIN orientations are in the higher symmetry.

inverted relief Topography in which areas of high relief, such as ANTICLINES, have become depressions due to enhanced EROSION of the uplands.

invisible gold GOLD invisible even under MICROSCOPIC examination because of its small grain size or inclusion in a SOLID SOLUTION in a MINERAL.

involute With edges that roll under or inwards.

inyoite ($CaB_3O_3(OH)_5.4H_2O$) A colourless, TRANSPARENT borate MINERAL.

iodargyrite (iodyrite) (AgI) A rare SUPERGENE ORE MINERAL of SILVER.

iodobromite ($Ag(Cl,Br,I)$) A rare SUPERGENE ORE MINERAL of SILVER.

iodyrite See IODARGYRITE.

iolite A GEM variety of CORDIERITE.

ion microprobe An instrumental analytical method for mapping the distribution of elements and determining their concentrations, similar in its usage to the ELECTRON MICROPROBE. A beam of ions bombards a polished section *in vacuo*. The ejected ionic species are identified and quantified in a MASS SPECTROMETER. The advantage of the technique is that isotopes, TRACE ELEMENTS and light elements can be determined to a high accuracy.

ion milling A preparation method for TRANSMISSION ELECTRON MICROSCOPY in which beams of argon ions or atoms are fired at a demounted, polished THIN-SECTION, eroding its surface to provide small, electron-transparent holes.

Iowan glaciation The fourth of the GLACIATIONS experienced by North America in the QUATERNARY.

IPOD See INTERNATIONAL PROGRAM FOR OCEAN DRILLING.

Iren'skiy 1 (Fillipovskiy) A PERMIAN succession on the eastern Russian Platform equivalent to the KUNGURIAN. 2 (Vyl'skiy) A PERMIAN succession in the Timan area of the former USSR equivalent to the KUNGURIAN.

iridescence A shimmering effect exhibited by some MINERALS caused by the DIFFRACTION of light by fine-scale lamellar intergrowths.

iridium (Ir) A very rare, PLATINUM-group, NATIVE ELEMENT.

iridium anomaly The high concentration of IRIDIUM found in rocks formed at the CRETACEOUS-TERTIARY boundary, and believed to be the result of a BOLIDE impact.

iridosmine (Ir,Os) A very rare natural alloy of IRIDIUM and osmium.

iris (rainbow quartz) A variety of QUARTZ exhibiting a chromatic reflection of light from FRACTURES.

IRM See ISOTHERMAL REMANENT MAGNETIZATION.

iron (Fe) A rare NATIVE ELEMENT, also found in METEORITES.

iron alum See HALOTRICHITE.

iron glance A specular variety of HAEMATITE.

iron hat See GOSSAN.

iron meteorite (siderite) A METEORITE comprising iron plus <20% nickel.

iron pan See CARSTONE.

iron pyrites See PYRITE.

ironstone A SEDIMENTARY ROCK containing iron-rich NODULES or layers, of actual or potential economic importance.

irregular fracture See UNEVEN FRACTURE.

irregular mullion A MULLION STRUCTURE comprising cylinders with irregular cross-sections.

Ischnacanthiformes An order of subclass ACANTHODII, class OSTEIICHTHYES, superclass PISCES; bony FISH of uncertain affinity. Range U. SILURIAN–U. DEVONIAN.

island arc An arc of volcanic islands formed above the dipping slab at a SUBDUCTION ZONE by melting mainly of the superjacent MANTLE.

island silicate See SOROSILICATE.

iso- Equal.

isoash map A COAL QUALITY MAP showing trends in the ASH content of a COAL.

isobar **1** A line joining points of equal PRESSURE. **2** A line joining points of equal PRINCIPAL STRESS magnitude.

isobath A line joining points of equal water depth.

isochore A line of equal drilled thickness of a rock unit. Cf. ISOPACHYTE.

isochromatic A line joining points of equal maximum SHEAR STRESS, obtained by photoelastic methods.

isochron **1** A line joining points of equal age, e.g. of OCEAN CRUST. **2** A line joining points of constant ratio of radioactive isotopes used in RADIOMETRIC AGE DATING.

isoclinal fold A FOLD with an INTERLIMB ANGLE of 0°–10°, i.e. parallel or nearly parallel limbs.

isocline A line joining points of equal magnetic INCLINATION.

isoclinic A line joining points of similar PRINCIPAL STRESS orientations.

isofacies map A map showing the limits of FACIES, which may include several rock types.

isogal A line joining points of equal GRAVITY ANOMALY.

isogam A line joining points of equal MAGNETIC ANOMALY.

isogon A line joining points of equal magnetic DECLINATION.

isograd A line joining locations of the same METAMORPHIC GRADE.

isogyre A black pattern in an INTERFERENCE FIGURE.

isoline See ISOPLETH.

isolith map A map showing the total thickness of beds of one lithology in a STRATIGRAPHIC succession composed of several lithologies.

isometric system See CUBIC SYSTEM.

isomorphism The phenomenon of two or more MINERALS of similar chemistry crystallizing with the same CRYSTAL SYMMETRY.

isomorphous Descriptive of substances which form a series of SOLID SOLUTIONS.

isopach **1** See ISOPACHYTE. **2** A line joining points of equal mean STRESS magnitude.

isopachyte (isopach) A line joining points of equal STRATIGRAPHIC thickness. Cf. ISOCHORE.

isopleth (isoline) A line joining points of equal abundance.

isopycnal A line joining points of equal pressure in water.

isoseismal A line joining points of equal EARTHQUAKE INTENSITY.

isostasy The study of the response of the Earth to the removal and imposition of large loads. Isostatic theory states that at the DEPTH OF COMPENSATION all PRESSURES exerted by the rocks above are equal. This condition is satisfied by the AIRY and PRATT HYPOTHESES, which propose different geometrical arrangements for the subsurface compensation of major Earth features.

isostatic anomaly The BOUGUER ANOMALY minus the effect of the compensation predicted by the AIRY or PRATT HYPOTHESES. Non-zero values indicate ISOSTATIC EQUILIBRIUM is not extant.

isostatic equilibrium The condition when a surface load, or lack of mass, is perfectly balanced by a mass deficiency, or mass excess, at depth.

isostatic rebound The recovery of ISOSTATIC EQUILIBRIUM after the removal of a load, e.g. after the removal of an ICE SHEET, the LITHOSPHERE has to rise to regain equilibrium.

isostructural Descriptive of MINERALS with similar chemical, physical and CRYSTALLOGRAPHIC characteristics.

isosulph map A COAL QUALITY MAP showing trends in the SULPHUR content of a COAL.

isotherm A line joining points of equal temperature.

isothermal Having the same temperature.

isothermal remanent magnetization (IRM) A very complex NATURAL REMANENT MAGNETIZATION acquired when a rock is struck by lightning.

isotopic abundance The ratio of the quantity of an isotope to the total amount of the element.

isotopic fractionation The phenomenon whereby the isotopic composition of a MINERAL is different from that of the fluid from which it PRECIPITATED.

isotopic stratigraphy STRATIGRAPHY based on the isotopic composition of MINERALS.

isotropic Having no order or PREFERRED ORIENTATION. Does not affect doubly polarized light.

isotropic fabric A FABRIC of a CLASTIC ROCK which has EQUANT CLASTS or no PREFERRED ORIENTATION of non-EQUANT CLASTS.

isotropic point The point where the PRINCIPAL STRESSES are equal.

isotypic Descriptive of a pair of ISOMORPHS whose relative sizes of ions are the same, but whose absolute sizes are different, so that no SOLID SOLUTION is possible.

isovol A line joining points of equal volatile content in COAL.

ISSC International Subcommission on Stratigraphic Correlation.

Isuan An ERA of the PRECAMBRIAN, 3800–3500 Ma.

itabirite See BANDED IRON FORMATION.

itacolumnite A micaceous SANDSTONE with interlocking CLASTS so that the rock bends when in the form of a thin slab.

italite A rare, coarse-grained, PLUTONIC IGNEOUS ROCK comprising LEUCITE and a small amount of GLASS.

IUGG International Union of Geophysics and Geodesy.

IUGS International Union of Geological Sciences.

IUGS classification An internationally adopted scheme for the classification of IGNEOUS ROCKS proposed by the IUGS in 1989.

Iuosuchanskaya A TRIASSIC succession in Siberia equivalent to the RHAETIAN.

Ivorian A STAGE of the CARBONIFEROUS, 353.8–349.5 Ma.

J

J In EARTHQUAKE SEISMOLOGY, an S WAVE generated by conversion from a P WAVE at the boundary between the inner and outer CORE and reconverted on exit.

J-type lead (Joplin-type lead) An ANOMALOUS LEAD which provides a negative age, probably due to its acquiring extra radiogenic lead during its history.

jacinth See HYACINTH.

Jackson An EOCENE succession on the Gulf Coast of the USA equivalent to the PRIABONIAN.

jacobsite ($MnFe_2O_4$) A rare black SPINEL.

jacupirangite An IGNEOUS ROCK comprising TITANAUGITE, AEGIRINE-AUGITE and ACCESSORY NEPHELINE.

jade A precious GEM consisting of JADEITE or NEPHRITE.

jadeite ($NaAlSi_2O_6$) A green PYROXENE found in METAMORPHIC ROCKS.

jamesonite ($Pb_4FeSb_6S_{14}$) A minor ORE MINERAL of lead.

jargoon A yellow or smoky variety of ZIRCON.

jarosite ($KFe_3(SO_4)_2(OH)_6$) A SECONDARY MINERAL found coating iron ORES.

jasper A GRANULAR, MICROCRYSTALLINE variety of QUARTZ, usually coloured red by HAEMATITE.

jasperoid A form of WALL ROCK ALTERATION, common in some EPITHERMAL DEPOSITS, in which fine-grained, HAEMATITE-stained SILICA is developed.

jaspillite See BANDED IRON FORMATION.

Jeffreys-Bullen model A model for the SEISMIC VELOCITY structure of the Earth, in which velocity varies only radially.

jeppeite ($(K,Ba)_2(Ti,Fe)_6O_{13}$) An OXIDE MINERAL found as an ACCESSORY MINERAL in LAMPROITES.

jet A lustrous variety of LIGNITE found as isolated masses in some bituminous SHALES, used in jewellery.

Jiangtangjiang (Chientangkiang) An ORDOVICIAN succession in China covering part of the HARNAGIAN, the SOUDLEYAN, LONGVILLIAN, MARSHBROOKIAN, ACTONIAN, ONNIAN and ASHGILL.

jimthompsonite ($(Mg,Fe)_{10}Si_{12}O_{32}(OH)_4$) A BIOPYRIBOLE.

JMA scale A scale of EARTHQUAKE INTENSITY (0–VII) used by the Japanese Meteorological Agency, suitable for local building styles.

joesmithite ($(Ca,Pb)Ca_2(Mg,Fe^{2+},Fe^{3+})_5$ $(Si_6Be_2O_{22})(OH)_2$) An AMPHIBOLE containing beryllium in the tetrahedral chain.

johannsenite ($CaMnSi_2O_6$) A PYROXENE found in association with RHODONITE and BUSTAMITE in METASOMATIZED LIMESTONES.

JOIDES (Joint Oceanographic Institutions for Deep Earth Sampling) An organization established to undertake deep-sea drilling, subsequently evolving into the DEEP SEA DRILLING PROGRAM and the INTERNATIONAL PROGRAM FOR OCEAN DRILLING.

Joint Oceanographic Institutions for Deep Earth Sampling See JOIDES.

joint A FRACTURE on which any SHEAR DIS-PLACEMENT is too small to be visible to the unaided eye.

joint set A group of JOINTS with a common orientation.

jökull A small ICE CAP.

jökullhlaup A catastrophic flood caused by a VOLCANIC ERUPTION beneath an ICE SHEET.

Joplin-type lead See J-TYPE LEAD.

Josephinite See AWARUITE.

Jotnian orogeny An OROGENY affecting the Baltic SHIELD in the PROTEROZOIC.

Jurassic The middle PERIOD of the MESO-ZOIC, 208.0–145.6 Ma.

juvenile bomb A PYROCLASTIC fragment of congealed MAGMA over 64 mm in size.

juvenile gas Gas originating in the interior of the Earth, not previously having been at the surface.

juvenile water Water derived from MAGMA.

K In EARTHQUAKE SEISMOLOGY, a P WAVE that has travelled through the CORE.

K-cycle A concept of land evolution involving the cyclic EROSION of soils on upper slopes during unstable climatic phases and soil development during stable phases.

K-feldspar See POTASSIUM FELDSPAR.

K-T boundary See CRETACEOUS-TERTIARY BOUNDARY.

Kachian A PALAEOCENE succession in the former USSR covering the upper part of the THANETIAN.

kaersutite (titanium hornblende) $(Ca_2(Na,K)(Mg,Fe^{2+},Fe^{3+})_4TiSi_6Al_2O_{22}(O,OH,F)_2)$ A titanium AMPHIBOLE.

Kaiatan An EOCENE succession in New Zealand covering the upper part of the BARTONIAN.

Kaihikuan A TRIASSIC succession in New Zealand covering the lower part of the LADINIAN.

kainite $(KMgSO_4Cl.3H_2O)$ An EVAPORITE MINERAL.

Kainozoic See CENOZOIC.

kaiwekite A VOLCANIC ROCK containing PHENOCRYSTS of OLIVINE, TITANAUGITE, BARK-EVIKITE and ANORTHOCLASE.

kakortokite NEPHELINE SYENITE containing red EUDIALYTE and a sodic PYROXENE, usually ACMITE.

kaliophilite A POLYMORPH of KALSILITE.

kalsilite $(KAlSiO_4)$ A rare FELDSPATHOID found in complex PHENOCRYSTS and in the MATRIX of certain LAVAS.

kamacite (Fe,Ni) An iron-nickel alloy found in METEORITES.

kame An irregular, undulating mound of bedded SANDS and GRAVELS deposited unevenly along the front of a stationary or decaying ICE SHEET.

kame terrace A terrace between a hillside and GLACIER formed by GLACIOFLUVIAL activity.

kamenitza A solution PAN or BASIN formed on a LIMESTONE surface by the DISSOLUTION of calcium carbonate.

Kamyshinian A PALAEOCENE succession in the former USSR covering part of the THANETIAN.

kandite A group of CLAY MINERALS including KAOLINITE, DICKITE, NACRITE, HALLOYSITE and meta-HALLOYSITE.

Kanev An EOCENE succession in the former USSR covering the YPRESIAN and part of the LUTETIAN.

Kansan glaciation The second of the GLACIATIONS of North America in the QUATERNARY.

kaolin (china clay) An important MINERAL product composed principally of KAOLINITE.

kaolinite $(Al_2Si_2O_5(OH)_4)$ A common CLAY MINERAL formed by the WEATHERING or HYDROTHERMAL ALTERATION of FELDSPARS and other aluminous silicate MINERALS.

kaolinitization See KAOLINIZATION.

kaolinization (kaolinitization) The

ALTERATION of a MINERAL (commonly MUSCOVITE, BIOTITE or FELDSPAR) by a DISSOLUTION-PRECIPITATION mechanism to AUTHIGENIC KAOLINITE.

Kapitean A NEOGENE succession in New Zealand covering parts of the MESSIAN and ZANCLIAN.

Karaganian A MIOCENE succession on the Russian Platform covering the lower part of the TORTONIAN.

Karakan An ORDOVICIAN succession in Kazakhstan covering the upper part of the LLANVIRN.

karat See CARAT.

Karatau The youngest PERIOD of the RIPHEAN, 1050–800 Ma.

Karelian orogeny An OROGENY affecting the Baltic SHIELD in the ARCHAEAN from ~2000–1900 Ma.

Karman-Prandtl velocity law A law expressing the velocity profile within a turbulent BOUNDARY LAYER as a logarithmic function of the distance from the bed: $U = \kappa^{-1}\sqrt{(\tau/\rho)}.\ln(y/k)$, where U = velocity, κ = VON KARMAN'S CONSTANT, τ = SHEAR STRESS, ρ = fluid DENSITY, y = height of point at which U is measured and k = COEFFICIENT OF ROUGHNESS.

Karoo basalts A lava pile erupted from a HOTSPOT beneath southern Africa during the period 190–150 Ma. Covers 2 000 000 km², and in places is 9 km thick.

karren (lapiés) Minor solutional features developed on carbonate rocks, formed mostly by DISSOLUTION.

karst A terrain with distinctive landforms and drainage (often underground), mainly originating from SOLUTIONAL EROSION and commonly developed on carbonate rocks or EVAPORITES.

Kashirskian A STAGE of the CARBONIFEROUS, 309.2–307.1 Ma.

Kasimovian An EPOCH of the CARBONIFEROUS, 299.9–295.1 Ma.

kataphorite See KATOPHORITE.

katatectic layer A layer of SOLUTION residue, commonly of GYPSUM and/or ANHYDRITE, at the top of a SALT DOME.

katazone A DEPTH ZONE of high temperature and very high HYDROSTATIC PRESSURE. See also EPIZONE, MESOZONE.

katophorite (cataphorite, kataphorite) $(Na_2Ca(Fe^{3+},Al)_5(AlSi_7)O_{22}(OH)_2)$ An AMPHIBOLE found in BASIC, ALKALINE IGNEOUS ROCKS.

katungite MELILITE-bearing, ultrapotassic, VOLCANIC rock, usually with OLIVINE and minor amounts of LEUCITE but no AUGITE.

kavir A PLAYA, continental SABKHA or other saline desert BASIN which may be flooded periodically.

Kawhia A JURASSIC succession in New Zealand comprising the TEMAIKAN, HETERIAN and OHAUAN.

Kazakhstania A continent situated between Siberia and GONDWANALAND from CAMBRIAN to DEVONIAN times.

Kazanskiy A PERMIAN succession on the eastern Russian Platform covering part of the UFIMIAN, the WORDIAN and part of the CAPITANIAN.

keatite (SiO_2) A synthetic tetragonal form of SILICA.

kegel A conical hill on a LIMESTONE landscape.

kegelkarst (cockpit karst, cone karst) A LIMESTONE landscape characterized by KEGELS interspersed by closed depressions, typical of KARST.

keilhauite A variety of SPHENE containing >10% RARE EARTH ELEMENTS.

Kelvin model See VISCOELASTICITY.

Kelvin-Voight model See VISCOELASTICITY.

kelyphitic rim A rim of one MINERAL around another in an IGNEOUS ROCK resulting from reaction of the enclosed MINERAL with other constituents of the rock.

Kenoran orogeny (Algoman orogeny) An OROGENY of PRECAMBRIAN age at ~2400 Ma which affected the Canadian SHIELD.

kentallenite A coarse-grained, BASIC IGNEOUS rock comprising OLIVINE, AUGITE and BIOTITE with minor PLAGIOCLASE and ORTHOCLASE in equal amounts.

kenyite A fine-grained IGNEOUS ROCK; an OLIVINE-bearing PHONOLITE with PHENO-CRYSTS of ANORTHOCLASE ± AUGITE in a GLASSY GROUNDMASS.

keratophyre A rock of the SPILITE suite made up of ALBITE, CHLORITE, EPIDOTE and iron-titanium oxides, formed by the ALTER-ATION of intermediate VOLCANIC ROCKS.

kermesite (pyrostibnite) (Sb_2S_2O) Antimony oxysulphide, a SECONDARY MINERAL forming by the ALTERATION of STIBNITE.

kernite ($Na_2B_4O_6(OH)_2.3H_2O$) A MINERAL important as a source of boron compounds.

kerogen A bituminous material, found in OIL SHALES and other SEDIMENTARY ROCKS, which is composed of organic matter and can yield oil on distillation.

kersantite A CALC-ALKALINE LAMPROPHYRE containing BIOTITE, PLAGIOCLASE (usually An_{10-50}) and AUGITE, ± DIOPSIDE and OLIVINE.

kesterite (Cu_2ZnSnS_4) A SULPHIDE MINERAL derived from the SPHALERITE structure by ordered substitution.

Ketilidian orogeny An OROGENY of PRO-TEROZOIC age affecting Greenland at ~1800–1600 Ma.

kettlehole A topographic depression left when DRIFT-covered ice melts.

Keuper 1 A TRIASSIC succession in Germany covering part of the LADINIAN and the CARN-IAN, NORIAN and RHAETIAN. **2** A traditional name for the U. TRIASSIC in Europe.

key-hole vugs Bubble-like FENESTRAE found in SANDS.

Kharakovian An OLIGOCENE succession in the former USSR equivalent to the RUPELIAN.

Khedalichenskaya A TRIASSIC succession in Siberia equivalent to the CARNIAN and NORIAN.

Kibalian orogeny See BUGANDO-TORO-KIBALIAN OROGENY.

kidney ore A form of HAEMATITE with a fibrous, radiating, internal structure and a RENIFORM, red, external surface.

kidney stone 1 A RENIFORM LIMESTONE PEBBLE or NODULE. **2** See NEPHRITE.

kieselguhr See DIATOMITE.

kieserite ($MgSO_4.H_2O$) An EVAPORITE MINERAL.

Kieslager deposit See BESSHI-TYPE DEPOSIT.

killas A term for DEVONIAN-CARBONIFEROUS, low-GRADE PHYLLITES in SW England.

kilobar The CGS unit of PRESSURE, equivalent to 100 MPa.

kimberlite A SERPENTINIZED, carbonated, commonly brecciated, PORPHYRITIC MICA-PERIDOTITE made up of PHENOCRYSTS of OLI-VINE and PHLOGOPITE in a fine-grained GROUNDMASS of OLIVINE, PHLOGOPITE, PYROPE, iron-titanium oxide, PEROVSKITE plus SERPENTINITE, CHLORITE and carbonates. Found in VOLCANIC PIPES and characteristically containing a wide range of XENOLITHS of CRUSTAL and MANTLE origin. The main source of DIAMONDS.

Kimmeridgian A STAGE of the JURASSIC, 154.7–152.1 Ma.

Kinderhookian A CARBONIFEROUS succession in the USA covering the lower part of the TOURNAISIAN.

Kinderscoutian A STAGE of the CARBON-IFEROUS, 322.8–321.5 Ma.

kinematic Referring to motion; used to describe phenomena related to the relative motion of an object.

kinematic indicator A STRUCTURE that can provide the direction and sense of DISPLACE-MENT of a FAULT or SHEAR ZONE.

kinematic sieving A process suggested as a partial explanation for INVERSE GRADING in which 'vibration-strain' during flow promotes the downward filtration of small grains between large ones.

kinematic symmetry axes The direction of FLOW, the plane of FLOW and the plane normal to these, assuming PLANE STRAIN.

kinematic viscosity The ratio of VISCOSITY to DENSITY, allowing comparison of the resistance to shape change between materials of different DENSITIES.

kink A type of FOLD with an angular PROFILE.

kink band A MICROSCOPIC- to MESOSCOPIC-scale localized band where the orientation of a STRUCTURE changes abruptly.

kink band boundary See KINK PLANE.

kink plane (kink band boundary) The edge of a KINK BAND.

kinking The FOLD process forming a KINK.

kipuka (dagala) An 'island' of land surrounded by LAVA.

kirschsteinite ($CaFeSiO_4$) An OLIVINE.

Kiruna-type deposit See VOLCANIC-ASSOCIATED MASSIVE OXIDE DEPOSIT.

Klabava An ORDOVICIAN succession in Bohemia equivalent to the ARENIG.

Klazminskian A STAGE of the CARBONIFEROUS, 295.1–293.6 Ma.

klippe An OUTLIER formed by THRUSTING.

kluftkarren See GRIKE.

knebelite (($Mn,Fe)_2SiO_4$) An OLIVINE usually formed as a METAMORPHIC product in iron-manganese ORE deposits.

knee twin See GENICULATE TWIN.

knoch and lochan Topography dominated by a mixture of eroded rock ridges and small BASINS, formed by GLACIAL EROSION.

Knott's equations Equations which define how SEISMIC WAVE energy is partitioned between reflected P and S WAVES when it impinges on a discontinuity; similar to ZOEPPRITZ' EQUATIONS.

kobellite ($Pb_2(Bi,Sb)_2S_5$) Lead-bismuth-antimony sulphide, found in GRANITE PEGMATITE VEINS.

Koenigsberger ratio (Königsberger ratio) The ratio of intensity of NATURAL REMANENT MAGNETIZATION to magnetization induced by the local GEOMAGNETIC FIELD.

Kogashyk An ORDOVICIAN succession in Kazakhstan covering the upper part of the ARENIG.

Kokhanskiy A PERMIAN succession on the eastern Russian Platform equivalent to the U. ASSELIAN.

Koln stone (Cullen stone) A MILLSTONE of MAYEN LAVA traded widely in NW Europe from the Iron Age.

komatiite An ULTRAMAFIC VOLCANIC ROCK with >18% MgO composed of OLIVINE and PYROXENE $\pm$ CHROMITE in a GLASSY or DEVITRIFIED GROUNDMASS. It has the morphological features of subaerial and submarine BASALTIC LAVA flows (i.e. PILLOW and HYDROCLASTITE STRUCTURE) and a distinctive SPINIFEX TEXTURE. Characteristic of ARCHAEAN terrains.

Komichanskiy A PERMIAN succession in the Timan area of the former USSR equivalent to the U. ARTINSKIAN.

Königsberger ratio See KOENIGSBERGER RATIO.

Konkian A MIOCENE succession on the Russian Platform covering parts of the TORTONIAN and MESSIAN.

Kopaly An ORDOVICIAN succession in Kazakhstan covering the lower part of the LLANVIRN.

Kopanina-Schichten A SILURIAN succession in Bohemia covering part of the GORSTIAN and the LUDFORDIAN.

kopje (koppie) A rocky hill probably formed by the exhumation of relatively unweathered rock and CORESTONES from within DEEP WEATHERING profiles.

koppie See KOPJE.

Korangian A CRETACEOUS succession in

New Zealand covering parts of the APTIAN and ALBIAN.

kornerupine $(Mg_3Al_6(Si,Al,B)_5O_{21}(OH))$ A rare magnesium-aluminium borosilicate, sometimes used as a GEM.

Kosov An ORDOVICIAN succession in Bohemia equivalent to the HIRNANTIAN.

koum See ERG.

Kraluv An ORDOVICIAN succession in Bohemia approximately equivalent to the RAWTHEYAN.

krennerite $(AuTe_2)$ A rare GOLD ORE MINERAL.

Krevyakinskian A STAGE of the CARBONIFEROUS, 303.0–299.9 Ma.

krotovina (crotovina) An animal burrow filled by later material.

kulaite An AMPHIBOLE-bearing NEPHELINE BASALT.

Kungurian A STAGE of the PERMIAN, 259.7–256.1 Ma.

kunzite A GEM variety of SPODUMENE with a clear lilac colour.

kupfernickel See NICCOLITE.

Kupferschiefer 1 A PERMIAN succession in NW Europe covering the upper part of the KUNGURIAN. **2** A COPPER-rich SHALE of PERMIAN age found in Germany, Poland, Holland and England.

Kuroko-type deposit A type of volcanic-associated MASSIVE SULPHIDE DEPOSIT, usually an ORE OF COPPER, zinc and lead ± GOLD and SILVER, formed in a BACK-ARC BASIN environment.

kurtosis A measure of the 'peakedness' of a frequency distribution.

kutnahorite $(CaMn(CO_3)_2)$ A manganese-rich DOLOMITE.

Kuyalnitskian (Cimmeridian) A PLIOCENE succession on the Russian Platform covering parts of the ZANCLIAN and PIACENZIAN.

kyanite (disthene) (Al_2SiO_5) A NESOSILICATE commonly formed by the REGIONAL METAMORPHISM of an ARGILLACEOUS rock.

kyanite group An industrial name for the SILLIMANITE MINERALS.

kylite See THERALITE.

L

L Letter used to indicate a phase of LIN-EATION formation, subscripted to denote each separate phase.

L wave See LOVE WAVE.

L-S tectonite A rock with both a LINEATION and a FOLIATION.

L-tectonite A rock with well-developed LINEATIONS.

labile Unstable.

labradorescence The brilliant play of colours shown by LABRADORITE.

labradorite A PLAGIOCLASE FELDSPAR, An_{50-70}.

Labyrinthodontia A subclass of class AMPHIBIA; AMPHIBIANS with a large head and limbs splayed laterally from a relatively stubby body. Range M. DEVONIAN–U. TRIASSIC.

laccolite See LACCOLITH.

laccolith (laccolite) A CONCORDANT minor intrusion with a flat floor and convex upper surface. Generally with a diameter up to ~8 km and a thickness from a few metres to hundreds of metres.

Lachlan Fold Belt The PALAEOZOIC (CAMBRIAN-CARBONIFEROUS) OROGENIC BELT that occurs in southeastern Australia.

lacuna See DISCONFORMITY.

lacustral 1 See LACUSTRINE. **2** See PLUVIAL.

lacustrine (lacustral) Referring to a lake.

ladder ripples (ladder-back ripples) INTERFERENCE RIPPLES comprising a set of long wavelength WAVE RIPPLES and a second set of shorter wavelength orthogonal to them, observed to form during a fall in water level.

ladder-back ripples See LADDER RIPPLES.

Ladinian A STAGE of the TRIASSIC, 239.5–235.0 Ma.

lag breccia (co-ignimbrite breccia) A coarse-grained deposit rich in LITHIC fragments which accumulated at the same time as an IGNIMBRITE near the VOLCANIC VENT by the accumulation of CLASTS too large to be transported away.

lag deposit See LAG GRAVEL.

lag fault (low-angle fault) A NORMAL FAULT with a DIP of <45°.

lag gravel (lag deposit) A residual accumulation of coarse particles from which the fine material has been WINNOWED away.

lagerstätte See EXCEPTIONAL FOSSIL DEPOSIT.

Lagomorpha An order of infraclass EUTHERIA, subclass THERIA, class MAMMALIA; the rabbits and hares. Range U. EOCENE–RECENT.

lagoon A body of water enclosed by a barrier or between a barrier and its associated coastline.

lahar A flow of volcanic debris and water, travelling at great speed, deposited as a poorly sorted mass.

Lahn-Dill iron deposit A SYNGENETIC, CONFORMABLE, iron-rich layer or lens, dominantly siliceous at the base and calcareous at the top. The principal MINERALS are HAE-MATITE, MAGNETITE, SIDERITE and LIMONITE. Probably of VOLCANIC-EXHALATIVE origin.

Lamé constant (λ) An ELASTIC CONSTANT equal to the BULK MODULUS less two thirds the SHEAR MODULUS.

lamellibranchs See BIVALVIA.

laminar flow A non-TURBULENT FLOW in which the mean flow velocity and instantaneous velocities at any point are exactly the same, characterized by the dominance of viscous forces over inertial forces, and REYNOLDS NUMBERS <500. Cf. TURBULENT FLOW.

laminar twinning See POLYSYNTHETIC TWINNING.

lamination A fine, discrete layer of rock 0.005–1.00 mm thick.

laminite A sediment with millimetric-scale LAMINATION, common in lake environments and useful in determining PALAEOCLIMATE.

lamp shell A colloquial name for the BRACHIOPODA.

lamprobolite See OXY-HORNBLENDE.

lamproite A potassium- and magnesium-rich MAFIC to ULTRAMAFIC ALKALINE LAMPRO-PHYRE-type rock of volcanic or HYPABYSSAL origin composed of Ti-rich PHLOGOPITE, CLI-NOPYROXENE, alkali AMPHIBOLE, OLIVINE, LEU-CITE and SANIDINE with ACCESSORY CHROME SPINEL, PRIDERITE, WADEITE, NEPHELINE, ILMENITE, SHCHERBAKOVITE, JEPPEITE, APATITE, PEROVSKITE, SPHENE and AMPHIBOLE.

lamprophyllite $(Na_3Sr_2Ti_3(Si_2O_7)_2 (O,OH,F)_2)$ A rare titanium NESOSILICATE found mainly in NEPHELINE SYENITES and their PEGMATITES.

lamprophyre A minor intrusion of MESO-CRATIC and MELANOCRATIC MINERAL composition containing BIOTITE or PHLOGOPITE $\pm$ AMPHIBOLE, with CLINOPYROXENE, OLIVINE and occasionally MELILITE in a GROUNDMASS of FELDSPARS or FELDSPATHOIDS.

lanarkite (Pb_2SO_5) A rare lead sulphate which occurs with ANGLESITE and LEADHILL-ITE in the oxidized zone of lead deposits.

Lancefieldian An ORDOVICIAN succession in Australia covering parts of the TREMADOC and early ARENIG.

land mammal age A unit based on the rich TERTIARY MAMMAL faunas of the Great Plains of North America, used to correlate continental deposits over long distances.

land plant A plant spending most of its life on land and commonly with a fluid-conducting vascular system. Such plants appear in the FOSSIL record from at least as early as the SILURIAN.

landfill See SANITARY LANDFILL.

Landon An OLIGOCENE succession in New Zealand comprising the WHAINGAROAN and DUNTROONIAN.

Landsat satellites A series of satellites used in REMOTE SENSING to provide images particularly useful for geological studies. Previously known as the ERTS SATELLITES.

landscape marble A LIMESTONE showing a pattern on a cut and polished surface similar to a landscape. Believed to arise from biogenic activity, the mixing of different coloured sediment by injection or the action of gas produced during the decay of bituminous sediments.

landslide (landslip) The rapid movement of a mass of soil downslope along a curved or planar FAILURE surface, without DEFOR-MATION of the SOIL STRUCTURE.

landslip See LANDSLIDE.

Lang topographic method An X-RAY DIF-FRACTION method for the direct imaging of microstructural CRYSTAL DEFECTS.

langbeinite $(K_2Mg_2(SO_4)_3)$ An EVAPORITE MINERAL.

Langhian A STAGE of the MIOCENE, 16.3–14.2 Ma.

langite $(Cu_4SO_4(OH)_6.2H_2O)$ A blue to green-blue, hydrated COPPER sulphate found as a SECONDARY MINERAL in COPPER deposits.

lansfordite ($MgCO_3.5H_2O$) An unstable, hydrated magnesium carbonate.

lapiés See KARREN.

lapilli PYROCLASTIC fragments between 2 mm and 64 mm in size.

lapis lazuli An ornamental stone comprising a mixture of LAZURITE, CALCITE, PYROXENES and other silicates.

Laplace's equation A relationship obeyed by all POTENTIAL FIELDS which states that in Cartesian coordinates the sum of the second derivatives of the field in three orthogonal directions is zero.

Laramide orogeny An OROGENY responsible for the formation of the Rocky Mountains of America in late CRETACEOUS and PALAEOCENE times.

lardalite See LAURDELITE.

Large Igneous Province See LIP.

large-ion lithophile (LIL) An element of large ionic radius and valency of 1 or 2 which becomes concentrated mainly in the potassium silicates of silicic melts during igneous FRACTIONATION.

larnite (Ca_2SiO_4) A rare MINERAL formed by the CONTACT METAMORPHISM of LIMESTONE.

larsenite ($PbZnSiO_4$) A rare OLIVINE.

larvikite (laurvikite) A variety of SYENITE in which the FELDSPAR shows blue IRIDESCENCE.

lateral ramp A RAMP that trends parallel to the transport direction of a THRUST SYSTEM.

lateral resistivity log An ELECTRIC GEOPHYSICAL BOREHOLE LOG which measures the RESISTIVITY of the formations to a considerable distance from the borehole.

lateral secretion The derivation of MINERAL-forming materials from the WALL ROCKS around a VEIN or other MINERAL deposit.

laterite A RESIDUAL DEPOSIT of iron and aluminium hydroxides formed by the WEATHERING of rocks in humid, tropical conditions.

lateritization The process whereby rock is converted to LATERITE by the extraction of SILICA.

laterolog An ELECTRIC GEOPHYSICAL BOREHOLE LOG which measures the RESISTIVITY of the formations within a thin circular disc around the borehole.

latite A VOLCANIC ROCK similar to CALC-ALKALINE ANDESITE in its QUARTZ content but with a higher K_2O content and K_2O/Na_2O ratio, so that PHENOCRYSTS of PLAGIOCLASE, BIOTITE ± SANIDINE are more abundant than PYROXENE and HORNBLENDE.

latitude correction The correction applied to GRAVITY data for the variation of GRAVITY with latitude using the GRAVITY FORMULA. Magnetic data require a GEOMAGNETIC CORRECTION for latitude and longitude.

Laue back-reflection method An X-RAY DIFFRACTION method in which the incident beam is directed at the sample through the centre of a flat piece of film. Back-diffracted radiation causes a pattern of spots on the film which can be interpreted with the aid of a net.

laumontite (($CaAl_2Si_4O_{12}$).$4H_2O$) A ZEOLITE found in the VEINS and cavities of many rock types, especially BASALTIC.

Laurasia The northern SUPERCONTINENT prior to the CONTINENTAL SPLITTING that formed the Atlantic Ocean, comprising North America, Greenland, Europe and Asia excepting Siberia and other blocks which joined during and after the TRIASSIC.

laurdelite (lardalite) A coarse-grained SYENITE resembling LARVIKITE but containing rhomb-shaped ALKALI FELDSPAR crystals and large NEPHELINES.

Laurentia A SUPERCONTINENT present during and subsequent to the VARISCAN OROGENY comprising the Canadian SHIELD, Greenland, parts of NW Europe and some other parts of North America.

laurvikite See LARVIKITE.

lautarite ($Ca(IO_3)_2$) Calcium iodate, found rarely in CALICHE.

lava Molten rock material at the surface.

lava cave A CAVE formed as molten rock solidified, possibly termed a PSEUDOKARST feature.

lava flow A dense mass of molten or partially molten rock moving as a stream on the surface.

lava levée A retaining wall of SCORIA at the side of a LAVA FLOW.

lava tube A hollow subsurface passage, up to 30m in width, 15 m in height and up to tens of kilometres long, in a solidified LAVA FLOW formed when LAVA withdrew from a distributory tunnel.

Law of accordant junctions (Playfair's law) 'Tributary rivers enter a main river at the same level as that river without any sudden drop.' Not always correct, as channel sizes may affect the form of the junction.

Law of Bravais 'The favoured low-energy crystal faces are those parallel to lattice planes with a high density of lattice points, as those planes cut the smallest area of each UNIT CELL and so break the smallest number of bonds in the structure.' A law governing CRYSTAL MORPHOLOGY.

Law of constancy of angles 'In a given compound the angles between corresponding faces (INTERFACIAL ANGLES) are always the same.'

Law of divides 'The nearer the divide, the steeper the slope, with all points on a single slope being interdependent.' A law explaining the smooth slopes in BADLAND areas.

Law of effective stress 'PORE FLUID PRESSURE reduces the NORMAL STRESS across a plane by the magnitude of the PRESSURE.'

Law of equal declivities 'Slopes on either side of a DIVIDE are interdependent.' An extension of the LAW OF DIVIDES, which explains why a BADLAND ridge always stands midway between two streams of equal elevation.

Law of mineral stability 'MINERALS are in thermodynamic equilibrium only in the environment in which they form.' An explanation for many EOGENETIC reactions.

Law of the Sea Convention An agreement that all deep-sea mining of the seafloor beyond 200 miles of coastal states would be under the jurisdiction of an INTERNATIONAL SEABED AUTHORITY, which might undertake such mining itself.

lawsonite $(CaAl_2(Si_2O_7)(OH)_2.H_2O)$ A SOROSILICATE commonly found in BLUESCHISTS.

Laxfordian orogeny An OROGENY affecting NW Scotland in the PRECAMBRIAN from 1800–1600 Ma.

layer silicate See PHYLLOSILICATE.

layer sulphides group A group of SULPHIDE MINERALS characterized by a STRUCTURE derived from the NICKEL ARSENIDE GROUP by the omission of a complete cation layer, with metal atoms in octahedral coordination and octahedra in the same layer sharing edges.

layer-parallel shear SHEAR parallel to the layering in the folding of a layered bed.

layer-parallel shortening SHORTENING parallel to the layering in the formation of CHEVRON FOLDS from KINK BANDS and BOX FOLDS.

layered anorthosite complex A very extensive intrusion, often with a retained CUMULATE TEXTURE and rich in CHROMITE, with much more abundant ANORTHOSITE and LEUCOGABBRO than normal for a basic intrusion. Unique to the GRANULITE-GNEISS BELTS of the ARCHAEAN.

layered igneous rock An IGNEOUS ROCK which displays MINERALOGICAL and/or chemical layering.

layover A type of distortion occurring in SIDE-SCAN SONAR or RADAR images of high relief terrains, where the top of a steep feature maps nearer the sensor than its base.

lazulite $((Mg,Fe^{2+})Al_2(PO_4)(OH)_2)$ A deep blue, strongly PLEOCHROIC, hydrated aluminium-magnesium-iron-(calcium) phosphate found in high-GRADE, aluminous METAMORPHIC ROCKS and GRANITE PEGMATITES.

lazurite $((Na,Ca)_8(AlSiO_4)_6(SO_4,S,Cl)_2)$ A rare TECTOSILICATE found in CONTACT META-MORPHOSED LIMESTONES.

leaching A process of PEDOGENESIS in which soluble MINERALS are removed from the soil.

lead-glance See GALENA.

lead-lead dating A RADIOMETRIC DATING method based on the proportion of radio-genic ^{207}Pb and ^{206}Pb, the former of which accumulates six times more rapidly than the latter.

leader A thin mineralized VEIN related to the main ORE-carrying VEIN and which aids its discovery.

leadhillite $(Pb_4SO_4(CO_3)_2(OH)_2)$ An hydrated lead carbonate and sulphate found in association with lead ORES.

leaky transform fault A TRANSFORM FAULT across which there is a component of EXTENSION so that movement along it forms a gap up which MANTLE material may penetrate.

lean ore A low GRADE ORE.

lebensspur See TRACE D'ACTIVITÉ ANIMALE.

lechatelierite (SiO_2) Natural fused SILICA or SILICA GLASS, the main constituent of FULGURITES.

lectotype See TYPE SPECIMEN.

lee side The downstream side of a body sheltered from the dominant flow direction. Cf. STOSS SIDE.

lee slope The downstream slope of a body sheltered from the dominant flow direction. Cf. STOSS SLOPE.

Lehmann discontinuity The boundary between the inner and outer CORE.

Leighton-Pendexter classification A largely outmoded classification scheme for LIMESTONES and DOLOSTONES.

Leiner A PERMIAN succession in NW Europe covering the upper part of the WORDIAN.

lenad A term for the FELDSPATHOIDS LEUCITE and NEPHELINE.

Lenan A CAMBRIAN succession in Siberia covering part of the ATDABANIAN, the LENIAN and part of the SOLVAN.

Lenian A STAGE of the CAMBRIAN, 554.0–536.0 Ma.

lenticular bedding A type of HETERO-LITHIC BEDDING characterized by lenses and ripples of SAND in a MUD MATRIX.

Leonard A PERMIAN succession in the Dela-ware Basin, USA covering part of the SAK-MARIAN, the ARTINSKIAN, KUNGURIAN and UFIMIAN.

lepidoblastic texture A METAMORPHIC TEXTURE comprising FOLIATION or SCHISTOSITY.

lepidocrocite $(\gamma FeO.OH)$ A SECONDARY MINERAL of iron, a POLYMORPH of GOETHITE.

lepidolite (lithia mica) $(K(Li,Al)_{2-3}(AlSi_3O_{10})(O,OH,F)_2)$ A pink or lilac, lithium-bearing MICA found in GRANITIC rocks and PEGMATITES.

lepidomelane An iron-rich variety of BIO-TITE found in IGNEOUS ROCKS.

Lepidosauria A subclass of class REPTILIA; the 'scaly' REPTILES, including the LIZARDS and SNAKES. Range U. PERMIAN–RECENT.

lepisphere A spherical, MICROCRYSTALLINE AGGREGATE of bladed crystals formed during the transformation of OPAL to QUARTZ.

Lepodocystoidea A class of phylum ECHINODERMATA close to the common ancestry of the CRINOIDEA and CYSTOIDEA. Range L. CAMBRIAN.

Lepospondyli A subclass of class AMPHIBIA; a group of extinct forms whose relation-ships are uncertain. Range TRIASSIC–PERMIAN.

leptite (leptynite) An EQUIGRANULAR METAMORPHIC ROCK comprising QUARTZ and FELDSPARS.

Leptostraca An order of subclass PHYLLO-CARIDA, class MALACOSTRACA, subphylum CRUSTACEA, phylum ARTHROPODA; small, marine CRUSTACEANS with a bivalved carapace and an abdomen of seven somites. Range RECENT.

leptynite See LEPTITE.

Letna An ORDOVICIAN succession in Bohemia covering part of the COSTONIAN and the HARNAGIAN.

leucite ($KAlSi_2O_6$) A FELDSPATHOID common in potassium-rich LAVAS.

leucitite A fine-grained, often PORPHYRITIC LAVA or minor intrusion composed of LEUCITE (30–50%) and CLINOPYROXENE with ACCESSORY NEPHELINE, iron-titanium oxides and APATITE. Occurs in continental RIFTS, ISLAND ARCS and other complex, post-TECTONIC, continental settings.

leucitophyre A fine-grained IGNEOUS ROCK, commonly a LAVA, comprising PHENOCRYSTS of LEUCITE and other MINERALS in a TRACHYTIC GROUNDMASS.

leuco- Of lighter colour.

leucocratic Descriptive of a rock with 0–30% dark MINERALS.

leucogabbro A GABBRO with a predominance of FELSIC MINERALS.

leucosapphire See WHITE SAPPHIRE.

leucosome A centimetric- to metric-scale, coarse-grained, quartzofeldspathic VEIN found in PELITIC and PSAMMITIC METAMORPHIC ROCKS, which may represent a MIGMATITIC product of a low melting point liquid segregated from the sediment during high-GRADE METAMORPHISM.

leucoxene A fine-grained, yellow-brown ALTERATION product of titanium-rich MINERALS, made up principally of RUTILE.

levée A raised bank beside a terrestrial or subaqueous channel formed by the rapid deposition of sediment from water escaping the channel, which fines and thins away from it. Levées allow water to rise above the level of the FLOODPLAIN, and when breached major flooding or channel AVULSION occurs.

Levy-Mises equations A fundamental law of PLASTICITY in which STRAIN RATES are proportional to DEVIATORIC STRESS.

levyne See LEVYNITE.

levynite (levyne) ($(Na,Ca)_2(Al,Si)_9$ $O_{18}.8H_2O$) A ZEOLITE found in cavities in BASALT.

Lewisian A division of the PRECAMBRIAN in Scotland overlain by the TORRIDONIAN and affected by the SCOURIAN and LAXFORDIAN events.

Lg wave An S WAVE of high amplitude trapped in the CRUST, which acts as a waveguide.

lherzolite A PHANERITIC rock similar to PERIDOTITE which contains 40–90% OLIVINE, >5% ORTHOPYROXENE and >5% CLINOPYROXENE with ACCESSORY PLAGIOCLASE, SPINEL ± GARNET. It yields BASALTIC MAGMA on PARTIAL MELTING, so is taken as a model for the composition of the UPPER MANTLE. Occurs in some OPHIOLITES and as XENOLITHS in ALKALI BASALTS, LAMPROITES and KIMBERLITES.

Liangshan A PERMIAN succession in China equivalent to the ASSELIAN.

Lias A term for the L. JURASSIC.

Liben An ORDOVICIAN succession in Bohemia covering the MID and LATE LLANDEILO and part of the COSTONIAN.

libethenite (Cu_2PO_4OH) Hydrated COPPER phosphate, found rarely in the oxide zones of metalliferous VEINS.

libolite A PITCH-like member of the ASPHALTITE group.

lichenometry A relative or absolute dating method for the exposure of a surface using the concentric growths of long-lived lichen, often used in the dating of GLACIAL deposits.

liddicoatite ($Ca(Li,Al)_3Al_6B_3Si_6O_{27}$ $(O,OH)_3(OH,F)$) A calcium-rich form of TOURMALINE.

life assemblage See BIOCENOSIS.

lift force The force experienced by a sediment grain, in a direction at right angles to the flow direction, generated by pressure differences over its surface.

light oil Oil with an API GRAVITY >30°.

light red silver ore See PROUSTITE.

light ruby silver See PROUSTITE.

lignite A soft, low RANK, earthy, brown-black COAL, sometimes with a massive SAP-ROPELIC form but more commonly composed of humic material with wood and plant remains in a finer-grained, organic GROUNDMASS.

LIL See LARGE-ION LITHOPHILE.

Lillburnian A MIOCENE succession in New Zealand equivalent to the SERRAVALLIAN.

limburgite An alkali-rich and/or SILICA-UNDERSATURATED, VOLCANIC ROCK or minor intrusion made up of OLIVINE and PYROXENE crystals in BASALTIC GLASS. Originally, the name given to alkali-poor KOMATIITES from southern Africa.

lime (CaO) A substance produced by the calcining of high purity LIMESTONE, with a wide variety of industrial uses.

limestone A rock comprising >50% calcium carbonate, since the CAMBRIAN partly or wholly of biogenic origin.

limestone pavement A glacially stripped platform of LIMESTONE dissected into blocks and runnels by solutional WEATHERING.

limiting depth The maximum depth at which the top of an anomalous body could lie and still give rise to an observed GEO-PHYSICAL ANOMALY.

limnic Descriptive of the environment of a freshwater lake.

limnic basin A freshwater COAL BASIN formed in a river DELTA in an intracratonic lake. Cf. PARALIC BASIN.

limnology The study of lakes.

limonite (brown iron ore) (FeO.OH. $n\mathrm{H_2O}$) A general term for a hydrated IRON OXIDE MINERAL.

linarite ($\mathrm{PbCuSO_4(OH)_2}$) Deep blue, hydrated lead-COPPER sulphate found as a SECONDARY MINERAL in the oxidized zone of COPPER and lead deposits.

line crystal defect A DISLOCATION in a CRYSTAL STRUCTURE. See also EDGE DISLO-CATION, SCREW DISLOCATION, STACKING FAULT.

line mapping Detailed geological mapping at scales of 1:2500 or more, used for complex areas, in which offsets are taken to an EXPOSURE from a surveyed base line.

lineage zone A BIOSTRATIGRAPHIC ZONE based on an evolutionary succession of species of a particular genus, so that STRATI-GRAPHIC gaps are precluded.

lineament A major, linear, topographic feature of regional extent of structural or volcanic origin, most easily appreciated from REMOTE SENSING data, e.g. a FAULT system.

lineation A repeated or PENETRATIVE linear STRUCTURE in a rock mass. Commonly used for a METAMORPHIC FABRIC of non-specific genesis, but also applied to features of FAULT PLANES, FOLDS, elongate crystal alignments in IGNEOUS ROCKS and CURRENT LINEATIONS, etc. in SEDIMENTARY ROCKS.

linguoid Shaped like a tongue.

linnaeite ($\mathrm{Co_3S_4}$) A cobalt ORE MINERAL with SPINEL structure.

linsen A SAND lens of LENTICULAR BEDDING in MUD.

LIP (Large Igneous Province) Massive crustal emplacement of predominantly MAFIC EXTRUSIVE and INTRUSIVE rocks which originated from MANTLE PLUMES.

liparite See RHYOLITE.

Lipopterna An order of infraclass EUTHERIA, subclass THERIA, class MAMMALIA; South American ungulates from rabbit- to camel-size, whose nostrils sit well back in the skull, suggesting the presence of a trunk. Range L. EOCENE–RECENT.

liptite A rare MICROLITHOTYPE OF COAL composed of EXINITE MACERALS.

liptobiolith A FOSSIL gum or resin, e.g. AMBER.

liptodetrinite A COAL MACERAL of the EXIN-ITE group composed of fragments of ALGIN-ITE, CUTINITE, RESINITE and SPORINITE.

liquation deposit An oxide-sulphide deposit forming as the result of LIQUID IMMISCIBILITY in a mixed sulphide-silicate MAGMA. The dense sulphides form globules that sink and accumulate at the base of the MAGMA.

liquefaction The sudden loss of SHEAR RESISTANCE associated with the collapse of the grain-supported framework in an UNDERCONSOLIDATED SEDIMENT. This collapse causes a temporary increase in the PORE FLUID PRESSURE as the suspension resediments from the base upwards until a more tightly packed, grain-supported state is achieved. PORE FLUIDS escape during this process, which may be initiated by EARTHQUAKES or other types of shock.

liquefied flow A type of SEDIMENT GRAVITY FLOW which is kept in motion by the buoyancy imparted to the particles by the escape of PORE FLUID.

liquefied natural gas (LNG) METHANE liquefied at −160°C and one atmosphere pressure, reducing its volume by a factor of over 600.

liquefied petroleum gas (LPG) Liquefied propane and BUTANE extracted from WET GAS.

liquid immiscibility The separation of an homogeneous liquid into two contrasting liquids. Responsible for the diversification of some iron-rich BASALTIC and highly ALKALINE MAGMAS and in the formation of LIQUATION DEPOSITS.

liquid limit An ENGINEERING GEOLOGY term for the minimum amount of water required to be mixed with sediment so that it will flow under standard conditions. Cf. PLASTIC LIMIT.

liquidus The temperature above which a MINERAL ASSEMBLAGE is entirely liquid. Cf. SOLIDUS.

Lissamphibia A subclass of class AMPHIBIA; the frogs, toads and APODA. Range TRIASSIC–RECENT.

listric Smoothly curving.

listric fan A set of SYNTHETIC, HORSETAIL LISTRIC FAULTS produced in an EXTENSIONAL FAULT SYSTEM as the SOLE FAULT migrates into the FOOTWALL.

listric fault A NORMAL FAULT whose DIP decreases downwards and whose HANGING WALL rotates as the FAULT slips.

lit-par-lit injection The injection of MAGMA or fluid along BEDDING, CLEAVAGE or SCHISTOSITY planes to produce MIGMATITE.

Litenschichten A SILURIAN succession in Bohemia covering the RHUDDANIAN, AERONIAN, TELYCHIAN, SHEINWOODIAN and part of the HOMERIAN.

litharge (γPbO) A red OXIDE MINERAL of lead.

lithia emerald See HIDDENITE.

lithia mica See LEPIDOLITE.

lithic Formed of rock.

lithic block A PYROCLASTIC fragment of initially solid rock over 64 mm in size.

lithic tuff A TUFF in which rock fragments are predominant.

lithiophilite (Li(Mn,Fe)PO$_4$) A phosphate MINERAL occurring in PEGMATITES.

litho- Pertaining to rock.

lithoclast A CLAST of an origin not associated with the main depositional system. Cf. INTRACLAST.

lithofacies A body of SEDIMENTARY ROCK characterized by specific and distinctive physical and chemical characteristics. Cf. BIOFACIES.

lithographic stone An ultra-fine-grained LIMESTONE suitable for printing plates.

lithographic texture A TEXTURE resembling a LITHOGRAPHIC STONE.

lithology A description of the MACROSCOPIC features of a rock type.

lithophile element (oxyphile element) An element with an affinity for oxygen, which thus occurs as an oxide or silicate rather than a sulphide or NATIVE ELEMENT.

lithophysa A centimetric-scale, rounded

mass found in GLASSY, and partly crystalline, FELSIC VOLCANIC ROCKS comprising concentric shells separated by voids. Possibly formed during rapid CRYSTALLIZATION alternating with gas expansion or by the chemical ALTERATION of SPHERULITES, with which they are found.

lithophysae Concentric shells of APHANITIC material encircling hollow cores which make up centimetric-scale masses in GLASSY ROCKS.

lithosome A sediment body deposited under uniform physical and chemical conditions.

lithosphere The upper shell of the solid Earth comprising the CRUST and UPPER MANTLE (VISCOSITY $>10^{21}$ Pa s) which deforms in a BRITTLE fashion when subjected to a STRESS of ~100 MPa. Its base is at a depth of 2–3 km under OCEAN RIDGES, increasing to up to 180 km beneath old OCEANIC CRUST. Beneath CRATONIC areas it is at least 250 km thick and possibly as much as 500 km.

lithostatic pressure (lithostatic stress, load pressure, load stress) The vertical STRESS due to the weight of overlying rocks, i.e. ρgz, where ρ = bulk DENSITY of overlying rock and any contained fluids, g = GRAVITY and z = depth. Cf. HYDROSTATIC PRESSURE.

lithostatic stress See LITHOSTATIC PRESSURE.

lithostratigraphy The subdivision and correlation of sequences of STRATA by means of rock type. Much less reliable than BIOSTRATIGRAPHY as rocks rarely persist laterally for great distances, but can be used in the absence of FOSSILS.

lithotypes of banded coal The four basic constituents of BANDED COAL that can be used for identification in hand specimen and classification: CLARAIN, DURAIN, FUSAIN and VITRAIN.

Little Ice Age The period from 1550–1850 AD with extended cold seasons and GLACIER expansion.

littoral The BEACH environment.

liver opal (menilite) OPAL with a colour resembling liver.

lizardite $(Mg_3Si_2O_5(OH)_4)$ A SERPENTINE MINERAL.

Lizzie A CARBONIFEROUS/PERMIAN succession in Queensland, Australia, covering parts of the U. CARBONIFEROUS and ASSELIAN.

Llandeilo An EPOCH of the ORDOVICIAN, 468.6–463.9 Ma.

Llandovery The lowest EPOCH of the SILURIAN, 439.0–430.4 Ma.

Llanvirn An EPOCH of the ORDOVICIAN, 476.1–468.6 Ma.

LNG See LIQUEFIED NATURAL GAS.

load cast (load structure) A type of SOLE MARK formed during WET-SEDIMENT DEFORMATION when, for example, SAND is deposited on MUD, into which it presses to leave bulbous projections.

load pressure See LITHOSTATIC PRESSURE.

load stress See LITHOSTATIC PRESSURE.

load structure See LOAD CAST.

loam A soil containing approximately equal proportions of SAND, SILT and CLAY.

local magnitude An EARTHQUAKE MAGNITUDE scale for crustal events within 600 km of the SEISMOGRAPH, based on wave amplitudes and a simple correction for distance.

Lochkovian **1** A STAGE of the DEVONIAN, 408.5–396.3 Ma. **2** A DEVONIAN succession in the former Czechoslovakia covering the LOCHKOVIAN and part of the PRAGIAN.

Lochkovium A DEVONIAN succession in Bohemia equivalent to the LOCHKOVIAN.

Lockportian A SILURIAN succession in North America covering part of the SHEINWOODIAN, the HOMERIAN and part of the GORSTIAN.

lode A mineralized body resulting from the extensive REPLACEMENT of pre-existing COUNTRY ROCK. VEIN is now the preferred term for all such bodies, regardless of their genesis.

lodestone   A stone rich in MAGNETITE which aligns in the GEOMAGNETIC FIELD, used in the past for determining the MAGNETIC NORTH direction.

loellingite (löllingite)   (FeAs$_2$) A SULPHIDE MINERAL of the DISULPHIDE GROUP found mainly in MESOTHERMAL VEIN DEPOSITS or PEGMATITES.

loess   SILT of AEOLIAN derivation, often forming extensive, thick deposits.

Logan Canyon   A CRETACEOUS succession on the Scotian shelf of Canada covering the APTIAN, ALBIAN and part of the CENOMANIAN.

logan stone   A large, exposed BOULDER, balanced so that it is easily rocked.

löllingite   See LOELLINGITE.

Lomnitz law   A law controlling CREEP, $\varepsilon_1(t) = A[(1 + at)^\alpha - 1]$, where $\varepsilon_1(t) =$ TRANSIENT CREEP, $t =$ time and A and α are constants for a particular RHEOLOGY.

long river profile   The elevation of a river channel plotted against distance downstream.

longitudinal coast   See PACIFIC-TYPE COAST.

longitudinal joint (bc joint)   A JOINT parallel to a FOLD AXIS and perpendicular to the folded layer at the FOLD HINGE.

longshore bar   See SWASH BAR.

longshore drift   The shore-parallel transport of sediment in one direction as the result of oblique WAVE action.

Longtanian   A STAGE of the PERMIAN, 250.0–247.5 Ma.

Longvillian   A STAGE of the ORDOVICIAN, 449.7–447.1 Ma.

Longwangmiao   A CAMBRIAN succession in China covering parts of the ATDABANIAN and LENIAN.

lonsdaleite   (C) A hexagonal POLYMORPH of DIAMOND found in IRON METEORITES.

looping   A method of transferring the absolute value of GRAVITY from a location where it is known to a new BASE STATION prior to a GRAVITY SURVEY. Consecutive alternate readings are taken at the two locations in order to account for the DRIFT of the GRAVIMETER taking the measurements.

loparite   A niobium-bearing variety of PEROVSKITE.

lopolith   A saucer-shaped igneous intrusion with upper and lower surfaces that are concave upwards, usually of MAFIC composition.

losing stream   A stream which loses water when flowing across permeable rocks. Cf. GAINING STREAM.

loughlinite   (Na$_2$Mg$_3$Si$_6$O$_{16}$.8H$_2$O) A poorly known MINERAL found in thin VEINS.

Louisiana   A DEVONIAN succession in the USA equivalent to the FAMENIAN.

Love wave   A SEISMIC SURFACE WAVE in which particle motion is horizontal and perpendicular to the direction of propagation. The wave travels by MULTIPLE REFLECTION in the surface low velocity layer.

low angle fault   See LAG FAULT.

low quartz   See ALPHA QUARTZ.

low velocity zone (LVZ)   A zone at the top of the ASTHENOSPHERE in oceanic areas in which P WAVES are slowed by ~10% compared with waves at higher levels. The zone is interpreted as a region where the melting point of MANTLE MINERALS is most closely approached so that there is about 0.1% PARTIAL MELTING.

low-grade metamorphism   METAMORPHISM at low to moderate temperature and pressure.

löweite   (Na$_{12}$Mg$_7$(SO$_4$)$_{13}$.15H$_2$O)   An hydrated sodium-magnesium sulphate; an EVAPORITE MINERAL.

Lowell-Guilbert Model   A type of PORPHYRY COPPER DEPOSIT in which a cylindrical zone of POTASSIC ALTERATION of the host STOCK is surrounded in turn by a phyllic zone of SERICITIZATION, a zone of INTERMEDIATE ARGILLIC ALTERATION and a zone of PROPYLITIC ALTERATION. The ORE may be in the host STOCK, COUNTRY ROCKS or both. Cf. DIORITE MODEL.

lower continental crust The CONTINENTAL CRUST between 10–12 km depth or the CONRAD DISCONTINUITY and the MOHOROVICIC DISCONTINUITY, formed of GRANODIORITIC rocks in the GRANULITE FACIES with a DENSITY of ~3.0 Mg m^{-3}. Cf. UPPER CONTINENTAL CRUST.

lower flow regime A FLOW REGIME characterized by increasing SHEAR STRESS on the bed as flow velocity increases. It is accompanied by RIPPLE formation and then DUNES, causing FLOW SEPARATION and thus an increased resistance to flow. Cf. UPPER FLOW REGIME.

Lower Greensand A CRETACEOUS succession in England covering the APTIAN and part of the ALBIAN.

lower mantle The MANTLE below 700 km down to the GUTENBERG DISCONTINUITY, characterized by very uniform physical properties. Probably composed of close-packed oxides with PEROVSKITE structure. Cf. UPPER MANTLE.

lower-stage plane bed A flat sediment bed forming at low SHEAR STRESSES from SAND coarser than 0.6 mm diameter and replacing the RIPPLES of finer sediment.

löwigite See ALUNITE.

LPG See LIQUEFIED PETROLEUM GAS.

LREE Light RARE EARTH ELEMENT.

Ludfordian A STAGE of the SILURIAN, 415.1–410.7 Ma.

Ludlow An EPOCH of the SILURIAN, 424.0–410.7 Ma.

Lugeon test See PACKER TEST.

Luisan A MIOCENE succession on the west coast of the USA covering parts of the LANGHIAN and SERRAVALLIAN.

lujavrite Textural variety of KAKORTOKITE

distinguished by slender PLAGIOCLASE and ACICULAR PYROXENE.

luminescence The emission of light by a material when irradiated by electromagnetic radiation of different wavelength.

lump ore See DIRECT-SHIPPING ORE.

lunate Crescent shaped.

lunette An arcuate DUNE ~20 m in height formed on the LEE SIDE of a DEFLATED LAGOON, lake BASIN or river bed in semi-arid areas.

Luning A TRIASSIC succession in Nevada, USA equivalent to the CARNIAN.

lustre A MINERAL property caused by the interference of light with the MINERAL surface. See ADAMANTINE LUSTRE, GREASY LUSTRE, PEARLY LUSTRE, RESINOUS LUSTRE, SILKY LUSTRE, VITREOUS LUSTRE.

lutaceous Formed from mud.

Lutetian A STAGE of the EOCENE, 50.0–42.1 Ma.

lutite A fine-grained SEDIMENTARY ROCK in which SILT makes up one third to two thirds of the total.

luxullianite A rock formed by the TOURMALINIZATION of GRANITE with the addition of boron, comprising SCHORL, QUARTZ and corroded, reddened FELDSPAR.

luzonite (Cu_3AsS_4) A rare ORE MINERAL of COPPER found in low- to medium-GRADE COPPER deposits.

LVZ See LOW VELOCITY ZONE.

Lycopsida A class of division TRACHAEOPHYTA, kingdom PLANTAE; the club mosses and related plants. Range DEVONIAN–RECENT.

lysocline The level in the ocean, above the CARBONATE COMPENSATION DEPTH, separating well-preserved (above) from poorly preserved (below) assemblages of a given calcareous, microFOSSIL group.

M-discontinuity See MOHOROVICIC DIS-CONTINUITY.

Ma Okou A PERMIAN succession in China equivalent to the KUNGURIAN and ZECHSTEIN.

maar A type of TUFF RING in which the centre of the CRATER has been affected by down-faulting or sagging, so that it lies below the surrounding ground surface.

Maastrichtian The highest STAGE of the CRETACEOUS, 74.0–65.0 Ma.

maceral The basic organic constituent of COAL, recognizable at MICROSCOPIC scale and made up of the remains of plant materials existing at the time of PEAT formation. Three maceral groups are recognized on the basis of their physical appearance: VITRINITE, EXINITE and INERTINITE.

machair A coastal area of calcareous, sandy soils covered by rich grassland.

mackinawite $((Fe,Co,Ni,Cr,Cu)_{1+x}S)$ A SULPHIDE MINERAL of the LAYER SULPHIDES GROUP with tetragonal PbO structure.

macrinite A COAL MACERAL of the INERTIN-ITE group resembling FUSINITE but without cell structure.

macrocrystalline Descriptive of a material in which crystals are visible to the unaided eye. Cf. MICROCRYSTALLINE.

macrophyric Descriptive of the TEXTURE of a medium- to fine-grained IGNEOUS ROCK with PHENOCRYSTS >2 mm long. Cf. MICRO-PHYRIC.

macropore A structural void in a soil.

macroscopic Referring to a STRUCTURE or feature of a scale of kilometres to hundreds of kilometres. Cf. MESOSCOPIC, MICROSCOPIC.

maculose Spotted.

Madagascar aquamarine A strongly DICHROIC, blue variety of BERYL, valued as a GEM.

Madagascar topaz See CITRINE.

made ground An area of land that has been constructed by man, often using a LANDFILL of natural materials.

Madeira topaz A brown, heated AMETHYST.

madupite A SILICA-SATURATED LAMPROITE with SANIDINE rather than LEUCITE, often carrying DIAMONDS.

Maentwrogian A STAGE of the CAMBRIAN, 517.2–514.1 Ma.

mafic A general term for FERROMAGNESIAN MINERALS. Cf. FELSIC.

mafurite Ultra-potassic rock comprising KALSILITE, OLIVINE and AUGITE.

maghemite (γ-Fe_2O_3) An OXIDE MINERAL of the MAGNETITE series.

magma A melt, generally containing suspended crystals and dissolved gases or volatiles, formed by total or PARTIAL MELTING of solid CRUSTAL or MANTLE rocks. Comprises polymers of interconnected, disordered Si-O tetrahedra with cations such as magnesium, iron, calcium, sodium and potassium in loose coordination with the oxygens and may range in composition from ULTRAMAFIC through BASALTIC, ANDES-ITIC, DACITIC and RHYOLITIC. Magma diversification can take place through the processes of MAGMATIC DIFFERENTIATION (including FRACTIONAL CRYSTALLIZATION, LIQUID IMMISCIBILITY and VAPOUR TRANSPORT), ASSIMILATION or MAGMA MIXING.

magma chamber A subsurface accumulation of MAGMA.

magma mixing A mechanism of MAGMA diversification, generally on a local scale, involving the mixing of two MAGMAS of contrasting composition to form a hybrid intermediate in composition.

magma pressure The HYDROSTATIC PRESSURE created by MAGMA.

magmatic differentiation The separation of an initially homogeneous MAGMA into two or more MAGMAS of contrasting composition by the processes of FRACTIONAL CRYSTALLIZATION, LIQUID IMMISCIBILITY or VAPOUR TRANSPORT.

magmatic segregation deposit See ORTHOMAGMATIC SEGREGATION DEPOSIT.

magnesia (MgO) An industrial product derived from MAGNESITE.

magnesia alum See PICKERINGITE.

magnesian limestone A LIMESTONE with a small proportion of DOLOMITE.

magnesiochromite ($MgCr_2O_4$) An OXIDE MINERAL with SPINEL CRYSTAL STRUCTURE.

magnesioferrite ($MgFe_2O_4$) An OXIDE MINERAL with SPINEL CRYSTAL STRUCTURE.

magnesite ($MgCO_3$) A CARBONATE MINERAL, the source of MAGNESIA.

magnetic anisotropy The phenomenon whereby a FERROMAGNETIC particle is more easily magnetized in one direction than others, so deflecting the original direction in the direction of ANISOTROPY.

magnetic anomaly The variation in the GEOMAGNETIC FIELD arising from variation in the magnetic properties of underlying rocks.

magnetic coercivity The MAGNETIC FIELD required to reduce the magnetization of a FERROMAGNETIC MATERIAL to zero.

magnetic dating A relative or absolute dating method making use of the SECULAR VARIATION of the GEOMAGNETIC FIELD, POLARITY REVERSALS or APPARENT POLAR WANDER, which cause REMANENT MAGNETIZATION directions to vary with time. Comparison of directions for different regions at the same time will demonstrate if they have moved relative to each other since the magnetization was acquired.

magnetic domain A small (~1 μm) volume of a FERROMAGNETIC material within which electron spins are coupled to produce a unidirectional MAGNETIC FIELD within it.

magnetic elements See GEOMAGNETIC ELEMENTS.

magnetic epoch A time interval of constant GEOMAGNETIC FIELD polarity, normally longer than 10 000 years.

magnetic equator (aclinic line) A line joining the locations of zero magnetic INCLINATION, which approximately follows the geographic equator.

magnetic field The FORCE experienced by a unit positive MAGNETIC POLE at the point of measurement.

magnetic flux MAGNETIC INDUCTION multiplied by the area of the surface it cuts.

magnetic gradiometer An instrument comprising a pair of PROTON, CESIUM VAPOUR or FLUXGATE MAGNETOMETER sensors separated by a short distance in the vertical or horizontal planes, used to measure directly the vertical or horizontal gradient of the GEOMAGNETIC FIELD.

magnetic induction MAGNETIC FIELD multiplied by the MAGNETIC PERMEABILITY of the surrounding medium.

magnetic lineations See OCEANIC MAGNETIC ANOMALIES.

magnetic log A GEOPHYSICAL BOREHOLE LOG in which the GEOMAGNETIC FIELD is measured down a borehole to provide information on the presence of magnetic rocks.

magnetic meridian See MAGNETIC NORTH.

magnetic moment For a dipole, the product of the MAGNETIC POLE strength and the distance between the poles.

magnetic north (magnetic meridian) The direction to the Earth's magnetic north pole. Cf. TRUE NORTH.

magnetic observatory A fixed installation which continuously monitors all the MAGNETIC ELEMENTS and thus records DIURNAL and SECULAR VARIATION of the GEOMAGNETIC FIELD.

magnetic permeability A constant describing the magnetic properties of the medium separating the causative body of a MAGNETIC ANOMALY and its point of observation.

magnetic pole 1 The points at which the lines of MAGNETIC FORCE surrounding a dipole converge (single poles very rarely exist in isolation). **2** The Earth's magnetic poles are those points where the MAGNETIC INCLINATION is ±90°. They are displaced from the geographic poles and are not antipodal.

magnetic polarity The direction of the GEOMAGNETIC FIELD. Polarity is normal when in the present-day configuration and reversed when the north and south MAGNETIC POLES interchange.

magnetic potential The scalar quantity which provides the MAGNETIC FIELD of a source in any direction when differentiated in that direction.

magnetic pyrites See PYRRHOTITE.

magnetic quiet zone A region with few or no MAGNETIC ANOMALIES, particularly an area of OCEANIC CRUST with no MAGNETIC LINEATIONS due to an absence of GEOMAGNETIC POLARITY REVERSALS or the loss of original magnetizations.

magnetic reversal (geomagnetic reversal, polarity reversal) A change in the orientation of the GEOMAGNETIC FIELD in which the south MAGNETIC POLE becomes the north pole and vice versa.

magnetic spectrometer An instrument that can be attached to a TRANSMISSION ELECTRON MICROSCOPE which allows the energy

lost by inelastically scattered electrons to be measured. This is characteristic of the elements present and allows microanalyses to be made.

magnetic storm A severe type of DIURNAL VARIATION in the GEOMAGNETIC FIELD in which the field varies greatly in amplitude over short time periods. Caused by the arrival in the ionosphere of charged particles generated during sunspot activity.

magnetic survey A survey undertaken with a MAGNETOMETER on land, at sea or in the air to search for magnetic rocks and MINERALS, archaeological artefacts or buried ferrous metals.

magnetic susceptibility A measure of the magnetic behaviour of a material in a MAGNETIC FIELD; a dimensionless constant of proportionality in the relationship between INTENSITY OF INDUCED MAGNETIZATION and the MAGNETIZING FORCE of the inducing MAGNETIC FIELD.

magnetic variometer (variometer) An early form of MAGNETOMETER based on a small dipole suspended in the GEOMAGNETIC FIELD.

magnetite (Fe_3O_4) An OXIDE MINERAL with the SPINEL CRYSTAL STRUCTURE; the most common FERRIMAGNETIC MINERAL.

magnetizing force The phenomenon responsible for the creation of a MAGNETIC FIELD.

magnetochronology A GEOCHRONOLOGICAL sequence based on GEOMAGNETIC POLARITY REVERSALS.

magnetogram A recording of temporal variations of the GEOMAGNETIC ELEMENTS.

magnetohydrodynamics A field of physics concerned with the motions of electrically conducting fluids in the presence of MAGNETIC FIELDS. Used to model motion in the outer CORE and the origin of the GEOMAGNETIC FIELD.

magnetometer An instrument for measuring a MAGNETIC FIELD. FLUXGATE, PROTON and CESIUM VAPOUR MAGNETOMETERS are used in MAGNETIC SURVEYS to measure the GEO-

MAGNETIC FIELD. SPINNER and CRYOGENIC MAGNETOMETERS measure the fields associated with rock samples in PALAEOMAGNETIC studies.

magnetostratigraphy The use of geomagnetic POLARITY REVERSALS recorded in stratal sequences for the purposes of correlation, particularly useful in the absence of FOSSILS and for correlating between marine and terrestrial sequences.

magnetotelluric survey (MT survey) An ELECTROMAGNETIC INDUCTION METHOD of GEOPHYSICAL EXPLORATION making use of naturally occurring electromagnetic fields to investigate the distribution of ELECTRICAL CONDUCTIVITY in the subsurface. The DEPTH OF PENETRATION extends to some tens of km, so that the method has application in HYDROCARBON surveys.

magnitude See EARTHQUAKE MAGNITUDE.

magnitude-frequency relationship The relationship between EARTHQUAKE MAGNITUDE and the number of events of a given MAGNITUDE, which shows that the number increases with decreasing MAGNITUDE.

major element An element present in a rock in high concentration, so that it controls the presence of MINERALS such as FELDSPAR and PYROXENE. Cf. TRACE ELEMENT.

major fold The larger, MACROSCOPIC FOLD in a complexly folded region, generally with a wavelength of kilometric scale. Cf. MINOR FOLD.

malachite ($Cu_2CO_3(OH)_2$) A bright green CARBONATE MINERAL, often found in the oxidized parts of COPPER ORES.

Malacostraca A class of subphylum CRUSTACEA, phylum ARTHROPODA characterized by a carapace covering the head and part of the trunk, which comprises a distinct thorax and abdomen, and terminates in a telson. Range CAMBRIAN–RECENT.

Malakovian A TRIASSIC succession in New Zealand equivalent to the NAMMALIAN.

malignite A MESOCRATIC variety of NEPHELINE SYENITE.

malleable A TENACITY descriptor indicating a MINERAL that can be hammered into a thin sheet.

Malm The youngest EPOCH of the JURASSIC, 157.1–145.6 Ma.

Malvinokaffric Province A biogeographic province extant in SILURIAN times, including much of South America and adjacent parts of south and north Africa.

mammillated (mammillary) With the form of portions of spheres.

Mammalia/mammals A class of vertebrate animals that suckle their young, the FOSSILS of which are classified by their teeth. Range late TRIASSIC–RECENT.

mammillary See MAMMILLATED.

man-made earthquake An EARTHQUAKE induced by human activity as a result of mining, the construction of reservoirs, explosion AFTERSHOCKS and fluid injection into deep boreholes. All these factors change the regional STRESS pattern or add/redistribute PORE FLUIDS and thus can trigger events.

manganepidote See PIEMONTITE.

manganese nodule A CONCRETION of ferromanganese oxides on the ocean floor, forming by the extraction of metals from seawater and the PORE FLUIDS of seafloor MUDS. Generally 5–200 mm in size and spheroidal, ellipsoidal or BOTRYOIDAL in form, sometimes with a rock core or no core at all and usually with internal MINERAL zoning. Not presently of significant economic importance.

manganese spar See RHODOCHROSITE.

manganite (γ–$MnO(OH)$) An ORE MINERAL of manganese found mainly in low-temperature HYDROTHERMAL VEINS.

manganoan cummingtonite See TIRODITE.

manganophyllite A variety of PHLOGOPITE or BIOTITE rich in manganese.

manganosite (MnO) A green OXIDE MIN-

ERAL of manganese, which becomes black on exposure.

manganotantalite $((Mn,Fe)Ta_2O_6)$ An ORE MINERAL of tantalum found mainly in GRANITE PEGMATITES.

Mangaorapan An EOCENE succession in New Zealand covering the middle part of the YPRESIAN.

Mangaotanian A CRETACEOUS succession in New Zealand covering parts of the CENOMANIAN and TURONIAN.

Mangapanian A PLIOCENE succession in New Zealand covering part of the PIACENZIAN.

mangerite A CHARNOCKITIC rock equivalent to HYPERSTHENE MONZONITE.

mangrove swamp An intertidal region of mudflats and mangrove vegetation found along sheltered, low-energy shorelines in tropical areas.

manjiroite $((Na,K)Mn_8O_{16}.nH_2O)$ A SECONDARY ORE MINERAL of manganese.

Manning equation A formula for estimating stream velocity (v) from CHANNEL ROUGHNESS, slope (s) and HYDRAULIC RADIUS (R): $v = kRsn$, where k = constant (1 in SI units), n = Manning roughness coefficient of the stream bed.

mantle The inner shell of the Earth between the MOHOROVICIC and GUTENBERG DISCONTINUITIES, with a silicate MINERALOGY distinct from the CRUST above and CORE below. Its composition is probably equivalent to a mixture of ~75% DUNITE and ~25% BASALT. The uppermost mantle beneath oceanic areas is the location of the LOW VELOCITY ZONE. The MINERALOGY of the mantle changes with depth to denser MINERAL phases as the pressure increases (e.g. OLIVINE → SPINEL → PEROVSKITE) and this is responsible for the TRANSITION ZONE between 400 and 700 km.

mantle bedding A uniform thickness of PYROCLASTIC material over all but the steepest topography.

mantle drag The FORCE exerted on the base of the LITHOSPHERE by movement of the ASTHENOSPHERE. If the ASTHENOSPHERE moves at a higher velocity than the LITHOSPHERE, the latter's velocity is enhanced and vice versa. This is unlikely to be an important mechanism of PLATE movement as the LOW VELOCITY ZONE at the top of the ASTHENOSPHERE probably would not allow efficient coupling between LITHOSPHERE and ASTHENOSPHERE.

mantle plume A persistent column of hot material, in the form of a vertical cylinder with a radius of ~150 km, rising to the CRUST from the MANTLE, possibly originating by localized streaming from the CORE-MANTLE boundary. Responsible for HOTSPOTS.

mantle transition zone A layer in the MANTLE between 400 and 700 km across which there is an increase in SEISMIC VELOCITY due to phase changes of the MINERALS present to more closely packed, denser forms.

mantled gneiss dome A STRUCTURE comprising a variably oriented, dome-shaped core of GRANITIC GNEISS overlain by supracrustal metasedimentary and metavolcanic rocks. May originate by the DEFORMATION of an UNCONFORMITY, by intrusion of the core material into the cover or from the buoyant ascent of low DENSITY GRANITIC BASEMENT into denser cover rocks.

manto A horizontal to subhorizontal, tubular OREBODY, more rarely a TABULAR OREBODY.

Maozhuang A CAMBRIAN succession in China covering parts of the LENIAN and SOLVAN.

Maping A CARBONIFEROUS/PERMIAN succession in China equivalent to the KASIMOVIAN, GZELIAN and ASSELIAN.

marble A metamorphosed LIMESTONE formed by RECRYSTALLIZATION during THERMAL or REGIONAL METAMORPHISM. It may form an attractive BUILDING STONE, although usage of the term marble by stonemasons also encompasses unmetamorphosed LIMESTONES.

marcasite (white iron pyrites) (FeS_2) A

relatively common SULPHIDE MINERAL, occasionally used as a GEM.

March analysis The analysis of the behaviour of passive markers in an homogeneous body that is deforming by VISCOUS FLOW. Elongate markers originally parallel to the axis of SHORTENING rotate symmetrically from this direction by an amount dependent on the STRAIN to form a BIMODAL PREFERRED ORIENTATION.

marekanite A RHYOLITIC PERLITE which has been broken down to rounded PEBBLES.

margarite $(CaAl_2(Al_2Si_2O_{10})(OH)_2)$ A BRITTLE MICA.

marginal basin See BACK-ARC BASIN.

marginal ore ORE which just repays the cost of exploitation.

marginal sea See BACK-ARC BASIN.

marialite $(Na_4(AlSi_3O_8)_3(Cl_2,CO_3,SO_4))$ A METAMORPHIC MINERAL of the SCAPOLITE series.

marine abrasion platform See SHORE PLATFORM.

marine band An horizon of marine origin within a succession of non-marine STRATA and representative of a brief TRANSGRESSION. Constitutes a useful MARKER BED.

marine snow The discarded mucus feeding sheets or strands of gelatinous zooplankton such as of PTEROPODS, SALPS and appendicularians. This traps suspended particulate matter and transports it to the seabed, although it itself is not preserved.

marker bed A distinctive BED useful in constructing a LITHOSTRATIGRAPHY.

markfieldite A variety of DIORITE with PORPHYRITIC TEXTURE and a GRANOPHYRIC GROUNDMASS.

Markowitz wobble A systematic change in position of the Earth's axis of rotation of unknown origin with a periodicity of 30 years and an amplitude of 25 marcs.

Marl Slate The extensive basal deposit of the PERMIAN ZECHSTEIN SEA in NE England, equivalent to the KUPFERSCHIEFER in continental Europe.

marl A friable, calcareous MUDSTONE.

marlstone Indurated MARL.

marmorization The thermal RECRYSTALLIZATION of LIMESTONE to produce MARBLE.

Marsdenian A STAGE of the CARBONIFEROUS, 321.5–320.6 Ma.

marsh gas NATURAL GAS (METHANE), of no commercial importance, produced by bacterial alteration of organic matter near the Earth's surface.

Marshbrookian A STAGE of the ORDOVICIAN, 447.1–444.5 Ma.

Marsupiala/marsupials An order of infraclass EUTHERIA, subclass THERIA, class MAMMALIA; MAMMALS in which development of the young takes place in an external pouch. Range U. CRETACEOUS–RECENT.

martite HAEMATITE or an intergrowth of HAEMATITE and MAGNETITE which replaces MAGNETITE along CLEAVAGE planes.

maskelynite A GLASS of PLAGIOCLASE composition occurring in colourless, ISOTROPIC grains in METEORITES, probably representing re-fused FELDSPAR.

mass deficiency The difference in mass between a body of relatively low DENSITY (e.g. a sedimentary BASIN) and the relatively high DENSITY COUNTRY ROCK which would otherwise occupy its space. It can be estimated by a GRAVITY SURVEY.

mass excess The difference in mass between a body of relatively high DENSITY (e.g. an OREBODY) and the relatively low DENSITY COUNTRY ROCK which would otherwise occupy its space. It can be estimated by a GRAVITY SURVEY.

mass extinction The EXTINCTION of a large number of FOSSIL groups in a limited period of time, such as occurred at the K–T BOUNDARY, possibly as a result of BOLIDE IMPACT.

mass flow A SLIDE of sediment downslope under the force of GRAVITY.

mass movement (mass wasting) The movement of mass takes place by SLIDE or flow processes under the influence of GRAVITY. Flow processes include SOLIFLUCTION, SOIL CREEP, DEBRIS AVALANCHES, EARTH FLOWS and MUD FLOWS. SLIDE processes include ROCK FALLS, ROCK SLIDES, PLANAR SLUMPS and ROTATIONAL SLUMPS. Additionally there are FROST HEAVE, FREEZE-THAW and CAMBERING movements.

mass solute transfer A DIAGENETIC process of transport of dissolved species from reaction sites in donor sediments where DISSOLUTION occurs to reaction sites in receptor sediments where PRECIPITATION takes place.

mass spectrometry The determination of the ratios and/or concentrations of isotopes in rocks or other materials. Ionized atoms of the sample are accelerated by a voltage through a MAGNETIC FIELD which curves their path by an amount dependent on ionic mass. Variation of the voltage allows ions of different masses to be focused on a detector. Particularly important in RUBIDIUM-STRONTIUM, POTASSIUM-ARGON and ACCELERATOR RADIOCARBON DATING and increasingly used for the determination of most MAJOR and TRACE ELEMENTS.

mass wasting See MASS MOVEMENT.

massicot (PbO) A rare yellow lead oxide, found as a SECONDARY MINERAL associated with GALENA.

massive sulphide deposit (volcanic-associated massive sulphide deposit) A large, usually STRATIFORM, CONFORMABLE OREBODY composed mainly of iron sulphide, usually PYRITE ± PYRRHOTITE, along interfaces between volcanic units or volcanic units and sediments. Usually underlain by a STOCKWORK which acted as the feeder for the mineralizing fluids. Probably SYNGENETIC in origin, forming by VOLCANIC-EXHALATIVE processes, such as at BLACK SMOKERS.

master joint A JOINT whose extent is considerably greater than others in the set and against which less prominent JOINTS terminate.

Mata A CRETACEOUS succession in New Zealand comprising the HAUMURIAN and PIRIPAUAN.

matrix 1 See GROUNDMASS. **2** The fine-grained material separating CLASTS in a SEDIMENTARY ROCK.

matrix-supported conglomerate A RUDITE with 30–85% CLASTS which are commonly not in contact. Cf. CLAST-SUPPORTED CONGLOMERATE.

maturation The processes whereby organic matter is transformed into oil and gas.

Matuyama A MAGNETOSTRATIGRAPHIC EPOCH of REVERSED POLARITY in the PLEISTOCENE, 2.42–0.71 Ma.

Mauretanian Orogeny An OROGENY affecting NW Africa in DEVONIAN times.

Maxillopoda A class of subphylum CRUSTACEA, phylum ARTHROPODA including the barnacles. Range U. SILURIAN–RECENT.

maximum octahedral shear failure criterion A FAILURE CRITERION in which FAILURE occurs when the octahedral SHEAR STRESS reaches a constant value dependent on the material.

maximum projection sphericity The ratio of the maximum and minimum cross-sectional areas of a particle, used in the description of the CLASTS of a SEDIMENTARY ROCK.

maximum strain energy of distortion failure criterion A FAILURE CRITERION in which FAILURE occurs when the STRAIN energy of distortion reaches a constant value dependent on the material.

Maxwell model See ELASTOVISCOUS DEFORMATION.

Maxwell substance A substance in which the STRAIN RATE is equal to the ratio of SHEAR STRESS to three times the VISCOSITY.

Mayan A CAMBRIAN succession in Siberia covering part of the MENEVIAN.

Mayen lava (Andernach lava, Niedermendig lava) A highly VESICULAR, grey

NEPHELINE TEPHRITE from the Mayen region of western Germany, extensively quarried for MILLSTONES from the Bronze Age to the 19th century.

Mazon Creek An EXCEPTIONAL FOSSIL DEPOSIT of U. CARBONIFEROUS age in Ilinois, USA containing many species of plants and animals preserved in SIDERITE NODULES.

meander belt An area of land occupied by a meandering channel between AVULSIONS. Submarine channels may also meander.

meander scroll The topography, comprising low, curved ridges of relatively coarse sediment parallel to a river channel, of exhumed POINT BARS resulting from differential EROSION of BEDS in a truncated EPSILON CROSS-STRATIFIED complex.

meandering stream A stream with planform SINUOSITY >1.3, occurring in alluvial, submarine and tidal environments.

measured reserve See ORE RESERVE.

mechanical infiltration of fines The introduction of CLAY and SILT grade sediment into coarse-grained, porous, permeable sediment where surface SEEPAGE accompanies ALLUVIATION in areas of low WATER TABLE, thus giving rise to a secondary MATRIX.

median valley See AXIAL RIFT.

medium oil Oil with an API GRAVITY of 22°–30°.

meerschaum See SEPIOLITE.

mega- Very large.

megablast A large PORPHYROBLAST in a coarse-grained IGNEOUS ROCK.

megabreccia A very coarse BRECCIA in which the CLASTS may exceed 1 km in length, which may have been formed by a LANDSLIDE.

megacryst A crystal in an IGNEOUS ROCK which is large compared to those in the MATRIX.

megaripple A RIPPLE with a wavelength >1 m.

megaturbidite (seismoturbidite) A very thick-bedded TURBIDITE >1 m in thickness, which is commonly laterally continuous over an entire depositional BASIN. Internal CROSS-STRATIFICATION or RIPPLE-LAMINATION may be developed, indicative of flow reflection from the BASIN margins. Believed to originate from very large volumes of sediment produced by FAILURE of the BASIN margin, perhaps in response to EARTHQUAKE activity.

meimechite An ULTRAMAFIC VOLCANIC ROCK with PHENOCRYSTS of OLIVINE.

meionite $(Ca_4(Al_2Si_2O_8)_3(Cl_2CO_3SO_4))$ A METAMORPHIC MINERAL of the SCAPOLITE series.

mela- Prefix attached to an IGNEOUS ROCK name signifying that it is of darker colour than usual.

melaconite A massive variety of TENORITE in the form of a black, earthy material formed by OXIDATION in COPPER VEINS.

mélange A metric- to kilometric-scale body of rock composed of chaotic blocks of COMPETENT STRATA in a finer-grained MATRIX. Sedimentary mélanges (OLISTOSTROMES) originate by avalanching, GRAVITY slumping or sliding; TECTONIC mélanges are the result of TECTONIC DEFORMATION in which SEDIMENTARY ROCK is dismembered.

melanite A black to dark brown, titanium-bearing ANDRADITE.

melanocratic Descriptive of a rock with 60–90% dark MINERALS.

melanosome A dark-coloured band rich in MAFIC and aluminous MINERALS, found between coarse-grained, quartzofeldspathic VEINS in REGIONALLY METAMORPHOSED PELITES and PSAMMITES. Represents a layer which has been highly shortened by DISSOLUTION or melting as the QUARTZ and FELDSPAR of the parent rock were removed along non-PENETRATIVE CLEAVAGES caused by high SHEAR STRESS during METAMORPHISM, leaving the MAFIC MINERALS behind.

melanterite $(FeSO_4.7H_2O)$ A green-blue, SECONDARY MINERAL of iron, formed by the

WEATHERING of PYRITE, MARCASITE and cupriferous PYRITE ORES.

Melekesskian A STAGE of the CARBONIFEROUS, 313.4–311.3 Ma.

melilite A group of FELDSPATHOID MINERALS including GEHLENITE, AKERMANITE and SODA MELILITE.

melilitite An ULTRAMAFIC VOLCANIC ROCK comprising MELILITE and PYROXENE.

melitolite An ULTRAMAFIC PLUTONIC rock comprising MELILITE, PYROXENE and OLIVINE.

melteigite An IJOLITE with 70–90% MAFIC MINERALS.

member A grouping of BEDS used in LITHO-STRATIGRAPHY. See also BED, FORMATION, GROUP, SUPERGROUP.

membrane polarization (electrolytic polarization) A mechanism of INDUCED POLARIZATION caused by the varying mobility of ions in PORE FLUID travelling through small pores in a rock. Charge build-up on either side of pores gradually disperses when the polarizing current is removed, creating a decaying voltage.

membrane tectonics The DEFORMATION of a PLATE resulting from its movement from regions of different radii of curvature on the Earth's surface. Radial tensional STRESSES are generated when moving from small to large radii of curvature regions and compressional STRESSES in the opposite direction. Suggested as a possible, but unlikely, mechanism behind CONTINENTAL SPLITTING.

meneghinite $(CuPb_{13}Sb_7S_{24})$ A rare ORE MINERAL of COPPER and lead.

Menevian A STAGE of the CAMBRIAN, 530.2–517.2 Ma.

menilite See LIVER OPAL.

Meotic (Sarmatian) A MIOCENE succession on the Russian Platform covering the upper part of the MESSINIAN.

Meramec A CARBONIFEROUS succession in the USA covering the lower part of the VISÉAN.

Mercalli scale A twelve point scale of

EARTHQUAKE INTENSITY devised in 1902. See also MODIFIED MERCALLI SCALE.

mercury (Hg) A naturally occurring liquid NATIVE ELEMENT.

mere A small lake of uncertain origin on GLACIAL outwash deposits and other superficial materials, possibly the result of THERMOKARST or SOLUTION processes.

Merioneth The youngest EPOCH of the CAMBRIAN, 517.2–510.0 Ma.

Merions A DEVONIAN succession in E. Australia equivalent to the PRAGIAN.

merocrystalline Descriptive of a rock containing both crystals and GLASS.

merokarst A LIMESTONE landscape with only partially developed KARST landforms.

meromictic The status of a lake which is permanently stratified with a well-developed EPILIMNION and HYPOLIMNION as a result of atmospheric disturbances being insufficient to break down the layering. Cf. HOLOMICTIC.

Merostomata A class of phylum ARTHROPODA including the king crabs and water scorpions. Range ORDOVICIAN–RECENT.

mesa A steep-sided, flat-topped PLATEAU or promontory surrounded by a flat EROSIONAL plain, forming as a result of PARALLEL RETREAT or protection from EROSION by a hard capping such as a CUIRASSE.

mesocratic Descriptive of a rock with 30–60% dark MINERALS.

mesogenesis See BURIAL DIAGENESIS.

mesolite $(Na_{16}Ca_{16}(Al_{48}Si_{72}O_{240}).64H_2O)$ A ZEOLITE found in cavities of VOLCANIC ROCKS.

Mesosauria An order of subclass ANAPSIDA, class REPTILIA; an aquatic group of REPTILES with long, needle-like teeth for catching small CRUSTACEANS. Range late CARBONIFEROUS–early PERMIAN.

mesoscopic Referring to a STRUCTURE or feature of a scale of metres. Cf. MACROSCOPIC, MICROSCOPIC.

mesosiderites An heterogeneous group of

STONY IRON METEORITES with a metal content possibly exceeding 40%, but not forming a continuous network.

mesosphere A largely outdated term for the mechanically strong layer beneath the ASTHENOSPHERE.

mesostasis The final fraction of a MAGMA to crystallize, in the spaces between existing crystals.

mesothermal deposit An EPIGENETIC MINERAL deposit, intermediate between EPI-THERMAL and HYPOTHERMAL DEPOSITS, formed at 200–300° and depths of 1200–4500 m and generally found associated with near-surface INTRUSIVE IGNEOUS ROCKS. Comprises FRACTURE fills and REPLACEMENT BODIES, often with well-developed ZONING.

mesotype An IGNEOUS ROCK with a COLOUR INDEX of 30–60.

Mesozoic An ERA comprising the TRIASSIC, JURASSIC and CRETACEOUS, 245.0–65.0 Ma.

mesozone A DEPTH ZONE of medium temperature and high HYDROSTATIC PRESSURE. See also EPIZONE, KATAZONE.

Messel Oil Shales An EXCEPTIONAL FOSSIL DEPOSIT of EOCENE age near Darmstadt, Germany containing a LACUSTRINE assemblage of articulated vertebrates, INSECTS and PLANTS.

Messinian The highest STAGE of the MIOCENE, 6.7–5.2 Ma, during which many EVAPORITE BASINS were developed in the Mediterranean region.

meta- Prefix indicating a metamorphosed variety.

meta-anthracite The highest RANK of ANTHRACITE, with at least 98% FIXED CARBON.

metabasite Any metamorphosed BASIC IGNEOUS ROCK.

metacinnabar $(Hg_{1-x}S)$ A high temperature form of CINNABAR.

metacryst See PORPHYROBLAST.

metal excess group A group of SULPHIDE MINERALS characterized by the presence of greater numbers of metal atoms than sulphur atoms.

metal factor parameter A measure used in the quantification of frequency domain INDUCED POLARIZATION data, defined as $2\pi10^5(\rho_0-\rho_\infty)/\rho_\infty\rho_0$, where ρ_0 and ρ_∞ correspond to APPARENT RESISTIVITIES measured at low and high alternating current frequencies respectively.

metalimnion The top of the HYPOLIMNION, a zone of rapid temperature change.

metallic lustre A LUSTRE similar to polished steel seen in some OPAQUE MINERALS.

metallogenic epoch A period of time during which there was abundant mineralization of the same type.

metallogenic province A region of the CRUST in which there is abundant mineralization of the same type.

metalloid A substance with both metallic and non-metallic properties, e.g. arsenic.

metaluminous Descriptive of an IGNEOUS ROCK in which Al_2O_3 exceeds $(CaO+Na_2O+K_2O)$.

metamorphic aureole (contact aureole) The zone of METAMORPHISM of the COUNTRY ROCKS around an intrusion, which may involve METASOMATISM as a result of heating GROUNDWATER as well as RECRYSTALLIZATION and the development of new MINERALS.

metamorphic banding Banding caused by the RECRYSTALLIZATION of original FABRICS to produce new planar FABRICS such as FOLI-ATION and GNEISSOSITY.

metamorphic core complex (core complex) Part of the lower continental CRUST adjacent to a sedimentary basin and owing its existence to ISOSTATIC uplift associated with crustal thinning.

metamorphic differentiation (metamorphic segregation) The separation of components or phases in a rock by META-MORPHIC processes.

metamorphic facies A group of rocks

that reached chemical equilibrium at the same pressure and temperature range of METAMORPHISM and characterized by particular MINERAL ASSEMBLAGES.

metamorphic grade The intensity of METAMORPHISM, an indicator of the temperature and pressure conditions extant.

metamorphic rock A rock which results from the partial or complete RECRYSTALLIZATION in the solid state under temperature and pressure conditions elevated with respect to the surface.

metamorphic segregation See META-MORPHIC DIFFERENTIATION.

metamorphic zone An area of METAMORPHIC ROCKS defined by the appearance of certain MINERAL ASSEMBLAGES.

metamorphism The processes by which rocks are changed by the solid-state application of heat, pressure and fluids, excluding WEATHERING and DIAGENESIS. See also AUTO-METAMORPHISM, DYNAMIC METAMORPHISM, REGIONAL METAMORPHISM, RETROGRADE META-MORPHISM, THERMAL METAMORPHISM.

metapedogenesis The human alteration of the properties of a soil, both deliberate and unintentional.

Metaphyta The plants. Range PRECAMBRIAN–RECENT.

metaquartzite A QUARTZITE formed by the METAMORPHISM of SANDSTONE. Cf. ORTHO-QUARTZITE.

metasilicate See CYCLOSILICATE.

metasomatism A METAMORPHIC process in which the chemical composition of a rock is changed significantly, usually as a result of fluid flow.

metastable Existing under conditions outside the normal range of stability.

Metatheria An infraclass of subclass THERIA, class MAMMALIA; the MARSUPIALS. Range late CRETACEOUS–RECENT.

Metazoa The animals. Range PRECAMBRIAN–RECENT.

meteoric diagenesis DIAGENESIS caused by rainfall-derived GROUNDWATER. Important in the formation of LIMESTONES as they are prone to subaerial exposure. ARAGONITE and magnesium CALCITE are unstable in METEORIC WATERS and are replaced by low-magnesium CALCITE.

meteoric water GROUNDWATER derived from rainfall or INFILTRATION.

meteorite An extraterrestrial body, derived from the Asteroid belt of the solar system, impacting the Earth's surface. Three main classes exist: IRON METEORITES, STONY IRON METEORITES and STONY METEORITES. These are subdivided into CHONDRITES and ACHONDRITES on the basis of the presence or absence respectively of CHONDRULES.

methane (CH_4) The lightest component of CRUDE OIL.

methane series (paraffin series) Straight-chain HYDROCARBONS with the general formula C_nH_{2n+2}, found in CRUDE OIL.

methanogenesis The production of METHANE during FERMENTATION.

meulière A SARSEN-like stone found in the TERTIARY of France.

Mexican onyx A TRANSLUCENT, veined, partly coloured variety of ARAGONITE.

mgal See MILLIGAL.

miarolitic cavity A small, crystal-lined cavity in an INTRUSIVE IGNEOUS ROCK resulting from the segregation of small gas pockets into an irregular cavity defined by surrounding crystals.

miarolitic fabric A FABRIC in an INTRUSIVE IGNEOUS ROCK formed by the alignment of MIAROLITIC CAVITIES.

miaskite A LEUCOCRATIC, BIOTITE NEPHELINE MONZOSYENITE.

mica schist A SCHIST rich in MICA, commonly MUSCOVITE.

micaceous Containing or resembling MICA.

micaceous iron ore A SPECULAR variety of HAEMATITE with a flaky HABIT reminiscent of the MICAS.

micas Sheet silicates characterized by a platy morphology and perfect basal CLEAVAGE in consequence of their atomic structure. The general formula is $X_2Y_{4-6}Z_8O_{20}(OH,F)$ where X is K or Na, Y is Al, Mg, Fe, Mn, Cr, Ti, Li, etc. and Z is Si, Al or Fe^{3+}. Common in IGNEOUS and METAMORPHIC ROCKS, and also found in SEDIMENTARY ROCKS.

Michel-Lévy chart A chart of standard colours used in measuring BIREFRINGENCE.

micrinite A COAL MACERAL of the INERTINITE group made up of very small rounded grains ~1 μm across.

micrite An abbreviation of **micr**ocrystalline cal**cite**; very fine-grained (<4 μm) carbonate making up the MATRIX in LIMESTONES.

micritic limestone A LIMESTONE with a MATRIX of MICRITE rather than a sparry cement.

micritization The degradation of coarse calcareous material by REPLACEMENT or reduction in grain size to MICRITE, often by biological activity.

micro- Extremely small; $\times 10^{-3}$ when attached to a unit.

microatoll (patch reef) An individual massive CORAL colony in a LAGOON.

microboring A submillimetric diameter boring up to 1 mm in length made by a micro-organism. The DISSOLUTION process used is important in the BIOEROSION of carbonates.

microcline $(KAlSi_3O_8)$ The low-temperature form of POTASSIUM FELDSPAR.

microcontinent A small fragment or remnant of CONTINENTAL CRUST up to about the size of Malagasy.

microcrack A MICROFRACTURE on which there has been no visible DISPLACEMENT. Cf. MICROFAULT.

microcrystalline Descriptive of material that is so fine-grained that its crystals can only be viewed microscopically. Cf. MACRO-CRYSTALLINE.

microdiorite A medium-grained IGNEOUS ROCK with the MINERAL ASSEMBLAGE and chemistry of DIORITE.

microearthquake See MICROSEISM.

microfabric A MICROSCOPIC scale FABRIC.

microfalaise A small cliff.

microfault A MICROFRACTURE on which there has been visible DISPLACEMENT. Cf. MICROCRACK.

microfelsitic texture A CRYPTOCRYSTAL-LINE TEXTURE in the GROUNDMASS of FELSIC IGNEOUS ROCKS formed by the DEVITRIFI-CATION of an originally GLASSY MATRIX.

microfracture A MICROSCOPIC discontinuity across which there has been separation. Observed prior to faulting with a PREFERRED ORIENTATION orthogonal to the least PRINCIPAL STRESS. Significant in DILAT-ANCY-DIFFUSION THEORY.

microgal (μgal) The CGS unit of GRAVITY ANOMALIES in MICROGRAVITY surveys, equal to 10^{-3} g.u. or 10^{-4} cm s^{-2}.

microgranite A medium-grained, MICRO-CRYSTALLINE IGNEOUS ROCK with the composition and TEXTURE of a GRANITE.

microgranodiorite A medium-grained, MICROCRYSTALLINE IGNEOUS ROCK with the composition and TEXTURE of a GRANO-DIORITE.

micrographic texture (micropegmat-ite texture) A MICROSCOPIC form of GRAPHIC TEXTURE.

microgravity A technique of measuring GRAVITY to MICROGAL accuracy using specialized GRAVIMETERS. Used in the search for subsurface voids, monitoring underground water movement and in measuring rates of NEOTECTONIC movement.

microlite 1 $(Ca_2Ta_2O_6(O,OH,F))$ A MINERAL found in PEGMATITES. **2** A very small crystal, usually in the GLASSY GROUNDMASS of a rapidly chilled LAVA, representing an initial stage of crystal NUCLEATION and growth.

microlithon A TABULAR body of rock defined by CLEAVAGE surfaces, formed during the BUCKLING of a layered sequence along the hinges of buckles. See also GLEIT-BRETTER.

microlithotype of coal A MACERAL association visible at a MICROSCOPIC scale, given the suffix -ite for distinction from a MACERAL and commonly containing several MACERAL types. The seven microlithotypes are CLARITE, DURITE, INERTITE, LIPTITE, TRIMACERITE, VITRINERTITE and VITRITE.

microlog A GEOPHYSICAL BOREHOLE ELECTRIC LOG in which small electrodes mounted on a pad are pressed against the wallrock. Provides information particularly on the MUDCAKE.

micropegmatite texture See MICROGRAPHIC TEXTURE.

microperthite A MICROSCOPIC intergrowth of ALBITE and POTASSIUM FELDSPAR.

microphyric Descriptive of the TEXTURE of a medium- to fine-grained IGNEOUS ROCK with PHENOCRYSTS <2 mm long. Cf. MACROPHYRIC.

micropiracy A type of RIVER CAPTURE in which small RILLS and gullies migrate back and forth across a hillslope, giving rise to even EROSION.

microplate A small tectonic PLATE with identifiable margins, which may subsequently become a DISPLACED TERRANE.

microporosity The POROSITY arising from the presence of pores <0.5 μm in diameter, generally within the ARGILLACEOUS MATRIX of a SEDIMENTARY ROCK.

microprobe See ELECTRON MICROPROBE.

Microsauria An order of subclass LEPOSPONDYLI, class AMPHIBIA; mainly terrestrial AMPHIBIANS with close similarities to the REPTILIA. Range late CARBONIFEROUS–early PERMIAN.

microscopic Descriptive of features visible under an optical microscope, with a size in the range 5 μm to 2 mm. Cf. MACROSCOPIC, MESOSCOPIC.

microseism (microearthquake) The dominant, naturally occurring seismic NOISE, taking the form of long-duration RAYLEIGH WAVES of 5–20 s period and amplitude 0.1–10 000 nm. Mainly generated by sea WAVES.

microseismic Referring to MICROEARTHQUAKES.

microspar A recrystallized MICRITE with crystals 4–20 μm in size. Cf. PSEUDOSPAR.

microspherulitic texture The TEXTURE of an IGNEOUS ROCK in which MICROSCOPIC SPHERULITES are distributed through the GROUNDMASS.

microstalactitic structure A gravitationally driven CEMENTATION FABRIC developed in the VADOSE ZONE on the lower surface of ALLOCHEMICAL grains.

microstructure The MICROSCOPIC features of a rock.

microsyenite A medium-grained, MICROCRYSTALLINE IGNEOUS ROCK with the composition and TEXTURE of a SYENITE.

microtonalite A medium-grained, MICROCRYSTALLINE IGNEOUS ROCK with the composition and TEXTURE of a TONALITE.

mictite (mixtite) A CLASTIC ROCK with a very wide range of grain sizes.

mid-ocean ridge An ACCRETIVE PLATE MARGIN often, but not always, situated at the median line of an ocean basin, at which new oceanic LITHOSPHERE is generated by MAGMATIC processes. Marked by a topographic rise 1000–4000 km in width which rises 2–3 km above the flanking ocean BASINS.

mid-ocean ridge basalt See MORB.

middle ground bar A BAR dividing a river channel in a DELTA, formed where the water is shallow and the flow is dominated by friction.

Midway An EOCENE succession on the Gulf Coast of the USA equivalent to the DANIAN and THANETIAN.

migmatite A METAMORPHIC ROCK injected with IGNEOUS MATERIAL.

migmatization The process of forming a MIGMATITE by PARTIAL MELTING of the parent rock under extreme METAMORPHISM.

migration **1** See SEISMIC MIGRATION. **2** Of oil and gas, the movement from the SOURCE ROCK into the RESERVOIR ROCK beneath the OIL or GAS TRAP, largely controlled by their BUOYANCY. **3** The lateral or downcurrent movement of a BEDFORM or channel.

Milankovich cycles Periodic perturbations in the Earth's orbit round the Sun caused by the effects of the other solar system planets. These affect the tilt of the Earth's spin axis and variations in the radius of its orbit with respect to the Sun and thus cause changes in the INSOLATION experienced. The three main periodicities are 22 000, 40 000 and 100 000 years. This has been related to the periodicity of GLACIATIONS.

milarite ($KCa_2AlBe_2(Si_{12}O_{30}).H_2O$) An hydrated aluminium-beryllium-calcium-potassium silicate found in VEINS in GRANITIC rocks and PEGMATITES.

Miller index An index used in CRYSTALLOGRAPHY to specify the orientation of a crystal face.

Miller-Bravais index A modified MILLER INDEX.

millerite (capillary pyrite) (NiS) A rare ORE MINERAL of nickel found as a low-temperature MINERAL in LIMESTONE, DOLOMITE, HAEMATITE, SERPENTINE and carbonate ORE VEINS.

milligal (mgal) The CGS unit of GRAVITY equal to 10^{-3} cm s^{-2} or 10^{-5} m s^{-2}.

Millstone grit A coarse SANDSTONE division in the British CARBONIFEROUS, approximately corresponding to the NAMURIAN.

millstone One of a pair of stones used for grinding, chosen for its roughness, hardness and low degree of contamination of the ground material. Rocks which have been used include LIMESTONE, TRAVERTINE, SANDSTONE, GRANITE, BASALT, RHYOLITE and LEUCITITE.

mima mound An earth mound up to 2 m in height and 20–50 m in diameter, generally found at a density of 50–100 ha^{-1}. May be erosional remnants or formed from deposition around vegetation, by frost sorting or by communal rodents.

mimetic growth The growth of a METAMORPHIC MINERAL which mimics preexisting features in shape or orientation, e.g. MICA growing in SLATE in an orientation controlled by the MINERAL arrangement in the ARGILLITE.

mimetic twinning TWINNING which closely imitates a higher symmetry.

mimetite ($Pb_5(AsO_4)_3Cl$) A SECONDARY ORE MINERAL of lead occurring in the oxidized zone of lead deposits, forming a SOLID SOLUTION with PYROMORPHITE.

Mindel glaciation The second of the GLACIATIONS affecting the Alps in QUATERNARY times.

Mindyallen A CAMBRIAN succession in Australia covering the lower part of the MAENTWROGIAN.

mineral A naturally occurring, homogeneous solid with a defined chemical composition and highly ordered atomic arrangement. Cf. MINERALOID.

mineral assemblage (mineral paragenesis) MINERALS coexisting in equilibrium.

mineral lineation A LINEATION formed by MINERALS of elongated crystal HABIT.

mineral paragenesis See MINERAL ASSEMBLAGE.

mineral wax See OZOCERITE.

mineralogical limit The concentration of a RESOURCE below which an element is no longer recoverable as a distinct MINERAL phase.

mineralogy The study of MINERALS.

mineraloid A naturally occurring substance which does not conform to the definition of a MINERAL, e.g. NATIVE MERCURY.

minette A CALC-ALKALINE LAMPROPHYRE with PHENOCRYSTS of BIOTITE and CLINOPY-

ROXENE ± HORNEBLENDE, with ALKALI FELD-
SPAR > PLAGIOCLASE.

minette ironstone A rock containing SID-
ERITE, BERTHIERINE (often OOLITIC) and CAL-
CITE (causing the ORE to be self-fluxing) with
SILICA >20%. Widespread in the MESOZOIC of
Europe.

miniripple A WAVE RIPPLE with a wavelength <10 mm.

minium (Pb_3O_4) A brown-red OXIDE MIN-
ERAL of lead found as an OXIDATION product
of GALENA and other lead MINERALS.

minnesotaite (($Fe,Mg)_3Si_4O_{10}(OH)_2$) A
rare, iron-rich analogue of TALC found in
PRECAMBRIAN IRON FORMATIONS.

minor fold A FOLD distinguishable at OUT-
CROP scale. Cf. MAJOR FOLD.

minor structure A STRUCTURE distinguishable at OUTCROP scale.

minverite A DOLERITE containing a brown,
soda-rich HORNBLENDE.

mio- Less.

Miocene An EPOCH of the NEOGENE, 23.3–
5.2 Ma.

miogeosyncline A GEOSYNCLINE with no
related magmatism.

mirabilite (Glauber salt) (Na_2SO_4.
$10H_2O$) An EVAPORITE MINERAL formed by the
HYDRATION of THERNARDITE.

mire See BOG.

Mirnyy A SILURIAN/DEVONIAN succession in
the Mirnyy Creek area of NE Siberia equivalent to the PRIDOLI and LOCHKOVIAN.

miscible Capable of mixing to form a
single liquid. Cf. IMMISCIBLE.

mise-à-la-masse A RESISTIVITY METHOD in
which one current electrode is sited within
the conducting body. The second is placed
at effectively infinity and EQUIPOTENTIAL
lines are mapped with a pair of potential
electrodes, providing more information
about the extent of the conductor than
standard surface methods.

mispickel An old name for ARSENOPYRITE.

Missisauga A CRETACEOUS succession on
the Scotian shelf of Canada covering the
BERRIASIAN, VALANGINIAN, HAUTERIVIAN and
part of the BARREMIAN.

Mississippi Valley-type deposit A type
of CARBONATE-HOSTED BASE METAL DEPOSIT.

Mississippian Sub-PERIOD of the CARBON-
IFEROUS, 362.5–349.5 Ma.

Missourian A CARBONIFEROUS succession
in the USA covering the lower part of the
KASIMOVIAN.

missourite A MELANOCRATIC PLUTONIC
IGNEOUS ROCK comprising CLINOPYROXENE,
OLIVINE and LEUCITE.

mixed mode fracture A combination of
tensile-, sliding- or tearing-type FRACTURES.

mixing length The mean size of an eddy.
See VON KARMAN'S CONSTANT.

mixtite See MICTITE.

mizzonite ($mCa_4(Al_6Si_6O_{24})CO_3 + nNa_4$
$(Al_3Si_9O_{24})Cl$) A SCAPOLITE group MINERAL
comprising a mixture of MEIONITE and
MARIALITE, found in metamorphosed LIME-
STONE and altered BASIC IGNEOUS ROCKS.

MKSA The **M**etre-**K**ilogram-**S**econd-
Ampere system of units.

MO See MOLECULAR ORBITAL.

Moberg An Icelandic name for HYALO-
CLASTITE formed by subglacial cooling of
BASALTIC MAGMA and altered to PALAGONITE.

mobile belt (orogenic belt) A large-
scale, linear belt of CONTINENTAL CRUST affected by TECTONIC activity over a given period
of geological time. Cf. CRATON.

Mocha stone See MOSS AGATE.

modal composition (mode) The composition of a rock sample in terms of the volu-
metric proportions of the MINERALS in it.

modderite (CoAs) A MINERAL with a distorted derivative of the nickel arsenide
structure.

mode **1** See MODAL COMPOSITION. **2** The
most common value in a set of numbers.

modified Griffith failure criterion An extension of the GRIFFITH FAILURE CRITERION which introduces frictional sliding between crack surfaces.

modified Mercalli scale The most widely used scale of EARTHQUAKE INTENSITY, modified to allow for changes in design and construction standards since being devised in 1902.

modulus of rigidity See RIGIDITY.

mofette 1 An opening of volcanic origin emitting carbon dioxide, nitrogen and oxygen. **2** SOLFATARA rich in carbon dioxide.

Mogi doughnut See SEISMIC GAP.

mogote A generally steep-sided, residual LIMESTONE hill in tropical KARST terrain.

Mohnian A MIOCENE succession on the west coast of the USA covering part of the SERRAVALIAN, the TORTONIAN and part of the MESSINIAN.

Moho See MOHOROVICIC DISCONTINUITY.

Mohorovicic discontinuity (M-discontinuity, Moho) The SEISMIC DISCONTINUITY between the CRUST and MANTLE where the P WAVE velocity increases to about 8.0 km s^{-1}.

Mohr circle A circle used in a MOHR DIAGRAM.

Mohr diagram A graph in which a state of STRESS or STRAIN is represented by circles.

Mohr envelope A line showing the relationship between SHEAR and NORMAL STRESS at FAILURE on a MOHR DIAGRAM.

Mohr locus A closed line representing the STRAIN on a general section of the STRAIN ELLIPSOID.

Mohs' scale A ten point scale of HARDNESS to which MINERALS are compared.

molasse The copious sediment derived from a newly elevated mountain range, i.e. post-orogenic continental sediment. Cf. FLYSCH.

molecular orbital (MO) A wave function which describes the behaviour of an electron in the presence of many nuclei.

Mollusca/molluscs A phylum characterized by a fleshy mantle and often a calcareous shell of from one to eight parts in the form of a coiled, hollow cone in which the soft parts can be viewed as various modifications of a hypothetical 'archimollusc' body plan. The 'archimollusc' possesses a ventral muscular foot beneath the visceral organs, which contain a gut running from a mouth with a rasping plate to a mantle cavity with paired gills. Range CAMBRIAN–RECENT.

molybdenite (MoS_2) The major ORE MINERAL of molybdenum.

monadnock An upstanding rock, hill or mountain on an otherwise flat plain.

monalbite ($NaAlSi_3O_8$) A MONOCLINIC, high-temperature form of ALBITE.

monazite (($Ce,La,Y,Th)PO_4$) A RARE EARTH phosphate occurring as an ACCESSORY MINERAL in GRANITE and concentrated as an ORE MINERAL in BEACH SAND.

monchiquite An alkaline variety of LAMPROPHYRE comprising Al-TITANAUGITE, BARKEVIKITE $\pm$ KAESURTITE, BIOTITE/PHLOGOPITE, $\pm$ OLIVINE, in a GLASSY GROUNDMASS containing NEPHELINE or ANALCIME.

mono- Single.

monocline An asymmetric FOLD with one limb dipping at a lower angle than the other.

monoclinic system A CRYSTAL SYSTEM whose members have three unequal axes, two of which intersect at an oblique angle and the third perpendicular to the plane of the other two.

monomict (oligomict) Descriptive of a CLASTIC SEDIMENTARY ROCK composed of a single MINERAL type. Cf. POLYMICT.

monomineralic Composed of a single MINERAL.

monophyletic Descended from a common ancestor. Cf. POLYPHYLETIC.

mortlake

Monoplacophora A class of phylum MOL-LUSCA with multiple paired gills, kidneys, gonads and shell attachment muscles, usually with only a slightly coiled shell. Range CAMBRIAN–RECENT.

Monorhina A subclass of class AGNATHA, superclass PISCES including lamprey-type FISH. Range U. SILURIAN–RECENT.

Monoskaya A TRIASSIC succession in Siberia covering the upper part of the NAMMALIAN.

Monotremata An order of subclass PROTO-THERIA, class MAMMALIA; egg-laying MAM-MALS. Range MIOCENE–RECENT.

monotropy The relationship between two POLYMORPHS in which only one is stable and the change to this form is irreversible. Cf. ENANTIOTROPHY.

montebrasite ((Li,Na)Al(PO$_4$)(OH,F)) A variety of AMBLYGONITE found in GRANITE PEGMATITES.

monticellite (CaMgSiO$_4$) An OLIVINE found mainly as a METAMORPHIC or METASO-MATIC MINERAL.

montmorillonite ((Al,Mg)$_8$(Si$_4$O$_{10}$)$_3$ (OH)$_{10}$.12H$_2$O) A CLAY MINERAL; the principal component of BENTONITE CLAYS.

monzodiorite An IGNEOUS ROCK intermediate in composition between MONZON-ITE and DIORITE.

monzogabbro A PLUTONIC rock intermediate in composition between GABBRO and MONZONITE.

monzogranite A GRANITE with equal proportions of ALKALI and PLAGIOCLASE FELDSPARS.

monzonite (syenodiorite) A medium to coarse-grained INTRUSIVE IGNEOUS ROCK containing PLAGIOCLASE ($Ab_{>50}$), <20% QUARTZ, AMPHIBOLE , ALKALI FELDSPAR (>10%) and/or PYROXENE. It grades into TONALITE with increased QUARTZ and DIORITE with <10% ALKALI FELDSPAR.

monzonorite A NORITE containing ORTHOCLASE.

mooihoekite (Cu$_9$Fe$_9$S$_{16}$) A rare SULPHIDE MINERAL.

moonmilk A white, CRYPTOCRYSTALLINE substance which forms SPELEOTHEMS, normally composed of CARBONATE MINERALS, which feels like cream cheese when wet and is a very fine powder when dry.

moonstone A variety of ALBITE or OLIGO-CLASE with an OPALESCENT play of colours, used as a semiprecious GEM.

moraine A depositional landform generated directly by a GLACIER.

Morarian orogeny An OROGENY of PRE-CAMBRIAN age affecting northern Scotland from ~1050–730 Ma.

MORB (mid-ocean ridge basalt) A type of THOLEIITIC BASALT found in OCEANIC RIDGES, characterized by very low K$_2$O and TiO$_2$; low Fe, P$_2$O$_5$, Ba, Rb, Sr, Pb, Th, U and Zr; and high CaO. Depleted in light RARE-EARTH ELEMENTS with respect to heavy RARE-EARTH ELEMENTS. Has not been contaminated and so retains the chemical signature of its MANTLE source.

mordenite (Na$_3$KCa$_2$(Al$_8$Si$_{40}$O$_{96}$).28H$_2$O) A ZEOLITE found in VEINS and cavities of IGNEOUS ROCKS.

morganite A pink GEM variety of BERYL.

morion A black variety of SMOKY QUARTZ.

Morrison A JURASSIC succession in Colorado, Idaho, Utah and Wyoming, USA, covering part of the OXFORDIAN, the KIMMERIDGIAN and part of the TITHONIAN.

Morrowan A CARBONIFEROUS succession in the USA equivalent to the BASHKIRIAN.

mortar structure (mortar texture) An optical MICROSTRUCTURE of large strained grains surrounded by smaller, RECRYSTAL-LIZED STRAIN-free grains. Typical of MYLON-ITES in MONOMINERALIC rocks.

mortar texture See MORTAR STRUCTURE.

Mortensnes A STAGE of the VENDIAN, 600–590 Ma.

mortlake See OXBOW LAKE.

morvan The intersection of two EROSIONAL surfaces.

mosaic evolution Evolution in which not all changes are seen in all representatives at the same time.

mosaic texture See SACCHAROIDAL TEXTURE.

Moscovian An EPOCH of the CARBONIFEROUS, 311.3–303.0 Ma.

moss agate (Mocha stone) A semi-precious GEM variety of AGATE with a moss-like patterning.

Mössbauer spectroscopy The recoilless emission and resonant absorption of gamma rays by the nuclei of solids, principally IRON, which provide information on the nature and environment of the atom.

mottramite $(Pb(Cu,Zn)VO_4OH)$ DESCLOISITE in which most of the zinc is replaced by COPPER.

Motuan A CRETACEOUS succession in New Zealand covering part of the upper ALBIAN.

mouldic fabric A type of grain DISSOLUTION POROSITY caused by the LEACHING of grains or replacive cements whose characteristic morphology is preserved in the new void.

moulin A vertical cylindrical shaft, 0.5–1.0 m wide and up to 25 m deep, through which surface meltwater flows into a GLACIER.

mound dune See NEBKHA.

mound spring An ARTESIAN SPRING occurring preferentially along a FAULT and which gives rise to a small mound.

mouth bar A BAR at the mouth of a river dividing the flow into channels.

moveout (stepout) The increase in ARRIVAL TIME with distance from the detector of a SEISMIC WAVE reflected from a discontinuity. See also DIP MOVEOUT, NORMAL MOVEOUT.

MSK scale A twelve point EARTHQUAKE INTENSITY scale similar to the MODIFIED MERCALLI SCALE, used in the former USSR.

MT survey See MAGNETOTELLURIC SURVEY.

mud Sediment whose particles have a size <62 μm.

mud ball See ARMOURED MUD BALL.

mud cake The solid part of a DRILLING MUD which is left on the wall of a borehole when the MUD FILTRATE has penetrated the adjacent formations.

mud clast See RIP-UP CLAST.

mud crack A vertical to subvertical shrinkage crack formed by the contraction of cohesive muddy sediment, which may be preserved if infilled by a different sediment.

mud filtrate The liquid part of a DRILLING MUD which can penetrate adjacent formations and displace GROUNDWATER or HYDROCARBONS.

mud pebble See ARMOURED MUD BALL.

mud ripple A surface BEDDING STRUCTURE on a MUDSTONE, and its SANDSTONE cast, similar in form and scale to the straighter-crested CURRENT RIPPLE. Grades into FLUTE MARK, and also probably of erosional origin.

mud volcano 1 A HOT SPRING which produces boiling mud. **2** A conical mound formed when liquid mud is forced to the surface as a result of COMPACTION or EARTHQUAKE activity. Cf. SAND VOLCANO.

mudflow A MASS FLOW of debris mixed with water with a high proportion of MUD. Very similar to a LAHAR, but richer in fine-grained material.

mudlump A small-scale landform in a DELTA, probably formed by the DIAPIRIC intrusion of plastic CLAYS through SANDS.

mudrock (mudstone) A SEDIMENTARY ROCK of MUD grade.

mudstone See MUDROCK.

mugearite A SILICA-poor sodic TRACHYANDESITE.

mullion structure A linear, cylindrical STRUCTURE comprising elongate rods or columns 20 mm–2 m across and up to 100 m long which are either complete or incomplete in section. Surfaces may be smooth or corrugated and define a LINEATION parallel to FOLD AXES. Most common in strongly deformed METAMORPHIC ROCKS. May form by BUCKLING of the surface between COMPETENT and INCOMPETENT BEDS, by the intersection of BEDDING and CLEAVAGE, by FOLD HINGES or by BOUDINAGE.

mullite ($Al_6Si_3O_{15}$) A rare aluminosilicate formed by the intense heating of ANDALUSITE, KYANITE or SILLIMANITE.

multiple See MULTIPLE REFLECTION.

multiple bars A set of up to 10 shore-parallel COASTAL BARS.

multiple faults (fault set) A group of FAULTS with similar orientations and DISPLACEMENT VECTORS, probably forming simultaneously in a common STRESS field.

multiple intrusion An igneous intrusion made up of successively emplaced MAGMAS of distinctive, similar composition distinguished by internal contacts and/or PHENOCRYST content and CRYSTALLINITY.

multiple reflection (multiple, seismic multiple) A SEISMIC WAVE that has been reflected more than once before being recorded, and which can obscure the arrival of primary reflections.

multiple twinning See POLYSYNTHETIC TWINNING.

multiplexing The process of interleaving multiple input information channels into a single output channel, used, for example, in sending the output of several GEOPHONES to a recorder. Cf. DEMULTIPLEXING.

multispectral scanner A REMOTE SENSING device carried by the first LANDSAT SATELLITES which imaged four spectral bands in the region 500–1100 nm with a PIXEL size of ~80 m.

multistorey sandbody A vertical succession of SANDSTONE BEDS deposited by the infilling of river channels with little intervening MUD, formed by the rapid migration of a channel network over an ALLUVIAL PLAIN whose sediment has little opportunity of preservation.

Multituberculata An order of subclass ALLOTHERIA, class MAMMALIA; small rodent-like MAMMALS. Range late JURASSIC–EOCENE.

Muschelkalk **1** A TRIASSIC succession in Germany covering the ANISIAN and part of the LADINIAN. **2** The traditional German name for the Middle TRIASSIC, during which the TETHYS Ocean transgressed over the region from Europe and N. Africa to China.

muscovite (white mica) ($KAl_2(Al Si_3O_{10})(OH)_2$) A very common MICA.

muskeg **1** The waterlogged marshland of NW Canada. **2** A PEAT-filled BASIN with *Sphagnum* moss.

Muskingum method An empirical method of FLOOD ROUTING using the continuity equation and a relationship between flow rate and temporary storage of water in the channel during flooding.

Myachkovskian A STAGE of the CARBONIFEROUS, 305.0–303.0 Ma.

mylonite A fine-grained, FOLIATED FAULT ROCK with a RECRYSTALLIZED TEXTURE with 50–90% MATRIX and a strong LINEATION caused by SHEAR in a major DUCTILE FAULT or SHEAR ZONE. See also PROTOMYLONITE, ULTRAMYLONITE.

mylonite gneiss A MYLONITE with a well-developed GNEISSOSITY.

mylonite schist A MYLONITE with a well-developed SCHISTOSITY.

mylonite zone A belt of MYLONITE, up to hundreds of kilometres long and kilometres thick, along a major FAULT ZONE.

mylonitization The process of forming a MYLONITE, generally by CRYSTAL PLASTIC STRAIN by DISLOCATION CLIMB and RECOVERY, with new MINERAL grains forming by DYNAMIC RECRYSTALLIZATION.

Myriapoda/myriapods A superclass of phylum UNIRAMIA. Terrestrial animals breathing through trachaea or the body wall with a head bearing a pair of antennae and mandibles and two pairs of feeding maxillae, and many trunk somites. Range SILURIAN–RECENT.

myrmekite (myrmekitic texture) A VERMICULAR intergrowth of QUARTZ and sodic FELDSPAR adjacent to a crystal of ALKALI FELDSPAR, probably of secondary origin when QUARTZ is released during the REPLACEMENT of ALKALI FELDSPAR by PLAGIOCLASE.

myrmekitic texture See MYRMEKITE.

Nabarro-Herring creep A CREEP mechanism in which atoms diffuse within the CRYSTAL LATTICE.

nacreous lustre A LUSTRE resembling that of a pearl.

nacrite ($Al_2Si_2O_5(OH)_4$) A CLAY MINERAL with the same composition as KAOLINITE but a different CRYSTAL STRUCTURE.

Nafe-Drake curve An empirical relationship between P WAVE velocity and DENSITY, allowing estimation of the latter from the former to an accuracy of $\sim$0.1 Mg m^{-3}.

Nagssugtoqidian orogeny An OROGENY affecting W. Greenland during the PRECAMBRIAN at $\sim$2600 Ma and $\sim$1900–1500 Ma.

nagyágite (AuTe.6Pb(S,Te)) A rare MINERAL found in HYDROTHERMAL VEINS.

Nammalian A STAGE of the TRIASSIC, 243.4–241.9 Ma.

Namurian A CARBONIFEROUS STAGE covering the SERPUKHOVIAN and part of the BASHKIRIAN.

nano- Extremely small; $\times 10^{-9}$ when attached to a unit.

nanoplankton Marine, PLANKTONIC FOSSILS of ultramicroscopic size, useful in correlation over wide distances.

nanoTesla (nT) The SI subunit used for MAGNETIC ANOMALIES, equal to 10^{-9} TESLA (V s m^{-2}).

naphthene series A component of CRUDE OIL with the general formula C_nH_{2n}.

napoleonite A DIORITE containing centimetric-scale, spheroidal structures comprising alternating shells of HORNEBLENDE and FELDSPAR.

nappe (decke) A body of rock, generally highly folded and with greater DUCTILE DEFORMATION than a THRUST SHEET, which has suffered considerable horizontal TECTONIC transport in an OROGENIC BELT.

Narizian An EOCENE succession on the west coast of the USA comprising part of the LUTETIAN and the BARTONIAN.

native element A MINERAL comprising a chemical element in an uncombined state or as an alloy with another element(s).

native metal A MINERAL comprising a metallic element in an uncombined state or as an alloy with another element(s).

natroalunite ($(Na,K)Al_3(SO_4)_2(OH)_6$) The sodium-rich equivalent of ALUNITE.

natrocarbonatite A CARBONATITE composed of sodium, calcium or potassium carbonates.

natrojarosite ($NaFe^{3+}_3(SO_4)_2(OH)_6$) An hydrated sodium-iron sulphate of the ALUNITE group found as a SECONDARY MINERAL in cracks in FERRUGINOUS rocks and ORES.

natrolite ($Na_2Al_2Si_3O_{10}.2H_2O$) A ZEOLITE, often found in ACICULAR form in AMYDALES in BASALTIC rocks.

natron ($Na_2CO_3.10H_2O$) An hydrated sodium carbonate commonly found in SODA LAKES.

natron lake A lake rich in NATRON, often occurring in continental RIFT VALLEYS.

natural arch A bridge or arch joining two

rock pillars, produced by WEATHERING or EROSION.

natural gas The gaseous constituents of PETROLEUM, i.e. METHANE, ethane, propane and *n*-butane.

natural remanent magnetization (NRM) The permanent magnetization of the FERRIMAGNETIC MINERALS in certain rocks that was acquired when they formed or at some later stage in their history. Its measurement forms the basis of the study of PALAEOMAGNETISM.

natural strain A linear measurement of shape change during DEFORMATION based on the integration of instantaneous STRAIN increments.

naujaite SODALITE-rich NEPHELINE SYENITE with MICROCLINE, ALBITE, ACMITE and a sodic AMPHIBOLE.

Nautiloidea/nautiloids A subclass of class CEPHALOPODA, phylum MOLLUSCA characterized by coiled shells whose internal compartments are defined by smooth septae. Range CAMBRIAN–RECENT.

Navarro A CRETACEOUS succession on the Gulf Coast of the USA equivalent to the MAASTRICHTIAN.

Navier-Stokes equations A series of equations of linear momentum for a moving, viscous, compressible fluid which relate STRESS and the state of STRAIN.

nebkha (coppice dune, shrub-coppice dune) An AEOLIAN bedform of wind-blown SAND collected within and behind, and stabilized by, vegetation.

Nebraskan glaciation The first of the major GLACIATIONS affecting North America in the QUATERNARY.

neck A volcanic PLUG.

necking A localized thinning of a STRUCTURE during EXTENSION.

Nectarian An ERA of the PRECAMBRIAN, 3950–3850 Ma.

Nectridea An order of subclass LEPOSPONDYLI, class AMPHIBIA; small AMPHIBIANS of sometimes bizarre morphology. Range PERMIAN–CARBONIFEROUS.

needle stone (rutilated quartz) A colloquial name for clear QUARTZ with ACICULAR inclusions of RUTILE or, rarely, ACTINOLITE.

Néel temperature The temperature at which the coupling between MAGNETIC DOMAINS in a FERRIMAGNETIC material breaks down, so that it shows simple PARAMAGNETIC behaviour.

negative crystal A PSEUDOMORPH comprising a hollow mould from which the original crystal has been removed by SOLUTION.

negative grading See INVERSE GRADING.

Neichiashan See AIJIASHAN.

nekron mud See GYTTJA.

nektonic Free-swimming.

nematath A linear series of OCEANIC ISLANDS.

nematoblastic Descriptive of a FOLIATION formed by prismatic MINERALS.

Nenetskiy A PERMIAN succession in the Timan area of the former USSR equivalent to the U. ASSELIAN.

neo- New.

neocatastrophism A theory concerning events of great magnitude and low frequency, applied to MASS EXTINCTIONS and GEOMORPHOLOGY.

Neocomian An EPOCH of the CRETACEOUS, 145.6–131.8 Ma.

neoformation of clays The AUTHIGENESIS of CLAY MINERALS, as the result of direct PRECIPITATION from a PORE FLUID, the NEOMORPHISM of one CLAY MINERAL POLYMORPH to another or the REPLACEMENT of a precursor MINERAL by a new CLAY MINERAL.

Neogene The younger PERIOD of the CENOZOIC, 23.3–1.64 Ma.

neoglacial A small-scale GLACIAL advance during the HOLOCENE after the maximum GLACIAL retreat of the present INTERGLACIAL.

Neognathae A superorder of subclass NEORNITHES, class AVES, comprising all modern BIRDS with the exception of the flightless PALAEOGNATHAE. Range CRETACEOUS–RECENT.

neomorphism All DIAGENETIC transformations between one MINERAL and itself or a POLYMORPH, including differences in size and shape but excluding pore filling, in which there is simultaneous reoccupation of the space occupied by the original and the new. Occurs by the processes of POLYMORPHIC TRANSFORMATION and RECRYSTALLIZATION.

Neornithes A subclass of class AVES; all BIRDS with the exception of the very ancient forms of the ARCHAEORNITHES. Range CRETACEOUS–RECENT.

neotectonics Late CENOZOIC DEFORMATION.

neoteny An evolutionary reduction in the rate of morphological development, which may lead to PAEDOMORPHOSIS.

neotype See TYPE SPECIMEN.

nepheline ((Na,K)AlSiO₄) A FELDSPATHOID, found mainly in PLUTONIC and VOLCANIC ROCKS and in PEGMATITES associated with NEPHELINE SYENITES.

nepheline monzonite An UNDERSATURATED MONZONITE containing ESSENTIAL NEPHELINE.

nepheline syenite A PLUTONIC IGNEOUS ROCK composed of NEPHELINE, ALBITE and MICROCLINE. High in alumina (~25%) and soda (~9%), so exploited in glass and ceramic manufacture.

nephelinite A fine-grained IGNEOUS ROCK with >10% MODAL NEPHELINE and little or no ALKALI FELDSPAR, in which the NEPHELINE content exceeds that of MAFIC MINERALS. Occurs in association with ALKALINE IGNEOUS ROCKS in continental settings.

nepheloid layer A water layer, generally on the sea bed of deep oceans and a few hundred to 1500 m thick, containing abundant suspended sediment.

nephelometer An instrument used for the identification of NEPHELOID LAYERS by the scattering of light by fine particles.

nephrite (kidney stone) A tough, compact variety of TREMOLITE providing much JADE.

neptunian dyke The sediment infill of a vertical fissure, possibly recording a period of deposition not preserved in a normal stratal sequence, or injected into the COUNTRY ROCK.

neptunian sill The sediment infill of a horizontal fissure, possibly recording a period of deposition not preserved in a normal STRATAL sequence, or injected into the COUNTRY ROCK.

neptunite (KNa₂Li(Fe,Mn)₂TiO₂(Si₄O₁₁)₂) A complex, rare INOSILICATE.

neritic (sublittoral) The area between the lower limit of the LITTORAL zone and the CONTINENTAL SHELF.

Nerminskiy A PERMIAN succession in the Timan area of the former USSR equivalent to the L. ARTINSKIAN.

nesosilicate (orthosilicate) A silicate in which the silicon tetrahedra share no corners with other tetrahedra.

nesting lineation A LINEATION on a SLICKENSIDE which is perfectly matched across the FAULT PLANE.

net slip The distance along a FAULT PLANE between the original position of some marker and its new position.

Nettleton's method A method of determining the *in situ* DENSITY of an isolated hill by correcting GRAVITY measurements over it for a range of REDUCTION DENSITIES. The reduced GRAVITY profile providing the least CORRELATION with the topography then provides the best DENSITY estimate.

neutral fold A FOLD which closes sideways.

neutral surface Of a FOLD, the surface within a folded layer, usually parallel to its boundaries, which separates regions of COMPRESSION and EXTENSION.

neutron activation analysis (NAA) An analytical method for determining a small number of MAJOR ELEMENTS and many TRACE ELEMENTS in silicate rocks. Bombardment by neutrons forms new isotopes as neutrons join the nuclei, which decay, emitting gamma rays whose energy is diagnostic of the original isotope.

neutron diffraction The scattering of thermal neutrons by atomic nuclei, the scattering being enhanced by atoms with a MAGNETIC MOMENT. The different isotopes of the same element scatter differently, allowing their distinction.

neutron log See NEUTRON-GAMMA RAY LOG.

neutron-gamma ray log (neutron log) A GEOPHYSICAL BOREHOLE LOG in which the WALLROCK is bombarded by neutrons, causing them to emit gamma rays, the intensity of which is a function of the POROSITY of the WALLROCK.

Nevadan orogeny An OROGENY of late JURASSIC to early CRETACEOUS age affecting the western USA and approximately equivalent to the COAST RANGE OROGENY in Canada.

névé (firn) Compacted snow which has survived a summer.

New Red Sandstone The continental rocks of the PERMIAN and TRIASSIC SYSTEMS in W. Europe.

Newton The SI unit of FORCE; the FORCE required to accelerate a mass of 1 kg by 1 m s^{-2}.

Newton's law of gravitation The FORCE of attraction between two masses is proportional to their product and inversely proportional to the square of the distance between them, the constant of proportionality being the GRAVITATIONAL CONSTANT.

Newtonian flow FLOW in which the SHEAR STRAIN RATE is a linear function of SHEAR STRESS.

Newtonian fluid A fluid whose VISCOSITY is constant, independent of any external SHEAR FORCE applied to it and which obeys

the NAVIER-STOKES EQUATION. Cf. NON-NEWTONIAN FLUID.

Ngaterian A CRETACEOUS succession in New Zealand covering parts of the ALBIAN and CENOMANIAN.

Niagaran A SILURIAN succession in North America comprising the ONTARIAN, TONA-WANDIAN and LOCKPORTIAN.

niccolite (kupfernickel, nickeline) (NiAs) A sulphide-like MINERAL with a NiAs structure; an ORE MINERAL of nickel.

nickel antimony glance See ULLMANNITE.

nickel arsenic glance See GERSDORFFITE.

nickel arsenide group A group of SUL-PHIDE MINERALS characterized by a structure in which metals and anions lie on interpenetrating simple and close-packed hexagonal sublattices respectively.

nickel bloom See ANNABERGITE.

nickel laterite deposit A LATERITE enriched in nickel, formed by intense tropical WEATHERING of rocks with trace amounts of nickel, e.g. PERIDOTITE, SERPENTINITE.

nickeliferous Containing nickel.

nickeline See NICCOLITE.

Niedermendig lava See MAYEN LAVA.

Niggli number A variant of the NORM classification.

nigrite A member of the ASPHALTITE group with a PITCH-like appearance.

Ningguo (Ningkuo) An ORDOVICIAN succession in China covering part of the TREMA-DOC and EARLY ARENIG.

Ningkuo See NINGGUO.

niobite See COLUMBITE.

nitratine (NaNO$_3$) A nitrate MINERAL found in CALICHE.

nitre (saltpetre) (KNO$_3$) A nitrate MIN-ERAL used in fertilizer manufacture.

nivation 1 WEATHERING and transport processes intensified by a late-lying snow patch. **2** Descriptive of landforms substantially modified by snow-patch related processes.

noble metal See PRECIOUS METAL.

nodal avulsion An AVULSION occurring persistently from one point.

nodal plane One of two orthogonal planes dividing compressional and rarefactional P WAVE ARRIVALS from an earthquake on the STEREOGRAM used in a FOCAL MECHANISM SOLUTION. One such plane is the FAULT.

nodule An irregular, spherical to ellipsoidal, flattened to cylindrical body, commonly composed of CALCITE, SIDERITE, PYRITE, GYPSUM and CHERT, common in soils and EVAPORITE DEPOSITS.

Noginskian The highest STAGE of the CARBONIFEROUS, 293.6–290.0 Ma.

noise An unwanted disturbance on a record, such as a SEISMOGRAM. Random SEISMIC NOISE includes MICROSEISMS, whereas coherent noise includes MULTIPLE REFLECTIONS and SURFACE WAVES.

non-associated gas PETROLEUM gas not accompanied by oil.

non-coaxial deformation Progressive DEFORMATION during which the lines of maximum and minimum ELONGATION rotate. Cf. ROTATIONAL DEFORMATION.

non-conformity (heterolithic unconformity) An UNCONFORMITY in which younger STRATA rest on an EROSION surface of non-bedded IGNEOUS ROCKS.

non-contacting conductivity measurement The measurement of ground ELECTRICAL CONDUCTIVITY using an ELECTROMAGNETIC INDUCTION METHOD so that no ground contact is required.

non-cylindrical fold A FOLD which displays different PROFILES along the HINGE LINE.

non-metallic lustre The LUSTRE of a MINERAL that reflects light but does not shine metallically. Includes VITREOUS LUSTRE, SILKY LUSTRE, RESINOUS LUSTRE, etc.

non-Newtonian flow FLOW in which the SHEAR STRAIN RATE is a complex function of SHEAR STRESS. Cf. NEWTONIAN FLOW.

non-penetrative Descriptive of a feature, such as CLEAVAGE, developed throughout only part of a rock at the scale of observation. Cf. PENETRATIVE.

non-polarizing electrode An electrode designed to avoid the accumulation of charge on a metal electrode by ELECTRODE POLARIZATION effects, constructed of a metal in a solution of one of its salts contained in a porous casing.

non-rotational strain The STRAIN experienced by a rock with a NON-COAXIAL DEFORMATION history.

non-sequence (paraconformity) A DIASTEM-type UNCONFORMITY with faunal or other evidence of a time gap.

nontronite $(Fe_2(Al,Si)_4O_{10}(OH_2.Na_{0.3}(H_2O)_4))$ A CLAY MINERAL of the MONTMORILLONITE group.

norbergite $(Mg_3(SiO_4)(F,OH)_2)$ A MINERAL of the HUMITE group.

nordmarkite An ALKALINE, QUARTZ-bearing SYENITE comprising MICROPERTHITE, AEGIRINE, sodic AMPHIBOLE and ACCESSORY QUARTZ.

Norian A STAGE of the TRIASSIC, 223.4–209.5 Ma.

norite A GABBRO containing ORTHOPYROXENE and LABRADORITE.

norm See NORMATIVE COMPOSITION.

normal drag FAULT DRAG in which the curvature of a marker is consistent with the sense of DISPLACEMENT on the FAULT. Cf. REVERSE DRAG.

normal fault A DIP-SLIP FAULT with a dominant component of DIP-SLIP DISPLACEMENT and the HANGINGWALL displaced downwards relative to the FOOTWALL.

normal grain growth An increase in the size of a polycrystalline AGGREGATE during RECRYSTALLIZATION.

normal magnetization A REMANENT MAGNETIZATION in the same sense as the present

GEOMAGNETIC FIELD. Cf. REVERSED MAGNETIZATION.

normal moveout MOVEOUT generated by a horizontal reflector. Cf. DIP MOVEOUT.

normal polarity The orientation of a past GEOMAGNETIC FIELD or a NATURAL REMANENT MAGNETIZATION in the same direction as the present GEOMAGNETIC FIELD. Cf. REVERSED POLARITY.

normal resistivity log An ELECTRIC GEOPHYSICAL BOREHOLE LOG which provides RESISTIVITY information on a thick shell of the WALLROCK.

normal stress The STRESS acting at right angles to a surface.

normal twin A TWINNING phenomenon in which the TWIN AXIS is normal to the COMPOSITION PLANE.

normal zoning ZONING in PLAGIOCLASE in which the zones become more sodic towards the outside of the crystal. Cf. REVERSED ZONING.

normally consolidated sediment A sediment compacted by the PRESSURE expected from the OVERBURDEN thickness. Cf. OVERCONSOLIDATED SEDIMENT, UNDERCONSOLIDATED SEDIMENT.

normative composition (norm) The composition of an IGNEOUS ROCK expressed as weight proportions of idealized anhydrous MINERALS that CRYSTALLIZE from MAGMA, calculated in a specified sequence from a chemical analysis of the rock.

nosean (noselite) $(Na_8(AlSiO_4)_6SO_4)$ A rare FELDSPATHOID found in PHONOLITES and related rocks and in VOLCANIC BOMBS.

noselite See NOSEAN.

notch A landform 1–5 m in depth at the base of a SEA CLIFF, platform or REEF flat, especially in LIMESTONE and on tropical coasts, characteristic of areas with a low tidal range.

nothosaurs Extinct, pisciverous REPTILES of TRIASSIC age with small heads, long necks and paddle-like limbs.

Notoungulata An order of infraclass EUTHERIA, subclass THERIA, class MAMMALIA; an order of South American herbivores from rabbit to bear size. Range U. PALAEOCENE–PLEISTOCENE.

novaculite A fine-grained to CRYPTOCRYSTALLINE rock composed of QUARTZ or other forms of SILICA, i.e. a CHERT.

NRM See NATURAL REMANENT MAGNETIZATION.

nT See NANOTESLA.

nubbin A centimetric-scale, rounded or elongate earth lump produced by HEAVE associated with the growth of needle ice.

nuclear explosion seismology (forensic seismology) The detection, identification and YIELD estimation of underground nuclear explosions.

nuclear precession magnetometer See PROTON MAGNETOMETER.

nucleation The first stage in PRECIPITATION, before the free energy barrier resulting from the developing interface between the crystal nucleus and the solution has been overcome and CRYSTAL GROWTH can take place.

nuée ardente A laterally mobile, TURBULENT, hot cloud of air, VOLCANIC GASES and suspended, fine-grained TEPHRA generated by volcanic activity. Typical of PELÉEAN ERUPTIONS. Probably generated by the upward CONVECTION of hot gas and TEPHRA from an active PYROCLASTIC flow. Gives rise to thin, fine-grained deposits with low-angle CROSS-BEDDING overlain by air-fall TEPHRA.

Nuevo Leon A CRETACEOUS succession on the Gulf Coast of the USA covering the BARREMIAN and part of the APTIAN.

nugget A lump of NATIVE GOLD, rarely of PLATINUM, generally found in ALLUVIAL DEPOSITS.

Nukumaruan A PLIOCENE/QUATERNARY succession in New Zealand covering parts of the PIACENZIAN and early QUATERNARY.

nummulitic limestone A LIMESTONE of

EOCENE age made up of nummulites (FORAMINIFERA) quarried for some early Egyptian monuments such as the pyramids.

nunatak An isolated rock peak projecting through an ICE SHEET.

Nusselt number An hydrodynamic parameter related to the RAYLEIGH NUMBER whose value controls the possibility of CONVECTION and is related to the proportion of conductive to convective heat transport.

nye channel The basal conduit of a GLACIER, incised into BEDROCK, through which meltwater discharges. Cf. RÖTHLISBERGER CHANNEL.

Nyquist frequency (folding frequency) Half the sampling frequency of a digitized signal; the highest frequency reliably restored from that signal.

O

obduction The process whereby OCEANIC LITHOSPHERE is transported onto CONTINENTAL CRUST.

oblate Shaped like an ellipsoid of revolution in which the polar is less than the equatorial radius. Cf. PROLATE.

oblique extinction (inclined extinction) The EXTINCTION of a MINERAL which takes place at an angle to its CLEAVAGE traces or margins.

oblique ramp A RAMP trending at an angle between the transport direction and the normal to it in a THRUST ZONE.

oblique slip DISPLACEMENT at an angle to the STRIKE of the DISPLACEMENT PLANE and the normal to it.

oblique slip fault A FAULT which has similar magnitudes of STRIKE-SLIP and DIP-SLIP DISPLACEMENTS.

obrution deposit A deposit in which organisms are smothered by a rapid sediment influx which kills and buries them simultaneously, possibly giving rise to an EXCEPTIONAL FOSSIL DEPOSIT.

obsequent stream A tributary of a SUBSEQUENT STREAM which flows in a direction opposite to the regional DIP of the land surface. Cf. ANTECONSEQUENT, INCONSEQUENT, INSEQUENT, SUBSEQUENT and RESEQUENT STREAMS.

observation well A WELL drilled or allocated for monitoring the extraction from a RESERVOIR.

obsidian A GLASSY ROCK of volcanic origin of INTERMEDIATE to ACID composition formed by the chilling of RHYOLITIC LAVA. Widely used in prehistoric times for the manufacture of small tools.

occult mineral A MINERAL expected to form from a GLASS if it had crystallized completely.

Ocean Drilling Program (ODP) The current phase, since 1984, of ocean drilling using the ship JOIDES Resolution.

ocean island basalt (OIB) QUARTZ THOLEIITE, ALKALI BASALT and NEPHELINITE found on OCEANIC ISLANDS. With respect to MORB, it is enriched in LIL elements, LREES compared to HREES and INCOMPATIBLE ELEMENTS. Probably formed by PARTIAL MELTING of enriched MANTLE.

ocean plateau An areally extensive region of ocean floor reaching to 2–3 km of the sea surface above the surrounding seafloor, probably of volcanic origin.

ocean ridge Any major ridge on the ocean floor, usually either a MID-OCEAN RIDGE or ASEISMIC RIDGE.

ocean trench (deep-sea trench, trench) The topographic expression of SUBDUCTION; the large linear or arcuate depression of the oceanic LITHOSPHERE at a SUBDUCTION ZONE.

Oceanic Anoxic Event A period of about a million years in the MESOZOIC when BLACK SHALE was deposited and there was enhanced removal of light carbon isotopes from the world's oceans.

oceanic crust Thin (~7 km), young (<200 Ma) CRUST of three layers: Layer 1 – the uppermost layer of sediments, Layer 2 –

PILLOW LAVAS underlain by DYKES and Layer 3 – GABBROIC and underlying ULTRABASIC rocks.

oceanic island An intraPLATE island believed to have originated by the passage of the oceanic LITHOSPHERE over a nearly stationary HOTSPOT in the MANTLE.

oceanic magnetic anomalies (magnetic lineations) Linear MAGNETIC ANOMALIES of alternating polarity with amplitudes of ~1000 nT, 10–20 km in width and of large lateral extent which run parallel to OCEANIC RIDGES and are symmetrical about the crest of the ridge. Believed to form by the process described by the VINE-MATTHEWS HYPOTHESIS.

oceanite A type of BASALTIC rock with a lower proportion of alkalis and a higher proportion of MAFIC MINERALS than normal BASALT, typically found as LAVA FLOWS on OCEANIC ISLANDS.

ocellar Descriptive of an IGNEOUS ROCK TEXTURE in which AGGREGATES of small crystals are distributed around larger crystals to give the impression of an eye-like form.

Ochoan A PERMIAN succession in the Delaware Basin, USA equivalent to the LONGTANIAN and CHANGXINGIAN.

ochre Red, yellow and brown IRON OXIDES formed by the WEATHERING of iron deposits, used as pigments.

octahedrite A variety of ANATASE occurring as tetragonal bipyramids.

Octocorallia A subclass of class ANTHOZOA, phylum CNIDARIA; CORALS represented by rare FOSSILS, including the earliest ANTHOZOANS. Range ORDOVICIAN–RECENT.

OD See ORDNANCE DATUM.

Odontognathidae Superorder of subclass NEORNITHES, class AVES; toothed, flightless BIRDS similar to modern divers. Range L. CRETACEOUS–U. CRETACEOUS.

odontolite See BONE TURQUOISE.

ODP See OCEAN DRILLING PROGRAM.

Oe See OERSTED.

Oersted (Oe) The CGS unit of MAGNETIC INDUCTION.

offlap A STRUCTURE in which successive wedge-shaped BEDS do not extend to the margin of the underlying BED, such as would be formed in a contracting sedimentary BASIN. Cf. ONLAP.

offscraping See UNDERPLATING.

offset **1** The horizontal DISPLACEMENT across a FAULT normal to the interrupted feature. **2** The horizontal distance from SHOT to receiver in a seismic survey.

offset well A WELL drilled close to another further to explore or exploit a RESERVOIR.

offshore bar See SUBMERGED BAR.

Ohauan A JURASSIC succession in New Zealand covering the upper part of the KIMMERIDGIAN.

Ohm's law 'The electric current flowing between two points is equal to the ratio of the potential difference to the resistance between them.' Relevant to ELECTRICAL RESISTIVITY surveying.

Ohre A PERMIAN succession in NW Europe covering the middle part of the CAPITANIAN.

OIB See OCEAN ISLAND BASALT.

oikocryst A large crystal grain which encloses several randomly oriented, smaller grains of another phase or phases. Responsible for POIKILITIC TEXTURE.

oil basin A sedimentary BASIN containing commercial accumulations of oil.

oil shale A bituminous, non-marine LIMESTONE or MARL (rarely marine SHALE) which contains substantial organic matter at a low level of MATURATION such that liquid oil has never been released from the KEROGEN. A very abundant RESOURCE of FOSSIL FUEL, but with great obstacles to its exploitation.

oil source rock A rock, generally SHALE, LIMESTONE or COAL, in which HYDROCARBON generation processes produce NATURAL GAS and CRUDE OIL. Contains large quantities of organic matter, a PELAGIC fauna and flora and phosphorus. ANAEROBIC conditions necessary for preservation of the organic matter indicate deposition below a THERMOCLINE.

oil trap A geometrical arrangement of STRATA that allows the accumulation of oil. Commonly an ANTICLINE but also associated with FAULTING, STRATIGRAPHIC arrangements, REEFS and PIERCEMENT STRUCTURES, e.g. SALT DOMES. The HYDROCARBONS tend to move upward due to their low DENSITY until trapped beneath an impermeable or semipermeable CAP ROCK.

oilfield A region of the CRUST containing a number of oil pools.

oilfield water The water found in HYDROCARBON-bearing RESERVOIR rocks.

Old Red Sandstone The continental FACIES of the DEVONIAN in the UK.

oldhamite ((Ca,Mn)S) A calcium-manganese sulphide, found in METEORITES.

Olenekian A TRIASSIC STAGE in Siberia comprising the MONOSKAYA and SYGYNKANSKAYA

oligo- Prefix indicating few or little.

Oligocene The youngest EPOCH of the PALAEOGENE, 35.4–23.3 Ma.

oligoclase A PLAGIOCLASE FELDSPAR, An_{10-30}.

oligomict See MONOMICT.

oligomictic lake A lake which is thermally almost stable and undergoes only rare mixing. Cf. POLYMICTIC LAKE.

oligotrophic lake A lake with a deficiency of nutrients and large amounts of dissolved oxygen in the bottom layers, which have only a small amount of organic matter.

olistolith A waterlain block of rock of PEBBLE to BOULDER size that differs considerably in its PETROGRAPHY, composition or TEXTURE from the surrounding rocks. See also OLISTOSTROME.

olistostrome A bed or layer of OLISTOLITHS, commonly formed at the base of submarine FAULT scarps subject to MASS WASTING.

olivenite (Cu_2AsO_4OH) A rare, green,

hydrated COPPER arsenate; a SECONDARY MINERAL found in COPPER deposits.

olivine basalt See ALKALI BASALT.

olivines A group of ORTHOSILICATE MINERALS with the general formula M_2SiO_4 where M is magnesium, IRON, manganese and calcium with minor amounts of nickel. The main natural olivines derive from the SOLID SOLUTION from FORSTERITE (Fo_{100}) (Mg_2SiO_4) to FAYALITE (Fa_{100}) (Fe_2SiO_4).

omphacite ((Ca,Na)(Mg,Fe,Al)Si_2O_6) A PYROXENE found in ECLOGITE.

oncoid (oncolith) A type of carbonate COATED GRAIN coated by microbial mats in which the laminae are irregular in thickness, relief and continuity.

oncolite A rock composed mainly of ONCOIDS.

oncolith See ONCOID.

onion-skin weathering See SPHEROIDAL WEATHERING.

onlap (overlap) A STRUCTURE in which successive wedge-shaped BEDS extend further than the margin of the underlying BED, such as would be formed in an expanding sedimentary BASIN. Cf. OFFLAP.

Onnian A STAGE of the ORDOVICIAN, 444.0–443.1 Ma.

onset time See ARRIVAL TIME.

Ontarian A SILURIAN succession in North America covering parts of the AERONIAN and TELYCHIAN.

ontogeny The development of an organism through various stages. Cf. PHYLOGENY.

onyx (SiO_2) A layered, MICROCRYSTALLINE variety of SILICA used for decorative purposes.

onyx marble (Oriental alabaster) Banded CALCITE or ARAGONITE used for decorative purposes.

ooid (oolith) A type of small (<2 mm) carbonate or iron COATED GRAIN with a cortex of concentric fine laminae, lacking biogenic features, and a nucleus, often a shell fragment or SAND grain.

oolite A rock composed mainly of OOIDS.

oolith See OOID.

oomicrite A LIMESTONE containing OOLITHS in a MICRITE MATRIX.

oosparite An OOLITIC LIMESTONE with a sparry cement.

ooze A PELAGIC MUD.

opal ($SiO_2.nH_2O$) An hydrated variety of SILICA made up of minute (~ 300 nm) spheres, including the GEM precious opal.

opal agate A variety of OPAL which is AGATE-like in structure.

opalescence The play of changing colours caused by the thin-film interference of light along planes of voids between the packed spheres forming the MINERAL.

opaque mineral A MINERAL appearing black in THIN SECTION in transmitted plane-polarized light.

open fold A FOLD with an INTERLIMB ANGLE between 70° and 120°.

open hole A borehole with no casing.

opencast mining (openpit mining) Mining by excavation from the surface.

openpit mining See OPENCAST MINING.

ophicalcite See FORSTERITE-MARBLE.

Ophiocistioidea A class of subphylum ELEUTHEROZOA, phylum ECHINODERMATA, seemingly intermediate between the ECHINOIDEA and HOLOTHUROIDEA. Range L. ORDOVICIAN–DEVONIAN.

ophiolite An association of ULTRABASIC/ ULTRAMAFIC–BASIC/MAFIC rock types believed to represent a section through the LITHOSPHERE of a BACK-ARC BASIN emplaced onto CONTINENTAL or older OCEANIC CRUST by OBDUCTION.

ophitic texture A type of POIKILITIC TEXTURE in slow-cooling BASIC IGNEOUS ROCKS in which EUHEDRAL, randomly oriented PLAGIOCLASE laths are wholly enclosed within PYROXENE plates. Cf. SUBOPHITIC TEXTURE.

Ophiuroidea (brittle stars) A class of subphylum ELEUTHEROZOA, phylum ECHINODERMATA; the brittle stars. Range L. ORDOVICIAN–RECENT.

Opoitian A PLIOCENE succession in New Zealand covering part of the ZANCLIAN.

optic axis The direction in an ANISOTROPIC crystal along which there is no BIREFRINGENCE.

optical emission spectroscopy An analysis method for MAJOR and TRACE ELEMENTS in which a powdered sample is mixed with GRAPHITE and vaporized in a carbon arc. The spectrum of the light emitted can be analyzed to provide the concentrations of elements present. Now largely superseded by other techniques.

orbicular Descriptive of spherical to subspherical masses comprising concentric shells of different composition in PLUTONIC IGNEOUS ROCKS.

orbital forcing The control of terrestrial processes by changes in the Earth's orbit. See MILANKOVITCH CYCLE.

orbital ripple A type of steep WAVE RIPPLE.

Ordian A CAMBRIAN succession in Australia covering part of the ATDABANIAN, the LENIAN and part of the SOLVAN.

Ordnance Datum (OD) SEA LEVEL DATUM in the UK, measured at Newlyn or Liverpool, to which all BENCH MARKS are related.

Ordovician A sub-ERA of the PALAEOZOIC, 510.0–439.0 Ma.

ore A metalliferous MINERAL or an AGGREGATE of metalliferous MINERALS mixed with GANGUE that can be exploited at a profit.

ore microscopy The technique of examining OPAQUE MINERALS in reflected light with a polarizing microscope.

ore mineral A MINERAL from which a useful metal can be extracted.

ore reserve (indicated reserve, inferred reserve, measured reserve) The calculable tonnage of ORE in an ORE-BODY, including that believed to be present.

ore shoot A thin, ribbon-like extension of an OREBODY.

orebody A volume of rock that can be exploited commercially for its metal content.

orendite A SILICA-SATURATED LAMPROITE with SANIDINE rather than LEUCITE, often carrying DIAMONDS.

Oretian A TRIASSIC succession in New Zealand covering parts of the LADINIAN and CARNIAN.

organic weathering The disintegration and decomposition of rock by the action of microorganisms, plants, animals and decaying organic matter.

Oriental alabaster See ONYX MARBLE.

Oriental almandine A deep red CORUNDUM of GEM quality.

Oriental amethyst Purple CORUNDUM or SAPPHIRE.

Oriental cat's eye See CYMOPHANE.

Oriental emerald GEM-quality CORUNDUM with the colour of EMERALD.

Oriental topaz A variety of CORUNDUM with the colour of TOPAZ.

oriented core A drilled rock sample oriented with respect to the present horizontal and TRUE NORTH.

oriented lake A lake with a PREFERRED ORIENTATION of its long axis.

Ornithischia An order of subclass ARCHOSAURIA, class REPTILIA; herbivorous DINOSAURS with a bird-like pelvis. Range late JURASSIC–end CRETACEOUS.

Orochenian A CRETACEOUS succession in the far east of the former USSR equivalent to the CONIACIAN, SANTONIAN, CAMPANIAN and MAASTRICHTIAN.

orocline Old term for an OROGENIC BELT

that appeared to have been bent into a horse-shoe shape.

orogen (orogenic belt) The total volume of rock deformed during an OROGENY.

orogenesis (orogeny, tectogenesis) The process of creation of a mountain belt by TECTONIC activity, generally by the collision of continental PLATES or MICROPLATES. Characterized by REGIONAL METAMORPHISM, igneous activity and vertical movements. Cf. EPEIROGENY.

orogenic andesite association An association of CALC-ALKALINE rocks found at SUBDUCTION ZONES.

orogenic belt See OROGEN.

orogenic cycle A term referring to the cyclicity of OROGENIC events with time.

orogeny See OROGENESIS.

Orowan-Elsasser convection A convective model for the Earth in which the tops of CONVECTION cells are the oceanic LITHOSPHERE and PLATES are driven by the expansion of MANTLE material at OCEAN RIDGES and the pull of descending LITHOSPHERE at SUBDUCTION ZONES.

orpiment (As_2S_3) A rare ORE MINERAL of arsenic found in VEINS and HOT SPRING deposits.

Orthida An order of class ARTICULATA, phylum BRACHIOPODA; BRACHIOPODS with biconvex, impunctate shells, broad, straight hinge lines and no brachidium. Range L. CAMBRIAN–PERMIAN.

orthite See ALLANITE.

ortho- Genuine, right, straight, upright.

orthoamphibole An AMPHIBOLE with an orthorhombic structure, e.g. ANTHOPHYLLITE.

orthoclase ($KAlSi_3O_8$) A common FELDSPAR.

orthoconglomerate A CLAST-SUPPORTED CONGLOMERATE deposited by aqueous currents. Cf. PARACONGLOMERATE.

orthoconic Descriptive of a shell in the form of a straight, tapering cone.

orthoenstatite An orthorhombic POLY-TYPE of ENSTATITE.

orthoferrosilite (FeSiO₃) An ORTHOPY-ROXENE.

orthogneiss A GNEISS formed from an IGNEOUS ROCK parent. Cf. PARAGNEISS.

orthogonals See WAVE RAYS.

orthomagmatic (orthotectic) Descriptive of the main stage of CRYSTALLIZATION from a MAGMA, when up to 90% of it may crystallize.

orthomagmatic segregation deposit (magmatic segregation deposit) An ORE deposit that has crystallized directly from a MAGMA, produced by FRACTIONAL CRYSTALLIZATION or LIQUATION.

orthophyric Descriptive of a fine- or medium-grained SYENITIC rock with closely packed ORTHOCLASE crystals.

orthopyroxene A PYROXENE crystallizing in the ORTHORHOMBIC SYSTEM ranging between ENSTATITE and FERROSILITE.

orthoquartzite (quartzarenite) An unmetamorphosed sedimentary QUARTZITE. Cf. METAQUARTZITE.

orthorhombic system A CRYSTAL SYSTEM whose members have three mutually perpendicular axes of different lengths.

orthosilicate See NESOSILICATE.

orthotectic See ORTHOMAGMATIC.

Osagean A CARBONIFEROUS succession in the USA covering parts of the TOURNAISIAN and VISÉAN.

osar See ESKER.

oscillatory flow An aqueous flow which reverses periodically.

oscillatory zoning ZONING within a crystal comprising alternating layers of the two end-members of a SOLID SOLUTION series.

ossicle A single ECHINODERM plate.

Osteichthyes A class of superclass PISCES; the bony FISH. Range L. DEVONIAN–RECENT.

Osteostrachi An order of subclass MONORHINA, class AGNATHA, superclass PISCES; small, jawless FISH with a flattened body covered by polygonal plates and a broad head covered by a bony shield. Range SILURIAN–DEVONIAN.

Ostracoda/ostracods (ostracodes) A class of subphylum CRUSTACEA, phylum ARTHROPODA; animals with a small, bivalved, calcareous carapace of vastly diversified morphology, useful in BIOSTRATIGRAPHY. Range CAMBRIAN–RECENT.

Ostracodermi/ostracoderms An old or informal term for the FOSSIL AGNATHA (FISH) of the ORDOVICIAN to CARBONIFEROUS.

ostracodes See OSTRACODA.

Ostwald step rule 'POLYMORPHIC transformations proceed in steps from the unstable form through a series of metastable states of decreasing free energy until the stable form is achieved.'

Otaian A MIOCENE succession in New Zealand covering the AQUITANIAN and part of the BURDIGALIAN.

Otamitian A TRIASSIC succession in New Zealand covering the upper part of the CARNIAN.

Otapirian A TRIASSIC succession in New Zealand equivalent to the RHAETIAN.

Oteke A JURASSIC succession in New Zealand covering part of the TITHONIAN.

otolith The ear-bone of a FISH.

ottrelite A manganese-bearing CHLORITOID found in SCHISTS as the result of CONTACT METAMORPHISM of certain ARGILLACEOUS rocks.

ouady See WADI.

oued See WADI.

out-of-phase component See IMAGINARY COMPONENT.

out-of-sequence thrust A THRUST FAULT in an IMBRICATE STRUCTURE which causes deviation from a PIGGYBACK THRUST sequence in which SHORTENING takes place on the lowest, latest THRUST.

outcrop The total area over which a particular rock unit occurs at the surface.

outcrop mapping See EXPOSURE MAPPING.

outlier An area of younger rocks surrounded by older rocks. Cf. INLIER.

overbank deposit Suspended sediment and fine BEDLOAD deposited by floodwater on a FLOODPLAIN when BANKFULL DISCHARGE is exceeded.

overburden Loose, unconsolidated material resting on BEDROCK; the unwanted rock overlying material of value, such as an OREBODY.

overconsolidated sediment A sediment which, sometime in its history, has been subjected to pressure greater than at present and so is stronger than anticipated. Cf. NORMALLY CONSOLIDATED SEDIMENT, UNDERCONSOLIDATED SEDIMENT.

overcoring (doorstopper technique) A type of *in situ* STRESS MEASUREMENT in which small STRAINS are produced after the release of STRESS around a cylinder at the end of a borehole.

overfold A FOLD in which the AXIAL SURFACE and both FOLD LIMBS DIP in the same direction.

overland flow The flow of water downslope across the soil surface.

overlap See ONLAP.

overlap integral A measure of the spatial overlap between orbitals in MOLECULAR ORBITAL theory.

overpressuring (geopressuring) The PRESSURE in excess of HYDROSTATIC PRESSURE in subsurface sediments, caused when the flow of fluids out of sediments is impeded. Common in actively subsiding sedimentary BASINS where COMPACTION is occurring.

overprinting The superimposition of a MESOSCOPIC to MICROSCOPIC younger STRUCTURE on an older one.

oversaturated Descriptive of an IGNEOUS ROCK containing free SILICA. See also SILICA SATURATION.

overstep A STRATIGRAPHIC relationship at an UNCONFORMITY where the oldest unit of the younger sequence is in contact with more than one of the older BEDS beneath. Implies tilting or FOLDING before deposition of the younger BEDS.

overstep propagation sequence A THRUST FAULT sequence in an IMBRICATE STRUCTURE in which successively later THRUSTS form closer to the HINTERLAND than existing STRUCTURES; earlier THRUSTS and related FOLDS can be truncated by later THRUSTS.

overthrust 1 DISPLACEMENT of a THRUST SHEET over the edge of an adjacent rock. 2 See THRUST.

overthrust fault See THRUST.

overvoltage (electrode polarization) An INDUCED POLARIZATION effect caused by the different rates of electronic and electrolytic current flow. The impedance of electrolyte flow by metallic MINERALS, in which flow is electronic, gives rise to a charge build-up which gradually decays when the driving potential difference is removed, causing a decaying voltage.

overwash fan A deposit formed on the landward side of a BEACH RIDGE and the BACKSHORE zones of a BARRIER ISLAND or low SPIT when it is breached by storm waves.

oxbow lake (mortlake) A curved lake isolated from a stream when an acute meander is cut off from the channel.

Oxfordian A STAGE of the JURASSIC, 157.1–154.7 Ma.

oxidation 1 The loss of electrons from an ion or atom. 2 The loss of hydrogen or the addition of oxygen through chemical reaction.

oxide minerals MINERALS formed by the combination of an element or elements with oxygen, common as ACCESSORY MINERALS.

oxygen isotope analysis The determination of the ratio of the oxygen isotopes ^{18}O and ^{16}O by MASS SPECTROMETRY, used in determining the provenance of artefacts and in OXYGEN ISOTOPE STRATIGRAPHY.

oxygen isotope stratigraphy The use of the oxygen isotope ratios of FORAMINIFERA for STRATIGRAPHIC purposes.

oxygen minimum layer A level in the ocean, generally between depths of 150 m and 1000 m, where the concentration of oxygen is at a minimum, the bounding limit being 0.2 ml l^{-1}. With a high influx of organic matter and the oxygen minimum layer at the seabed, ANOXIC sediments may accumulate.

oxy-hornblende (basaltic hornblende, lamprobolite) $(Ca_2(Na,K)_{0.5-1.0}(Mg, Fe^{2+})_{3-4}(Fe^{3+},Al)_{2-1}(Si_6Al_2O_{22})(O,OH,F)_2)$ An AMPHIBOLE found in VOLCANIC ROCKS.

oxyphile element See LITHOPHILE ELEMENT.

ozocerite (mineral wax) A dark PARAFFIN wax found in irregular VEINS.

P

P In EARTHQUAKE SEISMOLOGY, a P WAVE that does not travel through the CORE.

P foliation A feature of a FAULT GOUGE comprising a PHYLLOSILICATE PREFERRED ORIENTATION or layering at an acute angle to the SHEAR direction.

P shear A subsidiary FAULT in a SHEAR ZONE.

P wave A SEISMIC BODY WAVE which propagates by the vibration of particles forwards and backwards in the direction of propagation, i.e. as a series of compressions and rarefactions.

Pa See PASCAL.

Pacific-type coast (concordant coast, longitudinal coast) A smooth coastline which runs parallel to the trend of topography and geological STRUCTURE. Cf. ATLANTIC-TYPE COAST.

packer test (Lugeon test) A method of measuring the PERMEABILITY of STRATA in a borehole by isolating a section with an inflatable tube, filling it with water and timing the water loss per metre.

packstone A GRAIN-SUPPORTED, CLASTIC LIMESTONE containing some lime MUD.

paedomorphosis The retention of ancestral juvenile characters in a descendant adult.

pagodite See AGALMATOLITE.

pahoehoe A thin (0.1–2m) LAVA FLOW with a thick, smooth, wavy surface, formed by low-VISCOSITY LAVA.

paired metamorphic belt Juxtaposed low-temperature/high-pressure and high-temperature/low-pressure linear belts believed to be characteristic of the OCEAN TRENCH and ISLAND ARC environments, respectively, of a SUBDUCTION ZONE. There is now evidence that paired metamorphic belts may have a more complex TECTONIC history.

paisanite A sodic MICROGRANITE with RIEBECKITE as the main MAFIC MINERAL.

palaeo- (paleo-) Ancient.

palaeoautecology The study of the past ecology of individual organisms or taxonomic groups. Cf. PALAEOSYNECOLOGY.

palaeobiogeography The study of the distribution of ancient organisms.

palaeobiology See PALAEONTOLOGY.

palaeobotany The study of FOSSIL plants.

Palaeocene The oldest EPOCH of the CENOZOIC, 65.0–60.5 Ma.

palaeoclimatology The study of ancient climates.

palaeocurrent A current which influenced sediment deposition at some time in the past.

palaeoecology The study of the past interactions of organisms with each other and their environment. See also PALAEOAUTECOLOGY, PALAEOSYNECOLOGY.

palaeoflow An ancient flow, whose direction can be assessed from the geometry of the CROSS-STRATIFICATION it produces.

Palaeogene The older PERIOD of the CENOZOIC, 65.0–23.3 Ma.

palaeogeographical map A map showing the OUTCROP pattern at a particular time.

palaeogeography The study of ancient geography, e.g. past distributions of land and ocean.

Palaeognathae Superorder of subclass NEORNITHES, class AVES; large, flightless BIRDS. Range late CRETACEOUS–RECENT.

palaeohydrology The reconstruction of FLOW REGIMES for ancient river channels.

palaeomagnetic pole A location of the GEOMAGNETIC pole, assumed to be an AXIAL DIPOLE, in the past, calculated from PALAEOMAGNETIC data. Cf. VIRTUAL MAGNETIC POLE.

palaeomagnetism The study of the magnetic properties of rocks, particularly the NATURAL REMANENT MAGNETIZATIONS, which can provide information on the PALAEOMAGNETIC POLE position and the ancient intensity of the GEOMAGNETIC FIELD.

palaeontology (palaeobiology) The study of ancient organisms.

palaeopiezometry The measurement of the previous state of STRESS in a rock.

palaeosalinity An ancient salinity.

palaeosol A soil formed on a past landscape, documented for both the PRECAMBRIAN and PHANEROZOIC.

palaeosome The apparently older part of a composite rock body.

palaeosynecology The study of the past ecology of FOSSIL COMMUNITIES as a whole. Cf. PALAEOAUTECOLOGY.

palaeotemperature An ancient temperature, which can be determined from oxygen isotopes in sediments and estimated from some FOSSILS.

Palaeozoic An ERA comprising the CAMBRIAN to PERMIAN systems.

palagonite A VOLCANIC ROCK consisting of hydrated and chemically altered BASALTIC HYALOCLASTITE, the ALTERATION occurring at any temperature in the presence of seawater or METEORIC WATER.

palagonitization The process of ALTERATION of BASALTIC HYALOCLASTITE to PALAGONITE.

paleo- See PALAEO-.

palimpsest structure (relict structure) A relic of an original TEXTURE or STRUCTURE visible through a superimposed TEXTURE or STRUCTURE.

palingenesis See ANATEXIS.

palinspastic map A geological map in which DEFORMATION has been removed so that the rocks are displayed in their pre-TECTONIC configuration.

palinspastic reconstruction The reconstruction of the geometry of a set of rocks before DEFORMATION.

palladium (Pd) A very rare, PLATINUM-group NATIVE METAL.

pallasite A STONY METEORITE comprising metals and silicates.

palsa A PEAT mound above an ice lens in PERMAFROST. See also PINGO.

paludal Relating to a marsh.

palustrine In a PALUDAL environment.

palygorskite ($(MgO)_2(SiO_2)_2.4H_2O–Al_2O_3$ $(SiO_2)_5.6H_2O$) A CLAY MINERAL with a chain-like structure, found mainly in HYDROTHERMAL VEINS or in altered SERPENTINE or GRANITIC rocks.

palynology The study of carbonaceous microorganisms, including pollen, spores, DINOFLAGELLATES and ACRITARCHS.

palynomorph A micro-organism studied in PALYNOLOGY.

pan A closed depression, often one of a great number, in arid to semi-arid areas, caused mainly by DEFLATION, but also by SOLUTION and animal activity.

Pan-African orogeny An OROGENY during the PROTEROZOIC from 900–600 Ma affecting the Arabian and African SHIELDS.

pandemic distribution See COSMOPOLITAN DISTRIBUTION.

panfan An arid landscape of eroded hills and ridges, extensive, coalescing PEDIMENTS and infilled BASINS.

Pangaea (Pangea) The single SUPERCONTINENT comprising all the CONTINENTAL CRUST in the late PALAEOZOIC, which split into the SUPERCONTINENTS LAURASIA and GONDWANALAND at ~300 Ma.

Pangea See PANGAEA.

panidiomorphic Descriptive of an IGNEOUS ROCK with well-developed crystals.

panplain (planplain) An area of very subdued relief, formed by the coalescence of FLOODPLAINS and resulting from the lateral migration of streams.

pantellerite A type of PERALKALINE RHYOLITE with >12.5% NORMATIVE FEMIC MINERALS. Normally weakly PORPHYRITIC with PHENOCRYSTS of ALKALI FELDSPAR, sodic FERROHEDENBERGITE, FAYALITE, AENIGMATITE, AMPHIBOLE, QUARTZ and Fe-Ti oxides. Found in continental RIFTS and rarely on OCEANIC ISLANDS.

Panthalassa The ancestral Pacific Ocean which surrounded PANGAEA.

Pantotheria An order of infraclass TRITUBERCULATA, subclass THERIA, class MAMMALIA; egg-laying insectivores of shrew-like appearance. Range M.–U. JURASSIC.

para- Parallel to, resembling.

parabolic dune A crescentic DUNE whose arms are tethered by vegetation and point upwind.

paraconformity See NON-SEQUENCE.

paraconglomerate A MATRIX-SUPPORTED CONGLOMERATE, MIXTITE or pebbly MUDSTONE deposited by a SEDIMENT GRAVITY FLOW or as GLACIAL TILL. Cf. ORTHOCONGLOMERATE.

paraffin series See METHANE SERIES.

paragenesis A MINERAL association, generally expressed in terms of a time sequence.

paragenetic sequence The order of deposition or CRYSTALLIZATION of MINERALS, commonly used in connection with EPIGENETIC HYDROTHERMAL DEPOSITS.

paragneiss A GNEISS derived from DETRITAL SEDIMENTARY ROCK.

paragonite $(NaAl_2(AlSi_3O_{10})(OH)_2)$ A MICA ISOSTRUCTURAL with MUSCOVITE found mainly in SCHISTS, PHYLLITES, QUARTZ VEINS and fine-grained sediments.

paralic basin A sedimentary BASIN in a marginal marine environment. Cf. LIMNIC BASIN.

parallel extinction (straight extinction) The EXTINCTION of a MINERAL which takes place parallel to its CLEAVAGE traces or margins.

parallel fold A FOLD in which the layer thickness is constant everywhere perpendicular to the folded layer surface.

parallel retreat A type of slope evolution in which the form and angle of the slope remains constant as it is eroded.

parallel twin A TWINNING phenomenon in which the TWIN AXIS lies in the COMPOSITION PLANE.

paramagnetism The magnetic property of some materials which acquire a relatively weak magnetization when placed in a MAGNETIC FIELD in the same direction as that field which is lost when the field is removed.

paramorph A PSEUDOMORPH formed by the conversion of one POLYMORPH to another.

parasitic antiferromagnetism The magnetization arising in an ANTIFERROMAGNETIC material when the CRYSTAL LATTICE is slightly distorted and a SPONTANEOUS MAGNETIZATION is produced. Exhibited by HAEMATITE.

parasitic cone (adventive cone) A small cone on the flank of a VOLCANO.

parasitic fold (satellite fold) A SECOND ORDER FOLD with a pattern of asymmetry depending on its position with respect to FIRST ORDER FOLD HINGES.

parataxitic texture A EUTAXITIC TEXTURE with greatly elongated FIAMME, caused by secondary FLOW.

Paratethys A large seaway extending from north of the Alps to the east of the Aral Sea from OLIGOCENE to PLIOCENE times.

paratype A specimen selected to show additional characters to a HOLOTYPE.

parautochthon A large structural unit which has travelled a short distance from its site of origin. Cf. ALLOCHTHON, AUTOCHTHON.

parental magma A MAGMA capable of fractionating to yield a suite of more evolved, derivative liquids.

Pareora A MIOCENE succession in New Zealand comprising the OTAIAN and ALTONIAN.

pargasite ($NaCa_2Fe_4(Al,Fe)Al_2Si_6O_{22}(OH)_2$) An AMPHIBOLE found as a constituent of IGNEOUS and METAMORPHIC ROCKS.

parna An AEOLIAN CLAY (LOESS) deposit in the form of a DUNE or a thin, discontinuous, widespread sheet, possibly derived from the DEFLATION of unvegetated, saline lake floors or other soil or alluvial surfaces.

partial area model A concept proposed to account for the low ratio of RUNOFF to rainfall in CATCHMENTS where HORTONIAN OVERLAND FLOW is the dominant storm RUNOFF mechanism. Only small parts of the BASIN contribute to surface RUNOFF while elsewhere INFILTRATION CAPACITY is sufficiently high for complete INFILTRATION. Cf. VARIABLE AREA MODEL.

partial dislocation A crystal DISLOCATION whose DISPLACEMENT is a fraction of the UNIT CELL.

partial melting The incomplete melting of rock to produce a melt of different composition.

particle cluster A grouping of particles on a bed through mutual interference between the constituent CLASTS, characteristic of GRAVEL-bed rivers.

particulate flow A CATACLASTIC DEFORMATION MECHANISM.

parting The breaking along planes of weakness in MINERALS, e.g. TWIN PLANES.

Similar to CLEAVAGE, but only found in those crystals that are TWINNED or deformed.

partition coefficient (distribution coefficient) The ratio of the concentration of an element between a MINERAL and a coexisting melt. Used to quantify PETROGENETIC models based on TRACE ELEMENTS.

pascal (Pa) The SI unit of PRESSURE or STRESS in $kg\ m^{-1}s^{-2}$.

pascichnia TRACE FOSSILS comprising grazing traces.

passive continental margin (quiescent continental margin, trailing continental margin) A CONTINENTAL MARGIN which is not a PLATE margin and represents the site of CONTINENTAL SPLITTING prior to the SEAFLOOR SPREADING that carried it to its present position. Cf. ACTIVE CONTINENTAL MARGIN.

patch reef See MICRO-ATOLL.

patronite (VS_4) A SULPHIDE ORE MINERAL of vanadium.

patterned ground Symmetrical forms, such as circles, stripes and polygons, developed on soil, sediment and fractured BEDROCK by frost processes and CRYOTURBATION.

pause plane The surface in a CROSS-STRATIFIED BED which represents a time of minimum flow velocity.

paved bed A layer similar to an ARMOURED SURFACE but more stable and much coarser than the substrate. Forms as a LAG deposit of the material too coarse to be transported away.

pay streak A zone of concentrated heavy MINERALS in a PLACER DEPOSIT.

Payntonian A CAMBRIAN succession in Australia covering the upper part of the DOLGELLIAN.

PDR See PRECISION DEPTH RECORDER.

peacock coal An IRIDESCENT BITUMINOUS COAL or ANTHRACITE found in upper mine levels where GROUNDWATER has deposited a film of IRON OXIDE along FRACTURES.

peacock ore A popular name for BORNITE.

peak zone See ACME ZONE.

pearlspar A rhombohedral form of DOLOMITE or ANKERITE.

pearly lustre An IRIDESCENT, pearl-like LUSTRE shown on the CLEAVAGE surfaces of layer lattice silicates, e.g. TALC.

peat A mass of dark brown, partly decomposed, fibrous plant debris. The precursor of COAL, requiring substantial growth, standing water to prevent OXIDATION or bacterial destruction and an absence of introduced DETRITAL sediment.

peatland See BOG.

pebble A rounded rock fragment of 2–64 mm in diameter.

Péclet number A dimensionless number whose magnitude determines the relative importance of DIFFUSION over flow in a sediment.

pectolite $(Ca_2NaH(SiO_3)_3)$ A PYROXENOID INOSILICATE found commonly in cavities in BASALTIC rocks.

pediment A gently sloping (0.5°–7°), concave-up EROSION surface on the flank of a steep-sided hill or mountain and common at the base of MESAS or INSELBERGS. Possibly the product of SHEETFLOOD EROSION or BACKWEARING by PARALLEL RETREAT.

pediplain A landform created by coalescing PEDIMENTS.

Pedleian A STAGE of the CARBONIFEROUS, 332.9–331.1 Ma.

pedogenesis The process of soil formation, depending on the interplay between parent, climate, organisms, topography and time. The main processes involved are CALCIFICATION, ELUVIATION, FERRALITIZATION, GLEYING, HUMIFICATION, ILLUVIATION, LEACHING, PEDOTURBATION, PODSOLIZATION, RUBIFACTION, SALINIZATION and WEATHERING.

pedogenetic calcrete A CALCRETE originating in a soil, mainly by biologically induced CALCITE PRECIPITATION.

pedology The study of soils.

pedoturbation A PEDOGENETIC process involving the churning of soil by physical processes.

peel method A technique used in carbonate SEDIMENTOLOGY and PALAEONTOLOGY in which calcareous material is etched by weak acid to enhance relief and polyvinylacetate sheeting rolled on the surface after flooding with acetone. The resulting mould is peeled off along with a thin surface layer which can reveal structures and can be stained to show extra details.

pegmatite A very coarse-grained IGNEOUS ROCK with PHENOCRYSTS over 250 mm in length, usually of GRANITIC composition and forming at a late stage of CRYSTALLIZATION.

pegmatite deposit A PEGMATITE enriched in MINERALS of economic interest, often providing the following elements: lithium, beryllium, rubidium, cesium, niobium, tantalum and TIN.

Pel'skiy A PERMIAN succession in the Timan area of the former USSR equivalent to the U. SAKMARIAN.

pelagic Descriptive of the deep sea environment.

pelagite A PELAGIC sediment, generally decimetres thick, formed by the slow settling of calcareous and siliceous biogenic material from suspension.

pelecypods See BIVALVIA.

Pelé's hair See ACHNELITH.

Pelé's tears Cylindrical, spherical or tear-shaped drops of PYROCLASTIC, GLASSY BASALT 6–13 mm in length which solidified during flight.

peléean eruption A VOLCANIC ERUPTION in which ASH FLOWS <1 km^3 in volume, accompanied by a NUÉE ARDENTE, sweep down the flanks of a VOLCANO and deposit block and ASH FLOWS as radiating FANS.

Pelican Creek A PERMIAN succession in Queensland, Australia, covering the upper part of the WORDIAN.

pelite An aluminium-rich rock formed by the METAMORPHISM of CLAY-rich sediments.

pelitic With a similar chemical composition to SHALE.

pellet A small ovoid to spherical particle of MICRITE with no internal structure, 0.03–0.15 mm in diameter, possibly of faecal or ALGAL origin.

Pelmatozoa A subphylum of phylum ECHINODERMATA; ECHINODERMS with the aboral surface expanded to form a functional stalk and the water vascular system extended into elongate appendages. Range CAMBRIAN–RECENT.

pelmicrite A LIMESTONE composed of MICRITE and small, rounded AGGREGATES of sediment.

peloid A SAND-size grain of carbonate MUD.

pelsparite A LIMESTONE composed of PELLETS in a SPARITE MATRIX.

Pelycosauria An order of subclass SYNAPSIDA, class REPTILIA; originally small lizard-like REPTILES, radiating in the PERMIAN to large carnivorous, piscivorous and herbivorous forms, some with a dorsal 'sail'. Range U. CARBONIFEROUS–L. PERMIAN.

Penck and Brückner model A framework for understanding the PLEISTOCENE history of the Alps based on four main GLACIAL phases: GÜNZ, MINDEL, RISS and WÜRM. Now superseded by a more complex history.

pene- Almost.

penecontemporaneous Very shortly after or before.

penecontemporaneous dolomitization The formation of DOLOMITE from LIMESTONE by METASOMATIC ALTERATION very soon after deposition when the LIMESTONE was still unconsolidated. Cf. SUBSEQUENT DOLOMITIZATION.

peneplain A low-angle ground surface developing at a late stage in the CYCLE OF EROSION.

penetration twin A TWIN in which two crystals penetrate each other.

penetrative Descriptive of a feature, such as CLEAVAGE or FABRIC, developed throughout a rock at the scale of observation. Cf. NON-PENETRATIVE.

Peng Lai-Zhen A JURASSIC/CRETACEOUS succession in Sichuan, China, covering the TITHONIAN and L. CRETACEOUS.

pennantite $((Mn,Al)_6(Si,Al)_4O_{10}(OH)_8)$ A CHLORITE group MINERAL found in manganese ORES.

Pennsylvanian The younger sub-PERIOD of the CARBONIFEROUS, 322.8–290.0 Ma.

Pentamerida An order of class ARTICULATA, phylum BRACHIOPODA; BRACHIOPODS with normally smooth, biconvex, impunctate shells, often with two diverging plates in the brachial VALVE and a median septum in the pedicle VALVE. Range M. ORDOVICIAN–U. DEVONIAN.

pentlandite $((Fe,Ni)_9S_8)$ The most important ORE MINERAL of nickel.

Penutian A PALAEOCENE/EOCENE succession on the west coast of the USA covering parts of the THANETIAN and YPRESIAN.

peperite A BRECCIA comprising GLASSY LAVA fragments and sedimentary material formed by MAGMA flowing over or through wet sediment.

peralkaline A composition with an excess of $(Na_2O + K_2O)$ over Al_2O_3.

peraluminous A composition with an excess of Al_2O_3 over $(Na_2O + K_2O)$.

peramorphosis The appearance of ancestral adult characters in a descendant juvenile.

percentage contour map A subsurface map showing the percentage of one lithology in the total thickness of a STRATIGRAPHIC unit.

percentage frequency effect (PFE) A parameter quantifying the INDUCED

POLARIZATION effect in FREQUENCY DOMAIN surveys, defined as $100(\rho_0 - \rho_\infty)/\rho_\infty$, where ρ_0 and ρ_∞ are APPARENT RESISTIVITIES measured at <1 Hz and a few tens of Hz respectively.

perched aquifer A locally developed, water-saturated body located above the regional WATER TABLE due to the presence of an underlying impermeable layer.

perched water table The top surface of a PERCHED AQUIFER.

percussion mark A LUNATE scar on a PEBBLE formed by a sharp blow, such as would occur in a high-velocity flow.

perennial head The highest point along the course of a river from which flow always occurs.

perennial stream A stream or river with continuous flow. Cf. EPHEMERAL STREAM.

pergelation The formation of permanently frozen ground.

peri- Near, around.

periclase (MgO) An OXIDE MINERAL occurring in CONTACT METAMORPHOSED LIMESTONE.

pericline A FOLD with elliptical or circular OUTCROP in which DIPS vary around the STRUCTURE.

pericline twinning A type of lamellar TWINNING, shown by FELDSPARS in which the TWIN AXIS is $b[101]$.

peridot A clear green, GEM variety of OLIVINE.

peridotite An ULTRAMAFIC rock with 40–90% OLIVINE and PYROXENE, including HARZBURGITE, WEHRLITE and LHERZOLITE.

periglacial Referring to a wide range of cold, non-GLACIAL climatic and geomorphic conditions, irrespective of the proximity to a GLACIER.

perihelion The location in the Earth's elliptical orbit where it is nearest to the Sun. Cf. APHELION.

period A second order geological time unit.

Perissodactyla An order of infraclass

EUTHERIA, subclass THERIA, class MAMMALIA; the ungulate, odd-toed herbivores, e.g. horses. Range EOCENE–RECENT.

peristerite A very fine-scale intergrowth of two PLAGIOCLASE FELDSPARS in the range An_{2-15}, invisible to the naked eye but possibly producing IRIDESCENCE in MOONSTONE.

perknite A coarse-grained, ULTRAMAFIC IGNEOUS ROCK comprising PYROXENES and AMPHIBOLES but no FELDSPAR.

perlite An hydrated, silicic, volcanic GLASS with curved, concentric, 'onion-skin' cracks probably arising from volume changes associated with the HYDRATION. Used commercially for thermal insulation and as a rooting medium.

perlitic texture A TEXTURE of GLASSY and DEVITRIFIED IGNEOUS ROCKS comprising curved or spherical to subspherical cracks produced by contraction during cooling.

permafrost Rock, soil and sediment in which temperatures remain below 0°C for at least two consecutive winters and the intervening summer.

permanent strain The retained STRAIN resulting from DUCTILE DEFORMATION before FAILURE.

permanent water table The lowest level to which the WATER TABLE falls in a given locality.

permeability The coefficient linking flow rate to the pressure gradient in a medium in the DARCY EQUATION.

Permian The youngest PERIOD of the PALAEOZOIC, 290.0–245.0 Ma.

permineralization (petrification) The preservation of organic hard parts by MINERAL-bearing GROUNDWATER infiltrating porous matter after burial. The common preserving MINERALS are SILICA, CALCITE and IRON OXIDES.

permitted emplacement The EMPLACEMENT of an IGNEOUS BODY in which the COUNTRY ROCK is passively displaced. Cf. FORCEFUL EMPLACEMENT.

perovskite ($CaTiO_3$) An OXIDE MINERAL

found in NEPHELINE SYENITES and CARBONATITES.

perthite A MACROSCOPIC-scale intergrowth of POTASSIUM FELDSPAR and ALBITE.

perthosite A sodic SYENITE mainly composed of PERTHITIC FELDSPARS.

Petalichthydia An order of class PLACODERMI, superclass PISCES; FISH resembling ARTHRODIRES, but BENTHONIC in habit. Range L. DEVONIAN–U. DEVONIAN.

petalite ($Li(AlSi_4O_{10})$) A TECTOSILICATE important as an ORE MINERAL of lithium found in GRANITE PEGMATITES.

petrification See PERMINERALIZATION.

petro- Rock.

petrofabric The FABRIC elements of, usually, an IGNEOUS ROCK.

petrofabric diagram See FABRIC DIAGRAM.

petrogenesis All aspects of the formation of a rock.

petrography The systematic description of rocks in hand specimen and THIN SECTION.

petroleum All naturally occurring HYDROCARBONS: ASPHALT, BITUMEN, CRUDE OIL, GAS HYDRATES and NATURAL GAS.

petrology The study of all aspects of rocks.

petromict conglomerate (polymict conglomerate) A CONGLOMERATE with CLASTS of a variety of compositions. Cf. OLIGOMICT CONGLOMERATE.

Petrotsvet A CAMBRIAN succession in Siberia covering part of the TOMMOTIAN and the ATDABANIAN.

petzite (Ag_3AuTe_2) A steel-grey to black SILVER-GOLD telluride.

PFE See PERCENTAGE FREQUENCY EFFECT.

phacoidal structure A STRUCTURE in an IGNEOUS or METAMORPHIC ROCK in which lens-shaped fragments are present.

phacolith A metric- to kilometric-scale, concavo-convex lens of INTRUSIVE IGNEOUS ROCK between BEDS in an ANTIFORM.

phaneritic Descriptive of a TEXTURE of an IGNEOUS ROCK in which crystals are discernible by the unaided eye.

phanerocrystalline Descriptive of an IGNEOUS ROCK in which the crystals of all ESSENTIAL MINERALS are discernible to the unaided eye.

Phanerozoic Post-PRECAMBRIAN time.

pharmacolite ($CaHAsO_4.2H_2O$) Hydrated calcium arsenate, formed by the ALTERATION of ARSENOPYRITE and other arsenical ORES.

pharmacosiderite ($Fe_3(AsO_4)_2(OH)_3.5H_2O$) An hydrated IRON arsenate formed by the ALTERATION of arsenic ORE.

phase component system An ELECTROMAGNETIC INDUCTION METHOD in which the relative magnitudes of the REAL and IMAGINARY COMPONENTS of the secondary field generated by a conductor are measured. Cf. DIP-ANGLE SYSTEM.

phase diagram A graphical method of illustrating the equilibrium boundaries of different phases of a chemical system in terms of temperature, pressure and composition.

phase transformation The isochemical transformation of one MINERAL phase into another, e.g. OLIVINE → SPINEL with increasing pressure.

phase velocity The velocity of a specific crest or trough of a waveform. Cf. GROUP VELOCITY.

phenakite (Be_2SiO_4) A rare NESOSILICATE found in PEGMATITES.

phengite A high-SILICA variety of MUSCOVITE.

phenoclast A CLAST larger than 4 mm in diameter in a SEDIMENTARY ROCK.

phenocryst A large, generally EUHEDRAL, MINERAL grain within the fine-grained MATRIX of an IGNEOUS ROCK, probably reflecting slow cooling and growth prior to the cooling giving rise to the MATRIX. See also PORPHYRITIC TEXTURE.

phi (φ) unit A unit used to express CLAST

sizes in CLASTIC ROCKS according to $\phi = -\log_2 x$, where x is the grain size in mm. See also UDDEN-WENTWORTH SCALE.

Philip equation A theoretical equation which describes the INFILTRATION process.

phillipsite $(KCa(Al_3Si_5O_{16}).6H_2O)$ A ZEOLITE found in cavities in BASALT, PHONOLITE and related rocks, in SALINE LAKE deposits, in calcareous deep-sea sediments and in HOT SPRING deposits.

phlogopite $(KMg_3(AlSi_3O_{10})(OH)_2)$ A MICA found mainly in metamorphosed LIMESTONES and ULTRABASIC ROCKS.

Pholidota An order of infraclass EUTHERIA, subclass THERIA, class MAMMALIA; the pangolins. Range ?MIOCENE-RECENT.

phonolite A fine-grained, commonly PORPHYRITIC, FELSIC IGNEOUS ROCK with >10% MODAL FELDSPATHOID plus ALKALI FELDSPAR and minor sodium-rich AMPHIBOLE or PYROXENE. Occurs in association with ALKALI BASALT-TRACHYTE in ALKALINE volcanic provinces and OCEANIC ISLANDS.

phosgenite (horn lead) $(Pb_2CO_3Cl_2)$ A rare CARBONATE MINERAL.

phosphate rock An IGNEOUS or SEDIMENTARY ROCK with a high concentration of phosphate MINERALS, commonly the FRANCOLITE-APATITE series.

Phosphatocopida An order of class OSTRACODA, phylum ARTHROPODA; OSTRACODS with a phosphatic carapace. Range U. CAMBRIAN.

phosphorite (rock phosphate) A PHOSPHATE ROCK which occurs in beds from centimetres to tens of metres thick, composed of grains of CRYPTOCRYSTALLINE carbonate FLUORAPATITE or COLLOPHANE and detrital material. Forms under low latitude, marine conditions in shallow water. The major commercial source of phosphates.

photic zone (euphotic zone) The zone at the surface of the ocean, from a few metres to 150 m thick, through which light penetrates at sufficient intensity for photosynthesis. Cf. APHOTIC ZONE, DIPHOTIC ZONE.

photo- Pertaining to light.

photogeological map A geological map constructed from AERIAL PHOTOGRAPHY.

phreatic eruption The explosive ejection of TEPHRA consisting of MUD and LITHIC material, but no juvenile MAGMATIC material, from a VOLCANIC VENT, resulting from the explosive boiling of GROUNDWATER in response to indirect MAGMATIC heating.

phreatic gas Gas originating in the atmosphere or ocean.

phreatic zone (zone of permanent saturation) The zone beneath the WATER TABLE in which intergranular pores and fissures are completely filled with water at HYDROSTATIC PRESSURES in excess of atmospheric.

phreatomagmatic eruption (hydrovolcanic eruption) An explosive interaction between MAGMA and water, typically producing very fine-grained TEPHRA.

phreatoplinian deposit A very widespread, very fine-grained air-fall TEPHRA, resulting from the reaction of silicic MAGMA and water.

phyletic gradualism A theory of evolutionary development of species by gradual, continuous change. Cf. PUNCTUATED EQUILIBRIUM MODEL.

phyllarenite A LITHIC ARENITE whose rock fragments are dominantly of METAMORPHIC origin.

phyllic alteration ALTERATION found in the COUNTRY ROCKS of PORPHYRY COPPER and molybdenum deposits.

phyllite A REGIONALLY METAMORPHOSED, FOLIATED, PELITIC rock.

Phyllocarida A subclass of class MALACOSTRACA, subphylum CRUSTACEA, phylum ARTHROPODA, characterized by a large, bivalved carapace, an abdomen of seven somites and a telson with a simple paired appendage. Range CAMBRIAN-RECENT.

Phyllolepida An order of class PLACODERMI, superclass PISCES; freshwater FISH with extensive dermal armour whose plates have a concentric pattern of ridges. Range U. DEVONIAN.

phyllonite A DYNAMICALLY METAMORPHOSED rock rich in PHYLLOSILICATES in PREFERRED ORIENTATION, typically formed by RETROGRADE METAMORPHISM of SCHIST or GNEISS.

phyllosilicate (layer silicate, sheet silicate) A silicate with layers formed when each silicon tetrahedron shares three corners with other tetrahedra.

phylogeny The sequence of branching events involved in the evolution of a TAXON. Cf. ONTOGENY.

phyteral Plant material in COAL whose morphological form remains discernible.

phytogeomorphology The study of the relationships between plants and landforms.

phytokarst See BIOKARST.

phytolith A rock formed of plant material or by plant activity.

pi (π) diagram A STEREOGRAM on which PI POLES are displayed. Cf. BETA(β) DIAGRAM.

pi (π) pole On a STEREOGRAM, the best-fit GREAT CIRCLE through a GIRDLE DISTRIBUTION of POLES to a folded surface, i.e. the mean FOLD AXIS.

Piacenzian The higher STAGE of the PLIOCENE, 3.40–1.64 Ma.

pickeringite (magnesia alum) $(MgAl_2(SO_4)_4.22H_2O)$ An hydrated aluminium-magnesium sulphate, occurring in fibrous masses and formed by the WEATHERING of PYRITE-bearing SCHISTS.

picotite (chrome spinel) $((Fe,Cr)Al_2O_4)$ A form of HERCYNITE with appreciable chrome replacing aluminium.

picrite A dark-coloured VOLCANIC or minor intrusive ROCK with abundant OLIVINE (totalling >90% with other ferromagnesian MINERALS) and <10% PLAGIOCLASE.

piedmont Descriptive of a feature at the base of a mountain.

piedmontite See PIEMONTITE.

piemontite (manganepidote, piedmontite) $(Ca_2MnAl_2O(SiO_4)Si_2O_7)$ A SOROSILICATE ISOSTRUCTURAL with EPIDOTE.

piercement fold A FOLD caused by the DIAPIRIC intrusion of an EVAPORITE.

piercement structure A STRUCTURE arising from the piercing of STRATA by the DIAPIRIC intrusion of an EVAPORITE.

piezometric surface See POTENTIOMETRIC SURFACE.

piezometry The measurement of PRESSURE or STRESS.

piezoremanent magnetization A NATURAL REMANENT MAGNETIZATION acquired by a rock subjected to sudden impact, such as by a METEORITE, or prolonged STRESS. Of potential value in studying changes in seismically induced STRESS from variations in the local GEOMAGNETIC FIELD.

pigeonite $(\sim Ca_{0.25}(Mg,Fe)_{1.75}Si_2O_6)$ A PYROXENE found in rapidly cooled IGNEOUS ROCKS.

piggyback basin A SEDIMENTARY BASIN formed on a PIGGYBACK THRUST sheet.

piggyback propagation sequence An IMBRICATE STRUCTURE in which new THRUST segments develop closer to the FORELAND than existing THRUSTS, so that all the SHORTENING at a particular time is taken up on the lowest, latest THRUST.

piggyback thrust A THRUST in a PIGGYBACK PROPAGATION SEQUENCE.

pillar structure A water escape structure which characteristically penetrates flat laminae lacking DISH STRUCTURES.

pillow lava A VOLCANIC ROCK of BASALTIC composition comprising rounded, sack-like bodies, 0.2–2 m in diameter, separated from each other by fine-grained rinds. Forms when LAVA FLOWS come into contact with water.

pilotaxitic texture A TEXTURE comprising a felted mass of ACICULAR or lath-like crystals, which may show flow structure. See also HYALOPILITIC.

pin line A reference point on a BALANCED or RESTORED SECTION, the distance between two of which provides an estimate of the SHORTENING.

pinacoid A CRYSTAL FORM whose faces are parallel to two of the axes.

pinch-and-swell structure A repetitive thinning and thickening of a body of rock, possibly formed when a FAULT crosses a series of BEDS at different inclinations so that the FAULT movement creates a series of openings.

pinger A high-frequency, shallow penetration, marine SEISMIC SOURCE used in SEISMIC REFLECTION profiling.

pingo A conical hill in PERMAFROST formed above a body of ice. See also PALSA.

pinite A fine-grained, blue mixture of MUSCOVITE and CHLORITE formed by the ALTERATION of CORDIERITE, SPODUMENE, FELDSPAR, etc.

pinnate fracture (feather fracture) A minor FRACTURE which intersects a larger FRACTURE at an acute angle, probably formed due to the TENSILE STRESSES generated by frictional sliding on the main FRACTURE.

pipe 1 A subsurface channel from several millimetres to three metres in diameter, usually formed in soil or PEAT in which flow is concentrated in cracks rather than being absorbed by the MATRIX. **2** A vertical or subvertical, tubular OREBODY, often acting as a feeder to a MANTO.

pipe clay See BALL CLAY.

Piripauan A CRETACEOUS succession in New Zealand covering parts of the SANTONIAN and CAMPANIAN.

Pisces (fish) A superclass of aquatic, cold-blooded, vertebrate animals with persistent gills, first appearing in the FOSSIL record in the ORDOVICIAN.

pisoid (pisolith) A carbonate COATED GRAIN over 2 mm in diameter, with an origin similar to an OOID.

pisolite A rock composed of PISOIDS.

pisolith See PISOID.

pisolitic texture A TEXTURE suggestive of being formed by PISOIDS.

pistacite See EPIDOTE.

pitch 1 (rake) The orientation of a line, measured as an angle from the horizontal, in a specified non-vertical plane. PLUNGE may be derived from a series of pitch measurements using a STEREOGRAM. **2** See ASPHALT.

pitch coal (bituminous brown coal) A BRITTLE, lustrous BITUMINOUS COAL or LIGNITE with a CONCHOIDAL FRACTURE.

pitch lake See TAR PIT.

pitchblende A massive, AMORPHOUS, MICROCRYSTALLINE variety of URANINITE, the major ORE of uranium.

pitchstone An hydrated, RECRYSTALLIZED, silicic, volcanic GLASS, typically with an irregular FRACTURE and a dull, resinous appearance, sometimes with a SPHERULITIC TEXTURE.

pixel The smallest picture element recorded by a REMOTE SENSING device.

place value The economic relevance of the location of a MINERAL deposit which reflects the proportion of the value of the material represented by transport costs.

placentals See EUTHERIA.

placer deposit A sedimentary deposit of economic MINERALS concentrated by natural, mechanical processes. The concentration mechanism is usually GRAVITY-driven and accomplished by moving water in which the dense MINERALS sink. The placer MINERALS must be durable after being freed from the source rock, and the main ones are CASSITERITE, CHROMITE, COLUMBITE, COPPER, GARNET, GOLD, ILMENITE, MAGNETITE, MONAZITE, PLATINUM, RUBY, RUTILE, SAPPHIRE, XENOTIME and ZIRCON.

Placodermi A class of superclass PISCES; an extinct group of armoured, jawed FISH with a mobile, bony carapace over the head and shoulders. Range early–late DEVONIAN.

Placodontia An order of subclass EURYAPSIDA, class REPTILIA; heavily armoured REPTILES similar in appearance to turtles. Range TRIASSIC.

plagioclase A series of FELDSPARS with compositions in the range $NaAlSi_3O_8$ to $CaAl_2Si_2O_8$.

plagiogranite A GRANITE relatively rich in PLAGIOCLASE.

plagionite ($Pb_5Sb_8S_{17}$) A rare ORE MINERAL of lead.

planar crystal defect A CRYSTAL DEFECT in a structure comprising layer modules which are displaced irregularly in different directions relative to an adjacent layer.

planar slump A type of MASS MOVEMENT involving sliding along a planar surface.

planation surface A relatively flat plain resulting from EROSION.

planchéite ($Cu_8(Si_4O_{11})_2(OH)_4.H_2O$) A blue, hydrated, SECONDARY MINERAL of COPPER with INOSILICATE structure.

plane bed A flat sediment bed with any irregularities less than 2–3 grain diameters in height but many diameters in length, over which there is sediment transport.

plane bed lamination A thin, flat-bedded LAMINATION formed by the AGGRADATION of PLANE BEDS. Cf. GRAIN FALL LAMINATION.

plane strain STRAIN in three dimensions in which the intermediate PRINCIPAL STRAIN is 1, i.e. there is no EXTENSION or CONTRACTION in the intermediate PRINCIPAL STRAIN direction.

planeze A wedge-shaped LAVA FLOW on the slopes of a dissected VOLCANO protecting underlying material from EROSION.

planktonic Floating passively in surface water.

planplain See PANPLAIN.

plastering See UNDERPLATING.

plastic deformation See PLASTICITY.

plastic limit An ENGINEERING GEOLOGY term for the minimum amount of water mixed with a sediment necessary for it to deform plastically under standard conditions. Cf. LIQUID LIMIT.

plasticity (plastic deformation) DEFORMATION that causes permanent, continuous STRAIN without BRITTLE FAILURE or a significant change in volume arising from applied STRESS in excess of the YIELD STRESS.

plate A large segment of oceanic or continental LITHOSPHERE that is in relative motion with adjacent segments. Up to 12 major plates have been recognized and a large number of MICROPLATES.

plate boundary (plate margin) The lateral margin of a PLATE, represented by an OCEAN RIDGE (CONSTRUCTIVE PLATE MARGIN), SUBDUCTION ZONE (DESTRUCTIVE PLATE MARGIN) or TRANSFORM FAULT (CONSERVATIVE PLATE MARGIN).

plate boundary force The FORCE acting on the lateral margins of a PLATE: RIDGE PUSH, SLAB PULL and SUBDUCTION SUCTION.

plate margin See PLATE BOUNDARY.

plate tectonics The generally fully accepted theory that the solid Earth's surface is made up of a small number of large PLATES of LITHOSPHERE which are in relative motion and internally largely undeformed, the majority of the Earth's TECTONIC activity taking place at the PLATE MARGINS.

plateau Any large, relatively flat area at high altitude.

plateau lava One of a number of LAVA FLOWS making up a great volume which covers earlier topography, creating a new, nearly flat PLATEAU between 1000 km^2 and 300 000 km^2 in area.

plateau uplift A kilometric-scale UPLIFT of large PLATEAU regions of CONTINENTAL CRUST, originating in ISOSTATIC response to UNDERPLATING, TECTONIC thickening, thermal expansion and HYDRATION or phase changes in the lower CRUST.

platinum (Pt) A very rare NATIVE METAL ORE MINERAL.

playa lake (alkali flat, dry lake, inland sabkha, salina, salt flat) A continental, shallow, dried-up, BRINE lake in a SALINE LAKE depositional complex.

Playfair's law See LAW OF ACCORDANT JUNCTIONS.

Pleistocene The older EPOCH of the QUATERNARY, 1.64–0.01 Ma.

Pleistogene See QUATERNARY.

pleochroism (dichroism) A phenomenon exhibited by some crystals in THIN SECTION whose colour changes on rotation in plane polarized light.

pleonaste (ceylonite) A green variety of SPINEL intermediate in composition between SPINEL and HERCYNITE.

plesiosaurs Large, extinct, exclusively aquatic, marine REPTILES with powerful, paddle-like limbs and a variety of adaptations for macrophagy and microphagy. Range late TRIASSIC–CRETACEOUS.

Pleurocanthodii An order of subclass ELASMOBRANCHII, class CHONDRICHTHYES, superclass PISCES; freshwater sharks. Range late DEVONIAN–L. TRIASSIC.

plicate Folded, wrinkled.

Pliensbachian A STAGE of the JURASSIC, 194.5–187.0 Ma.

plinian eruption A continuous, high rate VOLCANIC ERUPTION of volatile-rich MAGMA which is torn apart by VESICULATION with the resulting mix of PUMICE or SCORIA, ASH and VOLCANIC GAS forming a column rising rapidly above the VOLCANIC VENT. When mixed with air the column increases in buoyancy and can rise to up to 50 km. Generates widely dispersed, sheet-like, air-fall deposits with good SORTING.

plinthite A thick, well-cemented horizon which forms by the ILLUVIAL ACCRETION of FERRICRETE or other hard soil crusts, often LATERITIC and protecting underlying materials from EROSION.

Pliocene An EPOCH of the NEOGENE, 5.2–1.64 Ma.

pliosaurs Predatory PLESIOSAURS with large heads.

plis de couverture FOLDS in a cover sequence above a basal DETACHMENT HORIZON or DÉCOLLEMENT.

plis de fond FOLDS in the basement below a DÉCOLLEMENT.

plucking (glacial plucking) The removal of large particles of an irregular rock surface by a GLACIER.

plug 1 A cylindrical feeder of a VOLCANO filled with solidified MAGMA and/or PYROCLASTIC material and subsequently exposed by DENUDATION. **2** Any small, vertical to subvertical mass of IGNEOUS ROCK in the form of a cylinder. Cf. STOCK.

plumbago An obsolete name for GRAPHITE.

plume 1 A curving trace on the surface of a FRACTURE. FRACTURES are initiated at the convergence of plumes and propagate with the FRACTURE surface approximately orthogonal to the plume. **2** See HOTSPOT. **3** A streak of effluent entering water.

plumose Like a feather.

plunge The angle between a linear STRUCTURE and a vertical plane. Cf. PITCH.

plus-minus method (Hagedoorn method) A SEISMIC REFRACTION interpretation method for REVERSED PROFILES over non-planar STRATA whose velocity is variable. Cf. GENERALIZED RECIPROCAL METHOD.

pluton A large, thick, IGNEOUS BODY with steep lateral contacts which was emplaced and CRYSTALLIZED beneath the surface, possibly now exposed as an irregular polygonal OUTCROP.

plutonic Originating at great depth.

pluvial 1 (lacustral) A period of increased moisture availability as the result of increased precipitation and/or reduced evaporation. **2** Relating to rain or other precipitation.

pneumatolysis ALTERATION by hot gas, excluding water, commonly fluorine, hydrofluoric acid and boron fluorides.

Occurs in the late stage cooling of an IGNEOUS ROCK and can affect the igneous material and COUNTRY ROCK.

Podolskian A STAGE of the CARBONIFEROUS, 307.1–305.0 Ma.

podsol (podzol) An ILLUVIALLY accumulated soil with an upper horizon from which aluminium and IRON oxides and hydroxides have been LEACHED.

podsolization (podzolization) A PEDO-GENETIC process in which iron, aluminium and organic matter move downwards to give a prominent ELUVIAL horizon with residual SILICA.

podzol See PODSOL.

podzolization See PODSOLIZATION.

poikilitic texture An INEQUIGRANULAR TEXTURE seen in GABBROIC and ULTRAMAFIC ROCKS in which an OIKOCRYST encloses several randomly oriented smaller grains of another phase or phases.

poikilo- Spotted.

poikiloblastic texture A TEXTURE similar to POIKILITIC TEXTURE, seen in METAMORPHIC ROCKS.

poikilotopic fabric A FABRIC of a SEDIMENTARY ROCK in which coarse cement crystals enclose smaller, detrital grains.

point bar A channel bar of MUD to coarse CONGLOMERATE forming on the convex side of a channel bend due to reduced flow velocity or FLOW SEPARATION. EPSILON CROSS-STRATIFICATION may develop as the channel MIGRATES.

point crystal defect A CRYSTAL DEFECT caused by departure of the structure from perfect regularity at a point.

point group A group of symmetry elements in a crystal.

Poisson solid An homogeneous elastic solid whose SHEAR MODULUS and LAMÉ'S CONSTANT are equal, giving a POISSON'S RATIO of 0.25.

Poisson's equation The relationship between GRAVITATIONAL and MAGNETIC POTENTIALS which allows the transformation of a MAGNETIC FIELD into a PSEUDOGRAVITY FIELD and a GRAVITY field into a PSEUDOMAGNETIC FIELD.

Poisson's number The reciprocal of POISSON'S RATIO.

Poisson's ratio The ratio of lateral STRAIN to longitudinal STRAIN in an elastic body due to uniaxial longitudinal STRESS.

polar wander The true or apparent motion of the north MAGNETIC POLE over the Earth's surface. See APPARENT POLAR WANDER, TRUE POLAR WANDER.

polar wobble Regular and irregular changes in the Earth's axis of rotation in space. The main regular change is annual with an amplitude of 100 marcs and caused by varying planetary gravitational attractions and changes in oceanic and atmospheric gas distributions. See also CHANDLER WOBBLE, MARKOWITZ WOBBLE.

polarity epoch (polarity interval) A period of constant polarity of the GEOMAGNETIC FIELD, usually >10 000 years.

polarity excursion A time interval when the GEOMAGNETIC FIELD attempted unsuccessfully to reverse polarity.

polarity interval See POLARITY EPOCH.

polarity reversal A relatively rapid change in the polarity of the GEOMAGNETIC FIELD in which the north MAGNETIC POLE becomes the south MAGNETIC POLE and vice versa.

polarity timescale A timescale based on reversals of the GEOMAGNETIC FIELD.

polarization A transverse waveform, such as light or a SEISMIC WAVE, may have random transverse vibration directions, but polarization causes all the vibrations to be in a single direction.

polder A flat, low-lying area of land reclaimed from the sea or a lake by artificial drainage.

pole figure A STEREOGRAM showing the distribution of CRYSTALLOGRAPHIC orientations, normally as poles to CRYSTALLO-GRAPHIC planes on an EQUAL-AREA PLOT.

pole of a plane The normal to a plane, a more convenient method of plotting plane data from a large number of measurements on a STEREOGRAM.

pole of rotation See EULER POLE.

polje A large, commonly flat-floored, closed depression in a KARST area, of equivocal origin.

pollucite ($CsAlSi2O_6.H_2O$) A rare FELDSPATHOID found in PEGMATITES.

poloidal field A field, such as the GEOMAGNETIC FIELD, with both radial and tangential components. Cf. TOROIDAL FIELD.

Poltava An OLIGOCENE/MIOCENE succession on the Russian Platform covering the CHATTIAN and part of the AQUITANIAN.

poly- Many.

polybasite ($Ag_{16}Sb_2S_{11}$) A rare ORE MINERAL of SILVER found in low- to medium-temperature SILVER VEIN deposits.

polyclinal fold A FOLD whose AXIAL SURFACE is variably inclined along and between the folded layers.

polygonal karst A KARST landscape of a closely packed assemblage of closed, polygonal depressions.

polygonization The formation of polygonal grains or SUBGRAINS. One of the processes contributing to RECOVERY by reducing the surface energy of grain boundaries. Occurs by the CLIMB and CROSS SLIP of DISLOCATIONS towards the boundaries at high temperatures.

polyhalite ($K_2Ca_2Mg(SO_4)_4.2H_2O$) An EVAPORITE MINERAL, a source of potassium.

polymict Descriptive of a CLASTIC SEDIMENTARY ROCK composed of more than one MINERAL type. Cf. MONOMICT.

polymictic lake A lake whose waters are in continuous circulation. Cf. OLIGOMICTIC LAKE.

polymorph One of two or more solid state forms of a chemical compound with different CRYSTAL STRUCTURES.

polymorphic transformation The transformation of a MINERAL into a POLYMORPH, a mechanism of NEOMORPHISM. See also RECRYSTALLIZATION.

polymorphism The ability of a chemical compound to exist as two or more POLYMORPHS, e.g. DIAMOND and GRAPHITE, CALCITE and ARAGONITE.

polyphyletic Descriptive of a group of organisms evolved from different ancestral stock as a result of CONVERGENT EVOLUTION. Cf. MONOPHYLETIC.

Polyplacophora A subclass of class AMPHINEURA, phylum MOLLUSCA; marine MOLLUSCS whose dorsal surface is covered by seven or eight calcareous plates. Range CAMBRIAN–RECENT.

polysomatic series A range of structures found in crystals made up of layer modules which have different compositions and atomic arrangements but which can fit together, e.g. the relationship between PYROXENES, AMPHIBOLES and PYRIBOLES.

polysynthetic twinning (laminar twinning, multiple twinning) A type of INVERSION TWINNING in which the two twin orientations are equally likely to be adopted and are often found as closely spaced lamellae.

polytypism A type of POLYMORPHISM in which a compound exists with two or more layer-like structures differing in their stacking sequences. Small differences in chemical compositions (<0.25 atoms per formula unit) are allowed.

polytypoid A compound showing extreme POLYTYPISM in which there are large differences in chemical composition (>0.25 atoms per formula unit).

ponor A SINKHOLE or SWALLOW HOLE found in a LIMESTONE area.

pontic Deposited in deep, still water.

pool and riffle sequence Alternating deeps (pools) and shallows (riffles) along a

river channel spaced at 5–7 times the channel width. The pools probably arise from EROSION of the outer bank and deposition on the inner bank of a meander.

pop-up The relatively uplifted part of the HANGINGWALL of a BACKTHRUST.

Porangian An EOCENE succession in New Zealand covering the middle part of the LUTETIAN.

porcellanite An archaeological term for a rock formed by the THERMAL METAMORPHISM of a soil horizon in BASALT.

pore fluid (pore water) A solution occupying the pore spaces in soil or rock, whose composition reflects its origin, subsequent mixing and DIAGENETIC interaction with the host sediment. Plays an important role in DIAGENESIS.

pore fluid pressure The PRESSURE exerted by PORE FLUIDS, comprising a NORMAL STRESS with no DEVIATORIC components and equal PRINCIPAL STRESSES, which can exceed HYDROSTATIC PRESSURE in OVERPRESSURED rocks. Causes FAILURE of rocks at lower differential STRESS than in its absence by lowering the applied NORMAL STRESS.

pore throat A restricted opening that connects adjacent pores.

pore water See PORE FLUID.

Porifera (sponges) A phylum intermediate between the PROTOZOA and the ANTHOZOA; sac-like organisms, attached to a substrate, with a central cavity opening upwards, which comprise a jelly colloid, sometimes containing calcareous or siliceous SPICULES which may be fossilised. Range L. CAMBRIAN–RECENT.

porosity The ratio of the fraction of voids to the volume of rock in which they occur.

porphyrite The central part of a PORPHYRY COPPER DEPOSIT with PORPHYRITIC TEXTURE.

porphyritic microgranite See GRANITE PORPHYRY.

porphyritic texture The TEXTURE of an IGNEOUS ROCK that contains PHENOCRYSTS.

porphyroblast (metacryst) A large EUHEDRAL to SUBHEDRAL MINERAL grain within a fine-grained MATRIX formed during METAMORPHIC RECRYSTALLIZATION.

porphyroclast A mass of the original rock remaining after METAMORPHISM.

porphyry An HYPABYSSAL rock containing PHENOCRYSTS, commonly of FELDSPAR.

porphyry copper deposit A large, low-GRADE STOCKWORK to DISSEMINATED DEPOSIT of COPPER which may also contain minor molybdenum, GOLD and SILVER, commonly in a GRANITIC COUNTRY ROCK.

porphyry gold deposit A DISSEMINATED and STOCKWORK DEPOSIT of GOLD in an intrusive IGNEOUS ROCK.

porphyry tin deposit A STOCKWORK TIN deposit, generally of low GRADE.

portal A location where a meltwater stream leaves the snout or front of an ice mass.

Portland stone A yellow-white OOLITIC LIMESTONE from the Isle of Portland, S. England, widely used as a BUILDING STONE.

Portlandian A regional British name for the highest STAGE of the JURASSIC (TITHONIAN).

portlandite ($Ca(OH)_2$) Calcium hydroxide, a rare, colourless, transparent MINERAL.

Post-Idamean A CAMBRIAN succession in Australia covering the lower part of the DOLGELLIAN.

post-vortex ripple A term for a very flat variety of WAVE RIPPLE.

pot earth See POTTER'S CLAY.

potash feldspar See POTASSIUM FELDSPAR.

potassic silicate alteration (potassium silicate alteration) A form of WALLROCK ALTERATION characterized by the formation of POTASH FELDSPAR and/or BIOTITE (BIOTIZATION) and ANHYDRITE.

potassic zone A zone characterized by the POTASSIC SILICATE ALTERATION of the host

STOCK in the LOWELL-GUILBERT MODEL of PORPHYRY COPPER DEPOSITS.

potassium feldspar (K-feldspar, potash feldspar) (KAlSi$_3$O$_8$) The general name for the potassium end-member of the ALKALI FELDSPAR series.

potassium silicate alteration See POTASSIC SILICATE ALTERATION.

potassium-argon dating A dating method based on the radioactive decay of ^{40}K to ^{40}Ar by electron capture. Potassium concentration is determined chemically and argon concentration by MASS SPECTROMETRY. Errors can occur by the presence of ^{40}Ar in the rock on formation and the subsequent loss of the gaseous argon.

potential The energy required to bring a unit quantity from infinity to the point of measurement against the ambient POTENTIAL FIELD.

potential field A field which obeys LAPLACE'S EQUATION, i.e. gravity, magnetic and electrical fields.

potentiometric surface (piezometric surface) An imaginary surface indicating the static HEAD OF WATER in an AQUIFER.

pothole 1 (swirlhole) A deep, circular hole in a river bed or CAVE stream caused by EROSION. **2 (aven, swallow hole)** A vertical shaft in a LIMESTONE area connecting a CAVE system with the surface.

potstone A massive variety of STEATITE.

potter's clay (pot earth) Any CLAY or earth that can be used in pottery making.

Pound Quartzite An EXCEPTIONAL FOSSIL DEPOSIT of VENDIAN age in the Flinders Ranges, Australia containing a diverse, shallow marine fauna preserved as casts and impressions.

Poundian The highest STAGE of the VENDIAN, 580–570 Ma.

powder diffractometer An instrument used in X-RAY DIFFRACTION ANALYSIS in which a powdered sample rotates in a focused beam to determine MINERAL characteristics from the angle between the incident and diffracted beams and the intensity of the latter.

powellite (CaMoO$_4$) A molybdate MINERAL with partial SOLID SOLUTION to SCHEELITE.

power spectrum The representation of a time or distance function as its power at the various frequencies or WAVENUMBERS present; the square of the amplitude spectrum.

power spectrum analysis See SPECTRAL ANALYSIS.

pozzolan A pumiceous ASH used mixed with lime to make cement by the Romans, nowadays used mixed with cement as a construction material.

Pragian 1 A STAGE of the DEVONIAN, 396.3–390.4 Ma. **2** A DEVONIAN succession in the former Czechoslovakia covering parts of the PRAGIAN and EMSIAN.

Prandtl number A dimensionless coefficient equal to the product of specific heat and KINEMATIC VISCOSITY divided by THERMAL CONDUCTIVITY.

Prandtl-Reuss equations A modification of the LEVY-MISES EQUATIONS incorporating a component of elastic behaviour.

prase A dull green, MICROCRYSTALLINE variety of SILICA.

Pratt hypothesis A model which suggests that ISOSTATIC COMPENSATION is achieved by varying the DENSITY of the outer layer of the Earth in inverse proportion to its elevation, i.e. mountain ranges are supposed to be made up of relatively low DENSITY rocks.

pre- Before.

Pre-Payntonian A CAMBRIAN succession in Australia covering the middle part of the DOLGELLIAN.

Precambrian The oldest (>570 Ma) ERA characterized by a paucity of organisms with hard parts capable of fossilization.

Precambrian iron formation A low-GRADE, METAMORPHIC iron deposit of PRECAMBRIAN age.

precious metal GOLD, SILVER or PLATINUM.

precipitate controlled replacement
The non-specific REPLACEMENT by highly concentrated, aggressive solutions regardless of the replaced MINERAL's composition, structure or surface free energy. Cf. HOST GRAIN CONTROLLED REPLACEMENT.

precipitation The deposition of an AUTHIGENIC MINERAL from a supersaturated PORE FLUID in either solid form by CRYSTALLIZATION or as a gel by FLOCCULATION followed by CRYSTALLIZATION, giving rise to total or partial CEMENTATION of the POROSITY of the COUNTRY ROCK. The two fundamental processes of precipitation are NUCLEATION and CRYSTAL GROWTH.

precision depth recorder (PDR) An ECHO SOUNDER of high precision.

preferred orientation A concentration of linear or planar, structural or FABRIC ELEMENTS in a particular attitude.

prehnite $(Ca_2Al(AlSi_3O_{10})(OH)_2)$ A green SECONDARY MINERAL found lining cavities in BASALTIC rocks.

prehnite-pumpellyite facies A META-MORPHIC FACIES characterized by the formation of PREHNITE, PUMPELLYITE and QUARTZ in BASIC IGNEOUS ROCKS under moderate pressure and low temperature, usually during BURIAL METAMORPHISM.

prescellite See PRESELITE.

preselite (prescellite) An archaeological term for the spotted DOLERITE of the Preseli Hills, SW Wales, used in the construction of Stonehenge.

pressure 1 (hydrostatic pressure, hydrostatic stress) A three-dimensional STRESS state in which the magnitude of STRESS is the same in all directions. **2** The FORCE per unit area acting on a surface.

pressure dissolution See PRESSURE SOLUTION.

pressure fringe A STRUCTURE at the margins of the more rigid body in a PRESSURE SHADOW formed by the growth of fibrous MINERALS.

pressure release A WEATHERING mechan-ism which causes spalling and EXFOLIATION of rocks which were once deeply buried by the release of PRESSURE when they are brought to the surface.

pressure shadow The region around a relatively rigid body in a deformed rock that has undergone EXTENSIONAL STRAIN, commonly containing new CALCITE or QUARTZ, whose growth is related to PRESSURE SOLUTION.

pressure solution (pressure dissolution) **1** The enhanced rate of transfer of material from a MINERAL grain into intergranular fluid caused by increasing STRESS as the external PRESSURE exceeds the PORE FLUID PRESSURE. **2** The process of STRAIN by DISSOLUTION, DIFFUSION through fluid on grain boundaries and redeposition of MINERALS. The DISSOLUTION is enhanced on surfaces normal to the maximum PRINCIPAL STRESS and deposition occurs preferentially on surfaces normal to the minimum PRINCIPAL STRESS.

pressure solution cleavage A CLEAVAGE developed by preferential solution at contact surfaces of grains or crystals when the external PRESSURE exceeds the PORE FLUID PRESSURE.

Priabonian The highest STAGE of the EOCENE, 38.6–35.4 Ma.

priderite $((K,Ba)(Ti,Fe^{3+})_8O_{16})$ A black-red OXIDE MINERAL found as an ACCESSORY in LEU-CITE-bearing rocks.

Pridoli The youngest EPOCH of the SILUR-IAN, 410.7–408.5 Ma.

Pridoli-Schichten A SILURIAN succession in Bohemia equivalent to the PRIDOLI.

primärumph A tectonically elevated DOME of rock which is ERODING at the same rate as the UPLIFT.

primary creep (transient creep) The initial stage of VISCOELASTIC STRAIN characterized by a concave-downwards STRAIN-time curve. See also SECONDARY CREEP, TERTIARY CREEP.

primary current lineation A low (2–3 grain diameters high), extensive, linear ridge of grains parallel to the flow direction on the surface of an UPPER STAGE PLANE BED. Results from high velocity SWEEP events in the VISCOUS SUBLAYER which entrains sediment and deposits it as flow-parallel ridges.

primary migration The first phase of the upward MIGRATION of HYDROCARBONS within and out of the SOURCE ROCK to the RESERVOIR ROCK. Cf. SECONDARY MIGRATION.

primary mineral A MINERAL formed at the same time as the rock bearing it. Cf. SECONDARY MINERAL.

primary porosity All the pore space initially present in a sediment at the time of deposition. Cf. SECONDARY POROSITY.

primary structure A STRUCTURE formed as a rock was formed, rather than by subsequent DEFORMATION, e.g. CROSS-BEDDING, SLUMP FOLDS.

Primates An order of infraclass EUTHERIA, subclass THERIA, class MAMMALIA; including the lemurs, monkeys, apes and man. Range PALAEOCENE–RECENT.

Primitive deposit A type of zinc- and COPPER-bearing MASSIVE SULPHIDE DEPOSIT which may contain GOLD and SILVER.

primitive circle The CYCLOGRAPHIC TRACE of a horizontal plane on a STEREOGRAPHIC PROJECTION.

principal axes of stress (stress axial cross) The normals to the PRINCIPAL PLANES OF STRESS.

principal finite strain FINITE STRAIN normal to the axis of maximum SHORTENING, along which it is assumed that CLEAVAGE is developed.

principal planes of stress The three mutually perpendicular planes in a STRESS system along which there are no SHEAR STRESSES.

principal strain axes The three axes of symmetry of a STRAIN ELLIPSOID.

principal strain planes The planes orthogonal to the three axes of a STRAIN ELLIPSOID.

principal strains The relative sizes of the three axes of a STRAIN ELLIPSOID, measured as STRETCHES or QUADRATIC ELONGATIONS.

principal stresses The values of STRESS along the principal axes of STRESS, termed the maximum, intermediate and minimum principal stresses.

Priscoan See HADEAN.

prismatic crystal A crystal with a prism as the main FORM, i.e. one dimension significantly greater than the others.

prismatic texture A TEXTURE of a METAMORPHIC ROCK characterized by PRISMATIC CRYSTALS.

Proanura An order of subclass LISSAMPHIBIA, class AMPHIBIA; extinct, frog-like AMPHIBIANS. Range L. TRIASSIC.

Proboscidea An order of infraclass EUTHERIA, subclass THERIA, class MAMMALIA; the elephants and their extinct relatives. Range ?MIOCENE–RECENT.

prochlorite See RIPIDOLITE.

prod-and-bounce mark A type of SOLE STRUCTURE or TOOL MARK. Prod marks are asymmetric, elongate, semicircular-triangular depressions in a sediment surface, the depression being broader and deeper at the downcurrent end, formed when a body in a sediment flow hits the sediment surface at an angle and momentarily stops. Bounce marks are more symmetrical depressions formed when a body impacts a bed at an acute angle and rises off it.

production well A WELL from which fluid is recovered.

proglacial lake A lake formed adjacent to the snout of a GLACIER.

progradation The forward movement of FACIES belts as a result of sediment supply.

Progymnospermopsida A class of division TRACHAEOPHYTA, kingdom plantae; the possible ancestors of the GYMNOSPERMOPSIDA. Range DEVONIAN–U. CARBONIFEROUS.

prokaryote The earliest type of primitive

organism with no cell wall around the nucleus. Cf. EUKARYOTE.

prolate Lengthened in the polar direction; rod- or spindle-shaped. Cf. OBLATE.

propagating rift An OCEAN RIDGE whose axis is growing in length while an adjacent, offset portion of the ridge (the doomed rift) is shortening. Often a propagating rift makes a small angle with the doomed rift and reflects a changed spreading direction.

propylitic alteration ALTERATION characterized by the development of CHLORITE, EPIDOTE, ALBITE and carbonate (CALCITE, DOLOMITE or ANKERITE).

propylitic zone A zone of PROPYLITIC ALTERATION around a PORPHYRY COPPER DEPOSIT in the LOWELL-GUILBERT MODEL.

Proterozoic The younger EON of the PRECAMBRIAN, 2500–570 Ma.

proto- First, foremost.

protocataclasite A cohesive FAULT ROCK with a random FABRIC and with 10–50% MATRIX. Cf. CATACLASITE, ULTRACATACLASITE.

protoclastic Descriptive of the structure of an IGNEOUS ROCK whose early crystals were broken or deformed by movement of the liquid fraction.

protoenstatite An orthorhombic POLYMORPH of ENSTATITE.

protomylonite A FOLIATED FAULT ROCK with 10–50% MATRIX. Cf. MYLONITE, ULTRAMYLONITE.

proton magnetometer (nuclear precession magnetometer, proton precession magnetometer) A MAGNETOMETER for measuring the strength of the GEOMAGNETIC FIELD by measuring the precessional frequency of protons which are returning to the GEOMAGNETIC FIELD direction after being deflected. The standard instrument for MAGNETIC SURVEYS and MAGNETIC OBSERVATORIES.

proton precession magnetometer See PROTON MAGNETOMETER.

protopyroxene An orthorhombic POLY-

MORPH of PYROXENE, not yet found naturally.

protore A MINERAL deposit which, by the action of further natural processes, may be upgraded to ORE.

Prototheria A subclass of class MAMMALIA of which the MONOTREMATA is the only order.

Protozoa Very small, primitive, unicellular animals including the FORAMINIFERA and RADIOLARIA. Range PRECAMBRIAN–RECENT.

proustite (light red silver ore, light ruby silver, ruby silver ore) (Ag_3AsS_3) An ORE MINERAL of SILVER found in VEIN deposits.

provenance The source area(s) of sedimentary material and the nature(s) of the rocks from which it is derived.

proximal Descriptive of a feature close to its source. Cf. DISTAL.

psammite A METAMORPHIC ROCK rich in QUARTZ.

psephite A coarse sediment or its METAMORPHIC form.

pseudo- False.

pseudobreccia An irregularly or partially DOLOMITIZED LIMESTONE in which the growth of some coarse crystals imparts a TEXTURE similar to a BRECCIA.

pseudobrookite $(FeTiO_5)$ An OXIDE MINERAL found in cavities in BASALT.

pseudogravity field The GRAVITY field derived from a MAGNETIC FIELD using POISSON'S EQUATION for the purposes of interpretation.

pseudokarren KARREN-type forms developed on non-carbonate rocks.

pseudokarst Landforms and landscapes resembling those of LIMESTONE areas but developed on non-carbonate rocks.

pseudoleucite A mixture of NEPHELINE, ORTHOCLASE and ANALCIME PSEUDOMORPHOUS after LEUCITE.

pseudomagnetic field The MAGNETIC FIELD derived from a GRAVITY field using

POISSON'S EQUATION for the purposes of interpretation.

pseudomalachite (tagilite) ($Cu_5(PO_4)_2(OH)_4.H_2O$) COPPER phosphate and hydroxide, found as a SECONDARY MINERAL, which resembles MALACHITE.

pseudomorph A SECONDARY MINERAL which has replaced another but maintained its shape.

pseudonodule A spherical body of SANDSTONE with an internal convolute LAMINATION set in a MUDSTONE, resulting from the sinking of SAND from an overlying bed into soft MUD.

pseudophite A massive variety of CHLORITE sometimes used as a substitute for JADE.

pseudospar Recrystallized MICRITE with grains >20 μm in size. Cf. MICROSPAR.

pseudosymmetry An arrangement of the atoms in a CRYSTAL STRUCTURE which is nearly, but not exactly, symmetrical about the TWIN AXIS or across the TWIN PLANE.

pseudotachylite A very fine-grained or GLASS-like FAULT ROCK, formed by rapid DISPLACEMENT and melting by SHEAR-generated heating.

pseudowollastonite ($CaSiO_3$) A high temperature (>120°C) form of WOLLASTONITE.

psilomelane A BOTRYOIDAL mass of manganese OXIDE MINERALS, of which ROMANECHITE is a major constituent.

pteropod ooze A PELAGITE composed primarily of the remains of PTEROPODS.

pteropods Animals with thin, calcareous shells forming gelatinous zooplankton which contributes to OOZES and MARINE SNOW.

Pterosauria/pterosaurs A class of subclass ARCHOSAURIA, class REPTILIA; flying REPTILES. Range JURASSIC–late CRETACEOUS.

Ptyctodontida An order of class PLACODERMI, superclass PISCES; small FISH with limited armour at the back of the head,

superficially resembling modern FISH. Range M. DEVONIAN–U. DEVONIAN.

ptygmatic fold A rounded, generally CONCENTRIC FOLD whose AMPLITUDE is large with respect to the thickness of the folded layer and whose WAVELENGTH is small, generally approaching an ELASTICAS FOLD PROFILE. Common in layers of high COMPETENCE in a less COMPETENT MATRIX.

Puaroan A JURASSIC succession in New Zealand covering part of the TITHONIAN.

pudding ball See ARMOURED MUD BALL.

puddingstone A colloquial term for coarse CONGLOMERATE, particularly the silicified variety from the EOCENE Reading Beds of Hertfordshire, England.

pulaskite A FELSIC ALKALI SYENITE mainly comprising ALKALI FELDSPAR plus subordinate FELSIC MINERALS and often a small amount of NEPHELINE.

pull-apart basin (rhombochasm) A small EXTENSIONAL BASIN, formed where two parallel STRIKE-SLIP FAULTS join.

pumice A light coloured, highly VESICULAR, ACID volcanic GLASS with a low DENSITY, used in construction because of its insulating properties. Cf. SCORIA.

pumpellyite ($Ca_2Al_2(Al,Mg,Fe)Si_2O_{10}(OH)_2(O,OH)_2$) A silicate MINERAL formed during moderate pressure and low temperature METAMORPHISM.

pumping test A method of measuring the properties of an AQUIFER by pumping from one WELL and monitoring the expansion of the CONE OF DEPRESSION in surrounding observation WELLS. The inclination of the cone is an indication of the HYDRAULIC GRADIENT, which depends on the pumping rate and the TRANSMISSIVITY and STORAGE COEFFICIENT of the AQUIFER.

Punctuated Aggradational Cycles A model for an ALLOCYCLIC mechanism of generating cyclic carbonate deposits on a platform.

punctuated equilibrium model A theory of evolutionary development of species involving morphological stasis and rapid evolutionary changes taking place at a rate too fast for STRATIGRAPHIC resolution. Cf. PHYLETIC GRADUALISM.

pure shear A shape change without volume change in which the STRAIN AXES do not rotate.

pure strain DEFORMATION that can affect each point in a body differently.

purple copper ore See BORNITE.

Pusgillian A STAGE of the ORDOVICIAN, 443.1–440.6 Ma.

puy A volcanic hill, often a PLUG, of the Massif Central, France.

pycnocline A DENSITY interface in a body of water, e.g. a THERMOCLINE.

pyralspite An acronym for the GARNETS PYROPE, ALMANDINE and SPESSARTITE. Cf. UGRANDITE.

pyrargyrite (dark ruby silver, ruby silver ore) (Ag_3SbS_2) An ORE MINERAL of silver found in VEIN deposits.

pyriboles (biopyriboles) A group of MINERALS with a MINERALOGY common to the PYROXENES and AMPHIBOLES, but also including certain non-classical chain silicates.

pyrite (iron pyrites) (FeS_2) The most common SULPHIDE MINERAL.

pyritization The REPLACEMENT of the hard parts of an organism by PYRITE. Cf. HAEMATIZATION.

pyro- From or by fire.

pyrochlore (($(Ca,Na)_2(Nb,Ta)_2O_6(O,OH,F)$)) An OXIDE MINERAL of niobium and tantalum found associated with ALKALINE IGNEOUS ROCKS.

pyroclastic ash Rapidly moving, gas-fluidized, high-density mass-flow of hot PYROCLASTIC debris.

pyroclastic flow See ASH FLOW.

pyroclastic rock A rock formed by the accumulation of material generated by the explosive fragmentation of MAGMA and/or

existing solid rock during a VOLCANIC ERUPTION.

pyroclastic surge Rapidly moving, turbulent, low density cloud of PYROCLASTIC ASH and gas.

pyrogenesis The intrusion and extrusion of MAGMA and its products.

pyrogenetic mineral A MINERAL crystallized from an anhydrous or near-anhydrous MAGMA.

pyrolite An hypothetical PERIDOTITIC rock proposed as representing the upper MANTLE composition from which BASALTIC MAGMA could be derived on PARTIAL MELTING.

pyrolusite (MnO_2) The most important ORE MINERAL of manganese.

pyrometamorphism Intense THERMAL METAMORPHISM at high temperature and pressure.

pyrometasomatic deposit See SKARN.

pyromorphite ($Pb_5(PO_4)_3Cl$) A SUPERGENE MINERAL found in the oxidized parts of lead VEINS.

pyrope (cape ruby, false ruby) ($Mg_3Al_2Si_3O_{12}$) A deep red to black magnesium GARNET, valued as a GEM when clear and transparent.

pyrophanite ($MnTiO_3$) An OXIDE MINERAL related to ILMENITE.

pyrophyllite ($Al_2Si_4O_{10}(OH)_2$) A rare PHYLLOSILICATE found in METAMORPHIC ROCKS, with many properties and uses in common with TALC.

pyrophyllite deposit A product of the HYDROTHERMAL ALTERATION of ACID VOLCANIC ROCKS.

pyrostibnite See KERMESITE.

pyroxene-hornfels facies A METAMORPHIC FACIES often formed adjacent to PLUTONS.

pyroxenes Silicate MINERALS with an internal structure of a single chain of linked silicate tetrahedra with cations occupying sites between oxygen ions at the edges of the chains. General formula: $A_{1-x}(B,C)_{1+x}T_2O_6$, where A is commonly Na or Ca, B is Mg or Fe^{2+}, C is Al or Fe^{3+} and T is Si or Al. Stable over a wide range of temperature and pressure and so used in GEOTHERMOMETRY and GEOBAROMETRY. Common in IGNEOUS and METAMORPHIC ROCKS.

pyroxenite An ULTRABASIC ROCK rich in PYROXENE.

pyroxenoids A group of SILICATE MINERALS with a formula similar to the PYROXENES, possessing a Si:O ratio of 1:3, but which do not crystallize with a PYROXENE structure, e.g. WOLLASTONITE, RHODONITE, PECTOLITE.

pyroxferroite $(Ca_{0.15}Fe_{0.85}SiO_3)$ A PYROXENOID found in lunar LAVAS.

pyroxmangite $((Mn,Fe)SiO_3)$ A PYROXENOID found in METAMORPHIC and METASOMATIC ROCKS.

pyrrhotite (magnetic pyrites) $(Fe_{1-x}S_x)$ A common, FERRIMAGNETIC SULPHIDE MINERAL.

Pytyr'yuskiy A PERMIAN succession in the Timan area of the former USSR covering part of the CAPITANIAN, the LONGTANIAN and the CHANGXINGIAN.

Q

Q See QUALITY FACTOR.

Q-mode analysis A technique of grouping organisms into communities on the basis of their relative similarity. Cf. R-MODE ANALYSIS.

QAP triangle A MODAL classification scheme for GRANITIC rocks based on the relative proportions of QUARTZ (**Q**), ALKALI FELDSPAR (**A**) and PLAGIOCLASE FELDSPAR (**F**).

QAPF classification A MODAL classification scheme for IGNEOUS ROCKS with a COLOUR INDEX <90 based on the relative proportions of QUARTZ to QUARTZ + total FELDSPARS (**Q**), ALKALI FELDSPARS to total FELDSPARS (**A**), PLAGIOCLASE to total FELDSPARS (**P**) and FELDSPATHOIDS to FELDSPATHOIDS + total FELDSPARS (**F**).

Qionghusi A VENDIAN/CAMBRIAN succession in China covering parts of the POUNDIAN and TOMMOTIAN.

quadratic elongation A linear measure of change in shape based on changes in line lengths: (deformed length/undeformed length)^2; the square of STRETCH.

quadrature component See IMAGINARY COMPONENT.

quality factor (Q) A descriptor of ANELASTIC ATTENUATION whereby the fraction $2\pi/Q$ of wave energy is absorbed per cycle. Varies from 100 to 5000 for P WAVES in the CRUST and MANTLE respectively.

quaquaversal Dipping radially away from a centre.

quartz (SiO_2) The most common SILICA MINERAL.

quartz diorite A coarse-grained, PLUTONIC, IGNEOUS ROCK comprising ESSENTIAL PLAGIOCLASE, normally with a small amount of ORTHOCLASE, plus BIOTITE and HORNBLENDE. The intrusive equivalent of DACITE.

quartz index An expression of the MINERALOGICAL maturity of a SANDSTONE in terms of the ratio of the percentage of (QUARTZ + CHERT) to (FELDSPAR + rock fragments + CLAY MATRIX).

quartz monzonite A GRANITIC rock with QUARTZ comprising 5–20% of the FELSIC MINERALS and ALKALI FELDSPAR making up 35–65% of the total FELDSPAR. The PLUTONIC equivalent of LATITE.

quartz porphyry A PORPHYRITIC MICROGRANITE, MICROGRANODIORITE or MICROTONALITE.

quartz topaz See CITRINE.

quartzarenite See ORTHOQUARTZITE.

quartzite 1 A METAMORPHIC ROCK usually formed by the METAMORPHISM of a SANDSTONE. **2** An outdated term for SANDSTONE.

quartzwacke A SANDSTONE comprising >95% QUARTZ CLASTS with 15–75% MATRIX.

quasi-equilibrium An equilibrium giving the impression of DYNAMIC EQUILIBRIUM because of its observation over only a relatively short period of time.

quasi-planar adhesion stratification See ADHESION LAMINATION.

quasi-plastic mechanism A PLASTIC DEFORMATION mechanism which can produce a MESOSCOPICALLY continuous STRAIN, including CRYSTAL PLASTICITY, DIFFUSIVE MASS TRANSFER and GRAIN BOUNDARY SLIDING.

Quaternary The most recent subPERIOD of the CENOZOIC 1.64 Ma–present.

quick clay A CLAY in which the delicate packing of particles can be completely destroyed on FAILURE to cause almost total loss of STRENGTH and, possibly, LIQUEFACTION.

quickflow A mixture of OVERLAND FLOW and subsurface STORMFLOW. Cf. BASEFLOW.

quicksilver A common name for MERCURY.

quiescent continental margin See PASSIVE CONTINENTAL MARGIN.

quiet zone See MAGNETIC QUIET ZONE.

R channels See RÖTHLISBERGER CHANNELS.

R-mode analysis A technique in which the distributions of individual TAXA are compared; those that co-occur are grouped and those which are mutually exclusive are placed in different communities. Cf. Q-MODE ANALYSIS.

R1 fault See RIEDEL FAULT.

R1 shear See RIEDEL SHEAR.

R2 fault A CONJUGATE RIEDEL FAULT.

R2 shear (P shear) A CONJUGATE RIEDEL SHEAR.

radar See GROUND-PENETRATING RADAR, REMOTE SENSING.

radial dykes A set of DYKES arranged radially around a PLUTON as a result of the STRESS field generated by the PLUTON.

radial joint A JOINT parallel to a FOLD AXIS which remains perpendicular to the folded layer on the FOLD LIMBS.

radiaxial Descriptive of a crystal which has grown in a cavity in a fan-like pattern approximately normal to the cavity wall.

radioactive dating See RADIOMETRIC DATING.

radioactivity log A GEOPHYSICAL BOREHOLE LOG making use of natural (GAMMA LOG) or induced (GAMMA-GAMMA LOG, NEUTRON LOG) radiation.

radiocarbon dating (carbon-14 dating) A method of RADIOMETRIC DATING based on the decay of ^{14}C, which is at a known, equilibrium concentration in living organisms, to ^{12}C. On death the proportion of ^{14}C decreases at a known rate. Normally suitable for dating materials younger than ~50 000 years, although ACCELERATOR RADIOCARBON DATING extends this range.

radiogenic Originating by radioactive processes.

radioisotope dating See RADIOMETRIC DATING.

Radiolaria A subclass of tiny, free-floating, marine, single-celled, round PROTOZOANS with siliceous endoskeletons, generally of OPALINE SILICA, arranged in radial or tangential elements. The skeletons make an important contribution to deep-sea sediments and CHERTS. Range CAMBRIAN–RECENT, with the greatest diversity in the CRETACEOUS.

radiolarian ooze A PELAGITE composed primarily of the remains of RADIOLARIA.

radiolarite CHERT containing abundant RADIOLARIA.

radiometric assay A measurement of the uranium concentration in buried bone, into which it is absorbed from percolating GROUNDWATER, from its radioactivity, generally for archaeological purposes. Cf. FLUORINE TEST.

radiometric dating (radioactive dating, radioisotope dating) Techniques of determining the age of rocks or FOSSILS from the relative proportions of a radioactive parent and its daughter decay product(s). Knowledge of the radioactive decay constant or HALF-LIFE allows the proportion to be converted into an age. Methods include ARGON-ARGON DATING, LEAD-LEAD DATING, POTASSIUM-ARGON DATING, RADIOCARBON

DATING, RUBIDIUM-STRONTIUM DATING, SAMARIUM-NEODYMIUM DATING, THORIUM-LEAD DATING and URANIUM-LEAD DATING.

radiometric surveying GEOPHYSICAL EXPLORATION methods for the location of radioactive elements, in practice uranium, thorium and potassium (^{40}K), using GEIGER COUNTERS, SCINTILLATION COUNTERS and GAMMA RAY SPECTROMETERS, the latter two of which can be used in AIRBORNE GEOPHYSICAL SURVEYS.

radon (Ra) The only radioactive gaseous element, produced by the decay of ^{238}U with a half-life of 3.8 days, and as such moves freely through pores, JOINTS and FRACTURES either as a gas or dissolved in PORE FLUIDS. Measured with a RADON EMANOMETER. Often indicative of buried uranium concentrations.

radon emanometer An instrument for measuring RADON from air samples from a shallow borehole. The filtered gas is dried and passed to an ionization chamber or SCINTILLATION COUNTER where ^{232}Ra is detected from its alpha particle activity.

Raibler Schichten A TRIASSIC succession in the Alps equivalent to part of the CARNIAN.

rain pit A RAINDROP-GENERATED STRUCTURE on a sediment surface.

rain-impact ripple A RAINDROP-GENERATED STRUCTURE produced when large raindrops are driven obliquely onto fine SAND by a strong wind.

rainbow quartz See IRIS.

raindrop impact erosion (rainsplash erosion) The movement of soil particles as the result of raindrop impact by rebound, undermining and downslope movement; most effective in the absence of surface RUNOFF.

raindrop-generated structure A sedimentary STRUCTURE formed by rain impacting on unconsolidated sediment. Impact marks are generally circular to elliptical, <10 mm in diameter and several millimetres deep with a raised rim. See also HAILSTONE-GENERATED STRUCTURE.

rainfall erosivity factor A term in the UNIVERSAL SOIL LOSS EQUATION.

rainsplash The mechanism of loosening soil particles by rain prior to their removal by surface RUNOFF.

rainsplash erosion See RAINDROP IMPACT EROSION.

raised beach A BEACH DEPOSIT stranded at altitude by a fall in sea level.

rake See PITCH.

Rakhmey An ORDOVICIAN succession in Kazakhstan covering the lower part of the ARENIG.

rammelsbergite (NiAs$_2$) A nickel-bearing MINERAL with LOELLINGITE structure.

ramp The part of a FAULT that cuts across DATUM surfaces, such as BEDDING, in THRUST and EXTENSIONAL FAULT SYSTEMS. Cf. FLAT.

Ramsaudolomit A TRIASSIC succession in the Alps equivalent to the ANISIAN.

Ramsdell notation A method of describing the UNIT CELL relationship in POLYTYPISM.

Randian An ERA of the PRECAMBRIAN, 2800–2450 Ma.

random fabric (isotropic fabric) A FABRIC with no PREFERRED ORIENTATION.

random structure (isotropic structure) A STRUCTURE with no PREFERRED ORIENTATION.

range The distance between the point of initiation of a SEISMIC WAVE and its point of detection.

rank The degree of COALIFICATION of a COAL.

rankinite (Ca$_3$Si$_2$O$_7$) Calcium disilicate, found in siliceous LIMESTONES subjected to high GRADE METAMORPHISM.

rapakivi granite A GRANITE exhibiting a RAPAKIVI TEXTURE, characteristic of the ANORTHOSITE-GRANITE associations of the PROTEROZOIC.

rapakivi texture A PORPHYRITIC TEXTURE in which centimetric-scale, rounded PHENO-

CRYSTS are surrounded by a rim of sodium-rich PLAGIOCLASE.

rapids A section of river channel in which flow is faster and more TURBULENT than elsewhere.

rare earth elements (REE) Elements with an atomic number between 57 and 71 plus scandium and yttrium.

ratio contour map A map showing the ratio of the total thickness of one lithological class to that of the remaining classes making up the total STRATIGRAPHIC unit section.

Raukumara A CRETACEOUS succession in New Zealand comprising the AROWHANAN, MANGAOTANIAN and TERATAN.

Rawtheyan A STAGE of the ORDOVICIAN, 440.1–439.5 Ma.

ray tracing (seismic modelling) A sophisticated method of interpreting SEISMIC data whereby a structure is assumed and the ARRIVAL TIMES from it computed. The model is progressively modified until the calculated arrivals simulate the observed arrivals.

Rayleigh number (Ra) A dimensionless parameter used in fluid dynamics whose magnitude determines when CONVECTION is initiated in a fluid. $Ra = d^3 g \Delta T \rho \alpha / K \eta$, where d = depth of fluid, g = GRAVITY, ΔT = super-ADIABATIC temperature gradient, ρ = DENSITY, α = volume coefficient of thermal expansion, K = THERMAL DIFFUSIVITY and η = KINEMATIC VISCOSITY.

Rayleigh wave A SEISMIC SURFACE WAVE whose particle motion is in the form of a vertical ellipse in the direction of propagation. Efficient in the transport of seismic energy and responsible for GROUND ROLL.

re-entrant angle The angle (>180°) between the members of a TWINNED crystal.

reaction rim A zone of SECONDARY MINERALS surrounding a PRIMARY MINERAL as the result of late-stage METASOMATISM or reaction with fluids or solids with which it is in contact.

reaction series See CONTINUOUS REACTION SERIES .

reactivation surface A surface which records the EROSION of a BEDFORM FORESET slope resulting from a rise or fall in the level of the flow. After the EROSION the BEDFORM reforms with the same orientation and shape and continues to migrate.

real component (in-phase component) The part of the secondary field that is in phase or 180° out of phase with the primary field in ELECTROMAGNETIC INDUCTION METHODS.

realgar (AsS) A red SULPHIDE MINERAL of the RING STRUCTURE GROUP found in VEINS, VOLCANIC-EXHALATIVE DEPOSITS and HOT SPRINGS.

reattachment point The outer margin of the SHEAR LAYER bounding a zone of FLOW SEPARATION where the freestream again impinges on the bed.

Recent See HOLOCENE.

Receptaculitids A group of green ALGAE, SPONGES or a separate phylum with a globular or pear-shaped form and internal radiating rods terminating distally in plate-like facets which make up an outer wall. Range ORDOVICIAN–PERMIAN.

recharge The precipitation reaching the WATER TABLE and replenishing GROUNDWATER supply.

recharge area The area over which RECHARGE is received.

reciprocal time The travel time in either direction between two points on a REVERSED SEISMIC PROFILE.

reclined fold (vertical fold) A FOLD whose HINGE LINE PLUNGES steeply.

reconstructive transformation The slow transformation of one POLYMORPH into another in which bonds must be broken and new bonds formed. Cf. DISPLACIVE TRANSFORMATION.

recoverability The percentage of total metal in an ORE that is present in the concentrate after processing.

recoverable strain See TEMPORARY STRAIN.

recovery 1 The change from a strained to an unstrained state during or after DEFORMATION, an important phenomenon in DISLOCATION CLIMB. **2** The percentage of the scheduled tonnage actually mined.

recrystallization A process of NEOMORPHISM in which the overall chemical composition is unchanged.

recumbent fold A FOLD whose HINGE LINE and AXIAL SURFACE are both horizontal.

red bed A sedimentary deposit whose CLASTS are coated with HAEMATITE, causing a red colour. The presence of such beds in the PROTEROZOIC is taken to indicate a major change from an oxygen-rich hydrosphere to an oxygen-bearing atmosphere.

red bed copper A copper variety of a SANDSTONE-URANIUM-VANADIUM BASE METAL DEPOSIT, whose COUNTRY ROCKS are often red.

red clay (brown clay) A red/brown PELAGITE primarily composed of CLAY MINERALS with minor QUARTZ, ASH, cosmic dust and FISH teeth, the product of very slow deposition in the central parts of ocean BASINS at >4 km depth and below the CALCITE COMPENSATION DEPTH.

red ochre A red, earthy variety of HAEMATITE used as a pigment.

red zinc ore See ZINCITE.

redox reaction OXIDATION coupled with CHEMICAL REDUCTION in a balanced exchange of electrons.

redruthite See CHALCOCITE.

reduced travel-time curve A means of displaying SEISMIC REFRACTION data by plotting the difference between the observed travel-time and the time which would have been observed if material of only one particular velocity had been present. Refractors with that velocity then plot horizontal to the distance axis.

reduction 1 See CHEMICAL REDUCTION. **2** The procedure used in processing GEOPHYSICAL survey data to remove all non-geological sources of variation.

reduction to the pole A method of processing MAGNETIC ANOMALY data so that the anomaly appears as it would be if located at the MAGNETIC POLE. Since the GEOMAGNETIC FIELD is vertical at the poles, this operation facilitates the interpretation of the anomaly.

reduzate A sediment accumulated under reducing conditions, typically rich in organic CARBON and iron sulphides.

REE See RARE EARTH ELEMENTS.

reedmergnerite ($NaBSi_3O_8$) A rare FELDSPAR found in unmetamorphosed DOLOMITIC OIL SHALES.

reef 1 A GOLD-bearing QUARTZ VEIN. **2** A PRECAMBRIAN GOLD-bearing CONGLOMERATE of southern Africa. **3** See CORAL-ALGAL REEF. **4** A ridge of rocks rising to near sea level.

reef knoll A large mass of REEF LIMESTONE WEATHERED out to form a rounded hill.

reef trap An OIL TRAP consisting of porous REEF LIMESTONE covered by impermeable STRATA, commonly SHALE or MUDSTONE.

Reeh calving A proposed mechanism for the splitting of icebergs from the front of GLACIERS and ICE SHEETS in which it is suggested that STRESS is greatest and sufficient for FRACTURE at a cross-section of floating GLACIER at a distance of about the ice thickness from the ice front.

reflectance 1 The strength with which the components of COAL, or other carbonaceous materials, reflect light, a measure of RANK. **2** The strength with which an OPAQUE MINERAL in polished section reflects light.

reflected light optics The study of OPAQUE MINERALS under the microscope in polished sections.

reflected seismic wave The SEISMIC WAVE returned to the surface by reflection at a boundary where the ACOUSTIC IMPEDANCE changes (a reflector). For a wave of the same type, the angle of reflection is equal to the angle of incidence. Forms the basis of a type of SEISMIC EXPLORATION, particularly for STRUCTURES forming potential OIL and GAS TRAPS.

reflection character analysis A technique of SEISMIC STRATIGRAPHY in which the changes in a SEISMIC REFLECTION wave shape from one record to another is studied in order to determine the nature of changes in STRATIGRAPHY, the fluid content and the properties of a RESERVOIR rock.

reflection coefficient The ratio of the amplitudes of an incident SEISMIC WAVE and the wave reflected from an ACOUSTIC IMPEDANCE contrast. For vertical incidence the reflection coefficient is given by $(A_2-A_1)/(A_2+A_1)$ where A_1 and A_2 are the ACOUSTIC IMPEDANCES of the layers through which the wave is travelling and the layer across the contrast respectively.

reflective beach (swell beach) A BEACH characterized by low frequency swell waves, low BEACH WATER TABLES, high percolation and low BACKWASH so that there is net onshore sediment movement.

refracted seismic wave The SEISMIC WAVE transmitted through a boundary where the ACOUSTIC IMPEDANCE changes (a refractor) according to the geometry predicted by SNELL'S LAW.

refractive index (RI) A most important optical property of a MINERAL, defined as the ratio of the velocity of light *in vacuo* to the velocity in the MINERAL.

refractometer An instrument used to measure REFRACTIVE INDEX.

refractory clay See FIRECLAY.

refractory gold NATIVE GOLD which occurs in MINERALS that are difficult to treat by amalgamation or cyanide.

refractory mineral A MINERAL resistant to decomposition by heat, pressure or chemical reaction.

refractory ore An ORE from which it is difficult to extract the valuable metal.

Refugian An EOCENE succession on the west coast of the USA equivalent to the PRIABONIAN.

reg (serir) A stony desert floor of GRAVELS underlain by mixed sediment sizes that have been removed from the surface by wind and water.

regelation The process whereby ice melts and refreezes as PRESSURE is raised and lowered.

regimes of diagenesis A genetic classification of the various stages of DIAGENESIS of carbonate and SILICLASTIC sediments during the evolution of a sedimentary BASIN, comprising EOGENESIS, MESOGENESIS and TELOGENESIS.

regional field The large-scale variation in a GRAVITY or MAGNETIC FIELD over a region which reflects the effects of relatively deep STRUCTURES. This field must be removed before interpreting any RESIDUAL ANOMALIES.

regional metamorphism Large-scale METAMORPHISM involving heat and pressure.

regolith The superficial layer of loose, unconsolidated material which overlies BEDROCK over much of the land surface, comprising unlithified *in situ* SAPROLITE, ASH, COLLUVIUM, ALLUVIUM or DRIFT.

regression A recession of the sea from a land area or from a shallow sea. Cf. TRANSGRESSION.

Reitzi A TRIASSIC succession in the Alps covering the lower part of the LADINIAN.

rejuvenation 1 Of a river, its stimulation to increased EROSIONAL activity. **2** Of a STRUCTURE, the topographic renewal of an OROGENIC BELT by UPLIFT after the main OROGENIC activity had ceased.

relative permeability The ratio of EFFECTIVE PERMEABILITY at a stated fluid saturation to the ABSOLUTE PERMEABILITY.

relative plate motion The relative motion between PLATES determined from the SPREADING RATE at an OCEAN RIDGE by matching OCEANIC MAGNETIC ANOMALIES of known age across it, by PALAEOMAGNETIC measurements, by using the direction of TRANSFORM FAULTS, by matching geological features, PALAEONTOLOGICAL correlations, PALAEOCLIMATIC data, etc.

relaxation time 1 The period of years

or decades after a major RIVER FLOOD over which a channel form reverts to its former state. **2** The time for REMANENT MAGNETIZATION to decay to e^{-1} of its original value. **3** The time taken for a disturbed system to reach equilibrium, or for a parameter to decrease to about 37% of its initial value.

releasing bend A curve in a STRIKE-SLIP FAULT across which there is EXTENSION parallel to the straight parts of the FAULT. Cf. RESTRAINING BEND.

relict structure See PALIMPSEST STRUCTURE.

Relizian A MIOCENE succession on the west coast of the USA covering parts of the BURDIGALIAN and LANGHIAN.

remanent magnetization The magnetization in a rock that is present in the absence of an external MAGNETIC FIELD. Principally NATURAL REMANENT MAGNETIZATION, but also includes magnetizations produced by laboratory processes.

remanié fossil (derived fossil) A FOSSIL surviving the EROSION of its enclosing rock and incorporated into a later sediment.

remote sensing The recording of images of parts of the Earth's surface using electromagnetic radiation, normally from an aircraft or satellite at sufficient height for a broad area to be covered. Includes passive AERIAL PHOTOGRAPHY and imagery from LANDSAT SATELLITES using the MULTISPECTRAL SCANNER and THEMATIC MAPPER, and the SPOT-1 SATELLITE, and also active techniques, such as RADAR, which shows the topographic texture of terrain in the presence of cloud cover, etc.

reniform Kidney-shaped.

Repettian A PLIOCENE succession on the west coast of the USA covering the middle part of the PIACENZIAN.

repichnia TRACE FOSSILS comprising locomotion traces.

replacement In DIAGENESIS, the growth of a chemically different AUTHIGENIC MINERAL within the body of an existing MINERAL, either of the COUNTRY ROCK, a detrital grain or an earlier AUTHIGENIC MINERAL.

replacement orebody An EPIGENETIC OREBODY formed by the REPLACEMENT of existing rocks, particularly carbonates, e.g. FLATS, SKARN DEPOSITS.

reprecipitation The PRECIPITATION of material following its DISSOLUTION, a mechanism for the NEOMORPHISM of SILICA MINERALS.

reptation The process of low velocity, AEOLIAN grain transport by a short trajectory following a grain impact. Cf. SALTATION.

Reptilia/reptiles A class of vertebrate animals with an egg typical of the AMNIOTA, which are not MAMMALIA or AVES, and which first appeared in the MISSISSIPPIAN.

resequent stream A stream following the original direction of drainage but which developed at a later time. Cf. ANTECONSEQUENT, INCONSEQUENT, INSEQUENT, OBSEQUENT and SUBSEQUENT STREAMS.

reserves 1 Of an economic MINERAL, the MINERAL known to be mineable under the technical and economic conditions at the time of assessment. *In situ* reserves refer to the amount that can probably be won; marketable reserves to the amount of recovered MINERAL that can ultimately be marketed after processing. **2** Of oil, several systems of categorization are in use. One variant used in the USA distinguishes between proved (or measured) reserves (those outlined by drilling), probable (or indicated) supply (likely to be present in the extension of existing OILFIELDS) and possible (or indicated) supply (from future discoveries in known productive FORMATIONS).

reservoir A subsurface rock containing commercially exploitable quantities of oil and/or gas because of its POROSITY and PERMEABILITY.

residual anomaly A GEOPHYSICAL ANOMALY, resulting from a relatively shallow source, in a GRAVITY or MAGNETIC FIELD, isolated by the removal of a REGIONAL FIELD.

residual breccia A deposit formed by the removal of all CARBONATE MINERALS from a rock by DISSOLUTION, causing collapse of the sedimentary FABRIC, occurring most often in the VADOSE ZONE where percolating rainwater is highly undersaturated with respect to CALCITE.

residual deposit An economic deposit formed by the action of WEATHERING and GROUNDWATER on PROTORE.

residual placer A PLACER DEPOSIT formed on a fairly flat surface immediately above a BEDROCK source by the chemical decay and removal of lighter materials.

residual strength The lower value of STRENGTH in a disturbed material.

resinite A COAL MACERAL of the EXINITE group made up of resinous material.

resinous lustre A LUSTRE like that of resin, as shown by SPHALERITE and NATIVE SULPHUR.

resistate mineral A MINERAL resistant to chemical WEATHERING.

resistivity log See ELECTRIC LOG.

resistivity method A GEOPHYSICAL EXPLORATION method for investigating the subsurface distribution of ELECTRICAL RESISTIVITY by introducing low frequency or direct electric current into the ground via spike electrodes and measuring the resulting voltage difference with a second electrode pair. The two principal techniques are CONSTANT SEPARATION TRAVERSING and VERTICAL ELECTRICAL SOUNDING. Widely used in HYDROGEOLOGY, ENGINEERING GEOLOGY, archaeology, etc.

resonance A term used in VALENCE BOND THEORY for the state in which pairs of electrons interact with other pairs to form a molecule in which the most stable structures dominate.

resorption The partial refusion or SOLUTION of a PHENOCRYST, which may give rise to a REACTION RIM.

resource The total amount of a particular commodity (element, MINERAL, oil) estimated for the world or a nation, comprising ORE RESERVES, uneconomic deposits and deposits as yet undiscovered (estimated by comparison with explored areas of similar geology).

restored section The arrangement of BEDS etc. in a section before DEFORMATION, essentially equivalent to a BALANCED SECTION.

restraining bend A curve in a STRIKE-SLIP FAULT across which there is COMPRESSION parallel to the straight parts of the FAULT. Cf. RELEASING BEND.

retardation time The time taken by a deforming VISCOELASTIC material to reach half the value of the STRESS divided by YOUNG'S MODULUS, equal to the ratio of the VISCOSITY to YOUNG'S MODULUS.

reticulated With a net-like STRUCTURE.

reticulite A very low DENSITY form of BASALTIC PUMICE formed in some HAWAIIAN ERUPTIONS.

retinite (burmite) A group of resins containing no succinic acid.

retrocharriage See BACKFOLD.

retrograde boiling Boiling in a system whose temperature is falling. Occurs in an acid MAGMA whose cooling promotes the CRYSTALLIZATION of anhydrous MINERALS so that the MAGMA becomes enriched in volatiles. If the increased vapour pressure exceeds the CONFINING PRESSURE, a rapidly boiling liquid separates, possibly with the development of CRACKLE BRECCIATION. Probably important in the development of PORPHYRY COPPER DEPOSITS.

retrograde metamorphism (diaphthoresis) METAMORPHISM which produces a lower GRADE from a higher GRADE.

return flow 1 THROUGHFLOW which EXFILTRATES from the soil profile at the base of a slope. **2** Flow in the opposite direction to the mean flow direction, such as might occur in the LEE SIDE of a DUNE.

reverse drag FAULT DRAG in which the curvature of a marker is contrary to the sense of DISPLACEMENT on the FAULT. Cf. NORMAL DRAG.

reverse fan See DIVERGENT FAN.

reverse fault A DIP-SLIP FAULT in which the HANGINGWALL moves upwards relative to the FOOTWALL.

reverse grading See INVERSE GRADING.

reverse kink bands See CONTRACTIONAL KINK BAND.

reversed magnetization A REMANENT MAGNETIZATION in the opposite sense to the present GEOMAGNETIC FIELD. Cf. NORMAL MAGNETIZATION.

reversed polarity The orientation of a past GEOMAGNETIC FIELD or a NATURAL REMANENT MAGNETIZATION in the opposite direction to the present GEOMAGNETIC FIELD. Cf. NORMAL POLARITY.

reversed profile A line of SEISMOMETERS with a SHOT at either end so that SEISMIC WAVES travelling in both directions are recorded. Employed in order to reveal true SEISMIC VELOCITIES and STRUCTURE in SEISMIC REFRACTION interpretation.

reversed zoning ZONING in PLAGIOCLASE in which the zones become more sodic towards the interior of the crystal. Cf. NORMAL ZONING.

reversing dune A DUNE with SLIP-FACES on opposing sides as a result of a BIMODAL wind regime.

reworking 1 The process of altering, by DEFORMATION and METAMORPHISM, deep, CRUSTAL material formed in a previous OROGENY by a subsequent orogenic event. **2** The EROSION, transfer and deposition of older sediment within a sedimentary BASIN.

Reynolds number (Re) A dimensionless parameter used in fluid dynamics which expresses the influence of viscous forces within a fluid to the inertial forces acting on the flow. $Re = \rho Ul/\upsilon$, where ρ = fluid DENSITY, U = mean flow velocity, l = length term and υ = KINEMATIC VISCOSITY. Its magnitude controls whether flow will be LAMINAR or TURBULENT.

Rhaetian (Rhetian) A STAGE of the TRIASSIC, 209.5–208.0 Ma.

Rhenanida An order of class PLACODERMI, superclass PISCES; small, shark-like armoured FISH. Range L.–U. DEVONIAN.

rheology The study of the FLOW and DEFORMATION of materials.

rheomorphism Post-compactional flow of a WELDED TUFF that has remained hot for long enough that it deforms in a DUCTILE fashion.

Rhetian See RHAETIAN.

rhexistasy The state of environmental and GEOMORPHIC instability which interrupts periods of BIOSTASY, probably resulting from climatic change or TECTONIC activity. Vegetation cover deteriorates, soils erode and the landscape denudes. Cf. BIOSTASY.

rhipidistians A group of SARCOPTERYGIIAN FISH from which the AMPHIBIA evolved.

rhizocretion An accumulation of MINERAL matter around the roots of a plant while living or after death.

rhizolite A rock showing evidence of having been formed largely by root activity.

rhizolith An organo-sedimentary STRUCTURE resulting from the accumulation and/or CEMENTATION around or within, or the REPLACEMENT of, plant roots by MINERAL matter.

rhodochrosite (dialogite, manganese spar) ($MnCO_3$) A rare ORE MINERAL of manganese found in some HYDROTHERMAL VEINS.

rhodolite A pale red or purple GARNET, compositionally two parts PYROPE to one of ALMANDINE.

rhodolith A NODULE of red ALGAE.

rhodonite ($MnSiO_3$) A PYROXENOID found in manganese-rich deposits.

Rhodophyta The CALCAREOUS red ALGAE. Range CAMBRIAN–RECENT.

rhodustalf See TERRA ROSSA.

rhomb-porphyry A medium-grained, INTERMEDIATE IGNEOUS ROCK comprising numerous PHENOCRYSTS of ANORTHOCLASE in a fine-grained GROUNDMASS, occurring in DYKES and minor intrusions.

rhomb-spar An obsolete name for DOLOMITE.

rhombochasm See PULL-APART BASIN.

rhombohedral packing One end-member of the different ways perfectly sorted spheres can be arranged. A sediment with such packing would have 26% POROSITY. Cf. CUBIC PACKING.

rhomboid ripple A surface sedimentary BEDFORM found in BEACH DEPOSITS indicative of the high power of the environment.

rhourd A large, pyramidal DUNE formed by the intersection of smaller DUNES in multi-directional winds.

Rhuddanian The lowest STAGE of the SILUR-IAN, 439.0–436.9 Ma.

Rhynchocephalia See RHYNCHOSAURIA.

Rhynchonellida An order of class ARTICUL-ATA, phylum BRACHIOPODA; BRACHIOPODS with biconvex, impunctate shells, usually with coarse ribs on each VALVE which meet in a zig-zag fashion. Range M. ORDOVICIAN–RECENT.

Rhynchosauria (Rhynchocephalia) An order of subclass ARCHOSAURIA, class REPTI-LIA; herbivorous REPTILES with powerful chopping teeth. Range U. TRIASSIC.

Rhynie Chert An EXCEPTIONAL FOSSIL DEPOSIT in Aberdeenshire, Scotland of L. DEVONIAN age containing LAND PLANTS and ARTHROPODS preserved in SILICA.

Rhyniopsida A class of division TRACHAEO-PHYTA, kingdom plantae; primitive VASCULAR PLANTS. Range M. SILURIAN–U. DEVONIAN.

rhyodacite An extrusive, commonly POR-PHYRITIC, IGNEOUS ROCK with PHENOCRYSTS of PLAGIOCLASE, SANIDINE and QUARTZ in a GLASSY to MICROCRYSTALLINE GROUNDMASS.

rhyolite (liparite) A fine-grained, ACIDIC VOLCANIC ROCK compositionally equivalent to GRANITE. APHANITIC or with PHENOCRYSTS of FELDSPAR and QUARTZ in a MATRIX of QUARTZ and FELDSPAR with sparse MAFIC MIN-ERALS. Most abundant in continental RIFTS as extensive IGNIMBRITES plus rarer DOMES and LAVA FLOWS, also occurring on OCEANIC ISLANDS and ISLAND ARCS.

rhythmite A finely laminated sediment in which two or three different lithologies are regularly repeated, common in GLACIAL lakes.

RI See REFRACTIVE INDEX.

ria A submerged, 'V'-shaped, coastal valley, narrowing landwards. Originally formed subaerially and inundated by a rise in sea level.

rib structure The saw-tooth cross-section of a PLUME STRUCTURE.

richness A PALAEOECOLOGICAL descriptor of the number of TAXA present in a com-munity.

Richter earthquake scale A logarithmic, EARTHQUAKE MAGNITUDE scale based on the amplitude of ground motion with a correc-tion for distance to the EPICENTRE.

richterite $((Na,K)_2(Mg,Mn,Ca)_6Si_8O_{22}(OH)_2)$ An alkali AMPHIBOLE found in contact METASOMATIC deposits.

rider A block separated by HORSETAIL FAULTS in an IMBRICATE EXTENSIONAL FAULT SYSTEM.

ridge push The horizontal FORCE applied to a PLATE at an OCEAN RIDGE due to the expansion of material brought to the sur-face from higher pressures at depth. An important PLATE-driving mechanism in OROWAN-ELSASSER CONVECTION.

ridges and runnels Alternating morpho-logical highs and lows with an amplitude of 0.1–1.5 m and a wavelength of 50–200 m. Run parallel or subparallel to the coastline on gently sloping, sandy BEACHES with a high TIDAL range, low to moderate WAVE energy and FETCH-limited WAVES. Develop as the TIDAL height varies and the SWASH zone moves to new locations on the BEACH face.

riebeckite $(Na_2Fe_3^{2+}Fe_2^{3+}Si_8O_{22}(OH)_2)$ A blue AMPHIBOLE, known as CROCIDOLITE when in ASBESTIFORM habit.

Riedel fault (R1 fault) A FAULT within a FAULT ZONE with the same sense as the zone but inclined at a low angle (<15°) to the main FAULT direction.

Riedel shear (R1 shear, synthetic shear) A SHEAR within a SHEAR ZONE with the same sense as the zone but inclined at a low angle (<15°) to the main SHEAR direction.

riegel A rock BAR extending across a GLACIAL trough, possibly formed where the erosive ability of the GLACIER wanes or the BEDROCK becomes harder.

riffles Topographic highs on the undulating longitudinal profile of a GRAVEL-bed river, spaced at 5–7 channel widths.

rift (rift valley) A major extensional GRABEN of large lateral extent, whose topographic expression may have been lost. Rifts in OCEANIC CRUST are found at the axes of slow-spreading OCEANIC RIDGES, and may be about 30 km in width. Continental rifts may connect to PLATE MARGINS and may represent sites of CONTINENTAL SPLITTING where new oceans grow between their edges and which become PASSIVE CONTINENTAL MARGINS. Other continental rifts are IMPACTOGENS.

rift valley See RIFT.

rigid body rotation Rotation without changes in length or angular relationships and without DISPLACEMENT of the centre of rotation. Used in concepts such as the STRAIN ELLIPSOID.

rigid-plastic Descriptive of a PLASTIC material which undergoes no DEFORMATION at STRESSES below the YIELD STRESS. Cf. ELASTIC-PLASTIC.

rigidity (modulus of rigidity, shear modulus) An ELASTIC CONSTANT indicating the degree of resistance to SHEAR STRAIN, defined as the ratio of SHEAR STRESS to SHEAR STRAIN in ELASTIC DEFORMATION.

rill A centimetric-scale channel that changes direction with every RUNOFF event.

rillenkarren (solution flutes) SOLUTION depressions found on steep or vertical surfaces with sharp ridges between the flutes.

rim syncline An annular SYNCLINE surrounding a SALT DOME formed by the collapse of STRATA over the region evacuated by the rising salt DIAPIR.

rimstone dam A type of SPELEOTHEM which forms a carbonate barrier that impounds CAVE pools.

ring complex (ring intrusion) An annular, TABULAR body of cylindrical shape, exposed at the surface with a ring-shaped outcrop, formed of almost any IGNEOUS ROCK and found in OCEANIC ISLANDS, at continental margins and in certain continental volcanic provinces.

ring dyke An arcuate, TABULAR body of IGNEOUS ROCK with a vertical axis and near-vertical or outward-dipping contacts.

ring intrusion See RING COMPLEX.

ring silicate See CYCLOSILICATE.

ring structure group A group of SULPHIDE MINERALS characterized by a structure comprising parallel chains of atoms.

ringwoodite ($(Fe,Mg)_2SiO_4$) A MINERAL with SPINEL structure and OLIVINE composition found in METEORITES.

rinnentaler See TUNNEL VALLEY.

rip current A strong, narrow current flowing seaward from a BEACH through the BREAKER ZONE.

rip-up clast (mud clast) An INTRAFORMATIONAL CLAST in a sedimentary sequence.

Riphean An ERA of the PRECAMBRIAN, 1650–800 Ma.

ripidolite (prochlorite) A CHLORITE group MINERAL found in METAMORPHIC ROCKS and some ORE VEINS.

rippability A measure of the facility with which rock can be broken by mechanical equipment; can be estimated from the P WAVE velocity.

ripple A flow-transverse BEDFORM generated at low SHEAR STRESSES above the threshold for sediment movement in COHESIONLESS SAND of grainsize <0.6 mm. Wavelength commonly <500 mm and amplitude <30 mm.

ripple form index (ripple index) The ratio of wavelength to height of a RIPPLE.

ripple index See RIPPLE FORM INDEX.

ripple mark The mark left by a RIPPLE on a sedimentary surface.

ripple symmetry index The ratio of the horizontal components of STOSS-SIDE length to LEE-SIDE length of a RIPPLE.

riser See STEP.

Riss Glaciation The third of the GLACIATIONS affecting the Alps in QUATERNARY times.

river capture A process in which one river undercuts the drainage area of another by more rapid incision, so enlarging its CATCHMENT at the expense of the other's.

river flood Water which arrives in a river channel sufficiently rapidly to produce a significant increase in DISCHARGE above BASEFLOW levels and forms a distinct peak in DISCHARGE.

RMQ See ROCK MASS QUALITY.

Robin effect A mechanism whereby the removal of rock at the edge of steps above a GLACIAL cavity is reinforced by a heat pump process resulting from the variable STRESS distribution in the ice and at the GLACIER bed as the GLACIER moves over an irregular surface.

rocdrumlin (tadpole rock) A DRUMLIN-like feature of BEDROCK moulded into a streamlined shape.

roche moutonée A hump of rock with one side moulded by ice and the other steepened; a product of GLACIAL EROSION.

rock crystal A colourless, crystalline variety of QUARTZ.

rock doughnut A doughnut-shaped rock formed when a veneer of CASE-HARDENED material is breached and exposes more easily eroded rock.

rock fall A rapid form of MASS MOVEMENT by falling under GRAVITY.

rock flour Rock debris in a GLACIER that has been ground down to a fine mixture of SILT and CLAY.

rock head The surface between BEDROCK and overlying unconsolidated material.

rock magnetism The study of the magnetic properties of naturally occurring magnetic MINERALS, including the processes by which they became magnetized. Important terrestrial magnetic MINERALS include iron oxides, hydroxides and sulphides, while IRON and nickel are found in METEORITES and on the Moon.

rock mass quality (RMQ) An engineering classification based on the number of major discontinuities (e.g. JOINTS) present, their orientation and spacing.

rock phosphate See PHOSPHORITE.

rock quality designation (RQD) A quantitative assessment of the intactness of material recovered from a borehole in SITE INVESTIGATION.

rock rose See DESERT ROSE.

rock slide A type of MASS MOVEMENT by sliding down a surface.

rockburst A rapid release of LITHOSTATIC PRESSURE in a deep mine, which can generate an EARTHQUAKE up to a MAGNITUDE of 4 or more.

rocksalt See HALITE.

rodding See RODS.

roddon A sinuous ridge of SILTY material above the general level of the PEAT FENS of East Anglia, England, which represents the remains of ancient river systems flowing between LEVÉES above the surrounding land level, or which was formed by PEAT wastage caused by drainage activities.

Rodentia An order of infraclass EUTHERIA, subclass THERIA, class MAMMALIA; the rodents. Range U. PALAEOCENE–RECENT.

rodingite A rock comprising GROSSULAR and PREHNITE ± WOLLASTONITE, DIOPSIDE and HYDROGROSSULAR, formed by calcium METASOMATISM of GABBRO or DOLERITE.

Rodinia The NEOPROTEROZOIC SUPERCONTINENT.

rods (rodding) Cylindrical STRUCTURES of a single MINERAL in a deformed rock, similar in form to BOUDINS and MULLIONS, but distinguished by their monomineralic composition. Commonly of QUARTZ, but rods of CALCITE and PYRITE have been described. Probably formed by the DEFORMATION of MINERAL segregations, the DETACHMENT and isolation of FOLD HINGES, where they define a LINEATION, or by the ELONGATION of CONGLOMERATE CLASTS.

rogen moraine A MORAINE field, with ridges 10–30 m high and spaced at 100–300 m, orthogonal to the direction of former ice advance, which probably formed beneath the GLACIER.

rollover anticline An ACCOMMODATION STRUCTURE accompanying the movement on a LISTRIC FAULT in an EXTENSIONAL FAULT SYSTEM.

romanèchite $(BaMn^{2+}Mn_8^{4+}O_{16}(OH)_4)$ An ORE MINERAL of manganese.

roméite $((Ca,Fe,Na)_2(Sb,Ti)_2(O,OH)_7)$ Hydrated calcium antimonite, sometimes with manganese and IRON, found in manganese deposits and PLACER DEPOSITS.

roof fault The FAULT at the top of the extensional DUPLEX formed when the SOLE FAULT migrates into the FOOTWALL of an EXTENSIONAL FAULT SYSTEM.

roof pendant A large mass of COUNTRY ROCK in the roof of an igneous intrusion.

roof rock See SEAL.

roof thrust The FAULT into which THRUST FAULTS merge at the top of a DUPLEX in an IMBRICATE STRUCTURE. Cf. SOLE THRUST.

room and pillar mining (bord and pillar mining, stoop and room mining) A mining method in flat-lying deposits in which MINERALS are extracted from rooms separated by a series of pillars which support the roof.

root cast See ROOT MOULD.

root mould (root cast) The cast of a root which has subsequently decayed.

root petrification The formation of a RHIZOLITH by the impregnation or REPLACEMENT of the organic matter of a root.

root tubule A RHIZOLITH comprising a cemented cylinder around a ROOT MOULD.

root zone Of a NAPPE, the area from which the NAPPE originated and from which it is now usually separated due to EROSION and/or DEFORMATION.

root-mean-square velocity The SEISMIC VELOCITY down to a reflector expressed as the square root of the sum of the products of layer thickness and squared velocity divided by the sum of the thicknesses. Usually estimated from NORMAL MOVEOUT measurements.

roscoelite $(KV_2(AlSi_3O_{10})(OH)_2)$ An ORE MINERAL of vanadium with a MICA structure.

rose diagram A type of circular histogram used to display directional data, e.g. the trends of GLACIAL STRIATIONS.

rose opal A red variety of OPAL.

Ross Orogeny An OROGENIC event in East Antarctica, at c. 500 Ma, associated with GONDWANALAND amalgamation.

Rossi-Forel Scale A ten point scale of EARTHQUAKE INTENSITY.

Rostroconchia A class of phylum MOLLUSCA, comprising a pseudobivalved shell whose two parts are connected dorsally by a continuous calcareous strip. Range CAMBRIAN–PERMIAN.

rotational deformation DEFORMATION in which the initial and final positions of the STRAIN AXES appear to have rotated with respect to an external reference frame. Cf. NON-COAXIAL DEFORMATION.

rotational fault A FAULT with a curved FAULT PLANE causing rotation of displaced material. See also LISTRIC FAULT.

rotational slip (rotational slump) A type of MASS MOVEMENT in COHESIVE sediments along a FAILURE plane with the shape of an arc of a circle, common where river valleys incise weak CLAY deposits.

rotational slump See ROTATIONAL SLIP.

rotational strain STRAIN resulting from a non-COAXIAL STRAIN path, described by VORTICITY.

Röthlisberger channels (R channels) A series of conduits at the bed of a GLACIER incised upwards into the ice, probably by the heat generated by the flow of TURBULENT water.

Rotliegendes 1 The older EPOCH of the PERMIAN, 290.0–256.1 Ma. **2** A PERMIAN succession in NW Europe equivalent to the ASSELIAN, SAKMARIAN and ARTINSKIAN.

roundness (angularity) A measure of the sharpness or otherwise of the corners of a CLAST, used in the description of a CLASTIC ROCK.

royalty The percentage of the revenue from mining or quarrying or HYDROCARBON recovery paid to the owner of the MINERAL rights.

RQD See rock quality designation.

rubefied Descriptive of a reddening of a soil or rock by the OXIDATION of IRON during WEATHERING.

rubellite A red to pink variety of TOURMALINE, sometimes used as a semi-precious GEM.

rubicelle A yellow or orange variety of SPINEL.

rubidium-strontium dating A RADIOMETRIC DATING technique based on the decay of ^{87}Rb to ^{87}Sr with a HALF-LIFE of ~50 Ga.

ruby A red GEM form of CORUNDUM.

ruby copper A variety of CUPRITE forming ruby-red TRANSPARENT crystals.

ruby silver ore See PROUSTITE, PYRARGYRITE.

ruby spinel (almandine spinel) ($MgAl_2O_4$) A red, GEM quality SPINEL.

rudaceous With the appearance of a RUDITE.

rudistid A cone-shaped BIVALVE resembling a solitary CORAL.

rudite A CLASTIC ROCK with >30% CLASTS of GRAVEL grade (>2 mm) with or without a MATRIX of SAND and/or MUD grade.

Rugosa (Tetracorallia) A subclass of class ANTHOZOA; solitary or colonial CORALS with bilaterally symmetrical, calcareous CORALLITES in which septae are inserted in four loci. Range CAMBRIAN–PERMIAN.

Runangan An EOCENE succession in New Zealand equivalent to the PRIABONIAN.

rundkarren Subsoil KARREN made up of rounded SOLUTION runnels.

runoff The flow of water on hillslopes, comprising BASEFLOW and QUICKFLOW.

runzelmarken (wrinkle marks) Parallel or reticulate ridges, <1 mm high and spaced at a few millimetres, on COHESIVE muddy sediment surfaces. Form in intertidal environments when strong wind STRESS affects the surface when covered by a very thin film of water, or by the action of subaqueous currents, or by sediment loading.

Rupelian The lower STAGE of the OLIGOCENE, 35.4–29.3 Ma.

Rustler A PERMIAN succession in the Delaware Basin, USA equivalent to the upper part of the LONGTANIAN.

rusty gold NATIVE GOLD with a stained surface.

rutilated quartz See NEEDLE STONE.

rutile (TiO_2) An ORE MINERAL of titanium and a common ACCESSORY MINERAL.

ruware A low, dome-shaped BEDROCK EXPOSURE protruding from a cover of ALLUVIUM or WEATHERED BEDROCK; a relict or incipient INSELBERG.

S

S 1 In EARTHQUAKE SEISMOLOGY, an S WAVE that does not travel through the CORE. **2** See SIEMEN.

S surface The planar, curved or irregular surface to which a FOLIATION is parallel.

S tectonite A rock with a tectonic FOLIATION. Cf. L-S TECTONITE.

S wave (secondary wave, shear wave, transverse wave) A SEISMIC BODY WAVE whose associated particle motion is at right angles to the direction of propagation. Its SEISMIC VELOCITY is generally ~0.58 of the P WAVE velocity in the same medium.

S wave splitting The phenomenon whereby cracks of common alignment within a rock cause S WAVES to be split into a fast wave polarized parallel to the cracks and a slow wave polarized at right angles to them. A means of observing DILATANCY and important in EARTHQUAKE PREDICTION.

s fold An ASYMMETRICAL FOLD with one LIMB shorter than the other whose profile defines an 'S' shape. Cf. Z FOLD.

Saale glaciation The second of the GLACIATIONS of N. Europe in the QUATERNARY.

sabkha (sebkha) A broad plain or SALT FLAT in an arid or semi-arid region containing EVAPORITES at a level dependent on the local WATER TABLE.

saccharoidal With the appearance of sugar.

saddle dolomite A coarsely crystalline DOLOMITE with pronounced lattice curvature, believed to form at temperatures between 50–100°C.

saddle reef A saddle- to triangle-shaped accumulation of MINERALS or COAL in the HINGE zone of a FOLD, forming in the gap produced when COMPETENT beds are interbedded, particularly when thin horizons of SHALE are present.

safflorite ($CoAs_2$) A MINERAL with a LOELLINGITE structure found in MESOTHERMAL VEIN deposits.

sagenite QUARTZ containing oriented needles of RUTILE in an attractive pattern.

St David's An EPOCH of the CAMBRIAN, 536.0–517.2 Ma.

Saint Venant equations FLOOD ROUTING equations which describe varying, unsteady flow in open channels.

Sakaraulian A MIOCENE succession on the Russian Platform covering parts of the AQUITANIAN, BURDIGALIAN and LANGHIAN.

Sakmarian A STAGE of the PERMIAN, 281.5–268.8 Ma.

Salado A PERMIAN succession in the Delaware Basin, USA equivalent to the middle part of the LONGTANIAN.

salar A BASIN of inland drainage in an arid or semi-arid region, dry for the majority of the time.

salcrete A surface or near-surface crust, principally of sodium chloride, which cements a SAND surface or other permeable soil, formed through the evaporation of moisture.

salic mineral A NORMATIVE MINERAL composed of SILICA and alumina.

salina See SOLAR POND.

salinity The total quantity of dissolved solids in sea water, usually expressed in parts per thousand when carbonate has been converted to oxide, bromide and iodide to chloride and all organic matter completely oxidized.

salinization A PEDOGENETIC process in which salts accumulate in the soil.

salps A gelatinous zooplankton contributing to MARINE SNOW.

salt diapir A STRUCTURE formed by the buoyant ascent of relatively low DENSITY EVAPORITE through a sediment OVERBURDEN by HALOKINESIS. Often constitutes an effective HYDROCARBON TRAP in combination with the deformed sediments.

salt dome A SALT DIAPIR with a dome-like shape.

salt flat A flat stretch of salt-encrusted ground.

salt heave A source of damage to man-made structures arising from the presence of soluble salts which increase in volume on HYDRATION.

salt marsh A vegetated, intertidal, MUD flat on a temperate, low energy coastline.

salt pan A deposit of salt developed in a SALT MARSH.

salt pillow The protuberance forming on a salt layer when HALOKINESIS is initiated.

salt pond See SOLAR POND.

salt weathering The WEATHERING resulting from physical changes associated with salt CRYSTALLIZATION, HYDRATION or thermal expansion, all of which can generate substantial PRESSURES.

saltation The process of BEDLOAD TRANSPORT comprising a series of ballistic jumps in which grains ascend steeply (>45°) from the bed and return at a low angle (<10°). Height and impact effects are dependent on the fluid characteristics, being greater in air than in water.

saltpetre See NITRE.

samarium-neodymium dating A RADIOMETRIC DATING technique based on the decay of ^{147}Sm to ^{143}Nd with a HALF-LIFE of ~2.5 x 10^5 Ma, which is resistant to the effects of ALTERATION and METAMORPHISM.

Samfrau fold belt A L. PALAEOZOIC OROGENIC BELT formed along the southern edge of GONDWANALAND and now found in Australia, Tasmania, Antarctica, S. Africa and Brazil.

sand MINERAL or rock grains of 0.625–2 mm diameter, often composed of QUARTZ.

sand flow cross-strata CROSS-STRATIFICATION of the deposits on the SLIP-FACE of a DUNE produced by grain flows of dry SAND, so that they show INVERSE GRADING.

sand rose A ROSE DIAGRAM showing the volume of SAND moved by the wind from different directions, usually over a one year period.

sand sea See ERG.

sand sheet A rippled to unrippled, flat to irregular area of AEOLIAN SAND where DUNES with SLIP-FACES are generally absent. Forms in ERGS where conditions are unsuitable for DUNE formation, i.e. a high WATER TABLE, periodic floods, surface CEMENTATION, the presence of vegetation or a significant component of coarse-grained particles.

sand volcano A mound of SAND/MUD, <30 cm high and about 25–50 mm in diameter with a conical depression at the apex from which fine-grained sediment in suspension with water is emitted. Develops by the extrusion of interstitial water as the sediments compact. Cf. MUD VOLCANO.

sandbed channel A channel of largely sandy material which is transported at a wide range of DISCHARGES, generally wider and shallower than channels in more COHESIVE sediment.

Sander's Symmetry Principle 'The symmetry of a FABRIC is the same or less than the kinetic symmetry producing the FABRIC.' Used to infer the kinematics of DEFORMATION from observed FABRICS.

sandstone A CLASTIC SEDIMENTARY ROCK with >25% by volume of CLASTS of SAND grade (0.625–2 mm diameter).

sandstone-uranium-type deposit An uranium-rich SANDSTONE-URANIUM-VANADIUM BASE METAL DEPOSIT.

sandstone-uranium-vanadium base metal deposit A terrestrial sedimentary deposit of probable FLUVIATILE origin laid down under arid conditions so that COUNTRY ROCKS are often red. Generally contains one or two metals in economic quantities.

sandstorm The elevation of SAND-grade particles by a strong wind so that visibility is reduced to <1000 m. Particles rarely rise >15 m above the surface. Cf. DUST STORM.

Sandugan A SILURIAN succession in the Mirnyy Creek area of NE Siberia equivalent to the TELYCHIAN and WENLOCK.

sandur A large outwash plain created by the meltwater from an ice mass.

sandwave **1** Any large-scale periodic BEDFORM. **2** A large-scale, linear-crested, two-dimensional periodic STRUCTURE with a large wavelength:height ratio. **3** A large-scale BEDFORM growing to equilibrium over longer time-scales than individual steady or quasi-steady flows. Formed by reversing flows of TIDAL frequency.

sanidine $(KAlSi_3O_8)$ A high-temperature form of POTASH FELDSPAR.

sanitary landfill (landfill) A land site where municipal solid waste is buried in a fashion such as to cause minimum disturbance or pollution to the environment. The cheapest method of disposal, but subject to problems of LEACHING and SEEPAGE of waste substances into GROUNDWATER.

sannaite An ALKALINE LAMPROPHYRE, similar to CAMPTONITE but with ALKALI FELDSPAR more abundant than PLAGIOCLASE.

Santonian A STAGE of the CRETACEOUS, 86.6–83.0 Ma.

sanukite A type of HYPERSTHENE ANDESITE,

black and GLASSY when fresh, commonly used for stone tools in prehistoric Japan.

saponite (bowlingite) $((Mg,Fe)_3(Al, Si)_4O_{10}(OH)_2(Ca,Na)_{0.3}(H_2O)_4)$ A swelling CLAY MINERAL of the MONTMORILLONITE group.

sapphir d'eau (water sapphire) An intense blue variety of CORDIERITE.

sapphire A blue GEM variety of CORUNDUM.

sapphirine $(Mg_{3.5}Al_{9.0}Si_{1.5}O_{20})$ A bluish, magnesium-aluminium silicate with some iron found in SILICA-poor METAMORPHIC ROCKS.

sapping The undercutting of the base of a cliff and subsequent FAILURE of the cliff face caused by wave action, lateral stream EROSION, GROUNDWATER outflow, etc.

saprolite A fine-grained CLAY material formed by the *in situ* DEEP WEATHERING of BEDROCK, particularly CRYSTALLINE IGNEOUS and METAMORPHIC ROCKS, under humid tropical and sub-tropical conditions.

sapropel An organic-rich deposit of NEOGENE age found in the Mediterranean and the Black Sea.

sapropelic coal A fine-grained, faintly stratified to homogeneous, massive COAL deposit. Generally dark coloured, tough and exhibiting CONCHOIDAL FRACTURE. The two main types are CANNEL COAL and BOGHEAD COAL.

Saratovan A PALAEOCENE succession in the former USSR covering parts of the DANIAN and THANETIAN.

Sarcopterygii A subclass of class OSTEICHTHYES, superclass PISCES; lobe-fin FISH. Range DEVONIAN–RECENT.

sard A brown variety of CHALCEDONY.

sardonyx A variety of MICROCRYSTALLINE SILICA composed of interlayered ONYX and SARD.

Sarka An ORDOVICIAN succession in Bohemia equivalent to the LLANVIRN.

Sarmatian See MEOTIC.

sarsen (gray wethers, grey wethers) Blocks of SILICA-cemented SANDSTONE (SIL-CRETE), BRECCIA or CONGLOMERATE believed to have formed on the surface or in the near-surface of S. England during warm phases of the TERTIARY and to have been broken down by frost action and mass movement. Used in the construction of Stonehenge, England and earlier structures. See also PUDDINGSTONE.

sassolite (H_3BO_3) The rare MINERAL boric acid.

sastrugi A small (<50 mm high), irregular ridge parallel to the wind direction, commonly with a concave upwind face, produced by wind EROSION of a moist or salt-cemented sandy surface.

satellite fold See PARASITIC FOLD.

satellite geodesy The use of satellites for geodetic observations, such as in SATELLITE LASER RANGING and SATELLITE RADIOPOSI-TIONING.

satellite imagery REMOTE SENSING from an artificial satellite.

satellite laser ranging (SLR) A technique involving the simultaneous measurement of the distance to two or more terrestrial stations from an artificial satellite by the travel time of a laser beam so as to determine the distance between the stations to an accuracy just sufficient to monitor CONTINENTAL DRIFT.

Satellite Probatoire pour l'Observation de la Terre (SPOT) A satellite used in REMOTE SENSING which produces stereo pairs of images.

satellite radiopositioning A technique of three-dimensional positioning by radio interferometry using the GLOBAL POS-ITIONING SYSTEM, accurate enough to measure rates of CONTINENTAL DRIFT and TEC-TONIC disturbance.

satin spar A fibrous variety of GYPSUM.

saturated Descriptive of an IGNEOUS ROCK with neither an excess nor deficiency of SIL-ICA. See also SILICA SATURATION.

saturated wedge A zone of soil saturation by THROUGHFLOW formed when INFIL-TRATION rates exceed the rate of percolation into lower horizons, with a wedge shape due to its extension upslope.

saturated zone The zone below the WATER TABLE saturated with GROUNDWATER.

saturation magnetization The maximum intensity of ISOTHERMAL REMAN-ENT MAGNETIZATION caused by a strong (>0.1 TESLA), external MAGNETIC FIELD, measured after removal of the field.

saturation overland flow RUNOFF from an area of soil on a shallow slope saturated by THROUGHFLOW which accumulates at the base of the slope.

Saucesian A MIOCENE succession on the west coast of the USA covering parts of the AQUITANIAN and BURDIGALIAN.

sauconite ($Na_{0.33}Zn_3(Si,Al)_4O_{10}(OH)_2$. $4H_2O$) A swelling CLAY MINERAL of the SMEC-TITE group.

Saurischia A class of order RHYNCHOCEPH-ALIA, subclass ARCHOSAURIA, class REPTILIA; DINOSAURS with a lizard-like pelvis. Range TRIASSIC–end CRETACEOUS.

Sauropoda Herbivorous members of the SAURISCHIA.

Sauropterygia An order of subclass EURYAPSIDA, class REPTILIA; the PLESIOSAURS and NOTHOSAURS. Range L. TRIASSIC–U. CRETACEOUS.

saussurite A fine-grained mixture of ZOISITE and other MINERALS formed by the ALTERATION of FELDSPAR.

Saxonian metallogenic epoch An EPOCH of EPIGENETIC mineralization from M. TRIASSIC to JURASSIC in the VARISCAN ORO-GENIC BELT of NW Europe.

saxonite A coarse-grained, ULTRABASIC ROCK comprising OLIVINE and ORTHOPYROX-ENE (commonly HYPERSTHENE), i.e. a HYPER-STHENE PERIDOTITE.

scabland A landscape of bare rock surfaces, thin soils and sparse vegetation underlain by flat BASALT flows and dissected by dry channels caused by GLACIAL floodwaters.

scanning electron microscopy (SEM) A technique of ELECTRON MICROSCOPY used in the investigation of surface structure and composition by irradiating a bulk sample with a scanning, focused beam of electrons and imaging backscattered and secondary electrons on a cathode ray tube scanned synchronously with the specimen.

scanning transmission electron microscopy (STEM) A form of SCANNING ELECTRON MICROSCOPY using transmitted electrons.

Scaphopoda A class of phylum MOLLUSCA; MOLLUSCS with a univalved, tusk-shaped shell open at both ends, a differentiated head region and radula but rudimentary gills. Range ORDOVICIAN–RECENT.

scapolite series (3NaAlSi$_3$O$_8$.NaCl–3CaAl$_2$Si$_2$O$_8$.CaCO$_3$) A group of METAMORPHIC TECTOSILICATES forming a SOLID SOLUTION between MARIALITE and MEIONITE.

scar A steep, rocky cliff in massively bedded LIMESTONE.

scar fold A FOLD formed by the FLOW of material around a BOUDIN.

scarpslope See ESCARPMENT.

scheelite (CaWO$_4$) A major ORE MINERAL of tungsten found in PNEUMATOLYTIC and HYDROTHERMAL VEINS and METAMORPHIC AUREOLES.

schiller The play of light on a crystal face at a certain angle of illumination.

schillerspar See BASTITE.

schist A rock exhibiting SCHISTOSITY, generally of high METAMORPHIC GRADE.

schistosity A FOLIATION produced by DEFORMATION in which TABULAR MINERALS, coarse enough to be visible to the unaided eye, have a PREFERRED ORIENTATION.

Schizomycophyta See BACTERIA.

schlieren Pencil-shaped, discoidal or blade-like inclusions made up of AGGRE-GATES with a greater concentration of MAFIC MINERALS than the COUNTRY ROCK.

Schlumberger configuration An electrode arrangement in the RESISTIVITY METHOD in which the potential electrodes are much closer together than the current electrodes.

Schmid factor See CRITICAL RESOLVED SHEAR STRESS.

Schmidt net A type of STEREOGRAPHIC NET using EQUAL AREA PROJECTION, in which areas are portrayed undistorted. Mainly used for the statistical analysis of orientation data. Cf. WULFF NET.

schollen Large blocks of COUNTRY ROCK found in TILL sheets, resulting from GLACIAL EROSION.

schorl The black, iron-rich variety of TOURMALINE.

schorlomite A black variety of ANDRADITE with Ti >Fe^{3+}, which is even richer in Ti than MELANITE.

Schottky crystal defect See VACANCY.

Schumann resonance frequencies The frequencies that are preferentially enhanced as SFERICS propagate in the Earth-ionosphere cavity.

schuppen structure An IMBRICATE STRUCTURE in an OROGENIC BELT.

scintillation counter (scintillation meter, scintillometer) An instrument used in RADIOMETRIC SURVEYING for measuring gamma rays by counting the flashes they produce on a screen coated with, e.g., zinc sulphide.

scintillation meter See SCINTILLATION COUNTER.

scintillometer See SCINTILLATION COUNTER.

scissor fault A FAULT with a reversal of DISPLACEMENT direction along the FAULT PLANE, so that there is pivoting about an axis of zero movement analogous to the opening of a pair of scissors.

Scleractinia (Hexacorallia) An order

of subclass ZOANTHARIA, class ANTHOZOA; CORALS secreting calcareous, commonly ARAGONITIC exoskeletons. Range M. TRIASSIC–RECENT.

Sclerospongia A class of phylum PORIFERA; SPONGES with an ARAGONITIC skeleton. Range ORDOVICIAN–RECENT.

sclerotinite A COAL MACERAL of the INERTINITE group made up of the coalified remains of FUNGAL material.

scolecite $(CaAl_2Si_3O_{10}.3H_2O)$ A ZEOLITE found in cavities in BASALTIC rocks.

scoria A rusty red to black, highly VESICULAR, MAFIC volcanic GLASS. Cf. PUMICE.

scoriaceous Descriptive of a LAVA or PYROCLASTIC ROCK containing empty cavities.

scorodite $(FeAsO_4.2H_2O)$ An hydrated iron-aluminium arsenate found as an ACCESSORY MINERAL formed by the ALTERATION of arsenic-bearing MINERALS.

Scorpionida An order of class ARACHNIDA, subphylum CHELICERATA, phylum ARTHROPODA; the scorpions. Range SILURIAN–RECENT.

scorzalite $((Fe,Mg)Al_2(PO_4)_2(OH)_2)$ A rare phosphate MINERAL occurring in METAMORPHIC ROCKS and PEGMATITES.

Scottish topaz A yellow, transparent variety of QUARTZ.

Scottville A PERMIAN succession in Queensland, Australia, covering the lower part of the WORDIAN.

scour mark A depression caused by EROSION at the base of a TURBIDITY CURRENT or on other erosional surfaces.

Scourian An OROGENIC episode affecting NW Scotland at 2900–2300 Ma.

scrap mica See FLAKE MICA.

scratch See TOOL TRACK.

scree (talus) A sloping accumulation of loose CLASTS of GRANULE grade or larger, generally in the form of a wedge, metres to hundreds of metres in height, at the base of a steep rock face from which the CLASTS fall as a result of WEATHERING and EROSION.

screw dislocation A CRYSTAL DEFECT in which the atoms are shifted relative to each other in a direction parallel to the DISLOCATION LINE.

scroll bar A sediment ridge deposited parallel to the contours of a POINT BAR in an alluvial channel.

Scyphozoa CNIDARIA with a polypoid or medusal form with few hard parts. Range U. PRECAMBRIAN–RECENT.

Scythian An EPOCH of the TRIASSIC, 245.0–241.1 Ma.

sea cliff A steep coastal slope caused by the interaction of marine and subaerial processes.

sea level datum The height of the mean sea level surface to which all elevations are related. Frequently used as the reference level in geophysical surveys.

seafloor spreading The hypothesis that CONTINENTAL DRIFT takes place through the growth of ocean basins between the continents. OCEANIC LITHOSPHERE is created at OCEANIC RIDGES, generating the source of OCEANIC MAGNETIC ANOMALIES according to the VINE-MATTHEWS HYPOTHESIS, and recycled into the MANTLE at SUBDUCTION ZONES.

seal (roof rock) The impervious capping which prevents the upward MIGRATION of HYDROCARBONS from a RESERVOIR, often comprising CLAYS, SHALES or EVAPORITES.

seamount An approximately conical, submarine peak which does not rise above sealevel. Often found in chains formed as the LITHOSPHERE passes over a HOTSPOT.

seat earth A thin horizon beneath a COAL SEAM containing FOSSIL rootlets which represents the soil in which the vegetation grew. Often rich in KAOLINITE formed by the LEACHING of cations by acidic soil waters and the ELUVIATION of CLAYS.

SEASAT A satellite which measured the height of the sea-level surface with great accuracy. This allowed the computation of the FREE-AIR ANOMALY map of the oceans.

sebkha See SABKHA.

second order fault (splay fault) A MINOR FAULT at an acute angle to a MAJOR FAULT with the same sense of slip.

second order fold A FOLD which is smaller than a FIRST ORDER FOLD and whose ENVELOPING SURFACE is folded by a FIRST ORDER FOLD.

secondary creep The second stage of VISCOELASTIC STRAIN characterized by steady-state VISCOUS FLOW and a constant STRAIN-time slope. See also PRIMARY CREEP, TERTIARY CREEP.

secondary enrichment See SUPERGENE ENRICHMENT.

secondary migration The MIGRATION of HYDROCARBONS, usually lateral, through or out of the RESERVOIR rock. Cf. PRIMARY MIGRATION.

secondary mineral A MINERAL formed after the formation of its enclosing rock, usually by the ALTERATION of a PRIMARY MINERAL.

secondary porosity The pore space generated in SEDIMENTARY ROCKS after deposition and during DIAGENESIS by the DISSOLUTION of detrital framework grains, MATRIX or earlier AUTHIGENIC cements, by the REPLACEMENT of other carbonates by DOLOMITE in carbonate rocks or by fracturing or shrinking associated with the dehydration of hydrous MINERALS. Cf. PRIMARY POROSITY.

secondary quartz QUARTZ forming as a cement in DIAGENESIS.

secondary wave See SHEAR WAVE.

sectile mineral A TENACITY descriptor indicating a MINERAL that can be cut by a knife but powders under a hammer.

secular variation The long-term, progressive and predictable change in the GEOMAGNETIC FIELD, probably arising from the changing circulation patterns in the fluid outer CORE. Cf. DIURNAL VARIATION.

sedarenite An ARENITE whose CLASTS are dominantly of sedimentary origin.

sedentary rudaceous deposit A RUDITE formed *in situ* by WEATHERING.

sedifluction The subaerial or subaqueous movement of unconsolidated sediment.

sedigraph A device which records the relative SETTLING VELOCITIES of particles in water so as to study the GRAIN SIZE DISTRIBUTION of a sediment.

sediment delivery ratio The ratio of gross EROSION to SEDIMENT YIELD for an entire drainage BASIN.

sediment drift An elongate, lobate mound of fine-grained SANDS and MUDS constructed by a CONTOUR CURRENT, i.e. a CONTOURITE.

sediment gravity flow An AGGREGATE of grains moving under the influence of GRAVITY mostly independently of the overlying medium, including GRAIN FLOWS, DEBRIS FLOWS, LIQUEFIED FLOWS and TURBIDITY CURRENTS, which overcome the friction between particles in different fashions.

sediment transport equation One of a number of equations designed to predict the SEDIMENT YIELD of a given flow from its SHEAR VELOCITY or SHEAR STRESS, the GRAIN SIZE DISTRIBUTION, the HYDRAULIC RADIUS of the channel and the characteristics of the fluid.

sediment yield The total quantity of sediment exported from a drainage BASIN over a given period of time.

sedimentary breccia A BRECCIA formed of SEDIMENTARY ROCKS.

sedimentary exhalative deposit See VOLCANIC-EXHALATIVE DEPOSIT.

sedimentary facies A SEDIMENTARY ROCK body with specific and distinctive characteristics, such as those defined by its biota, chemistry or physics. Also used in a genetic sense for sediment deposited by a particular process or in an environmental sense for sediment deposited in similar settings.

sedimentary rock A rock formed by the consolidation of sediment.

sedimentology The study of sediments and their deposition and accumulation.

seepage **1** An escape of gas or oil. **2** A diffuse flow of water from an AQUIFER, not sufficiently localized to be termed a SPRING.

seepage velocity The apparent velocity of GROUNDWATER calculated from the DARCY EQUATION, which is higher than the true velocity because of the effects of POROSITY and the TORTUOSITY of the flow path.

seiche A STANDING WAVE generated on the surface of an enclosed or partly enclosed body of water by winds, atmospheric gradients, TIDES, EARTHQUAKES or sediment SLIDES.

seif A linear DUNE, up to 100 m in height, whose orientation is controlled by the prevailing winds, commonly occurring in a series of parallel ridges.

seismic array **1** A group of SEISMOMETERS, whose output can be combined to improve the SIGNAL TO NOISE RATIO. In SEISMIC PROSPECTING arrays are generally arranged in linear patterns along survey lines. In EARTHQUAKE SEISMOLOGY and FORENSIC SEISMOLOGY arrays may be distributed over hundreds of square kilometres, and the direction of the causative EPICENTRE of the arrivals can be located from the different ARRIVAL TIMES at individual sensors. **2** Less commonly, a group of SEISMIC SOURCES, whose output can be combined to improve the SIGNAL TO NOISE RATIO.

seismic belt A linear grouping of EARTHQUAKE EPICENTRES, commonly along a PLATE BOUNDARY but also on a local scale.

seismic body wave See BODY WAVE.

seismic coda The concluding part of a SEISMOGRAM which follows the earlier, identifiable waves, and which can last a considerable time.

seismic creep A form of non-instantaneous DEFORMATION in which a residual STRAIN remains after removal of STRESS by an EARTHQUAKE, which decreases or increases over a period of time dependent on the initial STRESS, RIGIDITY and a time-dependent CREEP function.

seismic discontinuity An abrupt boundary between two media over which the ACOUSTIC IMPEDANCE changes significantly, e.g. MOHOROVICIC DISCONTINUITY, GUTENBERG DISCONTINUITY.

seismic event A sudden disturbance on or within the Earth that generates SEISMIC WAVES, such as an EARTHQUAKE, chemical or nuclear explosion, ROCKBURST or mine FAILURE, but excluding sources of MICROSEISMS.

seismic exploration (seismic prospecting) The use of seismic methods to locate and delineate geological STRUCTURES or bodies in the subsurface. The two principal techniques use SEISMIC REFLECTION and SEISMIC REFRACTION.

seismic facies analysis A technique of SEISMIC STRATIGRAPHY which determines the characteristics of SEISMIC REFLECTIONS (continuity, amplitude, frequency, INTERVAL VELOCITY) which distinguish them from adjacent reflections and uses them to infer the depositional environment and thus the presence of possible RESERVOIRS. See also REFLECTION CHARACTER ANALYSIS, SEISMIC SEQUENCE ANALYSIS.

seismic gap (Mogi Doughnut) A space in a SEISMIC BELT where no EARTHQUAKE has occurred over a stated period of time, commonly 10 years. Sometimes the anticipated site of a large EARTHQUAKE and so useful in EARTHQUAKE PREDICTION.

seismic head wave See HEAD WAVE.

seismic migration (migration) The process of displaying seismic reflectors on a SEISMIC REFLECTION SEISMOGRAM in their true position in space, rather than immediately below the SHOT-detector location. A necessary operation when STRATA are dipping.

seismic modelling See RAY TRACING.

seismic moment The scalar size of the set of couples that represent the FORCES causing FAILURE in an EARTHQUAKE or explosion; a measure of EARTHQUAKE 'size', equal to the product of the RIGIDITY, slip length along the FAULT PLANE and the slip area on the FAULT PLANE.

seismic multiple See MULTIPLE REFLECTION.

seismic noise The NOISE appearing on a SEISMIC RECORD.

seismic phase A grouping of seismic arrivals on a SEISMOGRAM representing, for example, a P WAVE.

seismic prospecting See SEISMIC EXPLORATION.

seismic record See SEISMOGRAM.

seismic reflection The SEISMIC WAVE or energy returned to a receiver from a SEISMIC SOURCE after reflection according to SNELL'S LAW at a discontinuity where there is a contrast in ACOUSTIC IMPEDANCE. The basis of an important form of SEISMIC EXPLORATION.

seismic refraction The SEISMIC WAVE or energy that has been refracted according to SNELL'S LAW at a discontinuity where there is a contrast in ACOUSTIC IMPEDANCE. The HEAD WAVE returning to the surface from a SEISMIC SOURCE after CRITICAL REFRACTION is the basis of an important form of SEISMIC EXPLORATION.

seismic sequence analysis A technique of SEISMIC STRATIGRAPHY in which a SEISMOGRAM is divided into units of common reflection and depositional characteristics by the identification of UNCONFORMITIES or changes in the seismic patterns. Each sequence thus represents a three-dimensional set of contemporaneous sediments in the same system of depositional processes and environment. See also REFLECTION CHARACTER ANALYSIS, SEISMIC FACIES ANALYSIS.

seismic shooting The detonation of a SHOT or other SEISMIC SOURCE so as to collect information for SEISMIC EXPLORATION.

seismic slip Slip on a FAULT PLANE accompanied by SEISMICITY. Cf. ASEISMIC SLIP.

seismic source Any mechanism which generates SEISMIC WAVES, e.g. EARTHQUAKES, chemical and nuclear explosions.

seismic spectrum Curve showing the amplitude of ground motion as a function of frequency or period, obtained by FOURIER ANALYSIS of the ground motion.

seismic stratigraphy (seismostratigraphy) The study of the STRATIGRAPHY and distinct, genetically related, depositional units from SEISMIC REFLECTION data by the analysis of non-structural information on SEISMOGRAMS. The techniques utilized are SEISMIC SEQUENCE ANALYSIS, SEISMIC FACIES ANALYSIS, REFLECTION CHARACTER ANALYSIS and the recognition of direct HYDROCARBON indicators, such as BRIGHT SPOTS.

seismic surface wave See SURFACE WAVE.

seismic tomography (tomography) The technique of mapping SEISMIC VELOCITY variations in a subsurface volume by the comparison of ARRIVAL TIMES at a network of SEISMOMETERS of SEISMIC WAVES traversing the volume in different directions.

seismic velocity The velocity with which a medium transmits SEISMIC WAVES, controlled by its ELASTIC MODULI and DENSITY.

seismic wave An ELASTIC WAVE travelling within the Earth, falling into one of the two general categories of BODY and SURFACE WAVES.

seismicity The distribution of EARTHQUAKE-induced seismic activity in time, location, MAGNITUDE and depth, of great significance to EARTHQUAKE PREDICTION and studies of the dynamic behaviour of the Earth.

seismogram A recording of the variation of ground displacement, velocity or acceleration with time, measured with a SEISMOMETER.

seismograph See SEISMOMETER.

seismology The study of SEISMIC WAVES within the Earth, including their generation by EARTHQUAKES or explosions, how they propagate and the information they provide about the structure and nature of the Earth's interior.

seismometer (seismograph) The general name for an instrument that monitors SEISMIC WAVES, e.g. GEOPHONE, HYDROPHONE.

seismostratigraphy See SEISMIC STRATIGRAPHY.

seismoturbidite See MEGATURBIDITE.

Selachii An order of subclass ELASMOBRANCHII, class CHONDRICHTHYES, superclass PISCES; the modern sharks and their ancestors. Range JURASSIC–RECENT.

selenite A variety of GYPSUM yielding broad, colourless, transparent CLEAVAGE folia.

self potential (spontaneous polarization) The phenomenon whereby natural electrical fields are generated in the subsurface by electrochemical reactions between electrically conductive MINERALS and GROUNDWATER.

self potential log A GEOPHYSICAL BOREHOLE LOG which measures the SELF POTENTIAL effects generated when PORE FLUIDS of different ionic concentration come into contact; particularly sensitive to SHALE horizons.

self reversal A rare mechanism by which a FERROMAGNETIC MINERAL can acquire a NATURAL REMANENT MAGNETIZATION of the opposite polarity to the ambient GEOMAGNETIC FIELD by interaction with the MAGNETIC FIELD of an existing MINERAL with NORMAL POLARITY.

self-potential method See SPONTANEOUS POLARIZATION METHOD.

SEM See SCANNING ELECTRON MICROSCOPY.

semi-metal The NATIVE ELEMENTS ARSENIC, ANTIMONY and BISMUTH, which are BRITTLE MINERALS with lower ELECTRICAL and THERMAL CONDUCTIVITIES than metals.

semifusinite A COAL MACERAL of the INERTINITE group, similar to FUSINITE but with a lower REFLECTANCE.

semseyite ($Pb_9Sb_8S_{21}$) A rare ORE MINERAL of lead.

Senecan A DEVONIAN succession in E. North America covering part of the GAVETIAN and the FRASNIAN.

Senonian A division of the CRETACEOUS, 88.5–65.0 Ma.

sepiolite (meerschaum) ($Mg_4(OH)_2$ $Si_6O_{15}H_2O+4H_2O$) A CLAY-like, hydrous, SECONDARY MINERAL of magnesium found in association with SERPENTINE.

septarian nodule (septarium) A NODULE with irregular internal cracks or VEINS.

septarium See SEPTARIAN NODULE.

Serevovinskiy A PERMIAN succession on the eastern Russian Platform covering parts of the LONGTANIAN and CHANGXINGIAN.

seriate texture A TEXTURE in an IGNEOUS ROCK in which there is a continuous range in crystal sizes.

sericite A fine-grained variety of MUSCOVITE.

sericitization A type of WALL ROCK ALTERATION whose products are dominantly SERICITE and QUARTZ.

series A third order CHRONOSTRATIGRAPHIC unit.

serir See REG.

serpentine asbestos See CHRYSOTILE.

serpentine group ($Mg_3Si_2O_5(OH)_4$) A group of common, hydrated PHYLLOSILICATES with three common POLYMORPHS: ANTIGORITE, LIZARDITE and CHRYSOTILE. Often found as an ALTERATION product of magnesium silicates, particularly OLIVINE, PYROXENE and AMPHIBOLE.

serpentinite A rock composed mainly of SERPENTINE MINERALS with minor BRUCITE, TALC, Fe-Ti oxide and Ca-Mg carbonates. Forms by SERPENTINIZATION with either a massive FABRIC with relict TEXTURES of the parent rock or a sheared FABRIC in which original FABRICS have been destroyed by intense PENETRATIVE DEFORMATION.

serpentinization The ALTERATION of ULTRAMAFIC ROCKS such as DUNITE, PERIDOTITE and PYROXENITE into SERPENTINITE.

Serpukhovian An EPOCH of the CARBONIFEROUS, 332.9–322.8 Ma.

Serravallian A STAGE of the MIOCENE, 14.2–10.4 Ma.

service reservoir See DISTRIBUTION RESERVOIR.

sessile Descriptive of a non-mobile organism.

settling velocity The terminal velocity at which a grain settles through a static fluid, dependent on the DENSITY, size, shape and bulk concentration of the particles and the DENSITY, VISCOSITY and RHEOLOGICAL properties of the fluid. May be measured to determine sediment grain size.

sferic A natural electromagnetic field in the audiofrequency range of $\sim 1-10^3$ Hz originating from thunderstorm activity and the source field of the MAGNETOTELLURIC and AFMAG techniques of GEOPHYSICAL EXPLORATION.

Sha-Xi-Miao A JURASSIC succession in Sichuan, China, equivalent to the BAJOCIAN, BATHONIAN and CALLOVIAN.

shadow zone A distance RANGE from a SEISMIC SOURCE in which some given type of SEISMIC WAVE does not emerge due to a subsurface refraction.

shakehole See DOLINE.

shale An ARGILLACEOUS rock with closely spaced, well-defined laminae.

shallow-focus earthquake An EARTHQUAKE with a depth of FOCUS shallower than 70 km. Cf. DEEP-FOCUS EARTHQUAKE, INTERMEDIATE-FOCUS EARTHQUAKE.

shallowing-upward carbonate cycle A phenomenon of sedimentation on CARBONATE PLATFORMS in which carbonate production exceeds the rate of subsidence.

Shannon-Weaver dominance diversity equation An equation used to calculate the DOMINANCE DIVERSITY (H) in a PALAEO-ECOLOGICAL study from $H = -\sum_{i=1}^{s} P_i \log Pi$, where P_i is the proportion of the ith species in the sample and s the total number of species.

Shaodong A CARBONIFEROUS succession in China equivalent to the FAMENNIAN.

shape fabric A planar or linear structure defined by PREFERRED ORIENTATION of the shapes of components of a material, requiring an ANISOTROPIC shape of the components and the alignment of the ANISOTROPY.

shard An abraded fragment of PUMICE.

shatter cone A centimetric- to metric-scale conical STRUCTURE believed to form during METEORITE impact.

shattuckite ($Cu_5(SiO_3)_4(OH)_2$) An INOSILICATE found in the oxidized zones of COPPER deposits.

shear DEFORMATION in which the angular relationship between material lines in a body change, i.e. a rotational STRESS or STRAIN.

shear band A narrow band of localized STRAIN developed in deforming, ANISOTROPIC, FOLIATED rocks, e.g. MYLONITE, PHYLLONITE, in which the DEFORMATION folds the existing METAMORPHIC FOLIATION.

shear direction The direction in which DISPLACEMENT takes place during SHEAR.

shear fault A FAULT on which DISPLACEMENT has been produced by SHEAR.

shear folding See FLOW FOLDING.

shear joint A JOINT forming by SHEAR at an acute angle to the maximum PRINCIPAL STRESS.

shear layer The zone of velocity gradient formed by the juxtaposition of two fluids of different velocity or DENSITY.

shear modulus See RIGIDITY.

shear plane The surface to which points undergoing SIMPLE SHEAR move parallel, i.e. the plane of a SHEAR ZONE.

shear resistance See SHEAR STRENGTH.

shear strain DEFORMATION involving a change in the intrinsic angular relationship between lines, defined as the tangent of the change in angle.

shear strength (shear resistance) The STRENGTH of a material subjected to SHEAR STRESS, a function of NORMAL STRESS.

shear stress The FORCE per unit area exerted tangentially to a given surface.

shear velocity (U^*) The velocity which expresses the velocity gradient of the lowest regions of a boundary layer (the lowest 15% of the flow depth). $U^* = \sqrt{(\tau/\rho)}$, where τ = SHEAR STRESS, ρ = fluid DENSITY.

shear wave See S WAVE.

shear zone A zone of DUCTILE DEFORMATION between two undeformed blocks that have suffered relative SHEAR DISPLACEMENT; the DUCTILE analogue of a FAULT.

sheath fold A highly NON-CYLINDRICAL FOLD with a strongly curved HINGE LINE within the FOLD AXIAL surface, formed during high STRAIN, DUCTILE DEFORMATION.

sheet erosion (sheetflood erosion) The removal of fine-grained, superficial material by sheets of flowing water rather than channelized streams, occurring, for example, in deserts where sporadic rainfall causes RUNOFF by SHEETFLOW on FANS and PEDIMENTS lacking incision.

sheet joint A FRACTURE parallel to the ground found in GRANITIC intrusions, possibly originating during cooling.

sheet silicate See PHYLLOSILICATE.

sheeted dyke complex A continuous series of BASALTIC DYKES in an OPHIOLITE which fed PILLOW LAVAS. Equated with layer 2 of the OCEANIC CRUST.

sheetflood The unconfined flow of water over the land surface resulting from channel overtopping or LEVÉE breaching, causing deposits of MUD, SILT, SAND or GRAVEL with a sharp base. Cf. CREVASSE SPLAY.

sheetflood erosion See SHEET EROSION.

sheetflow (sheetwash) OVERLAND FLOW over a smooth soil surface.

sheeting The process by which shells of rock split off along JOINTS running approximately parallel to the rock surface, resulting from PRESSURE RELIEF on DENUDATION.

sheetwash See SHEETFLOOD.

Sheinwoodian A STAGE of the SILURIAN, 430.4–426.1 Ma.

shell pavement An accumulation of shells left as a LAG deposit after the WINNOWING of fine-grained material.

Shemshinskiy A PERMIAN succession on the eastern Russian Platform covering the upper part of the UFIMIAN.

sherbakovite $(Na(K,Ba)_2(Ti,Nb)_2(Si_2O_7)_2)$ A dark brown, VITREOUS MINERAL found in PEGMATITE.

sheridanite $((Mg,Al)_6(Si,Al)_4O_{10}(OH)_8)$ A CHLORITE group MINERAL poor in both iron and silica, found in SCHIST and other METAMORPHIC ROCKS.

Shidert A CAMBRIAN succession in Siberia covering part of the MAENTWROGIAN and the DOLGELLIAN.

shield An extensive area of exposed BEDROCK with long-term TECTONIC stability, generally of PRECAMBRIAN age and forming the central core of a continent.

shield volcano A type of VOLCANO, 10–100 km in diameter, comprising many usually BASALTIC LAVA FLOWS and subordinate PYROCLASTIC ROCKS, with a circular to elliptical shape and flanks dipping at 2°– 8° away from a CRATER or line of CRATERS and/or a CALDERA. Often found in overlapping groups.

Shields beta (β) A dimensionless measure of SHEAR STRESS representing the ratio of the fluid FORCES promoting sediment movement to the submerged weight of a single layer of grains, or the GRAVITY forces opposing motion. $\beta = \tau/((\rho_s - \rho_f)gd)$, where τ = SHEAR STRESS per unit area of bed, ρ_s and ρ_f = DENSITIES of sediment and fluid respectively, g = GRAVITY and d = grain size.

Shields diagram A graphical representation of the threshold of sediment entrainment by a plot of SHIELDS BETA against particle REYNOLD'S NUMBER. Used to interpret the FORCES influencing the entrainment.

shipbourne geophysical survey GEOPHYSICAL EXPLORATION undertaken from a ship utilizing GRAVITY, MAGNETIC, SEISMIC or SIDE-SCAN SONAR techniques.

shock metamorphism METAMORPHISM at very high, transitory temperature and pressure generated by METEORITE impact.

shock remanent magnetization A NATURAL REMANENT MAGNETIZATION resulting from rapidly applied STRESS, such as in a METEORITE impact.

shoestring sand A STRATIGRAPHIC OIL TRAP comprising a long, narrow body of SANDSTONE enclosed by SHALE, probably originating as an OFFSHORE BAR or by a meandering river.

shonkinite A coarse-grained, FELDSPAR-rich SYENITE comprising PYROXENES and some OLIVINE.

shore platform (marine abrasion platform, wave-cut bench, wavecut platform) A low gradient, intertidal rock surface caused by the recession of a SEA CLIFF.

shortening (shortening strain) A linear measure of STRAIN defined as $(L_u–L_d)/L_u$, where L_u = undeformed length, L_d = deformed length.

shortening strain See SHORTENING.

shoshonite A POTASSIC, SILICA-poor TRACHY-ANDESITE.

shot An artificial SEISMIC SOURCE.

Shreve's magnitude A measure of STREAM ORDER which adds the order values at each confluence to provide the magnitude of the downstream link.

shrinkage crack A small fissure resulting from contraction of COHESIVE sediment.

shrub-coppice dune See NEBKHA.

shrub-coppice mound See NEBKHA.

SI Système International; the modern system of units based on m, kg, s, A, V.

sial An outmoded term for the UPPER CONTINENTAL CRUST derived from its **si**lica and **al**uminium-based composition. Cf. SIMA.

sichelwannen Metric-scale, crescentic marks cut into a flat or gently sloping rock surface by a GLACIER, the horns pointing away from the snout.

side-looking airborne radar (SLAR) An airborne RADAR system which scans sideways and detects RADAR backscattered from the ground surface in a similar way to SIDE-SCAN SONAR.

side-scan sonar A marine, sideways-scanning, acoustic method of GEOPHYSICAL EXPLORATION in which the seabed to either side of a ship's track is insonified by high frequency (6–110 kHz) sound beams. Seafloor features either reflect energy back to the ship or away from it, producing a characteristic SONOGRAPH.

siderite 1 (spathic iron) ($FeCO_3$) A CARBONATE MINERAL of IRON, sometimes of importance as an ORE. **2** See IRON METEORITE.

siderolite See STONY IRON METEORITE.

sideromelane The brown GLASS component of BASALTIC HYALOCLASTITE.

siderophile Descriptive of an element with an affinity with oxygen or sulphur and soluble in molten IRON.

siderophyllite An IRON- and aluminium-rich BIOTITE.

sief A linear or longitudinal DUNE, up to 20 km long, commonly in a group oriented in the direction of the prevailing wind.

siemen (S) The SI unit of electrical conductance, the inverse of resistance.

sieve deposit 1 A deposit originating from the deposition of fine sediment into existing coarse sediment because of the rapid DISCHARGE reduction of a surface flow due to percolation into the coarse sediment. **2** A coarse sediment lobe on a semi-arid, ALLUVIAL FAN originating from rapid DISCHARGE reduction due to the INFILTRATION of water.

signal to noise ratio (SNR) The ratio of the amplitude or energy of a particular signal to that of the unwanted background signal.

silcrete A nodular, highly indurated or massive, surface or near-surface layer composed of SILICA, resulting from the

CEMENTATION and replacive introduction of SILICA into soils, rock or sediment.

Silesian A CARBONIFEROUS succession in NW Europe equivalent to the SERPUKHOVIAN, BASHKIRIAN, MOSCOVIAN, KASIMOVIAN and GZELIAN.

silica minerals A group of MINERALS with the formula SiO_2. QUARTZ is the most common, but other POLYMORPHS include AGATE, CHALCEDONY, CHERT, COESITE, CRISTOBALITE, FLINT, JASPER, ONYX, OPAL, STISHOVITE and TRIDYMITE.

silica saturation The concentration of SILICA in an IGNEOUS ROCK relative to the other constituents which combine with SILICA to form silicate MINERALS. See OVERSATURATED, SATURATED, UNDERSATURATED.

silication The process by which a rock is converted into, or replaced by, silicates, in particular the conversion of carbonate rock to SKARNS.

siliceous ooze A PELAGIC, biogenic sediment comprising the skeletal remains of siliceous MICROFOSSILS.

siliciclastic (terrigenous) A CLASTIC ROCK descriptor indicating a rock whose CLASTS are predominantly of SILICATE MINERALS.

silicification **1** The ALTERATION by CEMENTATION or REPLACEMENT of sediment by AMORPHOUS, CRYPTOCRYSTALLINE, MICROCRYSTALLINE or MACROCRYSTALLINE SILICA. **2** A type of WALL ROCK ALTERATION involving an increase in QUARTZ or CRYPTOCRYSTALLINE SILICA in the altered rock.

silky lustre A LUSTRE with a silk-like appearance, due to the reflection of light from a fine AGGREGATE of parallel fibres.

sill A concordant, TABULAR or sheet-like IGNEOUS BODY from a few centimetres to hundreds of metres in thickness.

sillar A weakly consolidated, fine-grained ASH-FLOW TUFF.

sillénite (BiO_3) Cubic bismuth trioxide, occurring as a SECONDARY MINERAL. Cf. BISMITE.

sillimanite (fibrolite) (Al_2SiO_5) A NESO-SILICATE found in METAMORPHIC ROCKS rich in alumina.

sillimanite minerals A group of anhydrous, aluminium silicates comprising ANDALUSITE, KYANITE and SILLIMANITE. Used to make a MULLITE-SILICA mixture, which is highly REFRACTORY and has a small coefficient of thermal expansion.

silt A sediment with particles in the size range 4–62.5 μm.

siltstone A lithified SILT.

Silurian A PERIOD of the PALAEOZOIC, 439.0–408.5 Ma.

silver (Ag) A rare NATIVE METAL.

silver glance See ARGENTITE.

silver lead ore GALENA containing SILVER.

sima An outmoded term for the lower crust, which is rich in **si**lica and **mag**nesium. Cf. SIAL.

Simferopolian An EOCENE succession in the former USSR covering parts of the YPRESIAN and LUTETIAN.

similar fold A FOLD whose inner and outer arcs have the same curvature, so that all DIP ISOGONS are parallel to the AXIAL PLANE and the orthogonal thicknesses of the FOLD LIMBS are less than at the HINGE ZONE.

simple shear A shape change in which all particles of the deforming body move in parallel lines, the amount of movement depending on the distance of each particle from a given plane in which there are no DISPLACEMENTS.

Sinemurian A STAGE of the JURASSIC, 203.5–194.5 Ma.

singing sand (booming dune) DUNE SANDS which emit audible sounds when in motion, described as booming, roaring, squeaking, singing or musical. Probably arises from some unique combination of grain properties, such as high SORTING, uniform size and high ROUNDNESS.

sinhalite ($Mg(Al,Fe)BO_4$) A rare GEM.

Sinian The higher ERA of the PRECAMBRIAN, 800–570 Ma.

sinistral The sense of movement across a boundary, such as a FAULT, in which the side opposite the observer moves to the left. Cf. DEXTRAL.

sinkhole An approximately circular depression in LIMESTONE terrain into which water drains and collects.

sinter See GEYSERITE.

sinuosity A measure of the degree to which a curvilinear feature winds about the shortest route between two points, defined as the ratio of feature length to shortest distance.

sinusoidal fold A FOLD with a PROFILE approximating a sine curve.

Sirenia An order of infraclass EUTHERIA, subclass THERIA, class MAMMALIA; the dugongs and manatees. Range early EOCENE–RECENT.

site investigation A procedure which provides the information about a proposed site upon which decisions can be based regarding its suitability for engineering works; the primary means by which ground conditions and properties are considered during civil engineering design and construction work using the techniques of ENGINEERING GEOLOGY.

skarn (pyrometasomatic deposit) An irregularly shaped, REPLACEMENT ORE deposit, usually formed at high temperatures in metamorphosed, carbonate-rich sediments at the contacts with medium to large IGNEOUS BODIES. The OREBODIES are characterized by the development of calc-silicate MINERALS such as ACTINOLITE, ANDRADITE, DIOPSIDE and WOLLASTONITE, and provide IRON, COPPER, GRAPHITE, lead, molybdenum, TIN, tungsten, uranium and zinc.

skewness A statistical measure of the shape of a distribution which has more members at one end than the other. Used in the description of sedimentary GRAIN SIZE DISTRIBUTION.

skin depth In ELECTROMAGNETIC INDUCTION METHODS, the DEPTH OF PENETRATION, which is dependent on the product of the inverse square root of the frequency used, the ELECTRICAL CONDUCTIVITY and the PERMEABILITY of the medium.

skip mark A TOOL-MARK formed by the low-angle impact of a particle with a muddy sediment; a trail may be formed as the particle skips along the bed of a flow.

skutterudite (smaltite) $((Co,Ni)As_3)$ An ORE MINERAL of cobalt and nickel.

slab failure A form of WEATHERING common on hard rock cliffs, occurring when EROSION releases the lateral CONFINING PRESSURE and tension JOINTS open until the TENSILE STRENGTH of the rock at the edges of a slab is exceeded and it falls.

slab pull The mechanism whereby the gravitational pull of a downgoing slab in a SUBDUCTION ZONE exerts a lateral FORCE on the PLATE attached to the slab. An important PLATE driving mechanism in OROWAN-ELSASSER CONVECTION.

slack A flooded INTERDUNE.

slacking The process whereby LIGNITES and SUB-BITUMINOUS COALS crack and fall apart when they dry out after excavation.

slaking The disintegration of loosely consolidated material when water is introduced or it is exposed to the atmosphere.

SLAR See SIDE-LOOKING AIRBORNE RADAR.

slate A fine-grained, low-GRADE or REGIONALLY METAMORPHOSED MUDROCK with a well-developed PENETRATIVE CLEAVAGE. The CLEAVAGE is a FOLIATION in which submicroscopic PHYLLOSILICATE MINERALS are in well-developed, parallel alignment so that the rock splits into platy sheets.

slate belt A linear zone within an OROGENIC BELT in which PELITIC rocks are dominantly in the form of SLATE, characterized by the lowest GRADES of REGIONAL METAMORPHISM and a steeply dipping CLEAVAGE subparallel to the axis of the belt.

slaty cleavage The CLEAVAGE developed in a SLATE.

Slendoo A PERMIAN succession in Queensland, Australia, covering the middle part of the ARTINSKIAN.

slickencryst A fibrous MINERAL growth on the BEDDING PLANES within a FOLD LIMB, indicative of INTERLAYER SLIP.

slickenline (slickenside lineation, slickenside striation) A millimetric to metric scale LINEATION on a SLICKENSIDE parallel to the DISPLACEMENT direction of the FAULT.

slickenside A smooth or polished FAULT surface.

slickenside lineation See SLICKENLINE.

slickenside step A short, narrow offset at one end of the termination of a SLICKENLINE, usually approximately orthogonal to the SLICKENLINE.

slickenside striation See SLICKENLINE.

slickolite formation The process occurring during SLICKENSIDE formation in which solution of rock oblique to the FAULT PLANE forms oblique, incongruous STYLOLITES in the direction of compression.

slide 1 Any downslope MASS MOVEMENT of sediment under the influence of GRAVITY DEFORMATION. **2** A FAULT developed during major RECUMBENT FOLDING.

Slingram A system used in the COMPENSATOR ELECTROMAGNETIC METHOD.

slip A process of DISPLACEMENT within individual crystals during PLASTIC DEFORMATION.

slip clay A KAOLIN-rich CLAY used for glazes.

slip line A curve indicating the direction of maximum SHEAR STRESS at any point during PLASTIC DEFORMATION and, in a perfectly PLASTIC material, the direction of maximum STRAIN RATE. Forms an angle of 45° with the STRESS TRAJECTORY.

slip sheet A GRAVITY COLLAPSE STRUCTURE in the form of a detached sheet.

slip system The SHEAR PLANE and SHEAR DIRECTION within the plane of SIMPLE SHEAR, including all symmetrically equivalent combinations.

slip vector (displacement vector) The vector connecting a point to its original location before DISPLACEMENT.

slip-face The face on the LEE SLOPE of a DUNE where SAND accumulates at its ANGLE OF REPOSE (30°–34°).

slip-off slope A low gradient slope on the inside of a meander bend.

slope apron The region between the margin and floor of a BASIN.

slope gradient factor A term in the UNIVERSAL SOIL LOSS EQUATION.

slough channel A channel in a BRAIDED STREAM with relatively little flow and in which deposition is predominantly by the settling of fine material from SUSPENSION.

SLR See SATELLITE LASER RANGING.

slump A type of sediment SLIDE in which material moves downward as a unit or series of units, often with backward rotation about a horizontal axis parallel to the causative slope. Generally highly deformed internally.

slump fold A FOLD developed in a SLUMP.

small circle Any circle on a sphere that does not correspond to a circumference. Cf. GREAT CIRCLE.

Smallfjord A STAGE of the VENDIAN, 610–600 Ma.

smaltite See SKUTTERUDITE.

smaragdite A fibrous green variety of ACTINOLITE forming PSEUDOMORPHS after PYROXENE in ECLOGITIC rocks.

smectite clays (bentonite clays) A group of CLAYS comprising MONTMORILLONITE and ATTAPULGITE, exploited in DRILLING MUD for their swelling properties and for the increase in VISCOSITY they cause in DRILLING FLUIDS and as bonding agents.

smithsonite (calamine) ($ZnCO_3$) A zinc ORE MINERAL of SUPERGENE origin.

smokey quartz A brown to black variety of QUARTZ.

SMOW **S**tandard **M**ean **O**cean **W**ater, the international standard for $^{18}O:^{16}O$ ratios.

Snell's Law A law of optics that can be applied to SEISMIC WAVES impinging on an ACOUSTIC IMPEDANCE contrast: $(\sin i)/V$ is constant, where i = angle of incidence, refraction or reflection and V = the SEISMIC VELOCITY of the medium in which i is measured.

SNR See SIGNAL TO NOISE RATIO.

soapstone (steatite) A compact, massive variety of TALC.

soda lake A lake whose water has a high content of sodium, found in RIFT VALLEYS.

sodalite $(Na_8(AlSiO_4)_6Cl_2)$ A rare FELDSPA-THOID found mainly in NEPHELINE SYENITES and related rock types, in METASOMATIZED calcareous rocks and in cavities in EJECTA.

soft sediment deformation See WET SEDIMENT DEFORMATION.

soil creep A slow type of MASS MOVEMENT.

soil crust See CLAY PAN.

soil erodibility factor A term in the UNI-VERSAL SOIL LOSS EQUATION.

soil erosion The removal of material from the soil zone by natural processes, generally by wind, water and SOIL CREEP.

soil loss A term in the UNIVERSAL SOIL LOSS EQUATION.

soil mechanics The techniques of ENGIN-EERING GEOLOGY applied to soils.

soil profile (soil structure) The sequence through the soil zone from the surface to BEDROCK, comprising prominent horizons distinguished by different colours, structures and compositions.

soil structure See SOIL PROFILE.

Sokol'yergorskiy A PERMIAN succession on the eastern Russian Platform equivalent to the L. ASSELIAN.

solar luminosity The intensity of electro-magnetic radiation received from the Sun. Variation in this factor may influence the origin of ICE AGES.

solar pond (salina, salt pond) An arti-ficial pond in which evaporation of saline water by the Sun takes place, used as a guide to the processes of EVAPORATE BASIN evolution.

sole fault The DETACHMENT FAULT in an EXTENSIONAL FAULT SYSTEM.

sole mark (sole structure) A cast of a sedimentary structure on the base of a BED, often developed where relatively coarse-grained rocks overlie finer-grained, COHES-IVE sediment.

sole structure See SOLE MARK.

sole thrust The DETACHMENT FAULT in a THRUST BELT.

Solenoporaceae An heterogeneous group of red ALGAE. Range CAMBRIAN–TERTIARY.

solfatara A FUMAROLE in an active volcanic area whose escaping VOLCANIC GASES con-tain much SULPHUR, which commonly forms a bright yellow deposit.

solid Earth The CRUST of the Earth and its interior, i.e. the Earth excluding the hydro-sphere, atmosphere and biosphere.

solid solution The substitution of one ion for another in a CRYSTAL LATTICE with no change in STRUCTURE, i.e. the composition of MINERALS in the series varies continuously between the two end-members.

solidus The temperature below which a MINERAL ASSEMBLAGE is entirely crystalline. Cf. LIQUIDUS.

solifluction The slow, gravitationally driven, downslope movement of water-saturated, seasonally thawed materials.

Solikamsky A PERMIAN succession on the eastern Russian Platform covering the lower part of the UFIMIAN.

Solnhoffen Limestone An EXCEPTIONAL FOSSIL DEPOSIT of U. JURASSIC age in Bavaria, Germany containing a range of fauna, including articulated vertebrates, shrimps, limulines and INSECTS.

solution 1 An important process of chemi-cal WEATHERING in which a MINERAL in contact with a solvent is dissociated into its component ions. **2** A fluid containing ions, such as a PORE FLUID.

solution breccia See COLLAPSE BRECCIA.

solution cleavage A SPACED CLEAVAGE found in QUARTZITE and LIMESTONE which contains relatively soluble MINERALS, indicating the action of PRESSURE SOLUTION.

solution flutes See RILLENKARREN.

solution gas drive See DEPLETION DRIVE.

solution mining A mining method for soluble MINERALS in which water is introduced into the deposit and the dissolved MINERALS pumped to the surface.

solution transfer A DEFORMATION mechanism by DIFFUSION in the dissolved state.

solutional erosion The EROSION by SOLUTION of the component MINERALS of a rock.

Solva A CAMBRIAN succession in S. Wales equivalent to the SOLVAN.

Solvan A STAGE of the CAMBRIAN, 536.0–530.2 Ma.

solvus The line in a PHASE DIAGRAM separating a stable SOLID SOLUTION from two or more phases that form from the solution by unmixing.

somaite An ALKALINE IGNEOUS ROCK resembling ESSEXITE but with LEUCITE replacing NEPHELINE.

sonde A device lowered down a borehole for measuring or testing purposes.

sonic log See CONTINUOUS VELOCITY LOG.

sonobuoy A buoy used in marine SEISMIC EXPLORATION containing instruments for detecting seismic information and transmitting it to a recording ship.

sonograph A display of SIDE-SCAN SONAR data.

sorosilicate A CRYSTAL STRUCTURE classification in which the COORDINATION POLYHEDRA are Si tetrahedra which link by pairs sharing one corner.

sorting (grading) The standard deviation of the GRAIN SIZE DISTRIBUTION of a sediment, which is controlled partly by transport and depositional processes and is thus of diagnostic significance.

Soudleyan A STAGE of the ORDOVICIAN, 457.5–449.7 Ma.

sour crude oil CRUDE OIL with detectable hydrogen sulphide. Cf. SWEET CRUDE OIL.

source rock The rock in which HYDROCARBONS were generated.

South African jade See TRANSVAAL JADE.

SouthWest US and East AnTarctica hypothesis (SWEAT hypothesis) A controversial reconstruction of the PRECAMBRIAN SUPERCONTINENT in which it is proposed that the southwestern coast of North America and the east coast of Antarctica were juxtaposed in Late PROTEROZOIC times. The rifting of this continent would have resulted in its turning inside out (or 'EXTRAVERSION') and the drifting of its components into their more modern configuration.

Southland A MIOCENE succession in New Zealand comprising the CLIFDENIAN, LILLBURNIAN, and WAIAUAN.

sövite A coarse-grained, light-coloured, calcium carbonate CARBONATITE.

SP method See SPONTANEOUS POLARIZATION METHOD.

spaced cleavage A CLEAVAGE in which the CLEAVAGE PLANES are separated by uncleaved slices of rock.

Spanish topaz An orange-brown variety of QUARTZ.

spar A mining term for any white or light coloured MINERAL with a well-developed CLEAVAGE and a VITREOUS LUSTRE.

sparagmite An ARKOSE of PRECAMBRIAN age.

sparite Sparry CALCITE, occurring as the cement in some LIMESTONES and formed by nucleated PRECIPITATION and growth in a pore space.

sparker A marine SEISMIC SOURCE in which

a capacitor is discharged in the water, creating a plasma bubble which oscillates in the same fashion as an explosion.

sparry limestone A LIMESTONE with a SPARITE cement.

spastolith A grain composed of soft material which was deformed by mechanical COMPACTION during burial.

Spathian A STAGE of the TRIASSIC, 241.9–241.1 Ma.

spathic iron See SIDERITE.

spatter (driblet) Agglutinized masses of primary EJECTA, larger than LAPILLI, which were erupted as fluids.

spatter bank Bank formed by accretion of fluid LAVA squirted from a LAVA fountain.

spatter cone A low mound (<15 m high) formed by fountains of BASALTIC LAVA. Mainly composed of SPATTER but also including BOMBS and CINDERS.

specific form A crystal FORM whose faces bear a specific relationship to the symmetry operators of the crystal. They may be parallel or perpendicular to a rotation axis or a mirror plane. Cf. GENERAL FORM.

specific retention The water remaining in an area of soil at FIELD CAPACITY after draining under the influence of GRAVITY. Cf. SPECIFIC YIELD.

specific yield The water that has drained from an area of soil at FIELD CAPACITY under the influence of GRAVITY. Cf. SPECIFIC RETENTION.

spectral analysis (power spectrum analysis) The determination of the relationship between frequency or WAVENUMBER and energy for a waveform. The slope of linear segments of the power spectrum of a POTENTIAL FIELD can be used to determine the LIMITING DEPTH of its source.

spectral band A defined band of frequencies or WAVENUMBERS of a waveform.

spectrographic analysis The chemical analysis of a material by the matching of its spectral lines with those of elements.

spectrometer An instrument for measuring the spectrum of a waveform.

specular lustre See SPLENDENT LUSTRE.

specularite A platy, metallic variety of HAEMATITE.

speleology The scientific study of CAVES.

speleothem The generic name for chemical precipitates found in CAVES, including CAVE CORAL, CAVE DRAPERY, CAVE ONYX, CAVE PEARL, CURTAINS, FLOWSTONE, HELICTITES, MOONMILK, RIMSTONE DAMS, STALACTITES and STALAGMITES. Dominantly formed of CALCITE, but also, more rarely, GYPSUM and HALITE.

sperrylite ($PtAs_2$) A rare, white ORE MINERAL of platinum with a METALLIC LUSTRE.

spessartine ($Mn_3Al_2Si_3O_{12}$) A GARNET found mainly in GRANITE PEGMATITES, GNEISS, QUARTZITE and SCHIST and as LITHOPHYSAE in RHYOLITES, less commonly in SKARNS.

spessartite A CALC-ALKALINE LAMPROPHYRE containing HORNBLENDE, CLINOPYROXENE ± OLIVINE with PLAGIOCLASE in excess of ALKALI FELDSPAR.

sphalerite (zincblende) (ZnS) The major ORE MINERAL of zinc.

sphalerite group A group of SULPHIDE MINERALS characterized by a unit cube structure with metals at the corners and face centres and the four sulphur atoms coordinated so that each lies at the centre of a regular tetrahedron of metals and each metal is at the centre of a regular sulphur tetrahedron.

sphene See TITANITE.

sphenochasm A triangular gap of OCEANIC CRUST between two CRATONS whose faulted margins converge to a point, originating as one CRATON rotated with respect to the other. Cf. SPHENOPIEZM.

Sphenodontia An order of subclass LEPIDOSAURIA, class REPTILIA; small REPTILES with diverse dentition resembling, and related to, lizards. Range TRIASSIC–RECENT.

sphenopiezm A triangular region between two CRATONS which is being squeezed as one CRATON rotates with respect to the other. Cf. SPHENOCHASM.

Sphenopsida A class of division TRACHAEO-PHYTA, kingdom plantae; the horsetails and their relatives. Range M. DEVONIAN–RECENT.

spherical harmonic analysis The equivalent of FOURIER ANALYSIS in spherical coordinates.

sphericity A measure of how closely a grain resembles a sphere, defined as $^3\sqrt{(S^2/LI)}$, where S, L, I are the short, long and intermediate grain diameters respectively.

spheroid The ellipsoid whose flattening and polar and equatorial radii provide the best approximation to the GEOID.

spheroidal structure A STRUCTURE shown by some IGNEOUS ROCKS comprising large, rounded masses surrounding concentric shells of the same material. Probably a large-scale form of PERLITIC TEXTURE.

spheroidal weathering The process of mechanical spalling and surficial chemical WEATHERING of a BOULDER. Often associated with the DEEP WEATHERING of BASALT, DOLER-ITE and GRANITE, in which water promotes chemical WEATHERING within COOLING JOINTS.

spherule A small, spherical particle.

spherulite A subspherical STRUCTURE in a silicic VOLCANIC ROCK, 1–20 mm in diameter, comprising wholly or partly radially disposed, ACICULAR crystals, normally of ALKALI FELDSPAR. Probably the product of the DEVITRIFICATION of hydrated silicic GLASSES in RHYOLITIC, RHYODACITIC and TRACHYTIC GLASSY ROCKS.

spherulitic texture The TEXTURE of a rock containing many SPHERULITES.

spicular chert CHERT made up of SPONGE SPICULES.

spicule A small spine or needle.

spiculite A sediment or SEDIMENTARY ROCK composed of SPONGE SPICULES.

spider diagram A plot used to show variations in a variety of elements between two IGNEOUS ROCKS.

spike 1 An incongruous STYLOLITE growing in the direction of compression during SLICKOLITE FORMATION. **2** A mathematical waveform of infinitely small duration containing an infinite number of frequencies, used in the manipulation of waveforms, especially in SEISMOLOGY.

spilite A MAFIC VOLCANIC ROCK, usually PIL-LOW LAVA, consisting of ALBITE and CHLORITE. Probably represents a strongly altered BASALT.

spilite suite The association of SPILITE, KER-ATOPHYRE and QUARTZ KERATOPHYRE, believed to represent altered MAFIC, INTERMEDIATE and silicic VOLCANIC ROCKS respectively.

spillway A GLACIAL drainage channel.

spinel ($MgAl_2O_4$) The manganese end-member of the SPINELS.

spinels A group of isostructural OXIDE MIN-ERALS with the general formula M_3O_4.

spinifex texture A TEXTURE characterized by large, randomly oriented or subparallel, skeletal, plate or lattice OLIVINE grains, or ACICULAR PYROXENE grains. Formed by the rapid CRYSTALLIZATION of ULTRAMAFIC liquids and common in KOMATIITES.

spinner magnetometer A MAGNETO-METER for the determination of REMANENT MAGNETIZATIONS in which a sample is rotated within a magnetic detecting system. The generated signal's amplitude is proportional to the INTENSITY OF MAGNETIZATION and the phase directly related to its direction.

spinoidal decomposition EXSOLUTION without a NUCLEATION step, which can occur in a SOLID SOLUTION with an unstable composition.

Spiriferida An order of class ARTICULATA, phylum BRACHIOPODA; BRACHIOPODS with biconvex, usually impunctate shells with broad hinge lines. Range U. ORDOVICIAN–JURASSIC.

spit A long, narrow accumulation of BEACH DEPOSITS with one end attached to the shore and the other projecting into a large body of water, forming as a result of LONGSHORE DRIFT's carrying sediment beyond an abrupt change in coastline orientation.

splay fault See SECOND ORDER FAULT.

splendent lustre (specular lustre) LUSTRE of the highest intensity caused by an intense reflection of light.

splintery fracture A FRACTURE with a splinter-like appearance.

spodumene (LiAlSi$_2$O$_6$) A PYROXENE found in lithium-rich PEGMATITES.

sponges See PORIFERA.

spontaneous magnetization A natural magnetization exhibited by FERROMAGNETIC substances in the absence of an external MAGNETIC FIELD.

spontaneous polarization See SELF-POTENTIAL.

spontaneous polarization method (self-potential method, SP method) A GEOPHYSICAL EXPLORATION method making use of the naturally produced potential differences at the surface generated by electrochemical interactions between an OREBODY and GROUNDWATER at depths down to ~30 m.

sporinite A COAL MACERAL of the EXINITE group comprising coalified spores and pollen.

SPOT see SATELLITE PROBATOIRE POUR L'OBSERVATION DE LA TERRE.

spread A pattern of SEISMOGRAPHS which record a single SHOT.

spreading rate The rate of opening of an ocean by SEAFLOOR SPREADING. Cf. HALF SPREADING RATE.

spreiten TRACE FOSSIL structures comprising repeated, 'U'-shaped, blade-like, sinuous or spiral sedimentary LAMINAE caused by intense sediment reworking by animal locomotion, excavation or excretion.

spring A localized flow of water from an AQUIFER where the WATER TABLE intersects the surface.

spring sapping SAPPING by GROUNDWATER outflow at a SPRING.

spur-and-groove topography Undulating morphology on the front of a CORAL-ALGAL REEF.

Squamariaceae A group of red ALGAE. Range U. CARBONIFEROUS–RECENT.

Squamata An order of subclass LEPIDOSAURIA, class REPTILIA; the lizards and snakes. Range U. TRIASSIC–RECENT.

squid magnetometer See CRYOGENIC MAGNETOMETER.

Srbsko A succession in the former Czechoslovakia equivalent to the upper part of the DEVONIAN.

stable isotope An isotope that does not suffer radioactive decay.

stable sliding DISPLACEMENT on a FAULT at a constant or slowly varying STRAIN RATE. Cf. STICK-SLIP.

stable tectonic zone A region of the CONTINENTAL CRUST characterized by a lack of TECTONIC activity.

stack 1 A SEISMIC RECORD produced by adding together (STACKING) a number of different SEISMOGRAMS. **2** A coastal pillar of rock above the high tide level. Cf. STUMP.

stacking The process of adding together a number of different SEISMOGRAMS to produce a STACK.

stacking fault A PLANAR CRYSTAL DEFECT in which there is an error in the regular stacking sequence of layer modules which lies parallel to the plane of the layer.

stacking velocity The SEISMIC VELOCITY used to correct for NORMAL MOVEOUT prior to STACKING SEISMOGRAMS.

stadial A short cold period with a smaller volume of ice than in a GLACIAL.

stage 1 The water depth or elevation of the water surface at a location on a river system. **2** A fourth order CHRONOSTRATI-GRAPHIC unit.

stage hydrograph A graph of river STAGE.

stagnation deposit A deposit formed under conditions of stagnation, which may inhibit the decay of organic matter and give rise to an EXCEPTIONAL FOSSIL DEPOSIT.

staining A simple chemical technique used to distinguish MINERALS in THIN SECTION by painting it with an organic or inorganic reagent. Particularly useful in distinguishing the different CARBONATE MINERALS, whose optical properties are similar.

staircase trajectory See STAIRSTEP TRAJECTORY.

stairstep trajectory (staircase trajectory) A FAULT surface comprising FLATS and RAMPS.

stalactite A SPELEOTHEM composed of calcium carbonate which grows downwards from the roof of a CAVE.

stalagmite (dripstone) A SPELEOTHEM composed of calcium carbonate which grows upwards from the floor of a CAVE.

Standard Mean Ocean Water See SMOW.

standing wave A stationary surface WAVE.

stanniferous Containing TIN.

stannite (bell-metal ore, tin pyrites) (Cu_2FeSnS_4) An ORE MINERAL of TIN found mainly in HYDROTHERMAL VEIN deposits and more rarely in PEGMATITES.

star dune A pyramidal DUNE with three arms radiating from a high central dome.

starlite A GEM variety of ZIRCON produced by heating in a reducing environment to provide a blue colour.

stasis A period of no evolutionary change.

Stassfurt Evaporites A PERMIAN succession in NW Europe covering the middle part of the WORDIAN.

stassfurtite A massive variety of BORACITE.

static correction A constant time correc-tion applied to seismic data to compensate for the different elevations of SHOTS and detectors, and for the varying thickness and velocity of the WEATHERED LAYER.

staurolite (($Fe,Mg)_4Al_{17}(Si,Al)_8O_{44}(OH)_4$) A NESOSILICATE found in aluminium-rich rocks subjected to REGIONAL METAMORPHISM.

steady-state creep CREEP which takes place without any change in the physical state of the material.

steatite See SOAPSTONE.

steinkern The sediment infilling the space left by a shell after its removal by DISSOL-UTION, which provides an internal and external cast of the shell.

Steinmann trinity The association of SPIL-ITE, SERPENTINE and RADIOLARIAN CHERT found in deep-sea sediments.

STEM See SCANNING TRANSMISSION ELEC-TRON MICROSCOPY.

step A FAULT surface feature approximately orthogonal to its LINEATIONS and pointing either towards or opposite to the sense of motion of the block sliding over them form-ing WEAR GROOVES and GROWTH FIBRES respectively.

step faults A set of parallel NORMAL FAULTS which DIP and DOWNTHROW in the same direction.

Stephanian A CARBONIFEROUS STAGE in NW Europe equivalent to the KASIMOVIAN and GZELIAN.

stephanite (Ag_5SbS_4) A rare SILVER ORE MIN-ERAL commonly found in VEINS associated with other SILVER MINERALS.

stepout See MOVEOUT.

steptoe An isolated hill surrounded by LAVA.

stereogram See STEREOGRAPHIC PRO-JECTION.

stereographic net A type of protractor used in STEREOGRAPHIC PROJECTION which gives the CYCLOGRAPHIC TRACES of a complete set of GREAT CIRCLES about a common axis and the CYCLOGRAPHIC TRACES of a family of

SMALL CIRCLES about the same axis, which are used to graduate angles along the GREAT CIRCLES. See also SCHMIDT NET, WULFF NET.

stereographic projection (stereogram) A graphical method of showing three-dimensional geometrical data in two dimensions and of solving three-dimensional geometrical problems, particularly those involving the angular relationships between lines and planes.

stereoscope An instrument for viewing AERIAL PHOTOGRAPHS which overlap by ~60% that allows a viewer with binocular vision to obtain an apparently three-dimensional image of the topography.

Sterlitamakskiy A PERMIAN succession on the eastern Russian Platform equivalent to the U. SAKMARIAN.

stibnite (antimonite, antimony glance) (Sb_2S_3) The major ORE MINERAL of ANTIMONY found mainly in low-temperature HYDROTHERMAL VEIN or REPLACEMENT DEPOSITS.

stichtite ($Mg_6Cr_2CO_3(OH)_{16}.4H_2O$) A lilac or pink, hydrated magnesium-chromium carbonate found in SERPENTINITE and commonly associated with CHROMITE.

stick-slip FAULT behaviour comprising alternations of slow STRAIN RATE and rapid SLIP when the STRESS overcomes the static COEFFICIENT OF FRICTION on the FAULT PLANE. Cf. STABLE SLIDING. See also ELASTIC REBOUND THEORY.

stiffness See YOUNG'S MODULUS.

stilbite (desmine) ($CaAl_2Si_7O_{18}.7H_2O$) A ZEOLITE found mainly in AMYGDALES and cavities in BASALT, ANDESITE and related VOLCANIC ROCKS, but also as a HYDROTHERMAL MINERAL in crevices in METAMORPHIC ROCKS, in cavities in GRANITE PEGMATITES and in some HOT SPRING deposits.

stilpnomelane ($\sim K_{0.6}(Mg,Fe^{2+},Fe^{3+})_6Si_8Al(O,OH)_{27}.2-4H_2O$) A PHYLLOSILICATE found widespread in SCHISTS in association with CHLORITE, EPIDOTE, LAWSONITE and other MINERALS and in IRON ORE deposits.

Stinkschiefer See HAUPTDOLOMIT.

stishovite (SiO_2) A high pressure (>10 GPa) form of SILICA found in areas such as METEORITE impacts.

stock An IGNEOUS BODY smaller than a BATHOLITH with a subcircular section.

stockwork A closely spaced, interlocking network of veinlets, commonly in and around intermediate PLUTONIC igneous intrusions. An important control of various forms of disseminated mineralization.

Stokes law A law controlling the settling of a particle in a fluid: $v = (2r^2g\Delta\rho)/(9\eta)$, where v = settling velocity, r = radius, g = GRAVITY, $\Delta\rho$ = DENSITY CONTRAST with fluid, η = VISCOSITY of fluid.

Stokes surface A subhorizontal AEOLIAN BOUNDING SURFACE produced by DEFLATION over all or part of an ERG, which causes scouring of the COHESIVE or cemented SAND at the WATER TABLE.

stolzite ($PbWO_4$) A tungstate MINERAL found as a SECONDARY MINERAL in tungsten-bearing ORE deposits.

stone lattice See ALVEOLES.

stone pavement A flat desert area covered with a surface layer of rounded PEBBLES or GRAVEL, often a LAG deposit caused by the removal of fine particles by wind and wash EROSION.

stone-line An horizon of angular GRAVEL and COBBLES within the soil zone, often running parallel to the surface. Probably a LAG deposit originally formed at the surface by the removal of fine particles by wind and wash EROSION and subsequently buried.

stonefield See BLOCKFIELD.

stony iron meteorite (siderolite) A METEORITE comprising metal and silicate.

stony meteorite (aerolite) A METEORITE formed solely of rock-forming silicates.

stoop and room mining See ROOM AND PILLAR MINING.

stope An excavation within a mine made to extract ORE.

Stopes-Heerlen system A system for the

classification of COAL MACERALS on the basis of physical appearance, chemistry and biological affinity into three groups: VITRINITE, EXINITE and INERTINITE.

stoping The mechanical incorporation of blocks of WALL ROCK into a MAGMA during its ascent, an important mechanism in the EMPLACEMENT of PLUTONS.

storage coefficient (storativity) The volume of water given up per unit horizontal area of an AQUIFER and per unit fall of the WATER TABLE. Equal to SPECIFIC YIELD for an UNCONFINED AQUIFER, but dependent on the elastic compression for a CONFINED AQUIFER.

storativity See STORAGE COEFFICIENT.

stored deformation energy The potential energy within a deformed material, stored by the ELASTIC distortion of a CRYSTAL LATTICE or by imperfections in the lattice. The driving force in the process of RECOVERY and primary RECRYSTALLIZATION.

storm beach See DISSIPATIVE BEACH.

storm deposit (tempestite) A sediment body deposited as a result of one storm event.

storm setup The build-up of water against the coast during a storm.

storm surge A localized elevation of sea level by extreme meteorological events, caused by STORM SETUP and extremely low atmospheric pressure.

storm surge ebb The return of waters after a STORM SURGE.

stoss side The upstream side of a body where the fluid velocity and SHEAR STRESSES commonly increase as the flow moves over the body. Cf. LEE SIDE.

stoss slope The upstream slope of a body where the fluid velocity and SHEAR STRESSES commonly increase as the flow moves over the body. Cf. LEE SLOPE.

Strahler system A classification of STREAM ORDER whereby a stream of order $n+1$ is initiated at the confluence of two streams of order n, so that entry of a stream of lower order does not increase the order of the main stream.

straight extinction See PARALLEL EXTINCTION.

strain A fractional change in shape or internal arrangement of rock caused by TECTONIC activity.

strain axes The orientations of the PRINCIPAL STRAINS.

strain band A planar zone of high STRAIN in a set of asymmetrically folded layers corresponding to the superimposed short LIMBS of the FOLDS and bounded by the AXIAL SURFACES of the FOLDS.

strain ellipse The shape assumed by a unit circle after DEFORMATION which facilitates visualization of the shape change associated with the STRAIN.

strain ellipsoid The shape assumed by a unit sphere after DEFORMATION which facilitates visualization of the shape change associated with the STRAIN.

strain hardening 1 A phenomenon occurring during DISLOCATION CLIMB in which DISLOCATIONS and CRYSTAL DEFECTS impede the movement of a DISLOCATION in its own lattice plane so that DEFORMATION becomes increasingly more difficult. Cf. STRAIN SOFTENING. 2 The hardening of a FAULT PLANE caused by MINERALS precipitating on it.

strain increment Each small change in the shape of a body within a more extensive DEFORMATION sequence.

strain marker An object of known initial shape, geometry or shape distribution present in a deformed rock which allows STRAIN to be measured, e.g. FOSSILS, worm burrows, OOLITHS.

strain parallelepiped The shape taken by a unit cube after a vanishingly small STRAIN INCREMENT; a geometrical construction used in the analysis of INFINITESIMAL STRAIN.

strain path A line showing the successive states of STRAIN only during DEFORMATION. Cf. DEFORMATION PATH.

strain rate The rate of change of STRAIN with time, in units of time^{-1}.

strain softening (work softening) A phenomenon whereby DEFORMATION occurs at an increased rate as STRAIN increases, or a decrease in the applied STRESS is needed for constant rate DEFORMATION as STRAIN increases. Cf. STRAIN HARDENING.

strain superposition The application of successive STRAIN INCREMENTS to a rock body to produce the observed FINITE STRAIN.

strain-slip cleavage See CRENULATION CLEAVAGE.

strandflat A partly submerged, undulating, rocky lowland close to sea level, probably formed either by frost action on a shoreline or by marine and GLACIAL EROSION.

strandline A shoreline, commonly an ancient one.

strandplain A BEACH of multiple BEACH RIDGES formed when the rate of coastal sedimentation kept pace with a slow rate of sea level rise.

strata-bound mineral deposit A deposit restricted to a small STRATIGRAPHIC range in a group of STRATA.

stratification map A map showing the distribution of DUNES, INTERDUNES and AEOLIAN SAND SHEETS in a section through an AEOLIAN deposit, used in combination with DIP measurements to infer the nature of the BEDFORM(s) which deposited the sediment.

stratiform mineral deposit A deposit with a large lateral extent parallel to the principal planar STRUCTURE of the COUNTRY ROCK, such as the BEDDING or an igneous LAMINATION, but of limited thickness.

stratigraphic trap An OIL or GAS TRAP resulting from lateral and vertical variations in the STRATIGRAPHY of the RESERVOIR ROCKS in terms of thickness, TEXTURE, lithology and POROSITY. Cf. STRUCTURAL TRAP.

stratigraphy The study of layered SEDIMENTARY or METAMORPHIC ROCKS, especially their relative ages and correlation between different areas.

stratotype The type representative of a particular STRATIGRAPHIC unit or boundary.

stratovolcano (composite volcano) A VOLCANO comprising layers of LAVA and TEPHRA of dominantly ANDESITIC composition increasing in thickness towards a central vent.

stratum See BED.

stratum contour See STRUCTURE CONTOUR.

streak The colour of a MINERAL in finely powdered form, used in MINERAL identification.

streak plate A piece of unglazed white porcelain on which a MINERAL is rubbed so as to determine its STREAK.

stream flow (discharge) The volume of water passing a specified point in a specified period of time, measured by estimates of flow velocity and stream cross-sectional area or by use of a permanent structure which provides a set relation between STAGE and DISCHARGE.

stream order A topological classification of the links within a stream network. See also STRAHLER SYSTEM, SHRIEVE'S MAGNITUDE.

stream tin CASSITERITE occurring as detrital grains in a PLACER DEPOSIT.

streamer An array of HYDROPHONES towed behind a ship.

streamlines Lines depicting the motion of a fluid, drawn tangential to the flow direction at any point.

strength (yield strength) The maximum PRINCIPAL STRESS that a material can bear before FAILURE.

stress The FORCE per unit area acting on the surface of a solid plus the equal and opposite reaction of the material.

stress components The nine quantities which describe the state of STRESS at a point, the NORMAL and SHEAR STRESSES on the faces of an infinitesimally small cube around the point.

stress deviator A component of DEVIATORIC STRESS.

stress ellipsoid A construction showing the complete variation in STRESS with direction, comprising an imaginary ellipsoid with axes parallel and proportional to the directions and magnitudes of the three PRINCIPAL STRESSES, the length of any radius of which gives the magnitude of the NORMAL STRESS on the CONJUGATE plane to the radius.

stress invariants The three quantities I_{1-3}, defined $I_1 = \sigma_1 + \sigma_2 + \sigma_3$; $I_2 = -(\sigma_{12} + \sigma_{31} + \sigma_{23})$; $I_3 = \sigma_1 \sigma_2 \sigma_3$, where σ_1, σ_2, σ_3 are the maximum, intermediate and minimum PRINCIPAL STRESSES respectively.

stress trajectories Lines giving the orientations of the PRINCIPAL STRESSES.

stretch A linear measure of STRAIN; the ratio of deformed length to undeformed length, and equal to one plus the ELONGATION. Stretch is >1 for EXTENSION, <1 for CONTRACTION.

stretching direction The direction of maximum EXTENSION in a given plane, typically a CLEAVAGE plane.

stretching lineation A LINEATION in the STRETCHING DIRECTION produced by the ELONGATION of minute grains, grain AGGREGATES or FOSSILS on a CLEAVAGE surface, which can be used to determine the PRINCIPAL STRAIN AXES.

strewnfield An area in which a specific group of TEKTITES and MICROTEKTITES are found which are distinct in their chemistry and probably represent a particular impact event.

striation A marking on the surface of a PEBBLE or BEDROCK produced by ice movement.

strike The direction of a horizontal straight line constructed on an inclined planar surface, at a direction of 90° from the TRUE DIP direction.

strike fault A FAULT parallel to the STRIKE of the BEDROCK, BEDDING or CLEAVAGE.

strike line A straight line on a geological map parallel to the STRIKE of a rock unit or STRUCTURE. Cf. STRUCTURE CONTOUR.

strike separation The offset of a planar STRUCTURE in a horizontal direction parallel to a FAULT.

strike-slip The DISPLACEMENT parallel to the STRIKE of the DISPLACEMENT PLANE.

strike-slip duplex A part of a STRIKE-SLIP FAULT SYSTEM where there are several, subparallel bend segments of FAULTS.

strike-slip fault A FAULT with a dominant component of STRIKE-SLIP.

strike-slip fault system (wrench fault system) A set of related FAULTS on which the individual DISPLACEMENTS produce a net STRIKE-SLIP DISPLACEMENT on the system as a whole.

strike-slip tectonic regime A STRIKE-SLIP FAULT SYSTEM associated with a CONSERVATIVE PLATE MARGIN or within a continental PLATE.

strip mining A type of OPEN PIT MINING in which strips of land are worked, with the excavation being backfilled as mining proceeds.

stripping ratio The ratio of waste to ORE removed in OPEN PIT MINING.

stromatolite A laminated, calcareous, microbial structure, formed principally by CYANOBACTERIA and ALGAE. One of the oldest known FOSSILS, first occurring in the ARCHAEAN.

stromatoporoids Problematic, extinct organisms with a mesh-like skeleton forming sheet-like domes or discoidal or DENDROID masses, often forming REEFS with CORALS. Especially common in ORDOVICIAN to DEVONIAN carbonate sediments. Range CAMBRIAN–CRETACEOUS.

strombolian eruption A VOLCANIC ERUPTION characterized by continuous small explosions.

stromeyerite $((Ag,Cu)_2S)$ A rare ORE MINERAL of SILVER and COPPER.

stromotactis A type of cavity structure of uncertain origin with a smooth sediment floor and irregular roof, the infilling cement usually being of fibrous and/or DRUSY CALCITE.

strontianite (SrCO₃) A CARBONATE MIN-ERAL exploited as a source of strontium.

strontium isotope analysis A method of analysis by MASS SPECTROMETER in which strontium isotopes are used to distinguish the provenance of sources of OBSIDIAN, ALA-BASTER and MARBLE used in antiquity.

Strophomenida An order of class ARTICUL-ATA, phylum BRACHIOPODA; BRACHIOPODS with convex pedicle VALVES and, usually, a planar or concave brachial VALVE. The hinge is broad and straight and the pedicle for-amen usually closed. Range L. ORDOVICIAN–L. JURASSIC.

structural geology The branch of the Earth sciences dealing with rock STRUCTURES formed by DEFORMATION.

structural terrace A local flattening in an area of generally more inclined STRATA.

structural trap An OIL or GAS TRAP resulting from the STRUCTURE of the RESER-VOIR and SEAL rocks, commonly ANTICLINES, FAULTS and PIERCEMENT STRUCTURES such as SALT DOMES. Cf. STRATIGRAPHIC TRAP.

structure Any geological feature that can be defined geometrically.

structure contour (stratum contour) A curved line on a geological map which follows a constant height on a geological surface, i.e. parallel to STRIKE at each point on the line. Cf. STRIKE LINE.

structure contour plan A map showing the STRUCTURE CONTOURS of an OREBODY in horizontal or vertical plane projection for steeply dipping deposits, constructed for mining purposes.

structure grumeleuse A LIMESTONE TEX-TURE characterized by MICRITE clots sur-rounded by coarser, GRANULAR CALCITE or MICROSPAR. Probably the result of selective RECRYSTALLIZATION in which larger grains grow at the expense of smaller ones.

struvite (NH₄MgPO₄.6H₂O) Magnesium-aluminium phosphate, found in GUANO.

Stump A JURASSIC succession in Utah/Idaho, USA, covering parts of the CALLOVIAN and OXFORDIAN.

stump A coastal pillar of rock below the high tide level. Cf. STACK.

Sturtian The older PERIOD of the SINIAN, 800–610 Ma.

stylolite A surface within a rock along which DISSOLUTION by PRESSURE SOLUTION has occurred, common in carbonate rocks and often marked by seams of insoluble CLAY MINERALS which remained after dis-solved CALCITE had moved away.

sub- Under.

sub-bituminous coal A BROWN COAL, with a CALORIFIC VALUE of <19.3 MJ kg⁻¹ and a FIXED CARBON content of 46–60%.

subalkaline Containing no alkali MIN-ERALS other than FELDSPARS.

subarkose A SANDSTONE with insufficient FELDSPAR to be termed an ARKOSE.

subautomorphic See SUBHEDRAL.

subcretion See UNDERPLATING.

subcritical translatent stratification A WIND RIPPLE LAMINATION produced in fine to medium SAND by RIPPLES climbing at an angle below the angle necessary for the pres-ervation of STOSS SIDE deposits. Each RIPPLE deposits a millimetric-scale, planar, TABU-LAR, INVERSE GRADED lamina above an ERO-SION surface so that the LAMINATION is very well defined. Cf. SUPERCRITICAL TRANSLATENT STRATIFICATION.

subcrop **1** A subsurface OUTCROP, e.g. where a formation intersects a subsurface plane such as an UNCONFORMITY. **2** In min-ing, any near-surface development of a rock or OREBODY, usually beneath superficial material.

subduction The process of underthrusting an oceanic PLATE into the MANTLE at a DESTRUCTIVE PLATE MARGIN.

subduction complex See ACCRETIONARY PRISM.

subduction orogeny An OROGENY at a SUBDUCTION ZONE characterized by the formation of a PAIRED METAMORPHIC BELT and CALC-ALKALINE volcanic activity.

subduction suction (trench suction)
The mechanism whereby the overriding PLATE in a SUBDUCTION ZONE is thrown into tension, caused by the downgoing slab's descending at an increasing angle with depth near the surface.

subduction zone A region where oceanic LITHOSPHERE descends into the MANTLE, ideally characterized by the following set of features: LITHOSPHERIC bulge, TRENCH with ACCRETIONARY COMPLEX, FOREARC BASIN, sedimentary arc, volcanic ISLAND ARC, BACK-ARC BASIN and BACK-ARC RIDGE.

suberinite A COAL MACERAL of the EXINITE group made up of the corky cells of plants.

subgrain A region of a CRYSTAL LATTICE which differs in orientation from the surrounding MINERAL and is bounded by CRYSTAL DEFECTS, but without internal distortion.

subhedral (hypidiomorphic, subautomorphic) Exhibiting some traces of crystal form.

subjacent Bottomless.

sublittoral See NERITIC.

submarine canyon A steep-sided, canyon-like trench cut into the continental shelf and sometimes cutting across it. Probably related to a past or present river valley.

submarine fan (deep-sea cone, deep-sea fan, deep-water fan) A broad, convex-upwards, deep water, SILICICLASTIC or carbonate system with positive relief above the adjacent BASIN floor, developing from a localized source, such as a SUBMARINE CANYON. Generally of radial geometry from 10–3000 km wide.

submarine plateau Submarine prominences representing areas where water depth is shallower than normal. Such areas make up about 10% of the oceanic crust and are prime candidates to become EXOTIC TERRANES.

submerged bar See OFFSHORE BAR.

submerged forest An area of forest vegetation, usually a layer of PEAT with tree stumps in their growth positions, which has been inundated by the sea. Indicative of a sea level rise since growth.

submetallic lustre A LUSTRE of a MINERAL intermediate between METALLIC and NON-METALLIC.

submicroscopic Of a size smaller than MICROSCOPIC, imaged by techniques such as ELECTRON MICROSCOPY.

subophitic texture An OPHITIC TEXTURE in which the enclosure of PLAGIOCLASE crystals by PYROXENE is only partial.

subplinian eruption A less violent type of VOLCANIC ERUPTION than a PLINIAN ERUPTION, producing less widely dispersed TEPHRA.

subsequent dolomitization DOLOMITIZATION subsequent to the lithification of a LIMESTONE by the entry of magnesian fluids along JOINTS and FAULTS. Cf. PENECONTEMPORANEOUS DOLOMITIZATION.

subsequent stream A stream which follows a course controlled by the structure of the local bedrock. Cf. ANTECONSEQUENT, INCONSEQUENT, INSEQUENT, OBSEQUENT and RESEQUENT STREAMS.

subsolidus A chemical system below its melting point and in which reactions are solid state.

subsolvus granite A GRANITE which crystallized below the SOLVUS temperature and thus contains two types of ALKALI FELDSPAR, K-rich and Na-rich. Cf. HYPERSOLVUS GRANITE.

substage A fifth order CHRONOSTRATIGRAPHIC unit.

subsurface flow See THROUGHFLOW.

subsurface fluid migration The motion of PORE FLUIDS within a sediment along permeable CARRIER BEDS from regions of high pressure to regions of low pressure at a rate dependent on the sediment's PERMEABILITY, the type of fluid, the relative balance of opposing flows, the sedimentary geometry, the HYDRAULIC CONDUCTIVITY, the AQUIFER gradient and the overall hydrodynamic regime.

subsurface mapping The production of subsurface geological maps based almost totally on borehole data, including GEO-PHYSICAL BOREHOLE LOGGING.

subsurface stormflow See INTERFLOW.

subtidal Descriptive of the part of the TIDAL-FLAT environment below the normal level of mean low water of spring TIDES. Cf. SUPRATIDAL.

succinite See AMBER.

Suchanian A CRETACEOUS succession in the far east of the former USSR equivalent to the HAUTERIVIAN, BARREMIAN, APTIAN and ALBIAN.

sudoite $(Mg_2(Al,Fe^{3+})_3Si_3AlO_{10}(OH)_8)$ A rare PHYLLOSILICATE of the CHLORITE group found disseminated in some HAEMATITE deposits and as a HYDROTHERMAL MINERAL.

suevite A BRECCIA formed by SHOCK META-MORPHISM, whose angular fragments are set in a GLASS MATRIX.

suffosion The EROSION of unconsolidated surface sediment by slumping into an underground cavity produced by bedrock DISSOLUTION.

Sui Ning A JURASSIC succession in Sichuan, China, equivalent to the OXFORDIAN and KIMMERIDGIAN.

sulphate reduction An important, bacterially controlled process in fine-grained marine sediments during early DIAGENESIS, a major control on the fate of detrital IRON, since IRON sulphide is highly insoluble.

sulphide minerals Metal sulphide compounds which make up the single most important group of ORE MINERALS. Classified by CRYSTAL STRUCTURE into ten major groups: DISULPHIDE GROUP, GALENA GROUP, SPHALERITE GROUP, WURTZITE GROUP, NICKEL ARSENIDE GROUP, THIOSPINEL GROUP, LAYER SULPHIDES GROUP, METAL EXCESS GROUP, RING STRUCTURE GROUP, CHAIN STRUCTURE GROUP.

sulphur (S) A MINERAL occurring as a NATIVE ELEMENT, common in volcanic regions.

sunstone A TRANSLUCENT variety of AVEN-TURINE; OLIGOCLASE with a red glow formed by minute, platy inclusions of HAEMATITE in parallel orientation.

super- Over, above.

supercontinent A large grouping of continents, e.g. PANGAEA.

supercritical flow A water flow whose velocity is greater than the velocity of propagation of a long surface wave in still water.

supercritical translatent stratification A rare WIND RIPPLE LAMINATION occurring when rates of deposition are sufficiently high to preserve STOSS SIDE deposits. Lamina contacts are gradational. Cf. SUBCRITICAL TRANSLATENT STRATIFICATION.

supergene Descriptive of processes involving water percolating down from the surface.

supergene enrichment (secondary enrichment) An increase in the GRADE of mineralization by processes subsequent to the primary processes of formation, especially applicable to the enrichment of sulphide ORES, and also oxide and carbonate ORES. Metals are usually carried down into the ORE, leaving an upper GOSSAN, where they precipitate and increase the metal content, or GANGUE MINERALS are mobilized and carried away.

supergroup The largest LITHOSTRATI-GRAPHIC subdivision, comprising a series of GROUPS.

superimposed drainage (epigenetic drainage) A drainage pattern unrelated to the existing BEDROCK due to its initial development on a rock cover subsequently removed by DENUDATION.

superimposition of structures The DEFORMATION of existing STRUCTURES, with the exception of FAULTS.

Superior-type iron formation A type of BANDED IRON FORMATION comprising thinly banded rocks of the oxide, carbonate and silicate facies and free of clastic material.

The banding is of IRON-rich and IRON-poor CHERT layers of centimetric- to metric-scale and can be followed in a layer up to several hundred metres thick over wide areas, associated with QUARTZITE and BLACK SHALE.

superparamagnetism The phenomenon in which FERROMAGNETIC particles <0.5 mm in size can only retain a remanence for a few minutes.

superplasticity A DEFORMATION MECHANISM involving sliding along grain boundaries, accommodated by DIFFUSION, or by SOLUTION TRANSFER.

superplume Large streams of overheated material rising from the CORE-MANTLE boundary. The evidence for their occurrence in the geological record is equivocal.

superposition of strata The basic STRATIGRAPHIC principle that each BED in a stratified sequence is younger than the underlying BED and older than the overlying BED.

supra- Above, beyond.

suprafan The most rapidly AGGRADING part of a SAND-rich SUBMARINE FAN, which appears as a convex-up mound above the surrounding FAN.

suprastructure The part of an OROGENIC BELT deformed at a relatively shallow level. Cf. INFRASTRUCTURE.

supratidal Descriptive of the part of the TIDAL-FLAT environment above the normal level of mean high water of spring TIDES. Cf. SUBTIDAL.

surf The mass of broken, foaming water that forms as a wave breaks on the shore.

surf zone The area between the BREAKER ZONE and the SWASH ZONE, characterized by bores of shoreward-moving water from swirling or plunging breakers.

surface wave (seismic surface wave) A SEISMIC WAVE which travels along the top of a half-space, including RAYLEIGH WAVES and LOVE WAVES.

surface wave magnitude An EARTHQUAKE MAGNITUDE based on the amplitude of SURFACE WAVES. Cf. BODY WAVE AMPLITUDE.

surge An inflated, dilute, TURBULENT FLOW of gas and PYROCLASTS.

surtseyan eruption A VOLCANIC ERUPTION of MAFIC MAGMA into shallow water (1–2 km) causing explosive MAGMA-water interactions in an almost continuous series of blasts, producing characteristic clouds in the shape of a cock's tail. The TEPHRA produced is fine-grained and deposited as finely bedded BASE SURGE and air-fall BEDS, occasionally interbedded with massive BEDS similar to MUDFLOWS, and a TUFF CONE or RING is formed.

survivorship curve A graph of the number of survivors of a single age group against age obtained in a PALAEOECOLOGICAL study from the size-frequency distribution of individuals.

suspect terrane A TERRANE for which there are grounds for considering it to be a DISPLACED TERRANE.

suspended load The solid material transported in a fluid which is not in contact with the channel bed and is supported by the vertical component of velocity of TURBULENT fluid eddies.

suspension The transport of sediment within a fluid body, possible when upward fluid velocities exceed the SETTLING VELOCITY of the particles.

suture A surface within the LITHOSPHERE representing the contact between two PLATES or MICROPLATES at a CONTINENT-CONTINENT COLLISION.

Svecofennian orogeny An OROGENY affecting the Baltic SHIELD in the PROTEROZOIC at ~1700 Ma.

Sveconorwegian orogeny See DALSLANDIAN OROGENY.

swale 1 An area of low-lying, often marshy land. **2** A shallow trough between storm ridges on a BEACH.

swaley cross-stratification A variety of HUMMOCKY CROSS-STRATIFICATION in which hummocks are rare or absent and SWALES are preferentially preserved.

swallet (swallow hole) A vertical to steeply sloping shaft in a LIMESTONE area down which surface water disappears underground.

swallow hole See SWALLET.

swallowtail twinning TWINNING in which a crystal divides along a TWIN PLANE producing a TWIN with a 'V' shape.

swamp A waterlogged area with characteristic vegetation in both coastal and inland environments.

swash The shoreward-moving uprush of TURBULENT water formed as a wave breaks on a BEACH, with a velocity generally greater than the GRAVITY-driven BACKWASH, which moves coarse sediment onshore.

swash bar (longshore bar) A SAND ridge on a BEACH, probably formed by the steepening of a low gradient foreshore within the SWASH ZONE by constructional waves.

swash zone The portion of a BEACH alternately covered by SWASH and exposed by BACKWASH.

Swazian An ERA of the PRECAMBRIAN, 3500–2800 Ma.

SWEAT hypothesis See SOUTHWEST US AND EAST ANTARCTICA HYPOTHESIS.

sweep The inrush of water that penetrates downwards into the VISCOUS SUBLAYER of the TURBULENT BOUNDARY LAYER with a velocity higher than the time mean average and a vertical component towards the BED.

sweet crude oil CRUDE OIL low in SULPHUR. Cf. SOUR CRUDE OIL.

swell beach See REFLECTIVE BEACH.

swirlhole See POTHOLE.

syenite A PHANERITIC ALKALINE IGNEOUS ROCK comprising ALKALI FELDSPAR exceeding 67% of the total FELDSPAR and ACCESSORY QUARTZ or NEPHELINE which makes up 5–20% of the rock. Compositionally equivalent to TRACHYTE. Found in ALKALINE intrusive complexes.

syenodiorite See MONZONITE.

syenogabbro A coarse-grained IGNEOUS ROCK comprising ESSENTIAL ALKALI FELDSPAR and Ca-rich PLAGIOCLASE in equal proportions, AUGITE, BIOTITE and ACCESSORY APATITE.

syenoid A SYENITIC rock in which FELDSPATHOIDS take the place of ALKALI FELDSPAR.

Sygynkanskaya A TRIASSIC succession in Siberia equivalent to the SPATHIAN.

sylvanite (yellow tellurium) ((Au,Ag)Te$_2$) A rare ORE MINERAL of GOLD and SILVER found in low- to high-temperature VEIN deposits.

sylvine See SYLVITE.

sylvinite A mixture of SYLVITE and HALITE.

sylvite (sylvine) (KCl) An EVAPORITE MINERAL.

Symmetrodonta An order of infraclass TRITUBERCULATA, subclass THERIA, class MAMMALIA; small, probably predatory MAMMALS which may be ancestral to the marsupial and placental MAMMALS. Range JURASSIC-early CRETACEOUS.

sympatric speciation The evolutionary divergence of populations of the same area into distinct species.

symplectic texture A TEXTURE produced by the intergrowth of two MINERALS, e.g. GRAPHIC TEXTURE, OPHITIC TEXTURE, POIKILITIC TEXTURE.

symplectite A textural feature of some GABBROIC rocks in which bulbous, MYRMEKITE-like extensions of PLAGIOCLASE crystals contain VERMICULAR inclusions of ORTHOPYROXENE. Formed during the late stages of CRYSTALLIZATION or RECRYSTALLIZATION.

syn- Together, with, resembling.

synaeresis crack See SYNERESIS CRACK.

Synapsida A subclass of class REPTILIA; MAMMAL-like REPTILES. Range CARBONIFEROUS–JURASSIC.

syncline A FOLD with the younger rocks in its core.

synclinorium A large, composite SYNCLINE made up of smaller FOLDS. Cf. ANTILINORIUM.

syndiagenetic Descriptive of a MINERAL deposit formed during the early or late stages of DIAGENESIS of a sediment.

syneresis crack (synaeresis crack) A subaqueous shrinkage crack formed in response to DEWATERING of a MUD layer as it compacts under its own weight, as flocs break down or as some CLAY MINERALS shrink when overlying waters change SALINITY.

synform A FOLD with a downwards CLOSURE.

syngenetic deposit A deposit forming at the same time as its COUNTRY ROCKS.

synkinematic See SYNOROGENIC.

synneusis The moving together and attachment of crystals in a MAGMA. The mechanism behind the formation of GLOMEROPORPHYRITIC TEXTURE.

synoptic profile The degree of original relief of an organism such as a STROMATOLITE.

synorogenic (synkinematic, syntectonic) Descriptive of a process or event taking place at the same time as OROGENIC DEFORMATION.

synsedimentary fault See GROWTH FAULT.

syntaxial growth See SYNTAXIAL VEIN.

syntaxial vein (syntaxial growth) A VEIN filling which grows from the walls towards the centre, characterized by host and veinfill having different compositions, a lack of crystallographic continuity across the VEIN, an oblique contact of veinfill and vein wall, and single CRYSTALLOGRAPHIC veinfills and inclusions of WALL ROCK fragments which lie along the median line and in subparallel bands. Cf. ANTITAXIAL GROWTH.

syntaxy The cement on a crystal growing in CRYSTALLOGRAPHIC continuity with its substrate, forming a single crystal.

syntectite A rock formed by SYNTEXIS.

syntectonic See SYNOROGENIC.

syntectonic sediment A sediment deposited at the same time as TECTONISM affected the depositional area, e.g. the EROSION of a THRUST SHEET forming a sedimentary deposit in a FORELAND BASIN. The sediment need not show contemporaneous DEFORMATION.

syntexis MAGMA generation by the melting of at least two rock types. Cf. ANATEXIS.

synthetic Descriptive of a STRUCTURE or FABRIC with the prevailing orientation or VERGENCE. Cf. ANTITHETIC.

synthetic seismogram A computer-generated SEISMOGRAM obtained by using mathematical expressions which describe the propagation of SEISMIC WAVES through a particular STRUCTURE, used as an aid in the interpretation of seismic data.

synthetic shear See RIEDEL SHEAR.

Syrian garnet A variety of ALMANDINE of GEM quality.

system A second order CHRONOSTRATIGRAPHIC unit.

Syzranian A PALAEOCENE succession in the former USSR covering the lower part of the DANIAN.

T

T See TESLA.

T wave A P WAVE trapped in the ocean.

T-ΔT method A statistical technique for determining the ROOT-MEAN-SQUARE VELOCITY from SEISMIC REFLECTION data.

table mountain A large MESA.

table reef A large-scale CORAL-ALGAL REEF.

tabular Descriptive of a crystal or feature with a broad, flat, commonly rectangular, form.

Tabulata A subclass of class ANTHOZOA, phylum CNIDARIA; colonial CORALS with calcareous tubular CORALLITES, prominent tabulae and inconspicuous or absent septae. Massive, foliaceous, DENDROID, phaceloid or creeping in habit. Range L. ORDOVICIAN–U. PERMIAN.

tachylite A BASALTIC GLASS, a component of rapidly quenched BASALTIC MAGMA in the chilled margins of SILLS and on the rims of PILLOW LAVAS.

Taconian orogeny See TACONIC OROGENY.

Taconic orogeny (Taconian orogeny) An OROGENY affecting the northern parts of the APPALACHIAN FOLD BELT in the ORDOVICIAN–SILURIAN, corresponding to the CALEDONIAN OROGENY of Europe. See also ACADIAN OROGENY.

taconite BANDED IRON FORMATION suitable for the concentration of MAGNETITE and HAEMATITE by fine grinding and BENEFICATION.

tadpole plot A method of plotting data from a DIPMETER LOG with a dot on a depth-DIP graph and a tail pointing in the STRIKE direction.

tadpole rock See ROCDRUMLIN.

Taeniodonta An order of infraclass EUTHERIA, subclass THERIA, class MAMMALIA; a small group of North American herbivores of up to pig size of uncertain relationships. Range L. PALAEOCENE–U. EOCENE.

taenite (FeNi) A natural alloy of IRON and nickel found in METEORITES.

tafoni Pits and hollows on rock surfaces, particularly GRANITES in desert environments. Possibly form when salt WEATHERING and wind scour small depressions on the surfaces of exposed rock, which enlarge when moisture accumulates in them. Some may form when a CASE-HARDENED surface is breached, exposing the softer rock beneath to EROSION.

tagilite See PSEUDOMALACHITE.

Taitai A CRETACEOUS succession in New Zealand covering part of the early CRETACEOUS.

talc ($Mg_3Si_4O_{10}(OH)_2$) A PHYLLOSILICATE formed by the HYDROTHERMAL ALTERATION of ULTRABASIC ROCKS or the THERMAL METAMORPHISM of siliceous DOLOMITES.

talc schist A rock consisting mainly of TALC, formed by the REGIONAL METAMORPHISM of an ULTRABASIC IGNEOUS ROCK.

talnakhite ($Cu_9Fe_8S_{16}$) A SULPHIDE MINERAL of the SPHALERITE GROUP.

talus See SCREE.

taluvium A hillslope deposit comprising coarse rubble and finer material.

tamarugite ($NaAl(SO_4)_2.6H_2O$) An hydrated sodium-aluminium sulphate usually found as a SECONDARY MINERAL.

tangent modulus A variable parameter relating STRESS to STRAIN for a perfectly elastic material where the STRESS-STRAIN relationship is not necessarily linear.

tangent-arc method (Busk method) A method of constructing a vertical CROSS-SECTION through a set of FOLDS, which assumes that they are PARALLEL, by drawing normals to the FOLD surface at several points where DIPS are available and regarded as tangents to an arc of the FOLD surface. The normals intersect at the centre of curvature for that FOLD segment.

tangential longitudinal strain A FOLD MECHANISM recognized by a pattern of layer-parallel extensional STRAIN on the outer part of the folded layer and layer-parallel contractional STRAIN on the inner arc, so that the maximum PRINCIPAL STRAIN direction is tangential to the FOLD surface.

tangential velocity The velocity of a point on a rotating sphere, which varies from zero at the POLE OF ROTATION to a maximum at 90° from the pole. Cf. ANGULAR VELOCITY.

tantalite (($Fe,Mn)Ta_2O_6$) An ORE MINERAL of tantalum, found in PEGMATITES, GRANITIC rocks and ALLUVIAL DEPOSITS.

tanzanite A blue, GEM variety of ZOISITE.

taphonomy The study of the *post-mortem* history of an organism.

taphrogenesis Vertical movements of the CRUST causing the formation of major FAULTS and RIFT VALLEYS.

tapiolite (($Fe,Mn)(Ta,Nb)_2O_6$) A tantalate of IRON and manganese found in ALBITIZED GRANITE PEGMATITES or in PLACER DEPOSITS derived from them.

tar See ASPHALT.

tar mat A layer of plastic, MALLEABLE ASPHALT at or near the surface or at former surfaces now represented by UNCON-FORMITIES.

tar pit (pitch lake) A pool of BITUMEN which frequently contains the remains of animals trapped in it.

tar sand A bituminous SANDSTONE or LIME-STONE impregnated with oil too viscous and heavy to be extracted by conventional drilling.

Taranaki A NEOGENE succession in New Zealand comprising the TONGAPORUTIAN and KAPITEAN.

tarbuttite ($Zn_2PO_4(OH)$) An hydrated zinc phosphate found as a SECONDARY MINERAL in the oxidized zone of zinc ORES.

Tarkhanian A MIOCENE succession on the Russian Platform covering parts of the LANGHIAN and SERRAVALLIAN.

tarn A small lake in a CIRQUE.

Tastubskiy A PERMIAN succession on the eastern Russian Platform equivalent to the L. SAKMARIAN.

Tatar A JURASSIC succession in W. Siberia covering parts of the BATHONIAN and CAL-LOVIAN.

taxa Plural of TAXON.

taxon A group of organisms of any rank, e.g. family, genus, species.

taxonomic uniformitarian analysis A PALAEOENVIRONMENTAL ANALYTICAL technique used when direct comparison with modern analogues is not possible. The tolerance for various environmental parameters of each TAXON is estimated from that of a comparable living form and the value of each environmental parameter is estimated from the overlap of values from all TAXA or by comparing the FOSSIL community with the most comparable modern analogue.

Taylor A CRETACEOUS succession on the Gulf Coast of the USA equivalent to the CAMPANIAN.

Taylor number A dimensionless hydrodynamic parameter depending on the scale of a CONVECTION cell, its KINEMATIC VISCOSITY and the rate of rotation. If less than unity, rotation of the convecting system does not affect the pattern of the CONVECTION.

tear fault See WRENCH FAULT.

tectogene A narrow, linear belt of down-folded rocks possibly related to mountain building.

tectogenesis See OROGENESIS.

tectonic **1** Descriptive of a STRUCTURE produced by DEFORMATION. **2** Relating to a major Earth STRUCTURE and its formation.

tectonic attenuation The thinning of STRATA by EXTENSIONAL DEFORMATION.

tectonic earthquake An EARTHQUAKE resulting from TECTONIC activity. Cf. VOLCANIC EARTHQUAKE.

tectonic erosion The process in which EROSION results in the DETACHMENT of a sheet which then slides under GRAVITY.

tectonic province A region of the CONTINENTAL CRUST distinguished by a characteristic TECTONIC history.

tectonic release The mechanism whereby the outward PRESSURE from an explosion allows the regional STRESS to cause a small EARTHQUAKE beneath, and simultaneous with, the explosion.

tectonic style The aspects of FOLD PROFILE geometry used to characterize FOLDS of different TYPE and origin.

tectonic transport The direction of relative movement of a rock mass as the result of TECTONIC activity.

tectonics The processes responsible for TECTONIC activity.

tectonite See S-TECTONITE, L-S TECTONITE.

tectosilicate A silicate in which each silicon tetrahedron shares all four corners with others to form a three-dimensional framework.

tectosphere The stable, subcontinental zone in which TECTONICS originate.

tektite A GLASSY object probably formed by the melting of terrestrial material by the impact of an extra-terrestrial body, such as a METEORITE.

Teleostei An order of subclass ACTINOPTERYGII, class OSTEICHTHYES, superclass PISCES; the most recently evolved bony FISH with loosened bony articulations round the face and jaw bones so that the mouth can be used to suck prey from thick sediment. Range JURASSIC–RECENT.

telescoping A phenomenon observed in HYPOTHERMAL DEPOSITS where steep temperature gradients cause the superimposition of zones that would have been distinct at deeper levels.

teleseism A SEISMIC EVENT recorded at an EPICENTRAL ANGLE range of 30–100°.

telethermal deposit A deposit formed at a very low temperature at a great distance from the source of its causative HYDROTHERMAL SOLUTIONS.

telinite A COAL MACERAL of the VITRINITE group in which MICROSCOPIC cell structure is visible.

telluric bismuth (tellurobismuth, wehrlite) (Bi_2Te_3) A natural alloy of bismuth and tellurium.

telluric current An electric current induced in the Earth by MAGNETOTELLURIC field variations, which gives rise to a variable horizontal potential gradient of $\sim$–10 mV km^{-1}. The field can be used in GEOPHYSICAL EXPLORATION by seeking structures of anomalous ELECTRICAL CONDUCTIVITY which distort this gradient and is effective to depths of some 7 km.

tellurite (TeO_2) A TELLURIUM OXIDE MINERAL found in HYDROTHERMAL VEIN deposits.

tellurium (Te) A rare NATIVE METAL.

tellurobismuth See TELLURIC BISMUTH.

telogenesis DIAGENESIS which occurs when existing EOGENETIC or MESOGENETIC PORE FLUIDS are flushed from the host sediment by the ingress of METEORIC WATERS.

Telychian A STAGE of the SILURIAN, 432.6–430.4 Ma.

TEM See TRANSMISSION ELECTRON MICROSCOPY.

Temaikan A JURASSIC succession in New Zealand covering the BAJOCIAN, BATHONIAN and part of the CALLOVIAN.

Temnospondyli An order of subclass LABYRINTHODONTIA, class AMPHIBIA; AMPHIBIANS with FISH-like vertebrae seemingly designed to aid walking. Range L. CARBONIFEROUS–U. TRIASSIC.

temperature log A GEOPHYSICAL BOREHOLE LOG in which downhole temperature gradients are measured for GEOTHERMAL exploration purposes.

tempestite See STORM DEPOSIT.

Templetonian A CAMBRIAN succession in Australia covering the lower part of the SOLVAN.

temporary strain (recoverable strain) A transient shape change followed by reversion to the initial shape.

tenacity The resistance offered by a MINERAL to mechanical DEFORMATION or disintegration, depending on the nature of chemical bonding, the CRYSTAL STRUCTURE and the MICROSTRUCTURE of the MINERAL. See also BRITTLE MINERAL, DUCTILE MINERAL, ELASTIC MINERAL, FLEXIBLE MINERAL, MALLEABLE MINERAL, SECTILE MINERAL.

tennantite ($Cu_{12}As_4S_{13}$) An ORE MINERAL of COPPER, ISOMORPHOUS with TETRAHEDRITE and found mainly in low- to high-temperature HYDROTHERMAL ORE VEINS.

tenor See GRADE.

tenorite (CuO) A black, SUPERGENE COPPER MINERAL found mainly in the oxidation zone of COPPER deposits.

tensile Descriptive of a FORCE, STRESS or STRAIN which acts in a direction away from a reference point.

tensile fracture (extension fracture, tension fracture) A FRACTURE developed under TENSILE conditions perpendicular to the least (or most tensile) PRINCIPAL STRESS.

tensile strength The STRENGTH under TENSILE STRESS.

tensile stress The STRESS developed under TENSILE conditions.

tension The STRESS state produced by two equal FORCES acting in opposite directions from a reference point.

tension fracture See TENSILE FRACTURE.

tension gash A linear opening in a rock developed under TENSILE conditions.

tension joint See EXTENSION JOINT.

tepee An overthrust sheet of LIMESTONE with the form of an inverted 'V' in section, found in tidal areas, in CALCRETE and around PLAYAS as a result of DEFORMATION or desiccation related to fluctuating water levels or changes in the nature of chemical PRECIPITATION.

tephra A PYROCLASTIC material which was explosively fragmented during a VOLCANIC ERUPTION, comprising ASH-sized GLASS shards, PUMICE, SCORIA and LITHIC CLASTS and PHENOCRYST-derived crystals. Formed by the explosive decompression of MAGMA on reaching the surface as dissolved VOLCANIC GASES rapidly EXSOLVE.

tephrite An OLIVINE-free, ALKALINE VOLCANIC ROCK comprising PLAGIOCLASE, CLINOPYROXENE and FELDSPATHOIDS. The presence of OLIVINE causes gradation into a BASANITE, and of ALKALI FELDSPAR into a PHONOLITE. Occurs in OCEANIC ISLANDS and continental settings.

tephrochronology The study of widespread TEPHRA deposits in order to establish age relations.

tephroite (Mn_2SiO_4) The manganese end-member of the OLIVINES found mainly in iron-manganese ORE deposits and their associated SKARNS.

Teratan A CRETACEOUS succession in New Zealand covering part of the TURONIAN, the CONIACIAN and part of the SANTONIAN.

Terebratulida An order of class ARTICULATA, phylum BRACHIOPODA; BRACHIOPODS with biconvex, punctate shells with a rounded hinge and pedicle foramen. Range L. DEVONIAN–RECENT.

terminal fan A type of ALLUVIAL FAN in which the quantity of surface RUNOFF decreases down the FAN due to INFILTRATION and evaporation so that normally little or no water leaves the FAN as surface flow.

termite activity The contribution of termites to the GEOMORPHOLOGICAL development of a region.

terra rossa A red CLAY soil developed on LIMESTONE associated with KARST features in regions of strong seasonal variation in precipitation. Forms when LIMESTONE DISSOLUTION occurs in the wet season and CLAY and iron compounds are released. The red colour originates from the conversion of hydrated ferric oxides to HAEMATITE during the dry season.

terrace A nearly flat landform with a steep edge formed by a variety of processes.

terracette A small ridge or TERRACE which extends across a slope, usually at right angles to its direction. Possibly the expression of animal trackways or a consequence of soil instability on a steep slope.

terrain See TERRANE.

terrain correction See TOPOGRAPHIC CORRECTION.

terrane (terrain) A region of CRUST with well-defined margins, which differs significantly in TECTONIC evolution from neighbouring regions.

terrigenous See SILICICLASTIC.

Tertiary A PERIOD of the CENOZOIC comprising the PALAEOCENE to PLIOCENE, 65.0–1.64 Ma.

tertiary creep The third and final stage of VISCOELASTIC STRAIN, characterized by accelerating STRAIN which leads to FAILURE. See also PRIMARY CREEP, SECONDARY CREEP.

teschenite An ANDESINE-bearing GABBRO or DOLERITE comprising calcium-rich PLAGIOCLASE, CLINOPYROXENE and ANALCITE, with ACCESSORY Fe-Ti oxide, AMPHIBOLE, BIOTITE ± OLIVINE. Occurs as minor intrusions,

coarse PEGMATITIC SCHLIEREN or fine VEINS in DIFFERENTIATED minor intrusions.

tesla (T) The SI unit of MAGNETIC FIELD strength, with dimensions V s m^{-2}.

test A protective shell covering the soft parts of certain invertebrates.

Tethys An ocean which lay between GONDWANALAND and LAURASIA in pre-Neogene times.

Tetracorallia See RUGOSA.

tetradymite (Bi_2Te_2S) An ORE MINERAL of TELLURIUM, commonly found in GOLD-QUARTZ VEINS.

tetragonal system A CRYSTAL SYSTEM whose members have three mutually perpendicular axes, the vertical of which is shorter or longer than the equal length horizontal axes.

tetrahedrite (fahl ore, fahlerz, grey copper ore) ($Cu_{12}Sb_4S_{13}$) An ORE MINERAL of COPPER found mainly in low- to medium-temperature HYDROTHERMAL ORE VEINS.

Tetrapoda Vertebrates with four limbs.

Teuriamn A PALAEOCENE succession in New Zealand covering the DANIAN and part of the THANETIAN.

textural maturity The state of a CLASTIC ROCK with high ROUNDNESS and SORTING, inferred to have these properties because of a long history of transport, REWORKING and thus ABRASION.

texture The general character of a rock, shown by its component particles in terms of grain size and shape, degree of crystallinity and arrangement.

thalassostatic Descriptive of phenomena related to a period of static sea level.

thalweg A line connecting the points of deepest flow in successive downstream channel cross-sections, i.e. the planform pattern of maximum channel depth.

thanatocenosis (death assemblage, thanatocoenosis) An assemblage of FOSSILS brought together only after death. Cf. BIOCENOSIS.

thanatocoenosis See THANATOCENOSIS.

Thanetian The higher STAGE of the PALAEO-CENE, 60.5–56.5 Ma.

thaw lake An ORIENTED LAKE in a PERMA-FROST region.

Thecodontia An order of subclass ARCHO-SAURIA, class REPTILIA; very primitive ARCHO-SAURS. Range U. PERMIAN–U. TRIASSIC.

Thematic Mapper A powerful imaging system used for REMOTE SENSING from the second generation of LANDSAT SATELLITES, which records six spectral bands in the 450–2350 nm wavelength region with PIXELS ~30 m across and a thermal infrared channel with 120 m PIXELS.

thenardite (Na_2SO_4) An EVAPORITE MINERAL whose HYDRATION to MIRABILITE is accompanied by a significant increase in volume which assists in the WEATHERING of a rock.

theralite (kylite) The PHANERITIC equivalent of BASANITE.

Therapsida An order of subclass SYNAP-SIDA, class REPTILIA; REPTILES with an improved gait and a MAMMALIAN pattern of jaw-closing muscles. Range late CARBON-IFEROUS–early PERMIAN.

Theria A subclass of class MAMMALIA comprising the marsupials and placentals. Range end CRETACEOUS–RECENT.

thermal conductivity A measure of the ability of a material to conduct heat; the flow of heat across unit area in unit time under unit temperature gradient normal to the area. Unit: $W\ m^{-1}\ K^{-1}$.

thermal diffusivity A parameter which controls the rate of heat propagation through a material; the ratio of THERMAL CONDUCTIVITY to the product of DENSITY and specific heat.

thermal maturation The process by which organic matter is transformed into HYDROCARBONS by heating.

thermal metamorphism METAMORPHISM involving heat only.

thermal resistivity The inverse of THER-MAL CONDUCTIVITY; unit m K W^{-1}.

thermal subsidence Subsidence taking place as heated material is removed from the heat source and undergoes thermal contraction, forming a sedimentary basin.

thermobaric flow A form of MASS SOLUTE TRANSFER out of a compacting BASIN under the influence of heat and PRESSURE.

thermoclastis See INSOLATION WEA-THERING.

thermocline The layer within a body of water at which the rate of decrease of temperature with depth is at a maximum.

thermogravimetric thermal analysis A method of studying CLAY MINERALS in which the weight changes with increasing temperature are monitored.

thermohaline current A current driven by DENSITY differences caused by seasonal variation in temperature and/or SALINITY and responsible for deep oceanic circulation.

thermokarst Topographic depressions arising from the thawing of ground ice.

thermoluminescence dating A dating technique for material containing small amounts of radioactive substances. The radioactivity causes ionization and some of the electrons produced become trapped in CRYSTAL DEFECTS. Heating to ~500°C releases these electrons in the form of visible light and the amount of this thermo-luminescence per rad of ionizing radiation provides an estimate of the latest time the sample was heated to 500°. Used to date SPELEOTHEMS and ancient pottery and to determine if FLINT was heated in antiquity.

thermoremanent magnetization (TRM) A magnetization acquired when a FERROMAGNETIC substance heated above its CURIE TEMPERATURE cools in a MAGNETIC FIELD. Present in many IGNEOUS and META-MORPHIC ROCKS.

Theropoda Carnivorous forms of the SAURISCHIA.

thickness-grade plan A plan derived by UNDERGROUND MAPPING showing the product of GRADE and thickness of a deposit.

thin section A slice of MINERAL or rock ~30 µm thick mounted on a microscope slide which is transparent or translucent to light and can be used to study the optical properties of the material.

thiospinel group A group of SULPHIDE MINERALS characterized by a SPINEL structure based on a cubic close-packed anion sublattice with half the octahedral holes and one eighth of the tetrahedral holes occupied by cations.

thixotropy The behaviour shown by some QUICK CLAYS on DEFORMATION in which there is a partial RECOVERY of RESIDUAL STRENGTH at some time after FAILURE.

tholeiite (tholeiitic basalt) A SILICA-OVERSATURATED BASALT comprising calcium-rich PLAGIOCLASE, PYROXENE (in which ORTHOPYROXENE or PIGEONITE exceeds calcium-rich PYROXENE) and Fe-Ti oxide, possibly with AUGITE ± OLIVINE in a siliceous GROUNDMASS. Forms the most abundant BASALT group and makes up layer 2 of the OCEANIC CRUST.

tholeiitic basalt See THOLEIITE.

thompsonite $(NaCa_2(Al_5Si_5O_{20}).6H_2O)$ A ZEOLITE found in AMYGDALES in BASIC IGNEOUS ROCKS.

thorianite (ThO_2) An ORE MINERAL of thorium found in PEGMATITES and ALLUVIAL DEPOSITS.

thorite $(ThSiO_4)$ A thorium MINERAL isostructural with ZIRCON, found in SYENITES.

thorium-lead dating A RADIOMETRIC DATING method based on the decay of ^{232}Th to ^{208}Pb with a HALF-LIFE of 13.9 Ga.

threshold angle The angle of a THRESHOLD SLOPE.

threshold slope A graded or equilibrium hillslope whose inclination is controlled by the resistance of its soil cover to a dominant DEGRADATIONAL process.

threshold velocity The minimum velocity of wind or water required to move particles of material.

thrombolite A STROMATOLITE-like organism of uncertain origin with a MACROSCOPIC clotted FABRIC.

throughflow (subsurface flow) The downslope flow of water through soil approximately parallel to the ground surface, generated where the HYDRAULIC CONDUCTIVITY decreases with depth.

throw The vertical DISPLACEMENT of a DIP-SLIP FAULT.

thrust See THRUST FAULT.

thrust belt A TECTONIC zone in which most of the SHORTENING occurs on THRUST FAULTS, usually making up the external parts of an OROGENIC BELT.

thrust fault (thrust) A REVERSE FAULT with a low angle of DIP.

thrust fault system A set of related FAULTS on which individual DISPLACEMENTS give a net CONTRACTION in the system as a whole.

thrust sheet A STRUCTURE like a NAPPE, but with a lower intensity of internal DEFORMATION.

thrust zone A zone of THRUSTING between two undeformed blocks.

thufur A small (10–50 cm high, 0.3–1 m diameter) earth hummock in a PERIGLACIAL area above the treeline, resulting from differential freezing of the ground.

thulite A pink variety of ZOISITE containing manganese.

thuringite $((Fe^{2+},Fe^{3+},Mg)_6(Al,Si)_4O_{10}(OH)_8)$ A CHLORITE group MINERAL found in VEINS in METAMORPHIC ROCKS.

tidal bar A COASTAL BAR intermittently exposed by TIDES.

tidal bore (bore) A large WAVE which progresses up a funnel-shaped river or ESTUARY with the rising TIDE.

tidal bundle A sediment unit deposited during a spring/neap tidal cycle comprising SAND units transported by ebb and/or flood

TIDES, separated by NON-CONFORMITIES or PAUSE PLANES, commonly draped by fine-grained sediment which reflects fall-out during slack water.

tidal correction The correction applied to GRAVITY SURVEY data during REDUCTION for the variation in GRAVITY resulting from solid EARTH TIDES, whose effect is to cause a change of ~3 g.u. over a 12 hour period.

tidal current The horizontal water movement caused by the gravitational attractions of the Sun and Moon on the Earth linked to the vertical movements of TIDES.

tidal delta A fan-shaped accumulation of sediment at the mouth of a TIDAL INLET, formed as tidal flow expands and decelerates seawards.

tidal friction The phenomenon which causes a time lag between the oceanic TIDES and solid EARTH TIDES generated by frictional forces within the Earth.

tidal inlet A subaqueous channel between segments of a BARRIER ISLAND through which TIDAL CURRENTS flow, whose morphology is controlled by the relative magnitudes of the WAVE and TIDAL energy.

tidal palaeomorph An oversized, frequently meandering valley believed to represent a former tidal channel originating at a time of higher sea level.

tidal range The vertical height between consecutive high and low TIDES.

tidal rhythmite See TIDALITE.

tidal wave A term often applied to a TSUN-AMI, but which gives a false impression of its origin.

tidalite (tidal rhythmite) Stacked sets of laminated MUDSTONE or intercalated beds of SANDSTONE and MUDSTONE, the thicknesses of which depend on daily changes in current velocities associated with tidal processes. Used to determine length of year in the past.

tide The vertical water movement arising from the gravitational attraction of the Sun and Moon, the effect of the latter being predominant.

tiger-eye A fibrous, golden brown QUARTZ PSEUDOMORPHOUS after CROCIDOLITE and showing CHATOYANCY, used as an ornamental stone.

tight fold A FOLD with a small angle between its LIMBS.

tight sand A RESERVOIR rock with a PERMEABILITY so low as to allow production of no more than five barrels of oil per day without costly WELL stimulation.

tile clay A CLAY similar to BRICK CLAY, suitable for the manufacture of tiles.

till See BOULDER CLAY.

till fabric analysis The measurement of the orientations of CLASTS in TILL, whose plotting may indicate the direction of former GLACIER advance.

tillite A CONGLOMERATE of GLACIAL origin, formed from BOULDER CLAY.

tilloid A chaotic mixture of large blocks in a CLAY-rich MATRIX, formed by MUDFLOWS, LANDSLIDES and GLACIERS.

Tilodonta An order of infraclass EUTHERIA, subclass THERIA, class MAMMALIA; primitive, carnivore-like MAMMALS. Range U. PALAEOCENE–L. EOCENE.

tilt-angle The angle between the horizontal primary electromagnetic field and the resultant field, used to quantify the disturbance in certain simple ELECTROMAGNETIC INDUCTION METHODS.

tilted fold See INCLINED FOLD.

tiltmeter An instrument which measures ultra-long period, angular displacements of the Earth's surface, usually for the purposes of EARTHQUAKE PREDICTION.

time correction The STATIC or DYNAMIC CORRECTION applied to seismic data.

time-distance curve A plot of ARRIVAL TIME of SEISMIC WAVES against SHOT-SEISMOMETER separation, used in the interpretation of seismic data.

time-term method A method of interpretation of SEISMIC REFRACTION data in terms of an undulating refractor which makes use of the difference in travel time of a HEAD WAVE to and from a refractor and the time a wave would take to travel the same horizontal distance at the SEISMIC VELOCITY of the lower medium.

tin (Sn) A very rare NATIVE METAL.

tin girdle The region of SE Asia in which important deposits of TIN are found.

tin pyrites See STANNITE.

tincalconite ($Na_2B_4O_5(OH).3H_2O$) A MINERAL formed by the ALTERATION of borate MINERALS by HYDRATION or dehydration.

tinguaite An UNDERSATURATED, medium- to coarse-grained IGNEOUS ROCK comprising ESSENTIAL ALKALI FELDSPAR, NEPHELINE and AEGIRINE $\pm$ sodic AMPHIBOLE or BIOTITE. The HYPABYSSAL equivalent of PHONOLITE.

tinstone A popular name for CASSITERITE.

tip See TIP LINE.

tip fold A FOLD which maintains the STRAIN continuity between the end of a FAULT of finite DISPLACEMENT and unfaulted rock.

tip line (tip) The termination of a THRUST surface.

Tirek An ORDOVICIAN/SILURIAN succession in the Mirnyy Creek area of NE Siberia equivalent to the HIRNANTIAN and RHUDDANIAN.

tirodite (manganoan cummingtonite) ($(Mg,Mn,Fe)_7Si_8O_{22}(OH)_2$) A rare, yellow, manganese-rich, magnesium-bearing end-member of the AMPHIBOLES.

titanaugite Titanium-bearing AUGITE.

titaniferous iron ore See ILMENITE.

titanite (sphene) ($CaTiO(SiO_4)$) A common ACCESSORY MINERAL, sometimes exploited as an ORE of titanium.

titanium hornblende See KAERSUTITE.

titanomaghemite Titanium-bearing MAGHEMITE.

titanomagnetite Titanium-bearing MAGNETITE.

Tithonian The highest STAGE of the JURASSIC, 152.1–145.6 Ma.

titration A method of wet CHEMICAL ANALYSIS in which a solution of the unknown is gradually mixed with a suitable reagent of known concentration until some form of indicator, such as a colour change, is activated. The quantity of the reagent used can then be converted into the concentration of the unknown.

Tiverton A PERMIAN succession in Queensland, Australia, covering part of the ASSELIAN, the SAKMARIAN and part of the ARTINSKIAN.

toad's-eye tin A BOTRYOIDAL or RENIFORM variety of CASSITERITE which shows an internal concentric and fibrous structure.

toadstone A mining term for various types of altered IGNEOUS ROCK.

Toarcian A STAGE of the JURASSIC, 187.0–178.0 Ma.

todorokite ($(Mn,Ca,Mg)Mn_3O_7.H_2O$) An ORE MINERAL of manganese found as a SECONDARY MINERAL formed from the ALTERATION of other manganese-bearing MINERALS.

toe 1 The base of the working face of a mine or quarry. **2** The lowest, most DISTAL part of a FAN, DELTA, SCREE, DUNE, etc.

toe erosion EROSION at the base of a steep slope, such as a cliff.

toe method A mapping technique used in a mine or quarry in which the TOE is charted as a line and all features in a face are recorded and projected down to the reference level. The final map represents a single horizontal plane with all contacts shown in their true STRIKE direction.

toeset bed The basal, asymptotic part of CROSS-LAMINATED beds, usually of fine-grained sediment. Cf. FORESET BED, TOPSET BED.

Tolbonskaya A TRIASSIC succession in Siberia equivalent to the ANISIAN and LADINIAN.

Tolen An ORDOVICIAN succession in Kazakhstan covering part of the PUSGILLIAN, the CAUTLEYAN, RAWTHEYAN and HIRNANTIAN.

tombolo A BAR or SPIT of SAND linking an island to the mainland or another island, usually forming on the sheltered side of the island.

Tommotian The lowest STAGE of the CAMBRIAN, 570–560 Ma.

tommotiid A small, conical or tubular FOSSIL composed of phosphate, of uncertain taxonomic status, found in the TOMMOTIAN.

tomography See SEISMIC TOMOGRAPHY.

tonalite A PHANERITIC rock comprising PLAGIOCLASE, which makes up >90% of the FELDSPAR, and 20–30% QUARTZ, with ALKALI FELDSPAR absent or making up <10% of the rock. Associated with PLUTONS and BATHOLITHS of the DIORITE-GRANODIORITE-GRANITE association.

tonalite-trondhjemite-granodiorite suite See TTG SUITE.

Tonawandian A SILURIAN succession in North America covering parts of the TELYCHIAN and SHEINWOODIAN.

Tongaporutian A MIOCENE succession in New Zealand covering parts of the TORTONIAN and MESSIAN.

tonstein A KAOLIN-rich SEAT-EARTH possibly formed by the ALTERATION of volcanic ASH or ILLITE-rich CLAY.

tool mark A mark on a sediment surface made by a mobile object.

tool track (groove, scratch) An incongruous LINEATION on a FAULT surface caused by ASPERITY PLOUGHING.

topaz ($Al_2SiO_4(F,OH)_2$) A NESOSILICATE found in GRANITE, used as a GEM.

topazolite A honey-yellow, TRANSPARENT variety of ANDRADITE.

topographic correction (terrain correction) The correction applied to GRAVITY SURVEY data for the gravitational attraction of topography around an observation point, often calculated with a HAMMER CHART.

toponomy The mode of preservation of a TRACE FOSSIL.

toposequence See CATENA.

topotype A specimen from the same locality as the HOLOTYPE.

topset bed The upper, near-horizontal layers of a CROSS-LAMINATED BED, usually of coarse-grained sediment. Cf. FORESET BED, TOESET BED.

tor An exposure of bare rock with free faces on all sides resulting from differential WEATHERING, MASS WASTING and stripping, whose form is often controlled by JOINTS.

torbanite See BOGHEAD COAL.

torbernite (cuprouranite) ($Cu(UO_2)_2$ $(PO_4)_2.8-12H_2O$) A green SECONDARY MINERAL found where uranium ORE deposits have been OXIDIZED and WEATHERED.

Torian A CAMBRIAN succession in Siberia covering part of the MAENTWROGIAN.

toroidal field A MAGNETIC FIELD confined within a sphere with no radial components. Cf. POLOIDAL FIELD.

Torridonian The upper division of the PRECAMBRIAN in NW Scotland, generally of FLUVIAL origin.

torsion balance An early type of GRAVIMETER based on the oscillation of a beam suspended by a torsion fibre.

Tortonian A STAGE of the MIOCENE, 10.4–6.7 Ma.

total strain See FINITE STRAIN.

touchstone A piece of stone scratched across a PRECIOUS METAL or alloy to assess its purity from the STREAK.

tourmaline (($Na,K,Ca)(Li,Mg,Fe,Mn,Al)_3$ $(Al,Fe,Cr,V)_6(BO_3)_3(Si_6O_{18})(O,OH,F)_4$) A complex CYCLOSILICATE found in GRANITE

PEGMATITE and as an ACCESSORY MINERAL in IGNEOUS and METAMORPHIC ROCKS.

tourmalinization A type of WALL ROCK ALTERATION in which significant amounts of new TOURMALINE are formed. Usually associated with medium- to high-temperature mineralization.

Tournaisian The oldest EPOCH of the CARBONIFEROUS, 362.5–358.3 Ma.

Tournaisien A DEVONIAN/CARBONIFEROUS succession in France and Belgium covering part of the FAMENNIAN and the HASTARIAN.

tower karst (turmkarst) A KARST landscape characterized by steep, flat-topped towers of LIMESTONE rising from a flat plain.

town gas See COAL GAS.

trace The SEISMOGRAM for one SEISMOMETER.

trace d'activité animale (lebensspur) A TRACE FOSSIL in unconsolidated sediment.

trace element An element present in a rock in such low concentration, usually in the parts per million range, that it does not control the presence of MINERALS such as FELDSPAR and PYROXENE but occurs within them by ionic substitution for MAJOR ELEMENTS. Often indicative of the source of the rock. Cf. MAJOR ELEMENT.

trace fossil (biogenic sedimentary structure, ichnofossil) A STRUCTURE in sediment produced by the activity of an ancient organism. See also AGRICHNIA, CUBICHNIA, DOMICHNIA, ENDICHNIA, EPICHNIA, EXICHNIA, FODICHNIA, HYPICHNIA, PASCICHNIA, REPICHNIA.

Trachaeophyta A division of kingdom plantae; the VASCULAR PLANTS. Range CARBONIFEROUS–RECENT.

trachyandesite A VOLCANIC ROCK intermediate in composition between ANDESITE and TRACHYTE.

trachybasalt A VOLCANIC ROCK intermediate in composition between BASALT and TRACHYTE.

trachydacite A VOLCANIC ROCK with a lower Na_2O+K_2O content than TRACHYTE and with no PLAGIOCLASE PHENOCRYSTS.

trachyte A SILICA-SATURATED or OVERSATURATED, weakly PORPHYRITIC, ALKALINE VOLCANIC ROCK of INTERMEDIATE composition containing mainly SANIDINE or ORTHOCLASE PHENOCRYSTS, but also ALKALI FELDSPAR, CLINOPYROXENE, Fe-Ti oxides, AMPHIBOLE, FAYALITE, AENIGMATITE and BIOTITE in a GROUNDMASS often showing TRACHYTIC TEXTURE.

trachytic texture A TEXTURE in a VOLCANIC ROCK in which there is strong flow alignment of small ACICULAR ALKALI FELDSPAR crystals.

trachytoidal texture The TEXTURE of a PHANERITIC IGNEOUS ROCK in which FELDSPARS are in parallel to subparallel alignment and thus resemble a TRACHYTIC TEXTURE.

trailing continental margin See PASSIVE CONTINENTAL MARGIN.

transcurrent fault (wrench fault) A vertical STRIKE-SLIP FAULT.

transfer fault A steep FAULT with a STRIKE-SLIP DISPLACEMENT which transfers DISPLACEMENT from one DIP-SLIP plane to another; a small-scale equivalent of an oceanic TRANSFORM FAULT.

transform fault A special form of STRIKE-SLIP FAULT which joins the ends of CONSTRUCTIVE and DESTRUCTIVE PLATE MARGINS.

transformation twinning A type of TWINNING comprising two POLYMORPHS, caused by a change in CRYSTAL STRUCTURE under different temperature-pressure conditions during TWIN formation.

transformation-enhanced deformation DEFORMATION whose mechanism is enhanced by chemical reaction or phase changes.

transformational faulting (anticrack faulting) A sudden phase change in a MINERAL which takes place by rapid shearing of the CRYSTAL LATTICE along planes on which minute crystals of the new phase have grown. For example, the change from OLI-

VINE to SPINEL in the downgoing slab probably contributes to SUBDUCTION ZONE SEISMICITY.

transgranular displacement A DISPLACEMENT across a grain boundary during DEFORMATION. Cf. INTERGRANULAR DISPLACEMENT, INTRAGRANULAR DISPLACEMENT.

transgression An incursion of the sea over a land area or over a shallow sea. Cf. REGRESSION.

transient creep A slow DEFORMATION in rocks when they are subjected to STRESS at the temperatures and pressures found at or near the Earth's surface.

transition zone Any large-scale, diffuse boundary within the Earth, but usually the MANTLE TRANSITION ZONE between 400 and 700 km depth.

translation The DISPLACEMENT of the centre of a body to a new position after DEFORMATION.

translation gliding A crystal DEFORMATION mechanism whereby one or more rows of atoms are displaced laterally along a GLIDE plane.

translucent mineral A MINERAL in which the transmission of light is poorer than for a TRANSPARENT MINERAL, generally displayed by MINERALS with a NON-METALLIC LUSTRE.

transmission capacity The state of a soil which can accommodate the flow of no more water.

transmission coefficient The ratio of the amplitude of a SEISMIC WAVE transmitted across an ACOUSTIC IMPEDANCE contrast to the amplitude of the incident wave, described by ZOEPPRITZ' EQUATIONS.

transmission electron microscopy (TEM) A technique of ELECTRON MICROSCOPY used in the investigation of the MICROSTRUCTURE of crystalline specimens from the transmission of electrons through them. The electron beam is diffracted according to the BRAGG LAW, with a very small angle between incident and diffracted rays. This allows magnification of up to 10^6, providing resolution of ~0.3 nm.

transmissivity The product of HYDRAULIC CONDUCTIVITY and the thickness of a BED acting as an AQUIFER.

transparent mineral A MINERAL in which the transmission of light is more effective than for a TRANSLUCENT MINERAL, generally displayed by MINERALS with a NON-METALLIC LUSTRE.

transportation slope A slope which suffers neither EROSION nor deposition because the amount of material arriving from upslope is the same as that leaving downslope. Usually concave in form and situated at the base of a more extensive slope.

transposed foliation A FOLIATION developed by the DEFORMATION of a primary surface, commonly by folding accompanied by DIFFUSIVE MASS TRANSFER.

transposition The creation of a new FABRIC, commonly a FOLIATION, by DEFORMATION of an existing one.

transpression The compression associated with movement along a curved STRIKE-SLIP FAULT. Cf. TRANSTENSION.

transtension The tension associated with movement along a curved STRIKE-SLIP FAULT. Cf. TRANSPRESSION.

Transvaal jade (South African jade) A massive, light green variety of HYDROGROSSULAR, used as a substitute for JADE.

transverse bar A COASTAL BAR in a near-shore or intertidal zone normal to the shore. Its formation requires a restricted FETCH length and low energy wave conditions.

transverse coast See ATLANTIC-TYPE COAST.

transverse dune A DUNE whose long axis is normal to the main SAND-moving wind direction.

transverse wave See S WAVE.

trap **1** A STRUCTURE in which HYDROCARBONS are trapped. **2** Old term for a group of LAVA FLOWS, named after the Dutch for 'staircase' and descriptive of the characteristic form of a pile of LAVA FLOWS.

traverse mapping A method of GEOLOGI-CAL MAPPING for scales of 1:250 000–1:50 000 in which traverses are systematically spaced or follow primary topographic features.

travertine Calcareous material formed by PRECIPITATION from flowing fresh water at a HOT SPRING or WATERFALL after passage through calcareous rock or sediment, aided by biochemical activity.

Tremadoc The oldest EPOCH of the ORDOVICIAN, 510.0–493.0 Ma.

tremolite (grammatite) $(Ca_2Mg_5Si_8O_{22}(OH)_2)$ An INOSILICATE of the AMPHIBOLE group, found mainly in CONTACT and REGIONALLY METAMORPHOSED DOLOMITIC and low-GRADE ULTRABASIC ROCKS.

Trempealeauan A CAMBRIAN/ORDOVICIAN succession in the E. USA covering parts of the DOLGELLIAN and TREMADOC.

trench See OCEAN TRENCH.

Tresca's failure criterion 'FAILURE occurs when the maximum SHEAR STRESS reaches a critical constant value for the material.'

trevorite $(NiFe_2O_4)$ An OXIDE MINERAL of the MAGNETITE series.

Trias See TRIASSIC.

Triassic (Trias) The oldest PERIOD of the PALAEOZOIC, 245.0–208.0 Ma.

triaxial compression COMPRESSION with components along three orthogonal axes.

triaxial stress A STRESS system with components along three orthogonal axes.

triclinic system A CRYSTAL SYSTEM whose members have three unequal, mutually oblique axes; i.e. no symmetry other than a possible centre.

tridymite (SiO_2) A high-temperature form of SILICA, stable in the range 870–1470°C.

trigonal system A CRYSTAL SYSTEM whose members have three lateral axes of equal length intersecting at an angle of 60° to each other and reaching the centres of the three vertical faces, and a vertical axis of different length perpendicular to the other three.

Trilobita/trilobites A subphylum of phylum ARTHROPODA; ARTHROPODS characterized by trilobation of the exoskeleton into cephalon (head), thorax (body) and pygidium (tail). Range L. CAMBRIAN–U. PERMIAN.

trimacerite A lithotype of COAL with >5% of MACERALS from all three groups.

Trimerphytopsida A class of division TRACHAEOPHYTA, kingdom plantae; the ancestors of megaphyllous plants. Range L. DEVONIAN–U. DEVONIAN.

Trinity A CRETACEOUS succession on the Gulf Coast of the USA covering parts of the APTIAN and ALBIAN.

triphylite $(Li(Fe,Mn)PO_4)$ A lithium MINERAL found in PEGMATITES, ISOMORPHOUS with LITHIOPHILITE.

triple junction The point where three PLATE MARGINS intersect. Only certain configurations are possible, determined by the relative velocity vectors of the PLATES at the intersection. Quadruple junctions are inherently unstable.

triple point The point on a phase diagram at which the boundary curves of three phases meet and they can coexist.

tripoli A MICROCRYSTALLINE, soft, friable, porous siliceous material exploited as an ABRASIVE.

tripolite (infusorial earth) A variety of OPAL, a DIATOMITE with an earthy appearance.

tristanite An intermediate member of the potassic alkaline series.

trittkaren KARREN in the form of heel-shaped pits.

Trituberculata An infraclass of subclass THERIA, class MAMMALIA comprising the PANTOTHERIA and SYMMETRODONTA. Range U. JURASSIC–L. CRETACEOUS.

TRM See THERMOREMANENT MAGNETIZATION.

trochiform Descriptive of a coiled shell in which the sides are evenly conical and the base flat.

troctolite (troutstone) A GABBRO with OLIVINE and no PYROXENE.

troilite (FeS) An IRON SULPHIDE MINERAL mainly found in METEORITES.

trona ($Na_3H(CO_3)_2.2H_2O$) A MINERAL found in SALINE LAKE deposits.

trondhjemite A PHANERITIC, QUARTZ-ALBITE-rich PLAGIOCLASE rock, including PLA-GIOGRANITE within OPHIOLITE complexes, TONALITE within CALC-ALKALINE BATHOLITHS and forming part of a distinctive ARCHAEAN rock assemblage (BIMODAL TRONDHJEMITE and METAMORPHOSED BASALTIC rocks with no INTERMEDIATE ROCKS).

troostite A manganiferous variety of WIL-LEMITE.

trottoir A narrow organic REEF in the intertidal zone.

trough cross-bedding See TROUGH CROSS-STRATIFICATION.

trough cross-stratification (trough cross-bedding) CROSS-STRATIFICATION in which the lower bounding surface is highly curved and concave-up so that CROSS-SETS are scoop- or spoon-shaped, probably resulting from the migration of irregular, sinuous or LINGUOID DUNES.

trough line A line joining the topographically lowest points on a folded surface, normally parallel to the FOLD HINGE LINE.

troutstone See TROCTOLITE.

true dip The angle between the horizontal and the surface of a layer, measured in the vertical plane perpendicular to the STRIKE of the layer and thus the maximum DIP measurable. Cf. APPARENT DIP.

true north The direction towards the Earth's axis of rotation. Cf. MAGNETIC NORTH.

true polar wander The postulated movement of the Earth's surface, and hence its poles, relative to the plane of the ecliptic.

Tschermak's component The charge-coupled substitution of Al for divalent ions at B and for Si at T in PYROXENES.

tschermakite ($Ca_2Mg_3(Al,Fe)_2Al_2$ $Si_6O_{22}(OH)_2$) An AMPHIBOLE found in IGNEOUS and METAMORPHIC ROCKS.

Tselinograd An ORDOVICIAN succession in Kazakhstan covering part of the LATE LLAN-VIRN, the LLANDEILO and part of the COSTONIAN.

tsunami A long-wavelength water WAVE caused by sudden movement of the sea floor, such as caused by an EARTHQUAKE, sediment SLUMP or VOLCANIC ERUPTION, which travels at tens of m s^{-1}. Of low amplitude at sea, but growing to up to 20 m on reaching shallow water.

TTG suite (tonalite-trondhjemite-granodiorite suite) The suite of sodium-rich IGNEOUS ROCKS which formed extensively in the ARCHAEAN, and only form at the present day in those few places where young, hot oceanic CRUST is subducted.

Tubulidentata An order of infraclass EUTHERIA, subclass THERIA, class MAMMALIA; the aardvarks. Range MIOCENE–RECENT.

tufa (calc-tufa) A spongy, porous variety of TRAVERTINE associated with ALGAL colonies.

tuff A volcanic sediment.

tuff cone A STRUCTURE similar to a TUFF RING, but generally larger and steeper sided, with a maximum thickness of 100–350 m and a maximum DIP of 24–30°, and with a higher proportion of MUDFLOW deposits.

tuff ring A monogenetic VOLCANO, of any composition but often of BASALT, formed in a PHREATOMAGMATIC ERUPTION, with a low rim and a broad, flat CRATER. Comprises PHREATIC air-fall BRECCIAS and finely bedded air-fall and PYROCLASTIC SURGE beds, the maximum DIP of the TEPHRA being 3–12°. See also TUFF CONE.

tuffisite (tuffite) A fine-grained MATRIX in a volcanic BRECCIA, comprising finely divided fragments of MAGMA and grains of sediment, which appear to have been

FLUIDIZED and intruded along small cracks in blocks in the BRECCIA.

tuffite See TUFFISITE.

tugtupite ($Na_4BeAlSi_4O_{12}Cl$) A rare, intense pink, beryllium-bearing form of SODALITE.

Tumen A JURASSIC succession in W. Siberia covering the HETTANGIAN, SINEMURIAN, PLIENSBACHIAN and part of the TOARCIAN.

tundra A vast, level, treeless, marshy region, usually with a permanently frozen subsoil.

tungstates MINERALS containing tungsten in the form $(WO_4)^{2-}$, similar to the MOLYBDATES with which there is free substitution of molybdenum. Form two series, the WOLFRAMITE group and the SCHEELITE group.

tungstenite (WS_2) A SULPHIDE MINERAL of the LAYERED SULPHIDES GROUP.

tungstic ochre See TUNGSTITE.

tungstite (tungstic ochre) ($WO_3.H_2O$) An earthy, yellow or green, hydrated tungsten oxide, usually found as a SECONDARY MINERAL associated with WOLFRAMITE.

Tunicata SESSILE, filter-feeding animals of phylum CHORDATA with free-swimming larvae which may be ancestral to the PISCES.

tunnel erosion A form of PIPE EROSION initiated by an abnormally large HYDRAULIC GRADIENT.

tunnel valley (rinnentaler, tunneldale) A long, deep channel cut by a subglacial meltwater river.

tunneldale See TUNNEL VALLEY.

Turam EM method An ELECTROMAGNETIC INDUCTION METHOD for MINERAL exploration using a large fixed source and a pair of receiving coils about 10 m apart. EM components are measured at two frequencies, providing information on features with anomalous ELECTRICAL CONDUCTIVITIES in the subsurface.

turbidite A deposit from a waning TURBIDITY CURRENT.

turbidity current (turbidity flow) A generally turbulent, subaqueous DENSITY CURRENT of suspended sediment driven by GRAVITY, with the potential of flowing down slopes of <1° and even up-slope. Occurs in non-marine to marine, shallow to deep water environments, generated by storms, the entry of sediment-laden river PLUMES, shelf currents or slope FAILURE. Flow velocities are up to 10 m s^{-1} and the flow length may reach thousands of kilometres.

turbidity flow See TURBIDITY CURRENT.

turbidity meter An instrument using a light source which provides continuous measurement of SUSPENDED LOAD.

turbinate Descriptive of a coiled shell shaped like a top, but with a rounded base.

turbulence A series of quasi-random motions at different scales in a fluid; the root mean square of the instantaneous fluctuating velocity component in TURBULENT FLOW.

turbulence intensity The ratio of TURBULENCE to mean flow velocity in TURBULENT FLOW.

turbulent flow Flow with three-dimensional motion in which the velocity at any point fluctuates with time and STREAMLINES may not be parallel. Occurs at REYNOLDS NUMBER >200.

turgite See HYDROHAEMATITE.

turlough A seasonal lake, up to 5 km^2 in area, found in a glacially influenced KARST area, which fills and empties through SPRINGS and SINKHOLES.

turmkarst See TOWER KARST.

Turonian A STAGE of the CRETACEOUS, 90.4–88.5 Ma.

turquoise ($CuAl_6(PO_4)_4(OH)_8.4H_2O$) A blue-green SECONDARY MINERAL of COPPER valued as a GEM.

twin A composite crystal produced by TWINNING.

twin axis The axis of symmetry about

which an individual in a TWIN can be rotated to produce the orientation of the other.

Twin Creek A JURASSIC succession in Utah/Idaho, USA, covering part of the BAJOCIAN, the BATHONIAN and part of the CALLOVIAN.

twin gliding A crystal DEFORMATION mechanism whereby a DISPLACEMENT is taken up by each row in a CRYSTAL LATTICE.

twin law The fundamental elements along or about which a crystal is TWINNED in terms of a TWIN PLANE or TWIN AXIS.

twin plane The plane of symmetry in which an individual in a TWIN can be reflected to produce the orientation of the other.

twinning The formation of composite crystals (TWINS) in which the two individuals have different orientations in simple CRYSTALLOGRAPHIC relationship to each other. Described by the symmetry operation necessary to convert one orientation to the other. See also NORMAL TWIN, PARALLEL TWIN.

two-way travel-time (TWT) The time taken for a SEISMIC WAVE to return to the surface after undergoing a SEISMIC REFLECTION.

TWT See TWO-WAY TRAVEL TIME.

type locality 1 (type section) A locality selected as a standard for a STRATIGRAPHIC unit. **2** The locality of a TYPE SPECIMEN.

type section See TYPE LOCALITY.

type specimen (lectotype, neotype) A specimen (HOLOTYPE) on which the full description of a species is based.

typological method The STRATIGRAPHIC description of a unit in terms of its LITHOLOGY or FOSSILS.

tyuyamunite ($Ca(UO_2)_2(VO_4)_2.5-8.5H_2O$) A green, SECONDARY ORE MINERAL of uranium and vanadium.

U

Ubendian orogeny An OROGENY affecting central Africa at ~1800–1700 Ma.

Udden-Wentworth scale A scale for the subdivision of CLASTIC ROCKS on the basis of CLAST size, based on a ratio of two between successive class boundaries. See also PHI UNIT.

Udoteaceae A group of CALCAREOUS ALGAE characterized by internal medullary filaments which branch and curve outwards to form an exterior cortical layer. Range M. ORDOVICIAN–RECENT.

Ufimian A STAGE of the PERMIAN, 256.1–255.0 Ma.

ugandite A LEUCITE-bearing, SILICA-poor, potassic VOLCANIC rock rich in MAFIC minerals, principally OLIVINE, Ti-AUGITE and Ti-MAGNETITE.

ugrandite An acronym for the GARNETS UVAROVITE, GROSSULAR and ANDRADITE. Cf. PYRALSPITE.

uintaite (gilsonite) A brilliant black variety of ASPHALT occurring in rounded masses.

Uivakian orogeny An OROGENY affecting the NE Canadian SHIELD at ~3000 Ma.

Ulatizian An EOCENE succession on the west coast of the USA covering parts of the YPRESIAN and LUTETIAN.

ulexite ($NaCaB_5O_6(OH)_6.5H_2O$) A borate MINERAL found associated with BORAX, deposited from BRINE formed in enclosed BASINS in arid areas.

ullmannite (nickel antimony glance) (NiSbS) A SULPHIDE MINERAL found in HYDROTHERMAL VEINS.

Ulsterian A SILURIAN/DEVONIAN succession in E. North America covering the upper SILURIAN, LOCHKOVIAN, PRAGIAN and part of the EMSIAN.

ultimate strength The STRENGTH shown by a material which fails after DUCTILITY. Cf. BRITTLE STRENGTH.

ultrabasic rock An IGNEOUS ROCK comprising essential FERROMAGNESIAN MINERALS and FELDSPATHOIDS to the virtual exclusion of QUARTZ.

ultracataclasite A cohesive FAULT ROCK with a random FABRIC and with >90% MATRIX. Cf. CATACLASITE, PROTOCATACLASITE.

ultramafic rock A rock comprising >90% FERROMAGNESIAN MINERALS, composed of OLIVINE, ORTHOPYROXENE and CLINOPYROXENE ± AMPHIBOLE. Found in continents in association with GABBRO and in OPHIOLITE complexes.

ultramarine A pigment made from LAPIS LAZULI.

ultramylonite A foliated FAULT ROCK with 90–100% MATRIX.

ulvöspinel (Fe_2TiO_4) An OXIDE MINERAL with SPINEL structure found as EXSOLUTION bodies within MAGNETITE and also in lunar rocks.

umber A naturally occurring brown pigment composed of IRON and manganese oxides with minor SILICA, alumina and LIME, formed by the WEATHERING of rocks rich in FERROMAGNESIAN MINERALS.

unconfined aquifer An AQUIFER lacking an upper AQUICLUDE, so that the upper limit of saturation is the WATER TABLE.

unconformable Not following the underlying sequence of rock in STRUCTURE or age.

unconformity A break in the STRATIGRAPHIC record which represents a period of no sediment deposition. See also DIASTEM, DISCONFORMITY, NON-SEQUENCE, PARACONFORMITY.

unconformity trap A HYDROCARBON TRAP in which the SEAL is provided by impermeable material deposited on an UNCONFORMITY above permeable STRATA.

unconformity-related uranium deposit An EPIGENETIC uranium deposit in which most of the mineralization lies at or just below an UNCONFORMITY.

underclay A fine-grained sediment immediately beneath a COAL SEAM, generally with no BEDDING and strongly LEACHED, which causes a high KAOLINITE content.

underconsolidated sediment A sediment as it was deposited, before the commencement of DIAGENESIS etc. Cf. NORMALLY CONSOLIDATED SEDIMENT, OVERCONSOLIDATED SEDIMENT.

underfit stream A large valley channel, often meandering, which contains smaller alluvial channels, i.e. the current channel is too small to account for the valley channel. Possibly formed in response to climate change, RIVER CAPTURE, etc.

underflow GROUNDWATER flow in ALLUVIAL sediments parallel to and beneath a river channel.

underground mapping Mapping executed in tunnel drivage, working mines and abandoned mines, generally at a large scale.

underplating (offplating, plastering, subcretion) The addition of material to the base of a STRUCTURE, such as occurs in the growth of an ACCRETIONARY PRISM.

undersaturated Descriptive of an IGNEOUS ROCK deficient in SILICA so that MINERALS develop which are unstable in the presence of SILICA. See also SILICA SATURATION.

undersaturated oil pool An oil pool undersaturated with respect to gas, commonly arising from an increase in the depth of burial.

underthrusting The movement of an oceanic PLATE under the leading edge of a second PLATE at a SUBDUCTION ZONE.

Undillian A CAMBRIAN succession in Australia covering the middle part of the MENEVIAN.

undulose extinction The EXTINCTION of a MINERAL which takes place over a range of angles, which may be caused by the presence of MICROSCOPIC KINK BANDS.

unequal slopes A theory which considered the EROSIONAL development of a BADLAND divide between two slopes of unequal DECLIVITY, which acquire their smooth curves according to the LAW OF DIVIDES. See also LAW OF EQUAL DECLIVITIES.

uneven fracture (irregular fracture) A FRACTURE producing a rough and irregular surface.

uniaxial compressive strength The STRENGTH under uniaxial COMPRESSION.

uniaxial stress A STRESS system in which the PRINCIPAL STRESS is compressive and the INTERMEDIATE and MINIMUM PRINCIPAL STRESSES are zero.

uniclinal shifting The migration of a river in the direction of STRATAL DIP.

uniformitarianism (actualism) The present viewed as the key to the past. Essentially the converse of CATASTROPHISM.

unimodal Possessing one mode, i.e. a frequency distribution in which there is only one most common value.

Uniramia A phylum of mainly terrestrial ARTHROPODS characterized by single-branched limbs. The head bears a pair of antennae, mandibles and two feeding maxillae. Range ?CAMBRIAN–RECENT.

unit cell The fundamental building block of a crystal comprising an ordered arrangement of atoms repeated exactly in all directions.

universal gravitational constant See GRAVITATIONAL CONSTANT.

unloading The removal of STRESS, commonly gravitational, by EROSION, melting of ice, etc., followed by PRESSURE RELEASE effects.

unsaturated Descriptive of a MINERAL which cannot develop in stable equilibrium in the presence of free SILICA.

unsaturated zone See VADOSE ZONE.

uphole shooting A seismic technique in which SHOTS are detonated down a borehole and their arrivals monitored by a surface SEISMOMETER in order to determine SEISMIC VELOCITIES and assist in STATIC CORRECTIONS.

uplift 1 The upward movement of part of the Earth's surface, typically 500–1000 km in width. 2 A STRUCTURE caused by the upward movement of part of the Earth's surface.

Upper Greensand A CRETACEOUS succession in England covering the upper part of the ALBIAN.

Upper Sandugan A SILURIAN succession in the Mirnyy Creek area of NE Siberia equivalent to the WENLOCK.

upper continental crust The upper 10–12 km of the CONTINENTAL CRUST, previously termed the 'granitic layer'. Cf. LOWER CONTINENTAL CRUST.

upper flow regime The FLOW REGIME commencing when SHEAR STRESS on a bed abruptly decreases and BEDFORMS are washed out and replaced by a PLANE BED. FLOW SEPARATION no longer occurs and SHEAR STRESSES start to increase as STANDING WAVES and then ANTIDUNES form as velocity increases. Cf. LOWER FLOW REGIME.

upper mantle 1 The MANTLE above ~700 km depth, including the MANTLE TRANSITION ZONE. 2 The MANTLE above ~400 km depth, i.e. not including the MANTLE TRANSITION ZONE. Cf. LOWER MANTLE.

upper-stage plane bed A flat BED in SAND-grade sediment produced at FROUDE NUMBERS approaching unity, over which there is intense sediment transport, characterized by extensive parallel LAMINAE and PRIMARY CURRENT LINEATIONS.

upright fold A FOLD with a near horizontal HINGE LINE and a vertical AXIAL SURFACE.

upthrown Descriptive of the side of a FAULT with relative upward movement. Cf. DOWNTHROWN.

upthrust A REVERSE FAULT flattening upwards from a steep attitude to become a THRUST at a shallow level.

upward continuation The computation of how a POTENTIAL FIELD would appear at an elevation above the level of observation, used to enhance the anomalies of deep sources. Cf. DOWNWARD CONTINUATION.

upwelling The transport of subsurface waters to the surface at a wide variety of scales in oceans, ESTUARIES and lakes, usually in response to wind STRESS on the surface.

Uralian emerald A green variety of ANDRADITE used as a semi-precious GEM, found in NODULES in ULTRABASIC ROCKS.

Uralianeny An OROGENY in the former USSR coeval with the VARISCAN OROGENY.

uralite A blue-green AMPHIBOLE of generally ACTINOLITE composition formed by the ALTERATION of PYROXENE.

uralitization (amphibolitization) A form of ALTERATION affecting PYROXENE-bearing rocks in which the PYROXENE is replaced by fibrous AMPHIBOLES.

uraniferous calcrete A CALCRETE mineralized with uranium.

uraninite (UO_2) The major ORE MINERAL of uranium, found in PEGMATITES, high-temperature HYDROTHERMAL VEINS and PLACER DEPOSITS.

uranium-lead age dating A dating

method based on the radioactive decay of ^{238}U and ^{235}U to ^{206}Pb and ^{207}Pb respectively with HALF-LIVES of 4498 and 713 Ma. ZIRCON, which contains trace amounts of uranium, is the preferred MINERAL as its primary lead concentration is low and it is resistant to uranium LEACHING.

uranophane (uranotile) $(Ca(UO_2)_2$ $Si_2O_7.6H_2O)$ Hydrated calcium-uranium silicate, found in PEGMATITES as PSEUDO-MORPHS after URANINITE, as coatings on FRAC-TURES and in the oxidized zone of URANINITE VEIN deposits.

uranotile See URANOPHANE.

Urgonian A massive, landscape-dominating, U. BARREMIAN to L. APTIAN LIME-STONE facies in S. Europe.

Urodela An order of subclass LISSAMPHIBIA, class AMPHIBIA; the newts and salamanders. Range late JURASSIC–RECENT.

urtite An IJOLITE with 0–30% MAFIC MINERALS.

'uruq See DRAA.

Ururoan A JURASSIC succession in New Zealand equivalent to the PLIENSBACHIAN, TOAR-CIAN and AALENIAN.

Urutawan A CRETACEOUS succession in New Zealand covering part of the middle ALBIAN.

USLE See UNIVERSAL SOIL LOSS EQUATION.

Ust'kel'Terskaya (Induan) A TRIASSIC succession in Siberia covering the GRIES-BACHIAN and part of the NAMMALIAN.

uvala A series of complex closed depressions in a KARST area, often caused by coalescing DOLINES.

uvarovite $(Ca_3Cr_2Si_3O_{12})$ A rare green GAR-NET found in association with CHROMITE in SERPENTINE, in SKARNS and in METAMOR-PHOSED LIMESTONES.

V

vacancy crystal defect (Schottky crystal defect) The simplest type of POINT CRYSTAL DEFECT in which an atom is missing from its usual position in the regular arrangement of atoms.

vadose zone (unsaturated zone, zone of aeration) A shallow subsurface zone of INFILTRATION and percolation of rainfall, RUNOFF or melt water between the surface and the WATER TABLE. Up to tens of metres in thickness with intergranular pores and fissures unsaturated with water and containing air at atmospheric pressure.

Valanginian A STAGE of the CRETACEOUS, 140.7–135.0 Ma.

valence bond theory An approach to chemical bonding based on pairs of electrons considered as shared between atoms, localized on atoms or on separate atoms with opposed spins that are arranged among the atoms of a molecule to form so-called 'structures'. These structures have characteristic energy and interact to form a stable representation of a molecule.

valentinite (Sb_2O_3) A white antimony tri-oxide formed by the decomposition of other antimony ORES.

valley bulge (valley-bottom bulge) STRATA bulged up in a valley floor by ERO-SION's acting on rocks of different character, e.g. fine-grained, water-charged sediments extruded through COMPETENT rock.

valley meander A meander with a greater wavelength than the present river system, produced under conditions of higher RUNOFF.

valley-bottom bulge See VALLEY BULGE.

vallon de gélivation A small valley formed by the widening of JOINTS in the BEDROCK by frost action rather than FLUVIAL processes.

valve An exoskeletal unit which makes up a shell.

van Krevelen diagram A plot of KEROGEN composition in terms of oxygen:carbon ratio against hydrogen:carbon ratio.

vanadinite ($Pb_5(VO_4)_3Cl$) A rare SECOND-ARY MINERAL found in the oxidized parts of lead VEINS.

vapour phase crystallization The CRYSTALLIZATION of MINERALS in cavities or pores in IGNIMBRITES from hot VOLCANIC GASES.

vapour transport A mechanism of MAG-MATIC DIFFERENTIATION in which constitu-ents such as sodium, potassium and silicon become concentrated in a vapour phase which can migrate within, or escape from, a MAGMA body.

Varanger The lower EPOCH of the VENDIAN, 610–590 Ma.

variable source area model A model for RUNOFF processes in which the surface RUNOFF is produced by saturation-excess OVERLAND FLOW. Cf. PARTIAL AREA MODEL.

variation diagram A graph showing the chemical variation in a suite of rocks to reveal genetic and other relationships between them.

variation method A method of determining the energies of stable MOLECULAR ORBITALS by minimizing their energy.

variegated copper ore A popular name for BORNITE.

variolitic structure A form of SPHERULITIC TEXTURE comprising radiating fibres in a GLASSY ROCK at the chilled margin of an intrusion.

variometer See MAGNETIC VARIOMETER.

Variscan orogeny (Armorican orogeny, Hercynian orogeny) An OROGENY in late CARBONIFEROUS–early PERMIAN in Europe and eastern North America.

varisicite ($Al(PO_4).2H_2O$) A blue-green, massive MINERAL similar to TURQUOISE and used as a GEM, found mainly in deposits formed by the action of phosphatic METEORIC WATERS on aluminous rocks.

varve A thin layer of a LAMINITE and RHYTHMITE which represents an annual cycle of deposition.

varve dating The measurement and counting of VARVES in lake deposits so that a chronology can be established, often linked to RADIOCARBON DATING.

vascular plant A plant with a fluid-conducting vascular system.

vasques A series of wide (up to several decimetres), shallow, flat-bottomed pools forming a network of tiered, terrace-like steps on LIMESTONE coastal platforms in the intertidal zone, separated by narrow, lobed ridges 10–200 mm in height.

vaterite A METASTABLE form of CALCITE or ARAGONITE.

vector 1 A quantity which represents the direction and magnitude of a motion. **2** The relative DISPLACEMENT direction across a FAULT.

vegetation arc A band of dense vegetation in BROUSSE TIGRÉE, generally convex downslope on INTERFLUVES and convex upslope in shallow drainage ways.

vein A sheet-like or TABULAR, DISCORDANT, mineralized body formed by complete or partial infilling of a FRACTURE within a rock.

vein system A group of VEINS with a regular orientation.

velocity defect law A relationship describing the shape of the outer region of a TURBULENT BOUNDARY LAYER where the KARMAN-PRANDTL VELOCITY LAW does not apply. $(U_m-U)/U^* = -5.75 \log_{10} (y/\delta)$, where U_m = freestream velocity, U = mean velocity at a point at distance y from the wall to the SHEAR VELOCITY U^*, δ = BOUNDARY LAYER thickness or flow depth.

velocity log See CONTINUOUS VELOCITY LOG.

Vendian The upper PERIOD of the SINIAN, 610–570 Ma.

ventifact An object modified in shape by wind action, primarily SAND ABRASION. See also EINKANTER, DREIKANTER, ZWEIKANTER, YARDANG.

Venturian A PLIOCENE succession on the west coast of the USA covering the upper part of the PIACENZIAN.

verdite A green rock mainly formed of green MICA and CLAY used as an ornamental stone.

Vereiskian A STAGE of the CARBONIFEROUS, 311.3–309.2 Ma.

vergence The sense of SHEAR deduced from the asymmetry of minor STRUCTURES (FOLDS, FAULTS, etc.) in an OROGENIC BELT.

vermiculite ($(Mg,Ca)_{0.3}(Mg,Fe,Al)_3$ $(Al,Si)_4O_{10}(OH)_2.nH_2O$) A PHYLLOSILICATE formed mainly by the ALTERATION of BIOTITE.

vernadite (δ–MnO_2) A manganese MINERAL found in MANGANESE NODULES.

vertical electrical sounding (VES, electric drilling) A RESISTIVITY METHOD in which the current electrodes are expanded symmetrically so that the current penetrates progressively deeper into the subsurface. A GEOELECTRIC SECTION can be obtained by inverting the sounding data.

vertical fold A FOLD in which the HINGE LINE is vertical.

vertical seismic profile (VSP) A seismic

profile obtained by recording a surface SEIS-MIC SOURCE at a series of depths down a borehole.

very long baseline interferometry (VLBI) A method for determining the distance between two stations from the phase difference of the same signal received from a quasar, at an accuracy sufficient to monitor CONTINENTAL DRIFT.

very low frequency method See VLF METHOD.

VES See VERTICAL ELECTRICAL SOUNDING.

vesicle A gas-filled cavity in a MAGMA or VOLCANIC ROCK.

vesiculation The growth of VESICLES in a MAGMA, caused by the EXSOLUTION of VOL-CANIC GASES.

Veslyanskiy A PERMIAN succession in the Timan area of the former USSR covering part of the WORDIAN.

vesuvian eruption A VOLCANIC ERUPTION characterized by very violent activity with the ejection of large quantities of material.

vesuvianite (idocrase) $(Ca_{10}(Mg,Fe)_2Al_4(SiO_4)_5(Si_2O_7)_2(OH)_4)$ A CYCLOSILICATE formed by the CONTACT METAMORPHISM of impure LIMESTONE.

VHN See VICKERS HARDNESS NUMBER.

vibration magnetometer A MAGNETO-METER in which a sample is vibrated to produce an oscillating voltage in a detector of the same frequency as the vibration.

Vibroseis® A mechanical SEISMIC SOURCE in which a vibrator generates a wavetrain whose frequencies vary with time. Of low energy and thus environmentally advantageous. A conventional SEISMOGRAM is obtained by CORRELATION of the wavetrain with the received signal.

Vickers hardness number (VHN) A measure of the HARDNESS of an ORE MINERAL used in identification.

Vicksberg An OLIGOCENE succession on the Gulf Coast of the USA covering the lower part of the RUPELIAN.

Vine-Matthews hypothesis A model which explains the origin of linear OCEANIC MAGNETIC ANOMALIES in terms of a THERMAL REMANENT MAGNETIZATION acquired as newly generated oceanic LITHOSPHERE cools in the GEOMAGNETIC FIELD. As SEAFLOOR SPREADING continues, the TRM changes direction as the GEOMAGNETIC FIELD reverses polarity.

Vinice An ORDOVICIAN succession in Bohemia covering the lower part of the SOUDLEYAN.

violane A massive, violet-blue variety of DIOPSIDE.

violarite $(FeNi_2S_4)$ A SULPHIDE MINERAL of the THIOSPINEL GROUP.

Virgilian A CARBONIFEROUS succession in the USA covering parts of the KASIMOVIAN and GZELIAN.

viridine A green IRON- and manganese-bearing variety of ANDALUSITE.

virtual magnetic pole An estimate of the PALAEOMAGNETIC POLE based on palaeomagnetic measurement of a single sample and therefore potentially including a non-axial dipole component. Cf. PALAEOMAGNETIC POLE.

viscoelasticity DEFORMATION in which ELASTIC and VISCOUS components act in parallel. Applied STRESS produces an initial rapid STRAIN which increases asymptotically with time. On removal of the STRESS, the STRAIN is recovered exponentially towards the initial value.

viscosity The ratio of SHEAR STRESS to SHEAR STRAIN RATE in a fluid. Units Pa s.

viscosity coefficients The coefficients of proportionality in the tensor representation of the linear STRESS-STRAIN RATE relationship for a pure NEWTONIAN FLUID.

viscous flow (viscous strain) Behaviour in which there is a positive relationship between SHEAR STRESS and STRAIN RATE.

viscous remanent magnetization (VRM) A low stability REMANENT MAGNETIZATION acquired when a sample is held in a

MAGNETIC FIELD, such as the GEOMAGNETIC FIELD.

viscous strain See VISCOUS FLOW.

viscous sublayer The region of flow closest to the bed in a TURBULENT BOUNDARY LAYER dominated by viscous forces and in which velocity increases linearly with height above the bed. $\delta = 11.5 \, \upsilon/U^*$, where δ = thickness of viscous sublayer, U^* = SHEAR VELOCITY of fluid of KINEMATIC VISCOSITY υ.

Viséan An EPOCH of the CARBONIFEROUS, 349.5–332.9 Ma.

vishnevite $(Na,Ca,K)_{6-7}Al_6Si_6O_{24}(SO_4,CO_3,Cl)_{1.0-1.5}.1–5H_2O)$ A sulphate-bearing variety of CANCRINITE, found in NEPHELINE SYENITE.

visual roundness scale A method of estimating the ROUNDNESS of a CLAST by comparison with a template of twelve standard CLASTS of different ROUNDNESS and SPHERICITY.

vital effects Non-equilibrium fractionations of stable isotopes of carbon and oxygen in the rapid PRECIPITATION of biogenic carbonate.

vitrain A LITHOTYPE OF BANDED COAL comprising black, VITREOUS, BRITTLE material with a CONCHOIDAL FRACTURE that occurs in thin (6–8 mm) bands and forms the major constituent of BITUMINOUS COAL.

vitreous lustre A LUSTRE similar to broken glass.

vitric tuff A TUFF with more GLASSY fragments than LITHIC or crystal fragments.

vitrification The formation of a GLASS.

vitrified fort An ancient Celtic fort that had been burned so intensely, possibly deliberately, that the BUILDING STONE (SANDSTONE, CONGLOMERATE, GNEISS or IGNEOUS ROCK) is fused.

vitrinertite A bi-MACERAL MICROLITHOTYPE OF COAL comprising VITRINITE and INERTINITE, characteristic of high RANK COALS.

vitrinite A COAL MACERAL group mainly formed of the remains of trunks, branches, stems, roots and leaves of LAND PLANTS, the principal group in most COALS.

vitrite A MICROLITHOTYPE OF COAL composed of VITRINITE MACERALS.

vitroclastic structure A structure formed in volcanic ASH by the disruption of VESICULAR GLASSY ROCKS, so that most fragments have a concave outline.

vitrodurain A LITHOTYPE OF BANDED COAL intermediate between VITRAIN and DURAIN.

vitrophyre A PORPHYRITIC IGNEOUS ROCK with a GLASSY MATRIX; an intensely WELDED TUFF.

vitrophyric texture An IGNEOUS ROCK TEXTURE in which PHENOCRYSTS are embedded in a GLASSY MATRIX.

vivianite $(Fe_3(PO_4)_2.8H_2O)$ A rare SECONDARY MINERAL formed by the WEATHERING of phosphates in PEGMATITES.

VLBI See VERY LONG BASELINE INTERFEROMETRY.

VLF method (very low frequency method) An ELECTROMAGNETIC INDUCTION METHOD utilizing, as the primary field, low frequency signals in the waveband 15–25 kHz transmitted by large-scale navigational networks or military communication systems. Only a receiver is required and the surveying can be performed by a single operator.

vogesite A CALC-ALKALINE LAMPROPHYRE with PHENOCRYSTS of HORNEBLENDE, PYROXENE ± OLIVINE in a MATRIX in which ALKALI FELDSPAR is in excess of PLAGIOCLASE.

Voight model See VISCOELASTICITY.

volcanic arenite A LITHIC ARENITE with CLASTS dominantly of volcanic origin.

volcanic bomb (bomb) A mass of liquid LAVA thrown through the air which rotates and takes on a characteristic shape and structure.

volcanic breccia A non-sedimentary fragmental rock of volcanic origin.

volcanic earthquake An EARTHQUAKE in a volcanically active area with an origin different from a TECTONIC EARTHQUAKE. Possible origins include the modification of regional STRESS by MAGMA, VOLCANIC ERUPTION, the oscillation of liquid ± gas in a MAGMA CHAMBER and MAGMA flow in conduits compressed by HYDROSTATIC PRESSURE.

volcanic eruption   An expulsion of MAGMA ± broken rock from a VOLCANIC VENT, excluding the emission of gas from a FUMAROLE and hot water from a GEYSER. See also HAWAIIAN ERUPTION, PELÉEAN ERUPTION, PLINIAN ERUPTION, SUB-PLINIAN ERUPTION, STROMBOLIAN ERUPTION, SURTSEYAN ACTIVITY and VULCANIAN ERUPTION.

volcanic gas   The gas discharged into the atmosphere by MAGMATIC activity, either EXSOLVED from MAGMA or through FUMAROLES. Common gases are H_2, N_2, HCl, HF, CH_4 and NH_4.

volcanic hazard   The risk to life or property from volcanic activity, assessed for an area from historical records, the mapping of prehistoric deposits and comparison with the types of activity of VOLCANOES with similar evolution.

volcanic neck   A positive topographic feature formed when a VOLCANIC PLUG is exhumed by EROSION.

volcanic pipe   An approximately cylindrical conduit underlying a VOLCANIC VENT, 50 m–1 km in diameter and sometimes flaring towards the surface.

volcanic plug   A mass of solidified MAGMA or BRECCIA filling a VOLCANIC PIPE.

volcanic rock   A rock formed by the solidification of LAVA or PYROCLASTIC material.

volcanic vent   An orifice through which a VOLCANIC ERUPTION has occurred or is occurring.

volcanic-associated massive oxide deposit (Kiruna-type deposit)   A STRATIFORM oxide ORE similar to a MASSIVE SULPHIDE DEPOSIT, often characterized by MAGNETITE-HAEMATITE-APATITE in volcanic or volcanic-sedimentary terrains.

volcanic-associated massive sulphide deposit   See MASSIVE SULPHIDE DEPOSIT.

volcanic-exhalative deposit (sedimentary exhalative deposit)   A lenticular to sheet-like, MASSIVE SULPHIDE DEPOSIT often found at the interface between volcanic or volcanic and sedimentary BEDS. Believed to be formed by MAGMATIC fluids or the circulation of seawater.

volcanicity (vulcanicity, vulcanism)   Volcanic activity.

volcaniclastic   Descriptive of a CLASTIC ROCK containing volcanic material.

volcano   A location where MAGMA and volatiles issue through the CRUST and accumulate.

Volgian   A STAGE of the JURASSIC in the BOREAL realm approximately equivalent to the TITHONIAN.

Von Karman's constant   (κ) A constant, commonly taken as 0.4184, relating the mean size of eddies (l) in a flow to the depth of the flow (y): $l = \kappa y$.

Von Mises criterion   The condition whereby a general STRAIN at constant volume, excluding rotation, requires the simultaneous operation of five independent SLIP SYSTEMS.

vortex ripple   A steep variety of WAVE RIPPLE in which the wavelength of the ripple-mark is 65% of the orbital diameter of the water motion near the bed, although this relationship breaks down above a critical REYNOLDS NUMBER and the RIPPLE flattens.

vorticity   The average ANGULAR VELOCITY of lines between known points in a rock undergoing STRAIN, used to describe ROTATIONAL STRAIN.

Vrzhumskiy   A PERMIAN succession on the eastern Russian Platform covering parts of the CAPITANIAN and LONGTANIAN.

VSP   See VERTICAL SEISMIC PROFILE.

vug   A small, irregular cavity in an intrusive rock or carbonate sediment.

vulcanian eruption   A type of VOLCANIC ERUPTION comprising a discrete, powerful, cannon-like blast with the expulsion of LITHIC and SCORIA CLASTS, ASH and gas at high velocity from a VOLCANIC VENT, generating a short-lived ASH plume. Possibly occurs after a build-up of VOLCANIC GAS pressure in a vent trapped beneath viscous MAGMA.

vulcanicity See VOLCANICITY.

vulcanism See VOLCANICITY.

vulcanite A general name for a fine-grained IGNEOUS ROCK, usually forming a LAVA FLOW.

Vyatskiy A PERMIAN succession on the eastern Russian Platform covering part of the CHANGXINGIAN.

Vychegodskiy A PERMIAN succession in the Timan area of the former USSR covering the lower part of the UFIMIAN.

W

wacke A SANDSTONE with 15–75% mud MATRIX by volume.

wackestone A LIMESTONE comprising MATRIX-supported carbonate particles in MUD.

wad A fine-grained mixture of manganese oxide and hydrated OXIDE MINERALS.

wadeite ($K_2CaZrSi_4O_{12}$) A colourless to lilac, transparent MINERAL found in potassium-rich rocks.

wadi (ouady, oued, wady) A steep-sided watercourse with sporadic flow in an arid region.

wady See WADI.

Waiauan A MIOCENE succession in New Zealand covering the lower part of the TORTONIAN.

Waipawan A PALAEOCENE/EOCENE succession in New Zealand covering parts of the THANETIAN and YPRESIAN.

Waipipian A PLIOCENE succession in New Zealand covering the lower part of the PIACENZIAN.

wairakite ($Ca_8(Al_{16}Si_{32}O_{96}).16H_2O$) A ZEOLITE found in the deep parts of zones of HYDROTHERMAL ALTERATION.

wairauite (CoFe) A rare alloy found in PERIDOTITES.

Waitakian An OLIGOCENE succession in New Zealand covering the upper part of the CHATTIAN.

Wall A PERMIAN succession in Queensland, Australia, covering the lower part of the ARTINSKIAN.

wallrock The rock adjacent to a MINERAL deposit or igneous intrusion.

wallrock alteration The reaction of HYDROTHERMAL FLUIDS with WALLROCK, decreasing in intensity with distance.

Walther's law A concept which suggests that a CONFORMABLE vertical succession of SEDIMENTARY FACIES with no UNCONFORMITIES reflects their former lateral juxtaposition.

Wanganui A NEOGENE succession in New Zealand covering the OPOITIAN, WAIPIPIAN, MANGAPANIAN and part of the NUKUMARUAN.

waning slope A slope on which weathered SCREE is eroded to form a concave FOOTSLOPE, which develops, and extends uphill, when stream incision ceases to be the dominant slope-forming process. Cf. WAXING SLOPE.

want The missing part of a deposit resulting from EXTENSIONAL FAULTING.

Warendian An ORDOVICIAN succession in Australia covering the upper part of the TREMADOC.

Warepan A TRIASSIC succession in New Zealand equivalent to the NORIAN.

Warthe Glaciation The third of the major GLACIATIONS affecting northern Europe in the QUATERNARY.

wash 1 See ARROYO. **2** Loose debris.

wash load A fine-grained sediment that can be SUSPENDED at the lowest DISCHARGES.

Washita A CRETACEOUS succession on the

Gulf Coast of the USA covering parts of the ALBIAN and CENOMANIAN.

washout A lenticular body of CLASTIC SEDIMENT, commonly SANDSTONE, projecting from the roof of a COAL SEAM and replacing all or part of it. Formed by scour and fill by a stream.

washover (fan) Sediment deposited by overwash of a BARRIER ISLAND, characteristic of microtidal BARRIER ISLANDS where STORM SURGE-enhanced seas cannot pass through poorly developed TIDAL INLETS and break through the island.

washplain A nearly flat surface of ALLUVIUM covering deeply WEATHERED BEDROCK in an environment where seasonal floods cannot incise due to the lack of abrasive BEDLOAD and large sediment volume.

waste The unwanted material which is removed during mining and quarrying.

water drive The buoyant effect of underlying water which drives oil through permeable rocks during SECONDARY MIGRATION.

water gun A mechanical marine SEISMIC SOURCE in which a pulse of high pressure water is released into the sea, creating a vacuum into which the surrounding water implodes.

water lane A tract of concentrated SHEETFLOW in BROUSSE TIGRÉE with a uniformly dense tree cover.

water sapphire See SAPPHIR D'EAU.

water table The level of free-standing water within intergranular pores or fissures at the top of the PHREATIC ZONE, below which the pores of the host are saturated with water.

waterfall A stream falling over a precipice.

watershed (divide) The boundary of a drainage BASIN.

watten An area of tidal marshland and sand flats between the mainland and an OFFSHORE BAR or island, exposed only at low TIDE.

Waucoban A CAMBRIAN succession in the E. USA covering the TOMMOTIAN, ATDABANIAN and part of the LENIAN.

Waukesha Dolomite An EXCEPTIONAL FOSSIL DEPOSIT of L. SILURIAN age at Brandon Bridge, Wisconsin with a marine biota of ARTHROPODS, worms and CONODONTS whose soft parts are preserved in phosphate.

wave diffraction The fanning out of a WAVE as it passes through an opening, for example in a breakwater.

wave rays (orthogonals) Lines drawn normal to wave crests, between which wave energy is constant and which converge and diverge at the coast to concentrate energy on headlands and spread it in bays.

wave refraction The process by which water wave crests are bent until they become parallel to submarine contours according to SNELL'S LAW, caused by the wave velocity decreasing with decreasing depth.

wave-cut bench See SHORE PLATFORM.

wave-cut platform See SHORE PLATFORM.

wave-ripple A periodic BEDFORM generated by progressive GRAVITY WAVES, usually symmetrical to slightly asymmetrical and trochoidal in section with wavelengths of 10 mm–1 m. See also DECAYING RIPPLE, GROWING RIPPLE, ORBITAL RIPPLE, ROLLING GRAIN RIPPLE, VORTEX RIPPLE, POST-VORTEX RIPPLE.

wave-ripple cross-lamination A CROSS-LAMINATION formed by the MIGRATION and/or AGGRADATION of WAVE-generated RIPPLES, characterized by unidirectional cross-laminae, lensoid and complexly interwoven CROSS-SETS, irregular, undulating CROSS-SET bases and LAMINAE DISCORDANT with the external RIPPLE form.

wavefront The surface over which the phase of a wave is the same and which propagates according to HUYGEN'S PRINCIPLE.

wavelength dispersive analysis An ELECTRON MICROPROBE technique, using a crystal SPECTROMETER and gas-filled proportional counters, which provides low detection limits and a high resolution between elements.

wavellite $(Al_3(PO_4)_2(OH)_3.5H_2O)$ A SEC-ONDARY MINERAL found in low GRADE META-MORPHIC ROCKS.

wavenumber The number of cycles per unit distance for a spatial waveform.

waves Regular oscillations of the water surface of oceans and lakes resulting mainly from wind-generated differences in pressure.

wavy bedding A type of HETEROLITHIC BEDDING comprising rippled SAND with continuous MUD draping the RIPPLES.

waxing slope A convex slope at the crest of a cliff. Cf. WANING SLOPE.

way-up See YOUNGING.

way-up criteria (way-up indicators) Phenomena used to determine the YOUNG-ING direction of STRATA, such as SOLE MARKS, CROSS-STRATIFICATION and GRADED BEDDING.

way-up indicators See WAY-UP CRITERIA.

Weald Clay A CRETACEOUS succession in England equivalent to the HAUTERIVIAN and BARREMIAN.

wear groove A feature on a FAULT surface caused by scratching by a hard particle.

weathered layer An heterogeneous surface layer up to tens of metres thick with an anomalously low SEISMIC VELOCITY caused by the presence of open JOINTS and MICRO-FRACTURES and its unsaturated state. Corrections are required for its effects on SEISMIC WAVE travel times.

weathering The disintegration and decomposition of rock and sediment by near-surface mechanical and chemical processes.

weathering front A transition zone between unweathered BEDROCK and the overlying SAPROLITE of a DEEP WEATHERING profile, which may be at depths of up to 100 m in the tropics.

weathering index A measure of the intensity of chemical WEATHERING in which comparison is made between a material which is stable with respect to the material considered.

weathering rind A layer of partly WEATH-ERED rock a few centimetres thick which forms the surface of a BOULDER or OUTCROP. Often yellow, orange or red due to the OXI-DATION of iron MINERALS.

weathering series A sequence of common silicate MINERALS arranged in order of their susceptibility to chemical WEATHERING.

websterite A PYROXENITE with >95% ORTHOPYROXENE and CLINOPYROXENE (each >10%). Occurs within ULTRAMAFIC intrusions and as ULTRAMAFIC XENOLITHS in BASALT.

wehrlite 1 A PERIDOTITE with >95% CLINO-PYROXENE and OLIVINE (>40%) and minor (<5%) ORTHOPYROXENE. Occurs within ULTRAMAFIC intrusions and as ULTRAMAFIC XENOLITHS in BASALT. **2** See TELLURIC BISMUTH.

Weichert-Herglov inversion A method of deriving the SEISMIC VELOCITY-depth distribution within the Earth from the variation of ARRIVAL TIMES with RANGE for DIVING SEISMIC WAVES.

Weichsel Glaciation The fourth of the major GLACIATIONS affecting northern Europe in the QUATERNARY.

Weiss zone law (addition rule) A relationship defining the condition that a crystal plane is parallel to, or a face lies in, a given direction.

Weissenberg camera A technique of X-RAY DIFFRACTION ANALYSIS in which the X-ray beam is incident normal to the axis of rotation of a crystal and GONIOMETER. A cylindrical film concentric with the rotation axis oscillates through a large angle and screens limit the diffracted beams to those from the zero layer of the reciprocal axis. Different beam inclinations allow study of other lattice layers.

Weissliegendes A PERMIAN succession in NW Europe covering the lower part of the KUNGURIAN.

welded tuff A VOLCANIC ROCK made up of

explosively fragmented TEPHRA which retained sufficient heat on deposition for the GLASS shards and PUMICE CLASTS to deform and adhere to produce a compact, LAVA-like rock.

well A dug or drilled hole which yields a fluid.

well logging See GEOPHYSICAL BOREHOLE LOGGING.

well loss The part of the DRAWDOWN which results from water flowing into a WELL across the WELL face.

well shooting A method used to determine the SEISMIC VELOCITY of STRATA penetrated by a BOREHOLE by exploding a SHOT adjacent to it and measuring ARRIVAL TIMES at a string of GEOPHONES down the WELL.

Wengen A TRIASSIC succession in the Alps covering the middle part of the LADINIAN.

Wenlock An EPOCH of the SILURIAN, 430.4–424.0 Ma.

Wenner configuration A common electrode arrangement in the RESISTIVITY METHOD in which the distance between adjacent electrodes is constant.

Wentworth-Udden scale A scale used for CLAST size classification.

Werfen A TRIASSIC succession in the Alps equivalent to the SCYTHIAN.

wernerite Members of the SCAPOLITE SERIES intermediate in composition between MARIALITE and MEIONITE.

Werra A PERMIAN succession in NW Europe covering the upper part of the UFIMIAN.

western states-type deposit See SANDSTONE-URANIUM-VANADIUM BASE METAL DEPOSIT.

Westphalian A CARBONIFEROUS STAGE in NW Europe covering part of the BASHKIRIAN and the MOSCOVIAN.

westward drift The westwards motion, at ~0.2° a^{-1}, of most components of the GEOMAGNETIC NON-DIPOLE FIELD, probably originating in the passage of magnetic lines of force through the outer CORE.

wet chemical analysis A classical method of chemical analysis in which each element is determined by a distinct, appropriate method, usually involving COLORIMETRY or TITRATION.

wet gas NATURAL GAS containing significant amounts of gases other than METHANE. Cf. DRY GAS.

wet-sediment deformation (soft-sediment deformation) The post- or syn-depositional LIQUEFACTION and/or FLUIDIZATION of sediment, identified where STRAIN has altered the geometry of primary sediment LAMINATION, BEDDING, FABRIC or TEXTURE.

Wettersteinkalk A TRIASSIC succession in the Alps equivalent to the LADINIAN.

Whaingaroan An OLIGOCENE succession in New Zealand covering the RUPELIAN and part of the CHATTIAN.

whetstone (honestone) A sharpening stone for a metal blade, typically rectangular in shape. SCHIST and SANDSTONE were used in Britain from the Bronze Age; other utilized rocks include BASALT and some LIMESTONES.

whin stone A popular term for any dark, fine-grained IGNEOUS ROCK.

white copperas See GOSLARITE.

white corundum See WHITE SAPPHIRE.

white iron pyrites See MARCASITE.

white lead ore See CERUSSITE.

white mica See MUSCOVITE.

white nickel A popular name for CHLOANTHITE.

white sapphire (leucosapphire, white corundum) A pure, colourless variety of CORUNDUM used as a GEM.

white smoker A plume of relatively low-temperature HYDROTHERMAL FLUID from an ocean floor vent producing a white cloud of sulphate MINERALS. Cf. BLACK SMOKER.

white vitriol A popular name for GOSLARITE.

Whiterockian Series A SERIES of the ORDOVICIAN in North America.

whitings A suspension of ARAGONITE needles forming milky-white patches tens to hundreds of metres long in sub-tropical oceans. Of uncertain origin.

whitlockite $((Ca,Mg)_3(PO_4)_2)$ Calcium phosphate, found as a SECONDARY MINERAL in GRANITE PEGMATITES and in phosphate deposits.

Whitwellian A STAGE of the SILURIAN, 426.1–425.4 Ma.

Widmanstätten pattern A pattern of intersecting LAMELLAE on the polished, acid-etched surface of an IRON METEORITE, formed by the EXSOLUTION of TAENITE and KAMACITE from a nickel-IRON alloy during very slow cooling.

Wiener filter A filter that converts the input signal into an output which comes closest, by a least squares criterion, to some desired form.

Wilcox An EOCENE succession on the Gulf Coast of the USA covering the YPRESIAN and part of the LUTETIAN.

wildcat well A speculative exploratory WELL.

wildflysch A TURBIDITE-like, MASS FLOW MICTITE containing numerous exotic CLASTS.

willemite (Zn_2SiO_4) A NESOSILICATE sometimes occurring as an ORE MINERAL.

Wilson cycle A term referring to the sequence in the life of an ocean, from CONTINENTAL SPLITTING through ocean growth by SEAFLOOR SPREADING to its destruction following SUBDUCTION in a CONTINENT-CONTINENT COLLISION.

wind ripple A small-scale, AEOLIAN bedform created by wind acting on dry SAND or GRANULES. Wavelengths may reach 20 m and heights 1 m, and are significantly smaller than the dimensions of a DUNE. See also IMPACT RIPPLE, AERODYNAMIC RIPPLE.

wind ripple lamination A low-angle LAMINATION in a SAND SHEET deposit punctuated by subhorizontal EROSION surfaces.

window An area where EROSION has cut through a THRUST or RECUMBENT FOLD to expose the rocks beneath.

winnowing The removal of sedimentary particles by the wind or water currents.

wire-line logging See GEOPHYSICAL BORE-HOLE LOGGING.

Wisconsin Glaciation The fifth of the major GLACIATIONS affecting North America in the QUATERNARY.

withamite A variety of PIEMONTITE.

witherite $(BaCO_3)$ A rare MINERAL found associated with GALENA in HYDROTHERMAL VEINS.

within-plate basalt (WPB) A BASALT formed within a continental or oceanic PLATE.

Wolfcampian A CARBONIFEROUS/PERMIAN succession in the Delaware Basin, USA covering part of the GZELIAN, the ASSELIAN and part of the SAKMARIAN.

wolframite $((Fe,Mn)WO_4)$ The major ORE MINERAL of tungsten, found mainly in high-temperature ORE and QUARTZ VEINS in or near GRANITIC rocks, in medium-temperature ORE deposits and in PLACER DEPOSITS.

wollastonite $(CaSiO_3)$ A PYROXENOID mainly formed by the CONTACT METAMORPHISM of LIMESTONES.

Wonokan A STAGE of the VENDIAN, 590–580 Ma.

wood opal Wood fossilized by PETRIFACTION with OPAL.

wood tin A variety of CASSITERITE with a fibrous appearance.

Woodbine A CRETACEOUS succession on the Gulf Coast of the USA covering part of the CENOMANIAN.

Word A PERMIAN succession in the Delaware Basin, USA equivalent to the WORDIAN.

Wordian A STAGE of the PERMIAN, 255.0–252.5 Ma.

work hardening See STRAIN HARDENING.

work softening See STRAIN SOFTENING.

World-Wide Standardized Seismograph Network (WWSSN) A network designed and installed in the early 1960s to provide the means of monitoring underground nuclear explosions. The first global network utilizing common SEISMOMETER design and also used in the monitoring of EARTHQUAKES. Now superseded by the FEDERATION OF DIGITAL SEISMIC NETWORKS.

WPB See WITHIN-PLATE BASALT.

wrench fault See TRANSCURRENT FAULT.

wrench fault system See STRIKE-SLIP FAULT SYSTEM.

wrinkle mark See RUNZELMARKEN.

Wujiapping A PERMIAN succession in China equivalent to the LONGTANIAN.

wulfenite ($PbMoO_4$) An orange-red molybdate MINERAL found in the oxidized parts of LEAD VEINS.

Wulff net (equal angle net) A type of STEREOGRAPHIC NET in which equal angles on the surface of a sphere project as equal distances on the net. Used in calculating angular relationships between planes and lines. Cf. SCHMIDT NET.

Würm Glaciation The fourth of the major GLACIATIONS affecting the Alps in the QUATERNARY.

wurtzite (ZnS) A POLYMORPH of SPHALERITE found mainly in HYDROTHERMAL VEIN deposits.

wurtzite group A group of SULPHIDE MINERALS with a structure similar to the SPHALERITE GROUP, but with the arrangement of tetrahedra such that a lattice with hexagonal rather than cubic symmetry is formed.

WWSSN See WORLD-WIDE STANDARDIZED SEISMOGRAPH NETWORK.

Wyandot A CRETACEOUS succession on the Scotian shelf of Canada covering part of the SANTONIAN, the CAMPANIAN and part of the MAASTRICHTIAN.

wyomingite An alkaline VOLCANIC ROCK comprising LEUCITE, PHLOGOPITE and DIOPSIDE.

X

X-ray crystallography The study of CRYSTAL STRUCTURE using X-ray techniques.

X-ray diffraction analysis (XRD) A method of instrumental analysis making use of the DIFFRACTION of X-rays by crystalline materials. X-ray wavelengths are effective as they are of the same order of size as the atom spacings in crystals (~0.1 nm). X-rays are scattered by the electronic charges surrounding atomic nuclei and diffracted according to BRAGG'S LAW where they are detected by sensors at appropriate angles to provide information on the CRYSTAL STRUCTURE.

X-ray fluorescence analysis (XRF) A method of instrumental analysis making use of the FLUORESCENCE of crystalline materials by X-rays.

xalostocite A pale rose-pink variety of GROSSULAR.

xanthophyllite See CLINTONITE.

xeno- Foreign, strange or different.

xenoblastic texture A TEXTURE in a META-MORPHIC ROCK in which MINERAL grains have developed without showing crystal faces.

xenocryst A foreign crystal inclusion in an IGNEOUS ROCK. Cf. AUTOLITH.

xenolith A foreign rock inclusion in an IGNEOUS ROCK. Cf. AUTOLITH.

xenomorphic With no crystal form.

xenotime (YPO$_4$) Yttrium phosphate resembling ZIRCON, often with small amounts of cerium, erbium and thorium, found in GRANITES and PEGMATITES as an ACCESSORY MINERAL.

xenotopic fabric A FABRIC of a crystalline carbonate rock or cement, or an EVAPORITE, in which most of the crystals are ANHEDRAL.

Xinchang An ORDOVICIAN succession in China covering part of the TREMADOC.

Xiphosura A class of subphylum CHE-LICERATA, phylum ARTHROPODA; diverse ARTHROPODS with a large prosoma, a variable number of opisthosomal somites and a long pointed telson. Range SILURIAN–RECENT.

xonotlite (Ca$_6$Si$_6$O$_{17}$(OH)$_2$) A hydrous calcium silicate found mainly as small VEINS in SERPENTINE or CONTACT METAMORPHIC zones.

XRD See X-RAY DIFFRACTION ANALYSIS.

XRF See X-RAY FLUORESCENCE ANALYSIS.

Xuzhuang A CAMBRIAN succession in China covering parts of the SOLVAN and MENEVIAN.

Y shear A subsidiary FAULT in a SHEAR ZONE parallel to the SHEAR direction.

Yangyuan A CARBONIFEROUS succession in China equivalent to the TOURNAISIAN.

Yapeenian An ORDOVICIAN succession in Australia covering part of the LATE ARENIG.

yardang An elongate landform resembling the hull of an inverted boat sculpted by wind EROSION from weakly consolidated rocks.

yazoo A tributary stream running parallel to the main stream for some distance.

Yeadonian A STAGE of the CARBONIFEROUS, 320.6–318.3 Ma.

yellow ground Weathered and oxidized KIMBERLITE at the surface. Cf. BLUE GROUND.

yellow quartz See CITRINE.

yellow tellurium See SYLVANITE.

yellowcake Concentrated, precipitated and dried uranium oxide.

Yerkebidaik An ORDOVICIAN succession in Kazakhstan covering part of the COSTONIAN, the HARNAGIAN and part of the SOUDLEYAN.

Yichang (Ichang) An ORDOVICIAN succession in China comprising the XINCHANG and NINGGUO.

yield criterion The relation between the PRINCIPAL STRESSES at YIELD STRESS in the DEFORMATION of a PLASTIC material.

yield depression curve A graph of water extraction rate from a WELL against DRAWDOWN for different pumping rates. Used to assess the rate for minimum DRAWDOWN which uses least energy in overcoming HEAD LOSS.

yield point See ELASTIC LIMIT.

yield strength See STRENGTH.

yield stress The critical value of STRESS above which continuous, permanent STRAIN results in a PLASTIC material.

yield surface A representation of the YIELD CRITERION on a diagram showing the three PRINCIPAL STRESSES which define the STRESS conditions at which yield occurs.

Ynezian A PALAEOCENE succession on the west coast of the USA covering parts of the DANIAN and THANETIAN.

yoderite $((Al,Mg,Fe)_2Si(O,OH)_5)$ A rare iron-magnesium silicate found in SCHIST.

Yongningzhen A TRIASSIC succession in China covering part of the NAMMALIAN and the SPATHIAN.

Yoredale facies A British name for the CYCLOTHEM sequence LIMESTONE, SHALE, SANDSTONE, COAL produced in a sinking DELTA environment during the CARBONIFEROUS.

Young's modulus An ELASTIC MODULUS defined as the ratio of applied STRESS to the resultant EXTENSION along the axis of a plane-ended cylinder.

younging The property of STRATA by which they become STRATIGRAPHICALLY younger in a certain direction, which can be determined from WAY-UP CRITERIA. See also FACING.

Ypresian A STAGE of the EOCENE, 56.5–50.0 Ma.

yttrocerite A rare, massive, GRANULAR or earthy variety of cerian fluorite.

yu-stone A Chinese name for JADE of GEM quality.

yugawaralite $(Ca_2(Al_4Si_{12}O_{32}).8H_2O)$ A ZEOLITE found in networks and VEINS and as crystals in cavities in ANDESITE TUFFS which have been altered by waters from HOT SPRINGS.

Yumatin A PERIOD of the RIPHEAN, 1350–1050 Ma.

Z

z fold An ASYMMETRICAL FOLD with one LIMB shorter than the other whose profile defines a 'Z' shape. Cf. S FOLD.

Zahorany An ORDOVICIAN succession in Bohemia covering the upper part of the SOUDLEYAN.

Zanclian The lower STAGE of the PLIOCENE, 5.2–3.40 Ma.

Zechstein The younger EPOCH of the PERMIAN, 256.1–245.0 Ma.

Zechstein Sea A sea of restricted circulation covering parts of E. England, the Low Countries, Germany and North Sea in U. PERMIAN times, whose evaporation gave rise to thick EVAPORITE deposits.

Zechsteinkalk A PERMIAN succession in NW Europe covering the lower part of the UFIMIAN.

Zemorrian A MIOCENE succession on the west coast of the USA covering the OLIGOCENE and part of the AQUITANIAN.

zeolite facies A METAMORPHIC FACIES characterized by the development of SMECTITE ZEOLITES in BASIC IGNEOUS ROCKS at low temperatures and pressures.

zeolites Hydrated aluminosilicate MINERALS with a framework structure which encloses cavities occupied by large ions and water molecules, which both have considerable mobility, allowing ion exchange and reversible dehydration. Generally white to colourless in the pure state but often coloured by the presence of IRON OXIDES or other impurities. Occur in AMYGDALES and fissures in BASIC VOLCANIC ROCKS, as AUTHIGENIC MINERALS in SEDIMENTARY ROCKS, in TUFFS and in low GRADE METAMORPHIC ROCKS as a result of HYDROTHERMAL ALTERATION.

zeolitization A type of WALL ROCK ALTERATION in which STILBITE, NATROLITE, HEULANDITE and other ZEOLITES are developed.

zero length spring A special type of spring used in modern GRAVIMETERS which is pretensioned during manufacture so that the restoring force is proportional to its length, thus increasing the range and sensitivity of the instrument.

zeuge A TABULAR rock mass perched on a pinnacle of softer rock as the result of differential EROSION of the underlying rock.

Zhangxia A CAMBRIAN succession in China covering part of the MENEVIAN.

Zharyk An ORDOVICIAN succession in Kazakhstan covering the upper part of the ONNIAN.

Zi-Liu-Jing A JURASSIC succession in Sichuan, China, equivalent to the SINEMURIAN, PLIENSBACHIAN, TOARCIAN and AALENIAN.

zibar A coarse-grained, low relief (<10 m high), AEOLIAN depositional feature with no SLIP-FACES, found in regular spacings of up to 400 m and forming undulating surfaces on SAND SHEETS and in INTERDUNES.

zig-zag fold See CHEVRON FOLD.

zinc bloom See HYDROZINCITE.

zinc spinel See GAHNITE.

zincblende See SPHALERITE.

zincite (red zinc ore) (ZnO) An ORE

MINERAL of zinc found associated with FRANKLINITE, WILLEMITE and CALCITE in zinc deposits.

Zingg diagram A diagram used to plot the relative dimensions of a particle in terms of its maximum, intermediate and minimum dimensions.

zinkenite $(PbSb_2S_4)$ A steel-grey, lead-antimony sulphide occurring as columnar, often very thin, hexagonal crystals forming a fibrous mass in low- to medium-temperature VEIN deposits.

zinnwaldite $(K(Li,Al,Fe)_3(Al,Si)_4O_{10}(OH,F)_2)$ A grey to brown MICA similar to LEPIDOLITE containing appreciable iron, found in TIN VEINS and GRANITE PEGMATITES.

zircon $(ZrSiO_4)$ A NESOSILICATE, a common ACCESSORY MINERAL in IGNEOUS ROCKS and used as a GEM when transparent.

Zlichovian A DEVONIAN succession in the former Czechoslovakia covering the lower part of the EMSIAN.

Zoantharia A subclass of class ANTHOZOA; solitary or COLONIAL CORALS with polyps bearing multiples of six tentacles. Range M. TRIASSIC–RECENT.

Zoeppritz' equations Formulae which describe how the SEISMIC WAVE energy of a P or S WAVE is partitioned between reflected, transmitted and mode-converted waves incident on a planar discontinuity.

zoisite $(Ca_2Al_3O(SiO_4)(Si_2O_7OH))$ A SOROSILICATE MINERAL, an orthorhombic POLYMORPH of CLINOZOISITE, found in METAMORPHIC ROCKS.

zonal index fossil A FOSSIL with a wide geographic range and short temporal range used in BIOSTRATIGRAPHY.

zonation The subdivision of a STRATIGRAPHIC unit by means of FOSSILS.

zone 1 A BIOSTRATIGRAPHICAL subdivision, several of which comprise a STAGE. **2** A set of crystal faces all parallel to the ZONE AXIS.

zone axis The direction to which the crystal faces of a ZONE are parallel.

zone fossil A FOSSIL species characterizing a ZONE.

zone of aeration See VADOSE ZONE.

zone of fluctuation The zone between the highest and lowest levels reached by the WATER TABLE.

zone of oxidation 1 The zone in the upper part of a sulphide OREBODY in which copper, zinc and silver sulphides are soluble and the OREBODY is OXIDIZED and LEACHED of many of its valuable elements down to the WATER TABLE, leaving a GOSSAN. **2** Any zone in which OXIDATION occurs.

zone of permanent saturation See PHREATIC ZONE.

zone refining A process of MAGMA DIVERSIFICATION during PARTIAL MELTING in which a succession of zones of molten or partially molten MAGMA pass through a rock by causing melting at the front of each zone followed by CRYSTALLIZATION and deposition behind. Certain elements are preferentially included in the melt and carried forward by each zone.

zoned crystal A crystal with concentric zones of varying composition, reflecting a complex CRYSTALLIZATION history.

zoning of ore deposits The zoning of the MINERALOGY or chemistry of ORE, INDUSTRIAL or GANGUE MINERALS in a MINERAL deposit or district.

Zooxanthellae Symbiotic ALGAE found in the endoderm of REEF CORALS.

Zosterophyllopsida A class of division TRACHAEOPHYTA, kingdom plantae; the ancestors of microphyllous plants. Range U. SILURIAN–U. DEVONIAN.

zunyite $(Al_{13}Si_5O_{20}(OH,F)_{18}Cl)$ A colourless to grey to pink MINERAL found in VEINS, disseminated in PORPHYRIES and as an ALTERATION product of FELDSPAR.

zussmanite $(K(Fe,Mg,Mn)_{13}(Si,Al)_{18}O_{42}(OH)_{14})$ A pale green, TABULAR, hydrated iron-magnesium-potassium silicate found in METAMORPHOSED SHALE, siliceous IRON-STONE and impure LIMESTONE.

zweikanter A VENTIFACT with two curved surfaces intersecting at a sharp edge. Cf. EINKANTER, DREIKANTER.

Table 1 **The Stratigraphic Column** From Harland, W. B., Armstrong, R. L., Cox, A. V., Craig, L. E., Smith, A. G. and Smith, D. G. (1990), *A Geologic Timescale 1990*. Cambridge University Press, Cambridge.

Era	*Sub-era Period Sub-Period*	*Epoch*		*Stage*	*Age Ma*	*Stage abbr.*	*Intervals Ma*	
Cenozoic	Quaternary or Pleistogene 1.64	Holocene			0.01	Hol	0.01	
		Pleistocene			1.64	Ple	1.63	
	Neogene	Pliocene	2	Piacenzian	3.40	Pia	3.6	22
		Pli	1	Zanclian	5.2	Zan		
			3	Messinian	6.7	Mes	5.2	
				Tortonian	10.4	Tor		
		Miocene	2	Serravallian	14.2	Srv	5.9	
				Langhian	16.3	Lan		
	22 Ng	18.1 Mio	1	Burdigalian	21.5	Bur	7.0	
				Aquitanian	23.3	Aqt		
	Paleogene	Oligocene	2	Chattian	29.3	Cht	6.0	42
		12.1 Oli	1	Rupelian	35.4	Rup	6.1	
		Eocene	3	Priabonian	38.6	Prb	3.2	
			2	Bartonian	42.1	Brt	11.4	
				Lutetian	50.0	Lut		
		21.1 Eoc	1	Ypresian	56.5	Ypr	6.5	
		Paleocene	2	Thanetian	60.5	Tha	4.0	
65 Cz	TT 42 Pg	8.5 Pal	1	Danian	65.0	Dan	4.5	
Mesozoic		Gulf K₂ Senonian		Maastrichtian	74.0	Maa	9.0	
				Campanian	83.0	Cmp	9.0	
				Santonian	86.6	San	3.6	
		23.5 Sen		Coniacian	88.5	Con	1.9	

Geological time scale — Mesozoic

Era	Period	Series	Subseries	Stage	Abbr.	Age (Ma)	Dur.	Total
Mesozoic (Mz, 180)	Cretaceous (K)	K₁ (32 Gul) — Gallic (43.3 Gal)		Turonian	Tur	90.4	1.9	81
				Cenomanian	Cen	97.0	6.6	
				Albian	Alb	112.0	15.0	
				Aptian	Apt	124.5	12.5	
				Barremian	Brm	131.8	7.3	
		(59) — Neocomian (13.8 Neo)		Hauterivian	Hau	135.0	3.2	
				Valanginian	Vlg	140.7	5.7	
				Berriasian	Ber	145.6	4.9	
	Jurassic (J)	Malm (11.5) — J₃ Mal		Tithonian	Tth	152.1	6.5	62
				Kimmeridgian	Kim	154.7	2.6	
				Oxfordian	Oxf	157.1	2.4	
		Dogger (20.9) — J₂ Dog		Callovian	Clv	161.3	4.2	
				Bathonian	Bth	166.1	4.8	
				Bajocian	Baj	173.5	7.4	
				Aalenian	Aal	178.0	4.5	
		Lias (30.0) — J₁ Lia		Toarcian	Toa	187.0	9.0	
				Pliensbachian	Plb	194.5	7.5	
				Sinemurian	Sin	203.5	9.0	
				Hettangian	Het	208.0	4.5	
	Triassic (Tr)	Tr₃ (27)		Rhaetian	Rht	209.5	1.5	37
				Norian	Nor	223.4	13.9	
				Carnian	Crn	235.0	11.6	
		Tr₂ (6)		Ladinian	Lad	239.5	4.5	
				Anisian	Ans	241.1	1.6	
		Scythian (4) — Tr₁ Scy		Spathian	Spa	241.9	0.8	
				Nammalian	Nml	243.4	1.5	
				Griesbachian	Gri	245.0	1.6	
				Changxingian	Chx	247.5	2.5	

Era	System	Subsystem	Group		Group (Ma)	Stage	Age (Ma)	Abbr.	Duration	Total
Paleozoic			Zechstein	Zec	11	Longtanian	250.0	Lgt	2.5	45
	Permian (P)					Capitanian	252.5	Cap	2.5	
						Wordian	255.0	Wor	2.5	
			Rotliegendes	Rot	34	Ufimian	256.1	Ufi	1.1	
						Kungurian	259.7	Kun	3.6	
						Artinskian	268.8	Art	9.1	
						Sakmarian	281.5	Sak	12.7	
						Asselian	290.0	Ass	8.5	
	Carboniferous	Pennsylvanian (C₂)	Gzelian	Gze	5	Noginskian	293.6	Nog	3.6	33
						Klazminskian	295.1	Kla	1.5	
			Kasimovian	Kas	8	Dorogomilovskian	298.3	Dor	3.2	
						Chamovnicheskian	299.9	Chv	1.6	
						Krevyakinskian	303.0	Kre	3.1	
			Moscovian	Mos	8	Myachkovskian	305.0	Mya	2.0	
						Podolskian	307.1	Pod	2.1	
						Kashirskian	309.2	Ksk	2.1	
						Vereiskian	311.3	Vrk	2.1	
			Bashkirian	Bsh	12	Melekesskian	313.4	Mel	2.1	
						Cheremshanskian	318.3	Che	4.9	
						Yeadonian	320.6	Yea	2.3	
						Marsdenian	321.5	Mrd	0.9	
						Kinderscoutian	322.8	Kin	1.3	
		Mississippian	Serpukhovian	Spk	10	Alportian	325.6	Alp	2.8	40
						Chokierian	328.3	Cho	2.7	
						Arnsbergian	331.1	Arn	2.8	
						Pendleian	332.9	Pnd	1.8	
						Brigantian	336.0	Bri	3.1	
						Asbian	339.4	Asb	3.4	

Era	Period	Series	Epoch	Stage	Abbr	Duration	Age (Ma)	Period total
Paleozoic	Devonian	Visean (Vis) 17		Holkerian	Hlk	3.4	342.8	46
				Arundian	Aru	2.2	345.0	
				Chadian	Chd	4.5	349.5	
		D₃ 15		Famennian	Fam	4.5	367.0	
				Frasnian	Frs	10.4	377.4	
		D₂ 9		Givetian	Giv	3.4	380.8	
				Eifelian	Elf	5.2	386.0	
		D₁ 22		Emsian	Ems	4.4	390.4	
				Pragian	Pra	5.9	396.3	
			D	Lochkovian	Lok	12.2	408.5	
	Silurian	Pridoli (Prd)	S₄		Prd	2.2	410.7	31
		Ludlow (Lud) 13	S₃	Ludfordian	Ldf	4.4	415.1	
				Gorstian	Gor	8.9	424.0	
		Wenlock (Wen) 6.5	S₂	Gleedonian	Gle	1.4	425.4	
				Whitwellian	Whi	0.7	426.1	
				Sheinwoodian	She	4.3	430.4	
		Llandovery (Lly) 8.5	S₁	Telychian	Tel	2.2	432.6	
				Aeronian	Aer	4.3	436.9	
			S	Rhuddanian	Rhu	2.1	439.0	
	Ordovician	Bala — Ashgill (Ash)		Hirnantian	Hir	0.5	439.5	71
				Rawtheyan	Raw	0.6	440.1	
				Cautleyan	Cau	0.5	440.6	
				Pusgillian	Pus	2.5	443.1	
		Bala — Caradoc 4		Onnian	Onn	0.9	444.0	
				Actonian	Act	0.5	444.5	
				Marshbrookian	Mrb	2.6	447.1	
				Longvillian	Lon	2.6	449.7	
				Soudleyan	Sou	7.8	457.5	

Eon/Era	Period	Epoch/Series	Abbr.	Stage	Abbr.	Period total	Duration	Age (Ma)
Paleozoic	Ordovician (325 / Pz)		Bal	Harnagian	Har		4.8	462.3
			21 / Crd	Costonian	Cos		1.6	463.9
		Llandeilo / Llo (4.5)	Dyfed	Late	Llo 3		1.5	465.4
				Mid	Llo 2		1.6	467.0
				Early	Llo 1		1.6	468.6
		Llanvirn / Lln (7.5)	Dfd	Late	Lln 2		4.1	472.7
				Early	Lln 1		3.4	476.1
		Arenig / Canadian	Cnd		Arg		17.0	493.0
		Tremadoc			Tre		17.0	510.0
	Cambrian (€)	Merioneth / Mer (7)		Dolgellian	Dol	60	4.1	514.1
				Maentwrogian	Mnt		3.1	517.2
		St David's / StD (19)		Menevian	Men		13.0	530.2
				Solvan	Sol		5.8	536.0
		Caerfai / Crf (34)		Lenian	Len		18.0	554
				Atdabanian	Atb		6.0	560
				Tommotian	Tom		10.0	570
	Vendian / Sinian (230) (V)	Ediacara / Edi (20)		Poundian	Pou	40	10.0	580
				Wonokan	Won		10.0	590
		Varanger / Var (20)		Mortensnes	Mor		10.0	600
				Smalfjord	Sma		10.0	610
	Sturtian (Z)				Stu		190	800
	Riphean (850)	Karatau			Kar		250	1050
		Yurmatin			Yur		300	1350
		Buzyan / Rif			Buz		300	1650
	Animikean				Ani		550	2200
	Huronian				Hur		250	2450
	Randian				Ran		350	2800
	Swazian				Swz		700	3500

Isuan						3800	Isu	300
					Early Imbrian	3850	Imb	50
					Nectarian	3950	Nec	100
					Basin Groups 1-9	4150	BG1-9	200
	Hadean	760	Hde		Cryptic	4560	Cry	410

Table 2 **SI, Cgs and Imperial (Customary USA) Units and Conversion Factors**

Quantity	SI name	SI Symbol	cgs equivalent	Imperial (USA) equivalent
Mass	kilogram	kg	10^3 g	2.205 lb
Time	second	s	s	s
Length	metre	m	10^2 cm	39.37 in
				3.281 ft
Acceleration	metre s^{-2}	$m \, s^{-2}$	10^2 cm $s^{-2} = 10^2$ gal	39.37 in s^{-2}
Gravity	gravity unit	gu = $\mu m \, s^{-2}$	10^{-1} milligal (mgal)	3.937×10^{-5} in s^{-2}
Density	megagram m^{-3}	$Mg \, m^{-3}$	g cm^{-3}	3.613×10^{-2} lb in^{-3}
				62.421 lb ft^{-3}
Force	newton	N	10^5 dyne	0.2248 lb (force)
Pressure	pascal	Pa = N m^{-2}	10 dyne $cm^{-2} = 10^{-5}$ bar	1.45×10^{-4} lb in^{-2}
Energy	joule	J	10^7 erg	0.7375 ft lb
Power	watt	W = J s^{-1}	10^7 erg s^{-1}	0.7375 ft lb s^{-1}
				1.341×10^{-3} hp
Temperature	T	°C*	°C	(1.8T + 32)°F
Current	ampere	A	A	A
Potential	volt	V	V	V
Resistance	ohm	$\Omega = VA^{-1}$	Ω	Ω
Resistivity	ohm m	Ω m	$10^2 \, \Omega$ cm	3.281 ohm ft
Conductance	siemen	$S = \Omega^{-1}$	mho	mho
Conductivity	siemen m^{-1}	$S \, m^{-1}$	10^{-2} mho cm^{-1}	0.3048 mho ft^{-1}
Dielectric constant	dimensionless			
Magnetic flux	weber	Wb = Vs	10^8 maxwell	
Magnetic flux density (B)	tesla	T = Wb m^{-2}	10^4 gauss (G)	
Magnetic anomaly	nanotesla	nT = 10^{-9} T	gamma (γ) = 10^{-5} G	
Magnetizing field (H)	ampere m^{-1}	Am^{-1}	$4\pi 10^{-3}$ oersted (Oe)	
Inductance	henry	H = Wb A^{-1}	10^9 emu (electromagnetic unit)	
Permeability of vacuum ($\mu 0$)	henry m^{-1}	$4\pi 10^{-7} H \, m^{-1}$	1	
Susceptibility	dimensionless	k	4π emu	
Magnetic pole strength	ampere m	A m	10 emu	
Magnetic moment	ampere m^2	A m^2	10^3 emu	
Magnetization (J)	ampere m^{-1}	A m^{-1}	10^{-3} emu cm^{-3}	
Viscosity (J)	Pascal s	Pa s	10 poise (η)	

*Strictly, SI temperatures should be stated in Kelvin (K = 273.15 + °C). In this book, however, temperatures are given in the more familiar Centrigrade (Celsius) scale.

Table 3 SI, Very Large and Very Small Numbers

10^{-18}	atto	a	attometre (am)
10^{-15}	femto	f	femtometre (fm)
10^{-12}	pico	p	picometre (pm)
10^{-9}	nano	n	nanometre (nm)
10^{-6}	micro	μ	micrometre (μm)
10^{-3}	milli	m	millimetre (mm)
1	*no prefix*		metre (m)
10^{3}	kilo	k	kilometre (km)
10^{6}	mega	M	megametre (Mm)
10^{9}	giga	G	gigametre (Gm)
10^{12}	tera	T	terametre (Tm)

Table 4 Unit Conversion

1 mil	$= 10^{-6} \, m^3$
1 litre	$= 10^{-3} \, m^3$
1 tonne	$= 10^3 \, kg$
1 bar	$= 10^5 \, Pa$
1 Å	$= 10^{-10} \, m$

Table 5 Facts about the Earth

Equatorial radius	6378 km
Polar radius	6357 km
Volume	$1.083 \times 10^{21} \, km^3$
Surface area	$5.1 \times 10^{14} \, km^2$
Percentage surface area of oceans	71
Percentage surface area of continents	29
Average continental elevation	0.623 km
Average oceanic depth	3.8 km
Mass	$5.976 \times 10^{24} \, kg$
Mean density	$5.517 \, Mg \, km^{-3}$
Mass of atmosphere	$5.1 \times 10^{18} \, kg$
Mass of ice	$25–30 \times 10^{18} \, kg$
Mass of oceans	$1.4 \times 10^{21} \, kg$
Mass of crust	$2.5 \times 10^{22} \, kg$
Mass of mantle	$4.05 \times 10^{24} \, km$
Mass of core	$1.90 \times 10^{24} \, kg$
Mean distance from Sun	$1.496 \times 10^8 \, km$

Bibliography

I am aware that some of the books listed below are out of print. However, they should still be available in most geology libraries. They are included to give as comprehensive a coverage as possible.

General

Allaby, A. and Allaby, M. (eds.) (1990) *The Concise Oxford Dictionary of Earth Sciences*. Oxford University Press, Oxford.

Bott, M. H. P. (1982) *The Interior of the Earth*. Arnold, London.

Brown, G. C. and Mussett, A. E. (1993) *The Inaccessible Earth*. Allen & Unwin, London.

Brown, G. C., Hawkesworth, C. J. and Wilson, R. C. L. (1992) *Understanding the Earth*. Cambridge University Press, Cambridge.

Compton, R. R. (1985) *Geology in the Field*. John Wiley & Sons, New York.

Emiliani, C. (1992) *Planet Earth*. Cambridge University Press, Cambridge.

Finkl, C. W. (ed.) (1984) *The Encyclopedia of Applied Geology*. Van Nostrand Reinhold, New York.

Finkl, C. W. (ed.) (1988) *The Encyclopedia of Field and General Geology*. Van Nostrand Reinhold, New York.

Harben, P. W. and Bates, R. L. (1984) *Geology of the Nonmetallics*. Metal Bulletin Inc., New York.

Jacobs, J. A. (1987) *The Earth's Core*. Academic Press, London.

James, D. E. (ed.) (1989) *Encyclopedia of Solid-Earth Geophysics*. Van Nostrand Reinhold, New York.

Kearey, P. (ed.) (1993) *The Encyclopedia of the Solid Earth Sciences*. Blackwell Scientific Publications, Oxford.

Kennett, J. (1982) *Marine Geology*. Prentice Hall, Englewood Cliffs, NJ.

Le Roy, L. W. and Le Roy, D. O. (eds.) (1977) *Subsurface Geology*. Colorado School of Mines, Golden.

Lunine, J. I. (1998) *Earth: Evolution of a Habitable World*. Cambridge University Press, Cambridge.

Moseley, F. (1981) *Methods in Field Geology*. W. H. Freeman, San Francisco.

Murck, B. (1998) *Geology Today*. Wiley, Chichester.

Press, F. and Siever, R. (1982) *Earth*. W. H. Freeman, San Francisco.

Scrutton, R. A. and Talwani, M. (eds.) (1982) *The Ocean Floor*. John Wiley & Sons, Chichester.

Skinner, B., Porter, S. and Botkin, D. (1999) *The Blue Planet: An Introduction to Earth System Science*. Wiley, Chichester.

Tarbuck, E. and Lutgens, F. (2000) *Earth Science*. Prentice Hall, Englewood Cliffs, NJ.

Taylor, S. R. and McClennan, S. M. (1985) *The Continental Crust: Its Composition and Evolution*. Blackwell Scientific Publications, Oxford.

Van Andel, T. H. (1994) *New Views on an Old Planet: A History of Global Change* (2nd edn). Cambridge University Press, Cambridge.

Economic Geology

Armstead, H. C. H. (1978) *Geothermal Energy.* Spon, London.

Armstead, H. C. H. and Tester, J. W. (1987) *Heat Mining.* Spon, London.

Barnes, H. L. (ed.) (1979) *Geochemistry of Hydrothermal Ore Deposits.* John Wiley & Sons, New York.

Berger, B. R. and Bethke, P. M. (eds.) (1986) *Geology and Geochemistry of Epithermal Systems.* Society of Economic Geologists, El Paso, TX.

Brooks, J. and Fleet, A. J. (eds.) (1987) *Marine Petroleum Source Rocks.* Blackwell Scientific Publications, Oxford.

Brooks, J. (ed.) (1983) *Petroleum Geology and Exploration of Europe.* Geological Society Special Publication **12**. Blackwell Scientific Publications, Oxford.

Butler, E. W. and Pick, J. B. (1982) *Geothermal Energy Development.* Plenum, New York.

Collins, A. G. (1975) *Geochemistry of Oilfield Waters.* Developments in Petroleum Science **1**. Elsevier, Amsterdam.

Downing, R. A. and Gray, D. A. (1986) *Geothermal Energy – The Potential in the United Kingdom.* British Geological Survey, HMSO, London.

Edwards, R. and Atkinson, K. (1986) *Ore Deposit Geology.* Chapman & Hall, London.

Erickson, A. J. Jr. (ed.) (1984) *Applied Mining Geology.* American Institute of Mining, Metallurgical and Petroleum Engineers, New York.

Evans, A. M. (1993) *Ore Geology and Industrial Minerals.* Blackwell Scientific Publications, Oxford.

Freidrich, G. H., Genkin, A. J., Naldrett, A. J. *et al.* (eds.) (1986) *Geology and Metallogeny of Copper Deposits.* Springer-Verlag, Berlin.

Gray, P. H. J., Bowyer, G. J., Castle, J. F., Vaughan, D. J. and Warner, N. A. (eds.) (1990) *Sulphide Deposits – their Origin and Processing.* Institute of Mineralogy and Metallurgy, London.

Guilbert, J. M. and Park, C. F. Jr. (1986) *The Geology of Ore Deposits.* W. H. Freeman, New York.

Hobson, G. D. (ed.) (1980) *Developments in Petroleum Geology.* Applied Science Publishers, London.

Hutchison, C. S. (1983) *Economic Deposits and Their Tectonic Setting.* Macmillan, London.

Jenson, M. L. and Bateman, A. M. (1979) *Economic Mineral Deposits.* John Wiley & Sons, New York.

Kesler, S. E. (1994) *Mineral Resources: Economics and the Environment.* Prentice Hall, Englewood Cliffs, NJ.

Macdonald, E. H. (1983) *Alluvial Mining.* Chapman & Hall, London.

Magara, K. (1978) *Compaction and Fluid Migration – Practical Petroleum Geology.* Developments in Petroleum Science. Elsevier, Amsterdam.

Maynard, J. B. (1983) *Geochemistry of Sedimentary Ore Deposits.* Springer-Verlag, New York.

North, F. K. (1985) *Petroleum Geology.* Allen & Unwin, Boston.

Peters, W. C. (1987) *Exploration and Mining Geology.* John Wiley & Sons, New York.

Roberts, W. H. and Cordell, R. J. (eds.) (1980) *Problems of Petroleum Migration.* Studies in Geology **10**. American Association of Petroleum Geologists, Tulsa, OK.

Stach, E., Mackowsky, M-Th., Teichmuller, M. *et al.* (1982) *Stach's Textbook of Coal Petrology.* Gebrüder Borntraeger, Berlin.

Tarling, D. H. (1981) (ed.) *Economic Geology and Geotectonics.* Blackwell Scientific Publications, Oxford.

Titley, S. R. (1982) *Advances in Geology of Porphyry Copper Deposits*. University of Arizona Press, Tucson, AZ.

Ward, C. R. (1984) *Coal Geology and Coal Technology*. Blackwell Scientific Publications, Melbourne.

Wolf, K. H. (ed.) (1976) *Handbook of Strata-bound and Stratiform Ore Deposits*. Elsevier, Amsterdam.

Engineering Geology and Hydrogeology

Anderson, J. G. C. and Trigg, C. F. (1976) *Case Histories in Engineering Geology*. Elek Sciences, London.

Atkinson, B. K. (ed.) (1987) *Fracture Mechanics of Rock*. Academic Press, New York.

Brassington, R. (1998) *Field Hydrogeology*. Wiley, Chichester.

Chow, V. T. (ed.) (1964) *Handbook of Applied Hydrogeology*. McGraw-Hill, New York.

Clayton, C. R. I., Symonds, N. E. and Matthews, M. C. (1982) *Site Investigation – A Handbook for Engineers*. Granada, London.

Fetter, C. W. (1994) *Applied Hydrogeology* (3rd edn). Prentice Hall, Englewood Cliffs, NJ.

Francis, J. R. D. (1975) *Fluid Mechanics for Engineering Students*. Arnold, London.

Jaeger, J. C. and Cook, N. G. (1979) *Fundamentals of Rock Mechanics*. Chapman & Hall, London.

Price, M. (1985) *Introducing Groundwater*. Allen & Unwin, London.

Rahn, P. H. (1997) *Engineering Geology: An Environmental Approach*. Prentice Hall, Englewood Cliffs, NJ.

Wang, H. F. and Anderson, M. P. (1995) *Introduction to Groundwater Modeling: Finite Difference and Finite Element Methods*. Academic Press, London.

West, T. R. (1995) *Geology Applied to Engineering*. Prentice Hall, Englewood Cliffs, NJ.

Geochemistry

Brownlow, A. H. (1996) *Geochemistry*. Prentice Hall, Englewood Cliffs, NJ.

Faure, G. (1998) *Principles and Applications of Geochemistry*. Prentice Hall, Englewood Cliffs, NJ.

Gill, R. (ed.) (1997) *Modern Analytical Geochemistry*. Longman, London.

Henderson, P. (1982) *Inorganic Geochemistry*. Pergamon, Oxford.

Iler, R. K. (1979) *Chemistry of Silica*. Wiley-Interscience, New York.

Krauskopf, K. B. (1979) *Introduction to Geochemistry*. Dowden, Hutchinson & Ross, Stroudsburg, PA.

Lasaga, A. C. and Kirkpatrick, R. J. (eds.) (1981) *Kinetics of Geochemical Processes*. Reviews in Mineralogy **8**. Mineralogical Society of America, Washington, DC.

Lerman, A. (1979) *Geochemical Processes: Water and Sediment Environments*. John Wiley & Sons, New York.

Newman, A. C. D. (ed.) (1987) *Chemistry of Clays and Clay Minerals*. Mineralogical Society Monograph **6**.

Saxena, S. K. (1973) *Thermodynamics of Rock-forming Crystalline Solutions*. Springer-Verlag, Berlin.

Selley, R. C. (1997) *Elements of Petroleum Geology*. Academic Press, London.

Sosman, R. B. (1965) *The Phases of Silica*. Rutgers University Press, New Brunswick.

Stumm, W. and Morgan, J. J. (1981) *Aquatic Chemistry*. John Wiley & Sons, New York.

Vaughan, D. J. and Craig, J. R. (1978) *Mineral Chemistry of Metal Sulphides*. Cambridge University Press, Cambridge.

Geomorphology

Bird, E. C. F. (1984) *Coasts*. Basil Blackwell, Oxford.

Bowden, K. F. (1983) *Physical Oceanography of Coastal Waters*. Ellis Horwood, Chichester.

Büdel, J. (1982) *Climatic Geomorphology*. Princeton University Press, Princeton, NJ.

Coates, D. R. (1981) *Environmental Geology*. John Wiley & Sons, New York.

Cooke, R. U. and Doornkamp, J. C. (1974) *Geomorphology and Environmental Management*. Clarendon Press, Oxford.

Drewry, D. (1986) *Glacial Geologic Processes*. Arnold, London.

Eyles, N. (ed.) (1983) *Glacial Geology*. Pergamon Press, Oxford.

Gardner, R. and Scoging, H. (eds.) (1983) *Mega-geomorphology*. Oxford University Press, Oxford.

Goudie, A. S. (1973) *Duricrusts in Tropical and Subtropical Landscapes*. Clarendon Press, Oxford.

Goudie, A. S. (1981) *Geomorphological Techniques*. Allen & Unwin, London.

Goudie, A. S. (1986) The *Human Impact on the Environment*. Basil Blackwell, Oxford.

Goudie, A. S. (ed.) (1985) *The Encyclopaedic Dictionary of Physical Geography*. Blackwell Scientific Publications, Oxford.

Goudie, A. S. and Pye, K. (eds.) (1983) *Chemical Sediments and Geomorphology*. Academic Press, London.

Greeley, R. and Iverson, J. D. (1985) *Wind as a Geological Process*. Cambridge University Press, Cambridge.

Jennings, J. M. (1985) *Karst Geomorphology*. Basil Blackwell, Oxford.

Johnson, A. M. (1970) *Physical Processes in Geology*. Freeman Cooper, San Francisco.

King, C. A. M. (1972) *Beaches and Coasts*. Arnold, London.

La Fleur, R. G. (ed.) (1984) *Groundwater as a Geomorphic Agent*. Allen & Unwin, Boston.

Lerman, A. (ed.) (1978) *Chemistry, Geology and Physics of Lakes*. Springer-Verlag, New York.

Nickling, W. G. (ed.) (1986) *Aeolian Geomorphology*. Unwin Hyman, Boston.

Paterson, K. and Sweeting, M. M. (eds.) (1986) *New Directions in Karst*. Geo Books, Norwich.

Pethick, J. S. (1984) *An Introduction to Coastal Geomorphology*. Arnold, London.

Pond, S. and Pickard, G. L. (1983) *Introductory Dynamical Oceanography*. Pergamon, Oxford.

Richards, K. S. (1982) *Rivers: Form and Process in Alluvial Channels*. Methuen, London.

Richards, K. S., Arnett, R. R. and Ellis, S. (eds.) (1984) *Geomorphology and Soils*. Allen & Unwin, London.

Schrumm, S. A. (1977) *The Fluvial System*. Wiley-Interscience, New York.

Selby, M. J. (1985) *The Earth's Changing Surface*. Clarendon Press, Oxford.

Shepard, F. P. (1978) *Geological Oceanography*, Heinemann, London.

Syvitski, J. P. M., Burrell, D. C. and Skei, J. M. (1987) *Fjords: Processes and Products*. Springer-Verlag, New York.

Trenhaile, A. S. (1987) *The Geomorphology of Rock Coasts*. Clarendon Press, London.

Trudgill, S. (1985) *Limestone Geomorphology*. Longman, London.

Trudgill, S. T. (ed.) (1986) *Solute Processes*. John Wiley & Sons, Chichester.

Verstappen, H. T. (1983) *Applied Geomorphology*. Elsevier, Amsterdam.

Washburn, A. L. (1979) *Geocryology: A Survey of Periglacial Processes and Environments*. Arnold, London.

Geophysics

Aki, K. and Richards, P. G. (1999) *Quantitative Seismology: Theory and Methods*. W. H. Freeman, San Francisco.

Anstey, N. A. (1977) *Seismic Interpretation: The Physical Aspects*. IHRDC Publications, Boston.

Anstey, N. A. (1981) *Seismic Prospecting Instruments*. Gebrüder Borntraeger, Berlin.

Bolt, B. A. (1993) *Earthquakes – A Primer*. W. H. Freeman, San Francisco.

Brown, A. R. (1996) *Interpretation of Three-Dimensional Seismic Data*. Memoir **42**. American Association of Petroleum Geology, Tulsa, OK.

Bullen, K. E. and Bolt, B. A. (1985) *An Introduction to the Theory of Seismology*. Cambridge University Press, Cambridge.

Butler, R. F. (1992) *Paleomagnetism*. Blackwell Scientific Publications, Oxford.

Condie, K. C. (1997) *Plate Tectonics and Crustal Evolution* (4th edn). Pergamon, Oxford.

Dobrin, M. B. and Savit, C. H. (1988) *Introduction to Geophysical Prospecting* (4th edn). McGraw-Hill, New York.

Ellis, D. V. (1987) *Well Logging for Earth Scientists*. Elsevier, Amsterdam.

Fowler, C. M. R. (1990) *The Solid Earth: An Introduction to Global Geophysics*. Cambridge University Press, Cambridge.

Gibson, R. I. and Millegan, P. S. (eds.) (1999) *Geologic Applications of Gravity and Magnetics: Case Histories*. American Association of Petroleum Geology, Tulsa, OK.

Jacobs, J. A. (1984) *Reversals of the Earth's Magnetic Field*. Adam Hilger, Bristol.

Jones, E. J. W. (1999) *Marine Geophysics*. John Wiley & Sons, Chichester.

Kearey, P. and Brooks, M. (1991) *An Introduction to Geophysical Exploration*. Blackwell Scientific Publications, Oxford.

Khramov, A. N. (1987) *Paleomagnetology*. Springer-Verlag, Berlin.

Labo, J. (1986) *A Practical Introduction to Borehole Geophysics*. Society of Exploration Geophysicists, Tulsa, OK.

Lambeck, K. (1980) *The Earth's Variable Rotation*. Cambridge University Press, Cambridge.

Lay, T. (1995) *Modern Global Seismology*. Academic Press, London.

Lillie, R. J. (1999) *Whole Earth Geophysics: An Introductory Textbook for Geologists and Geophysicists*. Prentice Hall, Englewood Cliffs, NJ.

Lowrie, W. (1997) *Fundamentals of Geophysics*. Cambridge University Press, Cambridge.

McElhinny, M. W. and McFadden, P. L. (1999) *Paleomagnetism*. Academic Press, London.

Mörner, N. (ed.) (1980) *Earth Rheology, Isostasy and Eustasy*. John Wiley & Sons, Chichester.

Mussett, A. E. and Khan, M. A. (2000) *Looking into the Earth: An Introduction to Geological Geophysics*. Cambridge University Press, Cambridge.

O'Reilly, W. (1984) *Rock Magnetism*. Blackie, Glasgow.

Piper, J. D. A. (1987) *Palaeomagnetism and the Continental Crust*. Halsted Press, New York.

Poirier, J-P. (2000) *Introduction to the Physics of the Earth's Interior*. Cambridge University Press, Cambridge.

Reynolds, J. M. (1997) *An Introduction to Applied and Environmental Geophysics*. John Wiley & Sons, Chichester.

Rikitake, T. (1976) *Earthquake Prediction*. Elsevier, Amsterdam.

Serra, O. (1984) *Fundamentals of Well Log Interpretation* **1** & **2**. Developments in Petroleum Science **15A** & **15B**. Elsevier, Amsterdam.

Sharma, P. V. (1997) *Environmental and Engineering Geophysics*. Cambridge University Press, Cambridge.

Shearer, P. (1999) *Introduction to Seismology*. Cambridge University Press, Cambridge.

Sheriff, R. E. (1980) *Seismic Stratigraphy*. IHRDC Publications, Boston.

Sheriff, R. E. and Geldart, L. P. (1995) *Exploration Seismology*. Cambridge University Press, Cambridge.

Simon, R. B. (1981) *Earthquake Interpretation: A Manual for Reading Seismograms*. Kauffmann, Los Altos, CA.

Tarling, D. H. (1983) *Palaeomagnetism*. Chapman & Hall, London.

Telford, W. M., Geldart, L. P. and Sheriff, R. E. (1991) *Applied Geophysics*. Cambridge University Press, Cambridge.

Udias, A. (2000) *Principles of Seismology*. Cambridge University Press, Cambridge.

Weimer, P. and Davis, T. L. (eds.) (1996) *Applications of 3-D Seismic Data to Exploration and Production*. American Association of Petroleum Geology, Tulsa, OK.

Mineralogy and Crystallography

Bailey, S. W. (ed.) (1984) *Micas*. Reviews in Mineralogy **13**. Mineralogical Society of America, Washington, DC.

Bailey, S. W. (ed.) (1988) *Hydrous Phyllosilicates Exclusive of Micas*. Reviews in Mineralogy **19**. Mineralogical Society of America, Washington, DC.

Ballhausen, C. J. and Gray, H. B. (1965) *Molecular Orbital Theory*. Benjamin, New York.

Ballhausen, C. J. (1962) *Introduction to Ligand Field Theory*. McGraw-Hill, New York.

Berry, L. G. and Mason, B. (1983) *Mineralogy: Concepts, Descriptions, Determinations*. W. H. Freeman, San Francisco.

Bishop, A. C. (1967) *An Outline of Crystal Morphology*. Hutchinson, London.

Blackburn, W. H. and Dennen, W. (1997) *Encyclopedia of Mineral Names*. The Canadian Mineralogist Special Publication **1**. Mineral Association, Ottawa.

Brindley, G. W. and Brown, G. (eds.) (1984) *Crystal Structures of Clay Minerals and their X-ray Identification*. Mineralogical Society Monograph **5**.

Brown, W. L. (ed.) (1984) *Feldspars and Feldspathoids: Structures, Properties and Occurrences*. NATO ASI Series, Series C, **137**. Riedel, Dordrecht.

Cepeda, J. C. *Introduction to Rocks and Minerals*. Prentice Hall, Englewood Cliffs, NJ.

Clark, A. H. (1993) *Hey's Mineral Index* (3rd edn). British Museum of Natural History and Chapman & Hall, London.

Cox, P. A. (1987) *The Electronic Structure and Chemistry of Solids*. Oxford University Press, Oxford.

Cuilty, B. D. (1978) *Elements of X-ray Diffraction*. Addison Wesley, Reading, MA.

Deer, W. A., Howie, R. A. and Zussman, J. (1978) *Rock Forming Minerals*. Longman, London.

Deer, W. A., Howie, R. A. and Zussman, J. (1992) *An Introduction to the Rock Forming Minerals* (2nd edn). Longman, London.

Dent Glasser, L. S. (1977) *Crystallography and its Applications*. Van Nostrand Reinhold, New York.

Ernst, W. G. (1968) *Amphiboles*. Springer-Verlag, New York.

Farmer, V. C. (ed.) (1974) *The Infrared Spectra of Minerals*. Mineralogical Society Monograph **4**.

Faure, G. (1977) *Principles of Isotope Geology*. John Wiley & Sons, New York.

Fleischer, M. and Mandarino, J. A. (1991) *Glossary of Mineral Species 1991*. Mineralogical Record, Tucson.

Gaines, R. V., Foord, E., Mason, B. and Rosenweig, A. (1997) *Dana's New Mineralogy* (8th edn). John Wiley & Sons, Chichester.

Goldstein, J. I., Newbury, D. E., Echin, P. *et al*. (1981) *Scanning Electron Microscopy and X-ray Microanalysis*. Plenum, London.

Gottardi, G. and Galli, E. (1985) *Natural Zeolites*. Springer-Verlag, Berlin.

Gribble, C. D. and Hall, A. J. (1985) *A Practical Introduction to Optical Mineralogy*. Allen & Unwin, London.

Grim, R. F. (1968) *Clay Mineralogy*. McGraw-Hill, New York.

Hawthorne, F. C. (ed.) (1988) *Spectroscopic Methods in Mineralogy and Geology*. Reviews in Mineralogy **18**. Mineralogical Society of America, Washington, DC.

Hurbut, C. S. Jr. and Klein, C. (1985) *Manual of Mineralogy*. John Wiley & Sons, New York.

Kalló, D. and Sherry, H. S. (eds.) (1988) *Occurrence, Properties and Utilization of Natural Zeolites*. Akadémiai Kiadó, Budapest.

Klein, C. and Hurlburt, C. S. (1998) *Manual of Mineralogy*. John Wiley & Sons, New York.

Klug, H. P. and Alexander, L. E. (1974) *X-ray Diffraction Procedures*. John Wiley & Sons, London.

Ladd, M. F. C. and Palmer, R. A. (1985) *Structure Determination by X-ray Crystallography*. Plenum, New York.

Leake, B. E. (1978) 'The nomenclature of amphiboles'. *Mineralogical Magazine* **42**, 533–65.

McKie, D. and McKie, C. (1986) *Essentials of Crystallography*. Blackwell Scientific Publications, Oxford.

Morimoto, N. (1988) 'Nomenclature of pyroxenes'. *Mineralogical Magazine* **52**, 535–50.

Nesse, W. D. (1999) *Introduction to Mineralogy*. Wiley, Chichester.

Nickel, E. H. and Nichols, M. C. (1991) *Mineral Reference Manual*. Chapman & Hall, London and Van Nostrand Reinhold, New York.

Papike, J. J. (ed.) (1969) *Pyroxenes and Amphiboles: Crystal Chemistry*. Special Paper **2**. Mineralogical Society of America, Washington, DC.

Parker, S. P. (ed.) (1997) *Dictionary of Geology and Mineralogy*. McGraw-Hill, Chichester.

Phillips, F. C. (1971) *An Introduction to Crystallography*. Longman, London.

Poirier, J. P. (1986) *Creep of Crystals*. Cambridge University Press, Cambridge.

Putnis, A. (2000) *An Introduction to Mineral Sciences*. Cambridge University Press, Cambridge.

Putnis, A. and McConnell, J. D. C. (1980) *Principles of Mineral Behaviour*. Blackwell Scientific Publications, Oxford.

Ribbe, P. H. (ed.) (1982) *Orthosilicates*. Reviews in Mineralogy **5**. Mineralogical Society of America, Washington, DC.

Roberts, W. L., Rapp, G. P. Jr. and Weber, J. (1974) *Encyclopedia of Minerals*. Van Nostrand Reinhold, New York.

Rumble III, D. (ed.) (1976) *Oxide Minerals*. Reviews in Mineralogy **3**. Mineralogical Society of America, Washington, DC.

Sand, L. B. and Mumpton, F. H. (eds.) *Natural Zeolites: Occurrence, Properties, Use*. Pergamon, Oxford.

Smith, J. V. and Brown, W. L. (1988) *Feldspar Minerals* Vol. 1: *Crystal Structure, Physical, Chemical and Microtextural Properties*. Springer-Verlag, Berlin.

Tilley, R. (1999) *Colour and the Optical Properties of Materials*. Wiley, Chichester.

Urch, D. S. (1979) *Orbitals and Symmetry*. Macmillan, London.

Veblen, D. R. (ed.) (1983) *Amphiboles and Other Hydrous Pyriboles – Mineralogy*. Reviews in Mineralogy **9A**. Mineralogical Society of America, Washington, DC.

Veblen, D. R. and Ribbe, P. H. (eds.) (1983) *Amphiboles – Phase Relations*. Reviews in Mineralogy **9B**. Mineralogical Society of America, Washington, DC.

Velde, B. (1977) *Clays and Clay Minerals in Natural and Synthetic Systems*. Elsevier, Amsterdam.

Verma, A. R. and Krishna, P. (1966) *Polymorphism and Polytypism in Crystals*. John Wiley & Sons, New York.

Woodgate, G. K. (1980) *Elementary Atomic Structure*. Oxford University Press, Oxford.

Miscellaneous

Allum, J. A. E. (1986) *Photogeology and Regional Mapping*. Pergamon, Oxford.

Attendorn, H.-G. and Bowen, R. N. C. (1997) *Radioactive and Stable Isotope Geology*. Chapman & Hall, London.

Bauer, J. and Bouska, V. (1983) *A Guide in Colour to Precious and Semiprecious Stones*. Octopus Books, London.

Carter, D. J. (1986) *The Remote Sensing Sourcebook*. Kogan Page, London and McCarta Ltd, London.

Drury, S. A. (1987) *Image Interpretation in Geology*. Allen & Unwin, London.

Durrance, E. M. (1986) *Radioactivity in Geology*. Ellis Horwood, Chichester.

Ford, T. E. and Cullingford, C. H. D. (1976) *The Science of Speleology*. Academic Press, London.

Gillieson, D. (1996) *Caves: Processes, Development and Management*. Blackwell Science, Oxford.

Kempe, D. R. C. and Harvey, A. P. (eds.) (1983) *The Petrology of Archaeological Artefacts*. Clarendon Press, Oxford.

McGuire, W. J., Griffiths, D. R., Hancock, P. L. and Stewart, I. S. (2000) *The Archaeology of Geological Catastrophes*. Special Publication **171**, Geological Society of London.

Menard, H. W. (1986) *Islands*. Scientific American Library, New York.

Moore, C. A. (1973) *Handbook of Subsurface Geology*. Harper & Row, New York.

Murck, B. M. (1996) *Environmental Geology*. John Wiley & Sons, New York.

Ollier, C. D. (1969) *Weathering*. Oliver & Boyd, Edinburgh.

Roberts, J. L. (1982) *Introduction to Geological Maps and Structures*. Pergamon, Oxford.

Sabins, F. E. (1997) *Remote Sensing: Principles and Interpretations*. W. H. Freeman, San Francisco.

Tite, M. S. (1972) *Methods of Physical Examination in Archaeology*. Seminar Press, London.

Tritton, D. J. (1977) *Physical Fluid Dynamics*. Van Nostrand Reinhold, Wokingham.

Waltham, D. (1994) *Mathematics: A Simple Tool for Geologists*. Chapman & Hall, London.

Webster, R. A. (1979) *Practical Gemmology*. NAG Press, London.

Palaeontology

Aldridge, R. J. (ed.) (1987) *Palaeobiology of Conodonts*. Ellis Horwood, Chichester.

Alexander, R. McN. (1981) *The Chordates*. Cambridge University Press, Cambridge.

Anderson, O. R. (1983) *Radiolaria*. Springer-Verlag, New York.

Benton, M. J. and Harper, D. A. T. (1997) *Basic Palaeontology*. Addison-Wesley Longman, London.

Benton, M. J. (1997) *Vertebrate palaeontology* (2nd edn). Chapman & Hall, London.

Bolli, H. M., Saunders, J. B. and Perch-Nielsen, K. (eds.) (1985) *Plankton Stratigraphy*. Cambridge University Press, Cambridge.

Boucot, A. (1975) *Evolution and Extinction Rate*. Developments in Palaeontology and Stratigraphy. Elsevier, Amsterdam.

Brasier, M. D. (1980) *Microfossils*. Allen & Unwin, London.

Briggs, D. E. G. and Crowther, P. R. (eds.) (1992) *Palaeobiology – a Synthesis*. Blackwell Scientific Publications, Oxford.

Carroll, R. L. (1997) *Patterns and Processes of Vertebrate Evolution*. Cambridge University Press, Cambridge.

Clarkson, E. N. K. (1998) *Invertebrate Palaeontology and Evolution* (4th edn). Blackwell Science, Oxford.

Cortillot, V. (1999) *Evolutionary Catastrophes: The Science of Mass Extinction.* Cambridge University Press, Cambridge.

Dodd, J. R. and Stanton, R. J. (1981) *Paleoecology, Concepts and Applications.* John Wiley & Sons, Chichester.

Doyle, P. (1996) *Understanding Fossils: An Introduction to Invertebrate Palaeontology.* Wiley, Chichester.

Flügel, E. (ed.) (1977) *Fossil Algae, Recent Results and Developments.* Springer-Verlag, Berlin.

Fortey, R. A. (1992) *Fossils – The Key to the Past* (2nd edn). Natural History Museum, London.

Frankel, C. (1999) *The End of the Dinosaurs: Chicxulub Crater and Mass Extinctions.* Cambridge University Press, Cambridge.

Frey, R. W. (ed.) (1975) *The Study of Trace Fossils.* Springer-Verlag, Berlin.

Glaessner, M. F. (1984) *The Dawn of Animal Life – A Biohistorical Study.* Cambridge University Press, Cambridge.

Goldring, R. (1999) *Field Palaeontology.* Longman, London.

Hamilton, G. B. and Lord, A. R. (eds.) (1982) *A Stratigraphical Index of Calcareous Nannofossils.* Ellis Horwood, Chichester.

Haynes, J. R. (1981) *Foraminifera.* Macmillan, London.

House, M. R. (ed.) (1979) *The Origin of Major Invertebrate Groups.* Academic Press, London.

Laport, L. F. (ed.) (1974) *Reefs in Time and Space.* Special Publication **18**. Society of Economic Paleontologists & Mineralogists, Tulsa, OK.

Moore, R. C. (ed.) (1955,1969) *Treatise on Invertebrate Paleontology.* Geological Society of America and University of Kansas Press, Boulder and Lawrence.

Murray, J. W. (ed.) (1985) *Atlas of Invertebrate Macrofossils.* Longman, London.

Nielson, C. and Larwood, G. P. (eds.) (1985) *Bryozoa: Ordovician to Recent.* Olsen & Olsen, Fredensborg.

Nitecki, M. H. (ed.) (1984) *Extinctions.* University of Chicago Press, Chicago.

Purchon, R. D. (1968) *The Biology of the Mollusca.* Pergamon, Oxford.

Raup, D. M. and Jablonski, D. (eds.) (1986) *Patterns and Processes in the History of Life.* Springer-Verlag, Berlin.

Romer, A. S. (1966) *Vertebrate Palaeontology.* Chicago University Press, Chicago.

Rudwick, M. J. S. (1970) *Living and Fossil Brachiopods.* Hutchinson, London.

Sarjeant, W. S. (1974) *Fossil and Living Dinoflagellates.* Academic Press, London.

Schopf, T. J. M. (ed.) (1972) *Models in Paleobiology.* W. H. Freeman, San Francisco.

Schram, F. R. (1986) *Crustacea.* Oxford University Press, Oxford.

Smith, A. B. (1984) *Echinoid Palaeobiology.* Allen & Unwin, Hemel Hempstead.

Stanley, S. M. (1979) *Macroevolution, Pattern and Process.* W. H. Freeman, San Francisco.

Stanley, S. M. (1987) *Extinctions.* Scientific American Books, New York.

Tappan, H. (1980) *The Paleobiology of Plant Protists.* W. H. Freeman, San Francisco.

Tevesz, M. J. S. and McCall, P. L. (eds.) (1983) *Biotic Interactions in Recent and Fossil Communities.* Plenum, London.

Thomas, B. A. and Spicer, R. A. (1987) *The Evolution and Palaeobiology of Land Plants.* Croon Helm, London.

Tiffney, B. H. (ed.) (1986) *Geological Factors in the Evolution of Plants.* Yale University Press, New Haven, CT.

Walter, M. R. (ed.) (1976) *Stromatolites.* Developments in Sedimentology **20**. Elsevier, Amsterdam.

Wray, J. L. (1977) *Calcareous Algae.* Developments in Paleontology and Stratigraphy **4**. Elsevier, Amsterdam.

Petrology, Igneous and Metamorphic

Arndt, N. T. and Nisbet, E. G. (eds.) (1982) *Komatiites*. Allen & Unwin, London.

Atherton, M. P. and Tarney, J. (1981) *The Origin of Granite Batholiths*. Geochemistry Group of the Mineralogical Society. Shiva, Kent.

Barker, F. (ed.) (1979) *Trondhjemites, Dacites and Related Rocks*. Elsevier, Amsterdam.

Best, M. G. (1982) *Igneous and Metamorphic Petrology*. W. H. Freeman, New York.

Blatt, H. and Tracy, T. (1996) *Igneous, Sedimentary and Metamorphic Petrology*. W. H. Freeman, San Francisco.

Bowes, D. R. (ed.) (1990) *The Encyclopedia of Igneous and Metamorphic Petrology*. Van Nostrand Reinhold, New York.

Coleman, R. G. (1977) *Ophiolites*. Springer-Verlag, Heidelberg.

Cox, K. G., Bell, J. D. and Pankhurst, R. J. (1979) *The Interpretation of Igneous Rocks*. Allen & Unwin, London.

Crawford, A. J. (ed.) (1989) *Boninites*. Unwin Hyman, London.

Fitton, J. G. and Upton, B. G. J. (eds.) (1987) *Alkaline Igneous Rocks*. Geological Society Special Publication **30**. Blackwell Scientific Publications, Oxford.

Fry, N. (1991) *The Field Description of Metamorphic Rocks*. Wiley, Chichester.

Gupta, A. K. and Yaki, K. (1979) *Petrology and Genesis of Leucite-bearing Rocks*. Springer-Verlag, Berlin.

Hall, A. (1995) *Igneous Petrology*. Longman, London.

Kornprobst, J. (ed.) (1984) *Kimberlite 1: Kimberlites and Related Rocks*. Elsevier, Amsterdam.

MacKenzie, W. S. and Guilford, C. (1980) *Atlas of Rock-forming Minerals in Thin Section*. Longman, London.

MacKenzie, W. S., Donaldson, C. H. and Guilford, C. (1982) *Atlas of Igneous Rocks and their Textures*. Longman, London.

McBirney, A. R. (1993) *Igneous Petrology*. Jones & Bartlett, Boston & London.

Morse, S. A. (1980) *Basalts and Phase Diagrams*. Springer-Verlag, Berlin.

Nixon, P. H. (ed.) (1987) *Mantle Xenoliths*. John Wiley & Sons, Chichester.

Park, R. G. and Tarney, J. (eds.) (1987) *Evolution of Lewisian and Comparable Precambrian High Grade Terrains*. Geological Society Special Publication **27**. Blackwell Scientific Publications, Oxford.

Ringwood, A. E. (1975) *Composition and Petrology of the Earth's Mantle*. McGraw-Hill, New York.

Streckeisen, A. L. (1967) 'Classification and nomenclature of igneous rocks (Final report of an enquiry)'. *Neues Jahrbuch für Mineralogie, Abhandlung* **107**, 144–240.

Thorpe, R. S. (1982) *Andesites: Orogenic Andesites and Related Rocks*. Wiley & Sons, Chichester.

Thorpe, R. and Brown, G. (1991) *The Field Description of Igneous Rocks*. Wiley, Chichester.

Wager, L. R. and Brown, G. M. (1968) *Layered Igneous Rocks*. Oliver & Boyd, Edinburgh.

Wimmaenauer, W. (ed.) (1974) *The Alkaline Rocks*. John Wiley & Sons, New York.

Yardley, B. W. D. (1989) *An Introduction to Metamorphic Petrology*. Longman, London.

Yoder, H. S. (ed.) (1979) *The Evolution of the Igneous Rocks*. Princeton University Press, Princeton, NJ.

Plate Tectonics

Carey, S. W. (1976) *The Expanding Earth*. Elsevier, Amsterdam.

Cann, J. R., Elderfield, H. and Laughton, A. S. *Mid-Ocean Ridges: Dynamics of Processes*

Associated with the Formation of New Oceanic Crust. Cambridge University Press, Cambridge.

Condie, K. C. (1989) *Plate Tectonics and Crustal Evolution*. Pergamon, Oxford.

Cox, A. (1973) *Plate Tectonics and Geomagnetic Reversals*. W. H. Freeman, San Francisco.

Cox, A. and Hart, R. B. (1986) *Plate Tectonics: How it Works*. Blackwell Scientific Publications, Palo Alto, CA.

Davies, G. F. (1999) *Dynamic Earth: Plates, Plumes and Mantle Convection*. Cambridge University Press, Cambridge.

Davies, P. A. and Runcorn, S. K. (eds.) (1980) *Mechanisms of Continental Drift and Plate Tectonics*. Academic Press, London.

Gill, J. B. (1981) *Orogenic Andesites and Plate Tectonics*. Springer-Verlag, Berlin.

Kearey, P. and Vine, F. J. (1996) *Global Tectonics* (2nd edn). Blackwell Science, Oxford.

Le Grand, H. E. (1988) *Drifting Continents and Shifting Theories*. Cambridge University Press, Cambridge.

Park, R. G. (1988) *Geological Structures and Moving Plates*. Blackie, Glasgow.

Summerfield, A. (ed.) (1999) *Geomorphology and Global Tectonics*. John Wiley & Sons, Chichester.

Toksöz, M. N., Uyeda, S. and Francheteau, J. (eds.) (1980) *Ocean Ridges and Arcs*. Elsevier, Amsterdam.

Windley, B. F. (1995) *The Evolving Continents*. John Wiley & Sons, New York.

Sedimentology

Allen, J.R.L. (1982) *Sedimentary Structures, Their Character and Physical Basis*. Developments in Sedimentology **30**. Elsevier, New York.

Allen, J. R. L. (1985) *Principles of Physical Sedimentology*. Allen & Unwin, London.

Bathurst, R. G. C. (1971) *Carbonate Sediments and their Diagenesis*. Developments in Sedimentology **12**. Elsevier, Amsterdam.

Berner, R. A. (1980) *Early Diagenesis: A Theoretical Approach*. Princeton University Press, Princeton, NJ.

Blatt, H., Middleton, G. V. and Murray, R. C. (1980) *Origin of Sedimentary Rocks*. Prentice Hall, Englewood Cliffs, NJ.

Boggs, S. H. Jr. (1987) *Principles of Sedimentology and Stratigraphy*. Merrill, Columbus, OH.

Bradshaw, P. (1971) *An Introduction to Turbulence and its Measurement*. Pergamon, Oxford.

Brenchley, P. J & Williams, B. P. J. (1985) *Sedimentology: Recent Developments and Applied Aspects*. Geological Society Special Publication **18**. Blackwell Scientific Publications, Oxford.

Brodozikowski, K. and van Loon, A. J. (1991) *Glacigenic Sediments*. Elsevier, Amsterdam.

Brookfield, M. E. and Ahlbrandt, T. S. (eds.) (1983) *Aeolian Sediments and Processes*. Developments in Sedimentology **38**. Elsevier, Amsterdam.

Collinson, J. D. and Lewin, J. (eds.) (1983) *Modern and Ancient Fluvial Systems*. Special Publications of the International Association of Sedimentologists **6**. Blackwell Scientific Publications, Oxford.

Dans, R. A. (ed.) (1987) *Coastal Sedimentary Environments*. Springer-Verlag, New York.

Dowdeswell, J. A. and Scourse, J. D. (1990) *Glacimarine Environments*. Geological Society, London.

Duchafour, P. (1982) *Pedology, Pedogenesis and Classification*. Allen & Unwin, London.

Ethridge, F. C. and Flores, R. M. (eds.) (1987) *Recent Developments in Fluvial Sedimentology*. Special Publication **39**. Society of Economic Paleontologists and Mineralogists, Tulsa, OK.

Ethridge, F. C. and Flores, R. M. (eds.) (1981) *Recent and Ancient Nonmarine Depositional*

Environments: Models for Exploration. Special Publication **31**. Society of Economic Paleontologists and Mineralogists, Tulsa, OK.

Fitzpatrick, E. A. (1983) *Soils and their Formation, Classification and Distribution*. Longman, London.

Flügel, E. (1982) *Microfacies Analysis of Limestones*. Springer-Verlag, Berlin.

Frostick, L. E. and Reid, I. (eds.) (1987) *Desert Sediments: Ancient and Modern*. Geological Society Special Publication **35**. Blackwell Scientific Publications, Oxford.

Goudie, A. S., Livingstone, I. and Stokes, S. (eds.) (1999) *Aeolian Environments, Sediments and Landforms*. John Wiley & Sons, Chichester.

Ham, W. E. (ed.) (1962) *Classification of Carbonate Rocks*. Memoir **1**. American Association of Petroleum Geologists, Tulsa, OK.

Head, K. H. (1982) *Manual of Soil Laboratory Testing*. Pentech Press, London.

Kaplan, J. R. (ed.) (1974) *Natural Gases in Marine Sediments*. Plenum, New York.

Kirkland, D. W. and Evans, R. (1973) *Marine Evaporites: Origin, Diagenesis and Geochemistry*. Benchmark Papers in Geology. Hutchinson & Ross, Stroudsburg, PA.

Komar, P. D. (1976) *Beach Processes and Sedimentation*. Prentice Hall, Englewood Cliffs, NJ.

Larsen, G. and Chilingar, G. V. (eds.) (1967) *Diagenesis in Sediments*. Developments in Sedimentology **8**. Elsevier, Amsterdam.

Leeder, M. R. (1992) *Sedimentology: Process and Product*. Chapman & Hall, London.

Leggett, J. K. and Zuffa, G. G. (eds.) (1987) *Marine Clastic Sedimentology*. Graham and Trotman, London.

Lerche, I. and O'Brien, J. J. (1986) *The Dynamical Geology of Salt Related Structures*. Academic Press, London.

Lowe, D. and Waltham, T. (1995) *A Dictionary of Karst and Caves*. British Cave Research Association, Bridgwater.

Matter, A. and Tucker, M. (eds.) (1978) *Modern and Ancient Lake Sediments*. Special Publication of the International Association of Sedimentologists **2**. Blackwell Scientific Publications, Oxford.

Miall, A. D. (ed.) (1978) *Fluvial Sedimentology*. Canadian Society of Petroleum Geologists Memoir **5**.

Parker, A. and Sellwood, B. (eds.) (1983) *Sediment Diagenesis*. NATO ASI Series C, **115**, Reidel, Dordrecht.

Peryt, T. (ed.) (1983) *Coated Grains*. Springer-Verlag, Berlin.

Peterson, J. A. and Osmond, J. C. (eds.) (1961) *Geometry of Sand Bodies*. American Association of Petroleum Geologists, Tulsa, OK.

Pettijohn, F. J., Potter, P. E. and Siever, R. (1973) *Sand and Sandstone*. Springer-Verlag, Berlin.

Prothero, D. R. and Schwab, F. (1996) *Sedimentary Geology*. W. H. Freeman, San Francisco.

Reading, H. G. (ed.) (1996) *Sedimentary Environments: Processes, Facies and Stratigraphy*. Blackwell Scientific Publications, Oxford.

Reeder, R. J. (ed.) (1983) *Carbonates: Mineralogy and Chemistry*. Reviews in Mineralogy **11**. Mineralogical Society of America, Washington, DC.

Rubin, D. M. (1987) *Cross-bedding, Bedforms and Paleocurrents*. Concepts in Sedimentology **1**. Society of Economic Paleontologists and Mineralogists, Tulsa, OK.

Schlanger, S. O. and Cita, M. B. (eds.) (1982) *Nature and Origins of Cretaceous Carbon-rich Facies*. Academic Press, New York.

Scholle, P. A. and Schluger, P. R. (eds.) (1979) *Aspects of Diagenesis*. Special Publication **26**. Society of Economic Paleontologists and Mineralogists, Tulsa, OK.

Sieveking, G. de G. and Hart, M. B. (eds.) (1986) *The Scientific Study of Flint and Chert*. Cambridge University Press, Cambridge.

Soil Survey Staff (1992) *Key to Soil Taxonomy* (5th edn). Soil Management Support Services Technical Monograph **19**. Pochahontas Press, Blacksburg.

Sonnenfeld, P. (1984) *Brines and Evaporites*. Academic Press, London.

Stride, A. H. (ed.) (1982) *Offshore Tidal Sands: Processes and Deposits*. Chapman & Hall, London.

Tucker, M. E. (1991) *Sedimentary Petrology: an Introduction* (2nd edn). Blackwell Scientific Publications, Oxford.

Tucker, M. (1996) *Sedimentary Rocks in the Field*. Wiley, Chichester.

Van der Meer, J. J. M. (ed.) (1989) *Tills and Glaciotectonics*. Balkema, Rotterdam.

Walker, R. G. (ed.) (1984) *Facies Models*. Geoscience Canada Reprint Series **1**.

Wright, V. P. (ed.) (1986) *Paleosols: Their Recognition and Interpretation*. Blackwell Scientific Publications, Oxford.

Yalin, M. S. (1977) *Mechanics of Sediment Transport*. Pergamon, Oxford.

Stratigraphy

Audley-Charles, M. G. and Hallam, A. (1988) *Gondwana and Tethys*. Geological Society Special Publication **37**. Oxford University Press, Oxford.

Berggren, W. A. and van Couvering, J. A. (eds.) (1984) *Catastrophes and Earth History*. Princeton University Press, Princeton, NJ.

Brett, C. E. and Baird, G. C. (1997) *Paleontological Events: Stratigraphic, Ecological, and Evolutionary Implications*. Columbia University Press, Columbia.

Bruton, D. L. (ed.) (1983) *Aspects of the Ordovician System*. Palaeontological Contributions from the University of Oslo **295**. Universitetsforlaget, Oslo.

Doyle, P. and Bennett, M. (eds.) (1998) *Unlocking the Stratigraphic Record*. John Wiley & Sons, Chichester.

Ehlers, J. (1996) *Quaternary and Glacial Geology*. John Wiley & Sons, Chichester.

Eicher, D. L. and McAlester, A. L. (1980) *History of the Earth*. Prentice Hall, Englewood Cliffs, NJ.

Eicher, D. L. (1976) *Geologic Time*. Prentice Hall, Englewood Cliffs, NJ.

Glover, J. E. and Groves, D. I. (eds.) (1981) *Archaean Geology*. Special Publication of the Geological Society of Australia **7**.

Hallam, A. (1987) *Jurassic Environments*. Cambridge University Press, Cambridge.

Hambrey, M. J. and Harland, W. B. (1981) *Earth's Pre-Pleistocene Glacial Record*. Cambridge University Press, Cambridge.

Harland, W. B., Armstrong, R. L., Cox, A. V., Craig, L. E., Smith, A. G. and Smith, D. G. (1990) *A Geologic Timescale 1990*. Cambridge University Press, Cambridge.

Harris, A. L., Holland, C. H. and Leake, B. E. (eds.) (1979) *The Caledonides of the British Isles – Reviewed*. Geological Society Special Publication **8**. Scottish Academic Press, Edinburgh.

Hoffman, A. and Nitecki, M. H. (eds.) (1986) *Problematic Fossil Taxa*. Oxford University Press, New York.

Holland, C. H. (ed.) (1971) *Cambrian of the New World*. Wiley-Interscience, New York.

Holland, C. H. (ed.) (1974) *Cambrian of the British Isles, Norden and Spitzbergen*. John Wiley & Sons, Chichester.

House, M. R., Scrutton, C. T. and Bassett, M. G. (eds.) (1979) *The Devonian System*. Special Paper in Palaeontology **23**. The Palaeontological Association, London.

Imbrie, J. and Imbrie, K. P. (1979) *Ice-ages: Solving the Mystery*. Macmillan, London.

Kauffman, E. G. and Hazel, J. E. (eds.) (1977) *Concepts and Methods of Biostratigraphy*. Dowden, Hutchinson & Ross, Stroudsburg, PA.

Nilsson, T. (1983) *The Pleistocene, Geology and Life of the Quaternary Ice Age*. Riedel, Dordrecht.

Nisbet, E. G. (1987) *The Young Earth*. Allen & Unwin, Boston.

Odin, G. S. (ed.) (1982) *Numerical Dating in Stratigraphy*. John Wiley & Sons, Chichester.

Pomerol, C. (1982) *The Cenozoic Era*. Ellis Horwood, Chichester.

Prothero, D. R. (1990) *Interpreting the Stratigraphic Record*. W. H. Freeman, San Francisco.

Snelling, N. J. (ed.) *The Chronology of the Geological Record*. Memoirs of the Geological Society **10**. Blackwell Scientific Publications, Oxford.

Zapfe, H. (ed.) (1974) *The Stratigraphy of the Alpine-Mediterranean Triassic*. Springer-Verlag, Vienna.

Structural Geology and Geodynamics

Cosgrove, J. W. (ed.) (2000) *Forced Folds and Fractures*. Special Publication **169**, Geological Society of London.

Coward, M. P., Dewey, J. F. and Hancock, P. L. (eds.) (1987) *Continental Extensional Tectonics*. Geological Society Special Publication **28**. Blackwell Scientific Publications, Oxford.

Friedel, J. (1964) *Dislocations*. Pergamon, London.

Griggs, D. and Handin, J. (eds.) (1979) *Rock Deformation*. Geological Society of America Memoir **79**.

Hancock, P. L. (ed.) (1994) *Continental Deformation*. Pergamon, Oxford.

Hatcher, R. D. (1995) *Structural Geology: Principles, Concepts and Problems*. Prentice Hall, Englewood Cliffs, NJ.

Hobbs, B. E., Means, W. D. and Williams, P. F. (1976) *An Outline of Structural Geology*. John Wiley & Sons, New York.

Hsu, K. J. (ed.) (1982) *Mountain Building Processes*. Academic Press, London.

Hull, D. (1975) *Introduction to Dislocations*. Pergamon, Oxford.

Keller, E. A. and Pinter, N. (1996) *Active Tectonics: Earthquakes, Uplift and Landscape*. Prentice Hall, Englewood Cliffs, NJ.

Kroner, A. and Greiling, R. (eds.) (1984) *Precambrian Tectonics Illustrated*. Schweizerbart, Stuttgart.

Lawn, B. R. and Wilshaw, T. R. (1975) *Fracture of Brittle Solids*. Cambridge University Press, Cambridge.

McClay, K. R. and Price, N. J. (eds.) *Thrust and Nappe Tectonics*. Geological Society Special Publication **9**. Blackwell Scientific Publications, Oxford.

MacNiocaill, C. and Ryan, P. D. (eds.) *Continental Tectonics*. Special Publication **164**, Geological Society of London.

Means, W. D. (1978) *Stress and Strain*. John Wiley & Sons, New York.

Miyashiro, A., Aki, K. and Sengör, A. M. C. (1982) *Orogeny*. John Wiley & Sons, New York.

Moores, E. M. and Twiss, R. J. (1996) *Tectonics*. W. H. Freeman, San Francisco.

Nicholas, R. and Poirier, J. P. (1976) *Crystalline Plasticity and Solid-state Flow*. Wiley-Interscience, London.

Park, R. G. (1997) *Foundations of Structural Geology* (3rd edn). Chapman & Hall, London.

Paterson, M. S. (1978) *Experimental Rock Deformation – The Brittle Field*. Springer-Verlag, Berlin.

Phillips, F. C. (1971) *The Use of Stereographic Projection in Structural Geology*. Arnold, London.

Price, N. J. (1966) *Fault Development in Brittle and Semi-brittle Rock*. Pergamon, Oxford.

Ramsay, J. G. and Huber, M. I. (1987) *Techniques of Modern Structural Geology*. Academic Press, London.

Ramsay, J. G. (1967) *Folding and Fracturing of Rocks*. McGraw-Hill, New York.

Seyfert, C. K. (ed.) (1987) *The Encyclopedia of Structural Geology and Plate Tectonics*. Van Nostrand Reinhold, New York.

Skinner, B. J. and Porter, S. C. (1995) *The Dynamic Earth* (3rd edn). John Wiley & Sons, New York.

Suppe, J. (1985) *Principles of Structural Geology*. Prentice Hall, Englewood Cliffs, NJ.

Turner, F. J. and Weiss, L. E. (1963) *Structural Analysis of Metamorphic Tectonites*. McGraw-Hill, New York.

Twiss, R. J. and Moores, E. M. (1992) *Structural Geology*. W. H. Freeman, San Francisco.

Vita-Finzi, C. (1986) *Recent Earth Movements: An Introduction to Neotectonics*. Academic Press, London.

Volcanology

Cas, R. A. F. and Wright, J. G. (1987) *Volcanic Successions Modern and Ancient*. Allen & Unwin, London.

Chapin, C. E. and Elston, W. E. (eds.) *Ash Flow Tuffs*. Special Paper of the Geological Society of America **180**.

Decker, P. and Decker, D. (1997) *Volcanoes* (3rd edn). W. H. Freeman, San Francisco.

Fisher, R. V. and Schminke, H.-U. (1984) *Pyroclastic Rocks*. Springer-Verlag, Berlin.

Francis, P. (1976) *Volcanoes*. Penguin, Harmondsworth, Middlesex.

Le Bas, M. J. (1977) *Carbonatite – Nepheline Volcanism*. John Wiley & Sons, New York.

Macdonald, G. A. (1972) *Volcanoes*. Prentice Hall, Englewood Cliffs, NJ.

Sheets, P. D. and Grayson, D. K. (1979) *Volcanic Activity and Human Ecology*. Academic Press, London.

Simkin, T., Siebert, L., McLelland, L. *et al*. (1981) *Volcanoes of the World*. Hutchinson Ross, Stroudsburg, PA.

Sparkes, R. S. J., Carey, S. N., Gilbert, J., Glaze, L. S., Sigurdsson, H. and Woods, A. W. (1997) *Volcanic Plumes*. John Wiley & Sons, Chichester.

Tazieff, H. and Sabroux, J. C. (1983) *Forecasting Volcanic Events*. Elsevier, Amsterdam.

Williams, H. and McBirney, A. R. (1979) *Volcanology*. Freeman, Cooper & Co., San Francisco.